# Springer Series in Geomechanics and Geoengineering

Geomechanics deals with the application of the principle of mechanics to geomaterials including experimental, analytical and numerical investigations into the mechanical, physical, hydraulic and thermal properties of geomaterials as multiphase media. Geoengineering covers a wide range of engineering disciplines related to geomaterials from traditional to emerging areas.

The objective of the book series is to publish monographs, handbooks, workshop proceedings and textbooks. The book series is intended to cover both the state-of-the-art and the recent developments in geomechanics and geoengineering. Besides researchers, the series provides valuable references for engineering practitioners and graduate students.

**** Now indexed by SCOPUS, EI and Springerlink****

More information about this series at http://www.springer.com/series/8069

Ancuța Rotaru
Editor

# Critical Thinking in the Sustainable Rehabilitation and Risk Management of the Built Environment

## CRIT-RE-BUILT. Proceedings of the International Conference, Iași, Romania, November 7–9, 2019

Springer

*Editor*
Ancuța Rotaru
"Gheorghe Asachi" Technical University of Iași
Iași, Romania

ISSN 1866-8755 ISSN 1866-8763 (electronic)
Springer Series in Geomechanics and Geoengineering
ISBN 978-3-030-61117-0 ISBN 978-3-030-61118-7 (eBook)
https://doi.org/10.1007/978-3-030-61118-7



This Springer imprint is published by the registered company Springer Nature Switzerland AG
The registered company address is: Gewerbestrasse 11, 6330 Cham, Switzerland

# Prologue

This book presents contributions to the International Conference *Critical Thinking in the Sustainable Rehabilitation and Risk Management of the Built Environment—CRIT-RE-BUILT* that took place in Iași, Romania, on 7–9 November 2019. It attracted participants from many countries across four continents: Albania, Austria, Bulgaria, Czech Republic, France, Italy, Portugal, Romania, Slovenia, Spain (Europe), Indonesia, Jordan, Nepal, Thailand, Vietnam (Asia), Costa Rica, Mexico (North America), Morocco, South Africa (Africa).

During the conference, lively and fruitful discussions carried on sustainable rehabilitation of the built environment reminded me of the good old times I spent with dear colleagues far away: Prof. Wei Wu at BOKU, Vienna; Prof. Chavdar Kolev at the Higher School of Transport Sofia; Prof. Dashnor Hoxha at the University of Orléans; Prof. Paolo Simonini and Prof. Simonetta Cola at the University of Padua; Prof. Andrea Segalini at the University of Parma; Prof. Massimo Ramondini at the University of Naples Federico II; Prof. Paulo Carvalho and Prof. Luis Pais at the University of Beira Interior; Prof. Yoyong Arfiadi at the Atma Jaya University, Yogyakarta; Prof. Assal Haddad at the American University of Madaba; Prof. Ramesh Kumar Maskey at Kathmandu University; Prof. Toufik Cherradi at the University Mohammed V in Rabat; Prof. Wynand Steyn and Prof. Chris Roth at the University of Pretoria. I express a huge thanks to all mentioned colleagues for the opportunity to meet at their universities and discuss the topic before. In our dialogues, they turned out to be critical thinkers evaluating the evidence to decide what is accurate and relevant to the rehabilitation of the built environment. Each of those discussions paved the way to critical thinking on the built environment enabling us to see the angles before making judgements on its sustainable rehabilitation and risk management. A big thanks to all of them also for deciding to return the visit by participating in the CRIT-RE-BUILT Conference. Many thanks to all other conference attendees who proved that critical thinking is also about the sense of discovery consisting not in seeking new places, but in seeing the built environment topic with different eyes.

All authors of the papers issued in this book deserve heartfelt thanks for their contributions submitted in good time. Their papers identify the key ideas of the topic analysing the sources of information and evaluating arguments.

The CRIT-RE-BUILT Conference and its dedicated book of proceedings grew up based on the European project entitled *Rehabilitation of the Built Environment in the Context of Smart City and Sustainable Development Concepts for Knowledge Transfer and Lifelong Learning (RE-BUILT)* - Erasmus+ KA2 programme for higher education strategic partnerships. Hence, the eight European academic teams jointly working on the project committed to appoint reviewers for the papers published in the current volume. Reviewers contributed fundamentally to the elaboration of the book. In the listing below, the number written in brackets after each reviewer's name shows the number of reviewed papers.

Prof. Wei Wu (8) and Dr. Chong Peng (8) from BOKU, Vienna, Austria; Prof. Sébastien Rémond (9) and Prof. Dashnor Hoxha (6) from the University of Orléans, France; Prof. Léon Miguel (7), Prof. Mercader Pilar (5), Prof. Maria Victoria De Montes Delgado (2), Prof. Isidro Cortés Albalá (2) and Prof. Francisco Javier Guevara García (1) from the University of Seville; Dr. Assunta Pelliccio (4) and Prof. Maura Ibimbo (1) from the University of Cassino and Southern Lazio; Dr. Kaja Pogacar (11) and Prof. Miroslav Premrov (3) from the University of Maribor; Prof. Chavdar Kolev (14) and Prof. Nikolay Georgiev (7) from the Higher School of Transport, Sofia, Bulgaria; Prof. Vera Glisníková (13) and Dr. Jan Štefaňák (5) from Brno University of Technology, Czech Republic; Prof. Radu-Aurel Pescaru (7), Dr. Ioana Olteanu-Donțov (6), Dr. Alina-Mihaela Nicuță (6), Dr. Marius-Costel Balan (5), Prof. Ancuța Rotaru (3) and Prof. Petru Mihai (2) from "Gheorghe Asachi" Technical University of Iași, Romania. Many thanks to reviewers for their thoughtful comments and efforts towards improving the manuscript.

The book sets up eight chapters in relationship with sustainable rehabilitation and risk in the built environment. Most of the authors are old-established or new partners of the Faculty of Civil Engineering and Building Services, "Gheorghe Asachi" Technical University of Iași, Romania, the conference host. They witnessed skills which cover a broad palette of competences: observations, analysis, interpretation, reflection, evaluation, explanation, problem-solving, decision-making, etc. Each chapter starts with a broad state-of-the-art presentation comprising the latest ideas and methods in the field assessing and asserting the synthesized and evaluated levels of research, development and novelty through a critical thinking process. The distinct groups of authors of the eight initiatory presentations are partners under the framework of the *RE-BUILT* project. The papers show the critical thinking on topics which mirror the breath of the field and the importance of sustainable rehabilitation and risk management of the built environment in fundamental and applied research.

Going through one chapter or another, the authors invite the reader to look critically at how hazards could affect the built environment revealing that the built environment is more than the sum of its infrastructure, vulnerable areas or brownfields. Nature's transfer of surface, material, function, structure and process through the change of scale or range of conditions leads to new building materials, elements and structures. Using smart materials which can react to environmental conditions into smart cities already became common in the building industry. New buildings rise upon the old ones, so that old bricks carry with them the memories of how the city looked like before. The papers should allow the reader to catch a glimpse of these former areas in the angle of a street, gate, old façade or long-standing bridge.

The book has more than one address. It is an invitation to all who want to access the inside of the built environment topic walking around in it as long as it is a re-emergent subject that never ends its actuality and a path of human endeavour.

July 2020

Ancuța Rotaru

# Contents

## Transformation of the Built Environment for the Rehabilitation of Socially Disadvantaged City Districts



## Conservation, Rehabilitation and Integration of Cultural Heritage

# Hazard Risk Mitigation for a Sustainable Built Environment

# Hazard Risk Mitigation for a Sustainable Built Environment

Ancuţa Rotaru(✉), Radu-Aurel Pescaru, Ioana Olteanu-Donţov, Alina-Mihaela Nicuţă, Petru Mihai, Vasilică Ciocan, and Marius-Costel Balan

"Gheorghe Asachi" Technical University of Iaşi,
67 Prof.dr.doc.D.Mangeron Street, Iași 700050, Romania
ancuta.rotaru@tuiasi.ro

**Abstract.** The aim of this study is to collect and review resources that map the hazard risk mitigation strategy for transferring research knowledge in the field on sustainable rehabilitation of the built environment.

Natural disasters occur all over the world and impact on large or small land areas. This study makes a brief review on the nature of disasters, but also the impact and consequences of such hazards on the built environment. It also investigates the need to manage and mitigate the hazard risk for a sustainable rehabilitation and reconstruction of the built environment.

This paper presents the concept of hazard risk mitigation for a sustainable built environment, analysing the most dangerous types of environmental hazards causing important social and economic damage in Romania: earthquakes, landslides and subsidence, storms and tornadoes, highlighting imperative risk mitigation actions. It also screens temporary measures solving immediate issues, restoration of services, rehabilitation of structures, several case studies and scenario-based mitigation measures.

**Keywords:** Hazard risk · Mitigation strategy · Sustainable rehabilitation · Built environment

## 1 Introduction

### 1.1 Overview of the Term Disaster

The term 'disaster' refers to the impact of different complex hazards on vulnerable communities [14, 74]. It includes disasters associated with extreme natural events such as earthquakes, hurricanes or volcanic eruptions [36]. Hazards can be classified in a wide diversity of typologies, with the objective of optimization of preventive measures (Fig. 1).

A significant part of development assistance spent on the construction of infrastructure gets lost in seconds in the event of a hazard event [86]. The majority of human and economic losses from a hazard event occur either as a result of damage to the built environment or ineffective early warning and evacuation systems [65]. The negative


A. Rotaru (Ed.): CRIT-RE-BUILT 2019, SSGG, pp. 3–34, 2021.
https://doi.org/10.1007/978-3-030-61118-7_1

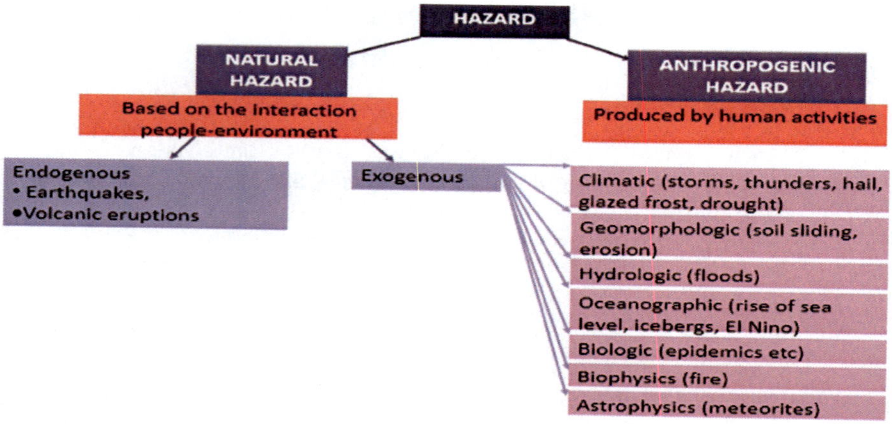


**Fig. 1.** Hazard typology

impact of hazards on communities can be mitigated by taking hazards into account when selecting sites, or designing/strengthening the infrastructure [76].

Hazard mitigation describes actions taken to reduce long-term risks caused by hazards or disasters, such as flooding, earthquakes, volcanic eruptions, tsunamis, cyclonic storms, wildfires, landslides, droughts dam failure, etc. [28, 66]. Each year, 130,000 people are killed, 90,000 are injured and 140 million are affected by natural disasters [2, 79]. Even though the costs of mitigating disasters continue to rise, governments as well as common citizens must find ways to reduce hazard risks. Hazard risk mitigation is an important component of the planning effort once communities plan for a sustainable built environment [75, 77, 81].

Hazard mitigation is essential after a disaster when repairs and reconstruction simply restore damaged property to pre-disaster conditions to get the community back to normal [15, 23]. However, the replication of pre-disaster conditions may result in a repetitive cycle of damage – reconstruction – repeated damage, the recurrent reconstruction becoming more expensive as the years go by. Hazard mitigation takes a long-term view of rebuilding for a sustainable built environment breaking this repetitive cycle. The implementation of such hazard mitigation actions leads to building stronger, safer and smarter to reduce future injuries and damage [16]. To design a disaster-resistant built environment is a successful way to mitigate losses of life and property [17].

A whole lot of techniques are applicable to mitigate the effects of hazards on the built environment [70]. The structure can be designed to resist hazard consequences depending on the hazard nature, location, construction type, and the specific requirements for the building. Later in the life cycle of the building, when incorporating disaster reduction measures into building design, additional opportunities to mitigate the risk from natural hazards may apply to protect lives and properties from damage [39, 41].

Hazard can be described by statements of 'what', 'where', 'when', 'how strong' and 'how often', demanding knowledge of fluctuation in both spatial conditions and temporal behaviour. The level of risk is the combination of the likelihood of something adverse occurring and the consequences if it does. The level of risk results from the intersection of hazard with the value of the elements at risk by way of their vulnerability (Fig. 2).

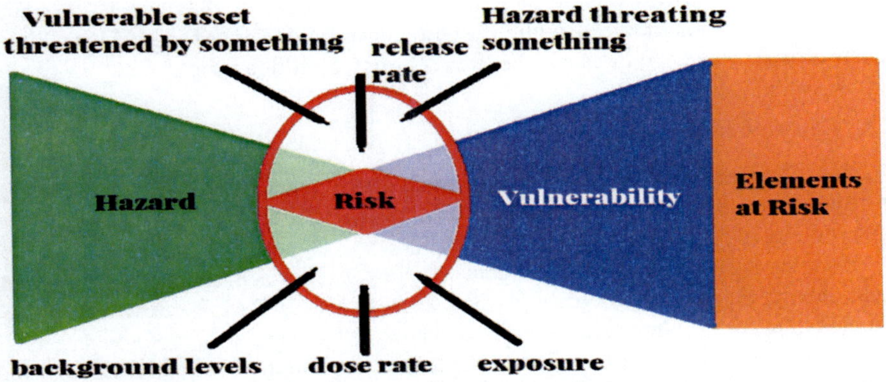


**Fig. 2.** Relationship between hazard, elements at risk, vulnerability and risk (after Alexander, 2002)

The wide discrepancy in losses has focused attention on the role and causes of vulnerability. The dominant view, referred to by scientists as the 'behaviourist' paradigm [2], attributes vulnerability to a lack of knowledge, insufficient preparedness and inappropriate adjustment to specific hazards. On the other hand, the 'structuralist' paradigm attributes vulnerability to disempowerment of the victims through political-economic structures that favour the elite at the expense of the mass of population. The denial of resources means that the affected populations can do little to improve their level of vulnerability. This view sees 'underdevelopment' in some countries as a product of 'development' in others [1].

## 2 Study on Several Types of Environmental Hazards

Natural hazards can be classified into: endogenous and exogenous hazards.

Endogenous hazards are generated by the energy coming from inside the planet such as volcanic eruptions and earthquakes.

Exogenous hazards are generated by climatic, hydrological, biological, etc., from where the following hazard categories result: geomorphological, climatic, hydrological, natural biological, oceanographic, biophysical and astrophysical.

Geomorphological hazards encompass a wide range of processes, such as collapses, landslides, avalanches. The most numerous landslides are recorded on the slopes with moderate inclinations, consisting of clays and alternations of clays, marls, tiles and sands. Anthropogenic causes, related especially to deforestation, play an important role.

Climatic hazards encompass phenomena and atmospheric processes that result in loss of human lives and environmental damage. The most common are the storms: 16

million storms occur annually on the earth resulting an average of 44,000 per day and 2,000 per hour. On the other hand, severe drought can lead to a reduction in water reserves from rivers, lakes and wells, which capture groundwater reserves, and causes a decrease of the harvests. Between 1876 and 1879, China was affected by drought killing 9 million people. Africa was hit by a severe drought between 1981 and 1984, which resulted in starvation of over 1 million people. Since 2011, East Africa has been hit by a new drought wave that threatens the lives of over 9.5 million people.

## 2.1 Earthquakes

The most frequent and violent seismic disasters occur in the 2 major belts of the world: the European and South Asian Alpine Belt and the Circumpacific Belt, where the degree of seismicity is the highest.

Building design will often be influenced by the level of seismic resistance desired. Even though the primary focus of earthquake design is initial life safety and getting people out of the building safely, this level can range from prevention of non-structural damage in occasional moderate ground shaking to prevention of structural damage and minimization of non-structural damage in occasional moderate ground shaking, and even avoidance of collapse or serious damage in rare major ground shaking. These performance objectives can be accomplished through a variety of structural engineering measures or structural components like shear walls, braced frames, moment resisting frames, and diaphragms, base isolation, energy dissipating devices such as viscoelastic dampers, elastomeric dampers, and hysteretic-loop dampers, and bracing of non-structural components [3, 13].

According to the geology of the earth there are areas on the globe with high seismicity where earthquakes are very frequent. The vulnerable countries developed sustainable measures in order to minimize losses by building adequate structures and preparing the population to react in case of a natural disaster. Considering the seismic hazard map, Romania is considered among the three most seismic countries from Europe due to Vrancea sesmic zone, one of the most active seismic areas in Europe.

In the list of most severe earthquakes occurred on the globe are the followings: China (1976), Indian Ocean (2004), Pakistan (2005), Haiti (2010), New Zealand (2011) and Japan (2011).

The great earthquake in Tangshan, China occurred on July 28, 1976 and was considered the most devastating earthquake of the 20th century, producing over 240,000 dead and 164,000 injured. On December 26, 2004, an earthquake with a magnitude of 9.15 on the Richter scale strike in the Sumatra-Andaman area. The earthquake, which lasted about 10 s, produced a tsunami wave that killed about 310,000 people on the coasts of Indonesia, Sri Lanka, South India and Thailand. The earthquake from Pakistan on October 8, 2005, with a magnitude of 7.6 on the Richter scale, had the epicentre in Kashmir and produced 75,000 dead and approximately 106,000 injured. The large number of damages revealed an inaccurate building configuration to earthquake, the country needing aids in the amount of $ 5.4 billion for the post disaster measures. The Haiti earthquake of January 12, 2010, with a magnitude of 7.0 on the Richter scale, with the epicentre in Port-Au-Prince, had 59 replications with magnitudes between 4.2 and 5.9 on the Richter scale, Fig. 3.

**Fig. 3.** Earthquakes in Haiti (2010), New Zeeland (2011), Japan (2011) (from left to right)

The need of efficient anti-seismic restrictions in the design code is sustained by the losses occurred: over 200,000 dead and 3 million people affected by this event. A total amount of over $ 300 million was needed to restore order after the disaster.

Less than 6 months after, the Darfield earthquake of September 4, 2010 with an intensity of 7.1° on the Richter scale, which did not result in significant damage, occurred on February 22, 2011, in Christchurch, an earthquake with a magnitude of 6.3 on the Richter scale, which was ranked second in the number of victims in New Zealand (185 dead, 238 missing and 164 injured). Old masonry buildings were mainly affected, but also the collapse of new buildings such as the one of the Local Television, Fig. 3. Land liquefaction, produced as a side effect of this earthquake, was another factor that led to the increase of the recorded damages, which were estimated at a total of 16 billion dollars [49, 60].

The most powerful earthquake in Japan so far, with a magnitude of 9° on the Richter scale occurred on March 11, 2011, Fig. 3. Strong ground accelerations were recorded in the prefectures of Miyagi, Iwate, Fukushima and Ibaraki to the east, where a huge landslide occurred. Initial research on the dimensions of this slip revealed that the difference between two points on opposite sides of the ridge reached 30 m. The maximum displacement of the land exceeded 100 cm in the Sendai area and 50 cm in the area from Tohoku to Kanto. It produced about 15,000 dead and injured, over 250,000 buildings damaged and was close to producing a nuclear disaster, affecting reactors at the Fukushima power plant [71, 72].

More recent events occurred in Nepal on April 25th, 2015, respectively on February 5th, 2016 in Taiwan. The Taiwan earthquake revealed severe irregularities in the execution process.

Earthquakes destroy engineering constructions in different ways, such as: huge inertia forces induced in structures due to the seismic motion; physical changes of the foundation soil properties (consolidations, settlements, liquefiers); fires; landslides; changing the topography of the land; earthquake-induced waves (tsunamis or tides); direct displacement of the fault on the ground level [59].

Even though Romania's territory is subjected to different natural calamities such as floods, drought, earthquakes and landslides, the event that caused the largest number of dead and required the most financial support to recover from is the March, 4th, 1977 earthquake. Even though not all the earthquakes had serious effects, the seismic risk in

Romania is important [48], being considered the third country in Europe, after Italy and Turkey, according to the number of people lost due to earthquakes in the 20th century. Bucharest is ranked as the tenth capital in the world in terms of seismic risk, being the city with the largest seismic exposure in Europe.

The most active crustal seismic regions in Romania are the Făgăraș and Banat seismic zones, which were highlighted by: the last major earthquake in Făgăraș occurred in January 1916 (estimated magnitude 6.4), respectively in 1991 in the Banat seismic zone with magnitudes of 5.6-5.7. The most powerful crustal earthquake in Romania is considered to be that of October 26, 1550, in the Făgăraș area, having an epicentral intensity of 9 on the MSK scale and a magnitude of 7.2 on Richter scale Lungu et al. [42].

The strongest earthquakes in our country, which cause great material damage, are recorded in Vrancea seismic region, located in the region of the Carpathian curvature. Such earthquakes, which caused numerous human lives losses, were recorded on November 26th, 1802, November 10th, 1940 and March 4th, 1977. Figure 4 shows the epicentres in Romania from 1984 - January 2013 according to the National Institute of Earth Physics, INFP.

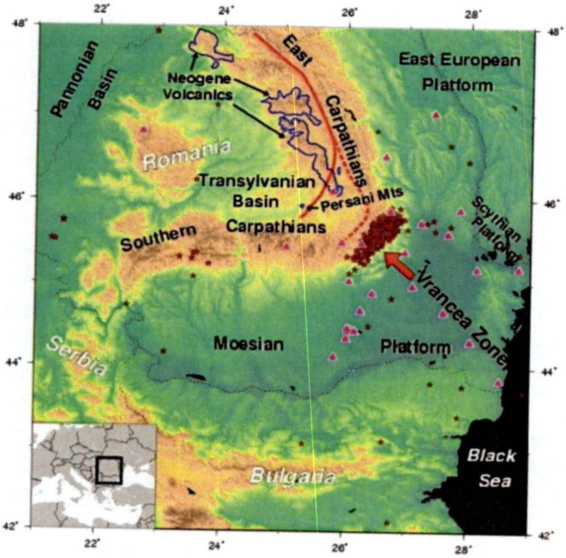


**Fig. 4.** Earthquakes epicentres recorded in Romania between 1984 and 2018 (Catalogue, INFP)

Earthquakes from Vrancea which occur at big depths can reach magnitudes of 7.6–7.7° on the Richter scale, while the crustal earthquakes are much smaller, rarely reaching 5.2–5.4° on the Richter scale [42]. In terms of effects, deep earthquakes are more important. Due to the great depth at which they occur, the remarkable properties of the focal mechanisms that determine a preferential directivity of the breaking process, as well as the nature of the soil, the effects of Vrancea earthquakes are felt at great distances from the epicentre [58, 73].

As a rule, the most serious consequences were reported not in the area near the epicentre, but on both sides, towards the North-East, in the direction of Focșani - Iași - Chisinau and to the South-West, in the direction of Bucharest - Zimnicea - Sofia. A characteristic of earthquakes from Vrancea region is that they feel weaker within the Carpathian Arch Bălat et al. [6].

The shallow crustal earthquakes have local effects, much limited in area, although in some cases they can be felt quite strongly in their epicentre areas, especially in the area of Focșani, Râmnicu Sărat [68]. Currently, in Romania, several national networks for seismic activity recording exist: - the national network for constructions RNSC, of the National Institute for Research and Development in Construction and Sustainable Territorial Development (URBAN-INCERC); CNRRS (National Centre for Seismic Risk Reduction) network, which works in partnership with the RNSC network; the INCDFP network (National Institute for Research and Development for Earth Physics) and GEOTEC network (Institute of Geological and Geophysical Studies).

**Earthquake Risk Mitigation Actions.** The fundamental problem or earthquake risk mitigation is not to find any solution, but to find a "best" solution. What is best usually depends on the circumstances, value and priorities of individuals, enterprise, authorities.

In an ideal world, all buildings have to be strong enough to withstand to any seismic action and damages must not occur. The main drawback of this idea is the higher cost of the structure, which is required in order to guarantee the safety and the stability of the building.

In the real world is not possible to provide maximum safety with no damages for all kind of buildings because the financial resources are limited. Thus, some compromises have to be made in order to keep the cost on a reasonable level.

The legislation from different countries adapt the earthquake risk mitigation according with the importance of the buildings. For vital buildings (like hospitals), no damage is accepted because victims have to be treated in them after a major earthquake.

A second category consist in buildings which are considered very important for the society. National museums, historical monuments and similar buildings have to be preserved and only minor damage is accepted in case of a big earthquake.

For regular buildings, in order to decrease the price of the structure, the best solution consists in the protection of human lives and goods. Damage is accepted in the structure but the stability of the building has to be guaranteed in order to avoid human losses and goods damage.

Last category of buildings involves low importance structure when the value of goods is small. In this case, a large financial investment in the structure is not rational because exceeds the value of the stored goods.

In case of an earthquake there are some general recommendation that people should consider so that the number of injures and human losses to be significantly reduced. Among these are: identifying the safest place to stay; create a correct mitigation plan with your colleagues and family in case of an earthquake; stop, if possible, gas and energy supply [25, 32]; do not use elevators or stairs; if you are in a public place, try not to run in order to avoid panic which can lead to accidents. The most common surviving technique in case of an earthquake is "duck, cover and hold".

Unfortunately, these measures do not prevent structures to be damaged or even to collapse in case of a seismic action. For this, the government of each country has a specific department which is responsible for the seismic risk mitigation [48]. In many countries, due to the change of the code, the authorities started to evaluate the seismic vulnerability of buildings designed before the current restrictions. In Romania for example, there is a code that combines geometric figures with color, which shows the vulnerability state of a structure. The most dangerous constructions are marked with a red circle. This means that urgent rehabilitation works should be performed on the building. Unfortunately, not a lot is happening, because there is not a plan regarding on who should finance the works: the authorities or the owners.

The changes into the new codes represents an important action from the seismic risk mitigation. Adopting the new code provisions should reduce earthquake damage risk.

Another approach on the seismic risk mitigation refers to developing more accurate hazard maps. Research all around the world is focusing on probabilistic maps, instead on the old deterministic ones, emphasizing their benefits. A probabilistic seismic map considers more seismic action, with different characteristics, meaning that is based on more magnitudes, more epicentre depths, different frequencies, in contrast with the classical one which considered only one dangerous event. Also, researchers started to connect different hazards on the same map in order to offer a multi hazard perspective on the area. The main benefit of this is that secondary effects of an earthquake can be known ahead and complex measures can be considered. For example, if an earthquake occurs in an area vulnerable to landslides, when the seismic risk mitigation plan is done, the secondary hazard is also considered and terrain reinforcing works could be performed prior to the seismic action in order to diminish losses.

The potential damage of the critical facilities and infrastructure would be significantly reduced through actions such as: retrofitting public facilities [34], requiring bracing of generators, elevators or other vital equipment in hospitals, reviewing bridges state to determine their susceptibility to collapse and retrofitting the problematic ones, use of flexible piping and installing shutoff valves and emergency connector where necessary (FEMA, 2013).

Raising awareness through the population and educating people at a young age regarding seismic risk, would, on long term, reduce damage. This can be reached by the help of the insurance industry, by developing outreach programs on earthquake risk and mitigation activities in homes, schools and offices; and educating house owners. In the same direction, people responsible for a building, from the design stage to the demolishing should be made aware of proper design and building requirements.

Making a guide for the public with retrofitting measures for structural and non-structural elements can help owner to identify problems in homes or workplaces and take the necessary measures to repair the damage by calling an authorized worker or, if it is not very complicated, doing it by themselves.

Sometimes, earthquake insurance can also be an effective method to mitigate financial loss, even though it does not cover life loss or injury, and typically only partial offsets primary financial loss (SSC Report 99-04, 1999).

## 2.2 Landslides and Subsidence

Landslides and subsidence present a threat to life and livelihood throughout the world, ranging from minor disruption to social and economic catastrophe.

Landslides generate a small but important component of the spectrum of hazard and increasing risk that faces mankind. For landslides, the 'adverse something' might be a large rockslide and its 'likelihood' expressed as the probability of its occurrence [21]. Similarly, the consequences will depend on what is affected by the landslide, the degree of damage it causes and the costs incurred. It is the intersection of humanity with landslide activity that has recast a natural land-forming process into a potential hazard [20].

Ground subsidence can result from mining, sinkholes, underground fluid withdrawal, hydrocompaction, and organic soil drainage and oxidation [53]. Subsidence mitigation can best be achieved through careful site selection, including geotechnical study of the site. In subsidence-prone areas, foundations must be appropriately constructed, and utility lines and connections must be stress-resistant [54]. When retrofitting structures to be more subsidence-resistant, shear walls, geo-fabrics, and earth reinforcement techniques such as dynamic compaction can be used to increase resistance to subsidence damage and to stabilize collapsible soils [55].

Two hundred years or so of science and practice related to slope stability problems have transformed the landslide and subsidence issue from an 'act of God' into a comprehensible geophysical process [62].

Work on landslide 'hazard' assessment is site-based and driven by development projects and engineering concerns. Conventionally, this is approached by stability analysis of the site, determined from the balance of shear stress and strength and expressed as a factor of safety [26]. Recognition of the natural variability of factors controlling stress suggests that the factor of safety is realistically evaluated in probabilistic terms [61]. The major challenge for site-based stability analysis is the conversion of the factor of safety or equivalent stability assessment into a useful expression of hazard that can then be used as a component of risk assessment [18]. This involves employing the factor of safety along with temporal variability in triggering factors to determine the probability of failure per unit of time. Probability of occurrence, in turn, needs to be qualified by a statement of expected behaviour of the failure in terms of its impact characteristics [56].

Predicting the nature of the landslide, particularly for first-time failures, is another challenge for landslide hazard science [19].

Gravity-driven movement of earth material can result from water saturation, slope modifications, rainfall occurring after wildfire, and earthquakes [50, 78].

Techniques for reducing landslide risks to structures include selecting non-hillside or stable slope sites; constructing channels, drainage systems, retention structures, and deflection walls; planting groundcover; and soil reinforcement using geo-synthetic materials, avoiding the cut and fill sites [82]. Design for the direct effects of a landslide is not cost-effective.

Landslide and subsidence risk mitigation actions (Fig. 5). Ground subsidence can result from mining, sinkholes, underground fluid withdrawal, hydro-compaction, and organic soil drainage and oxidation. Subsidence mitigation can be achieved through careful site selection, including geotechnical study of the site [33].

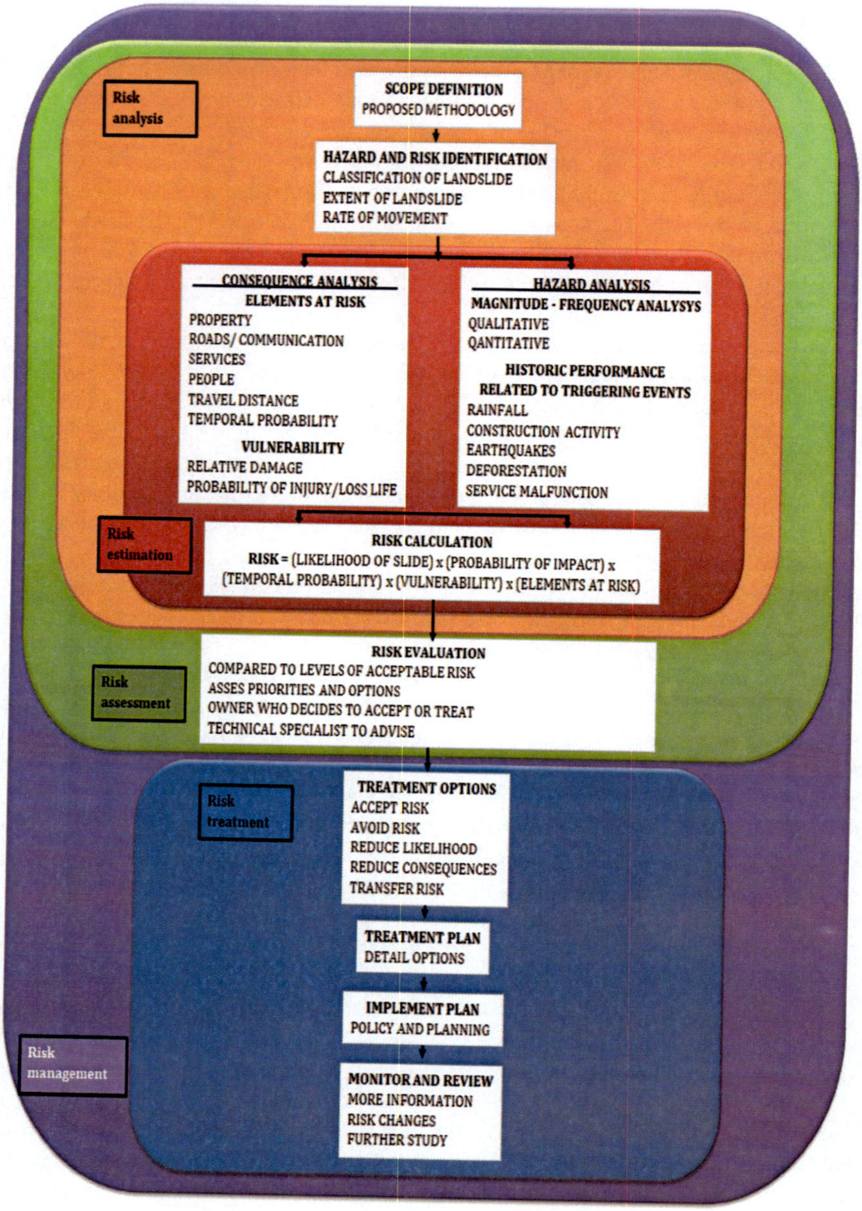


**Fig. 5.** Landslide risk mitigation actions

In subsidence-prone areas, foundations must be appropriately constructed, and utility lines and connections must be stress-resistant. When retrofitting structures to be more subsidence-resistant, shear walls, geo-fabrics, and earth reinforcement techniques such as dynamic compaction can be used to increase resistance to subsidence damage and to stabilize collapsible soils [5, 29].

## 2.3 Wind (Storms and Tornadoes)

Winds are generated by naturally convective movements of air masses in the nature and the moderate (speed) winds do not create hazard risk to the built environment. In the context of intense urbanization with an increasing number of high rise buildings or flexible, lightweight large buildings or constructions (like membrane structures, cable supported roofs) even the medium speed winds can produce important lateral forces and dynamic actions (as galloping, buffeting, vortex shedding, fluttering, look in effect etc.) that have to be taken into consideration in the design process.

In addition to the presented direct influence, the indirect risk is given by debris raised and carried by winds. First involves the structure, second involves the façade of the building and the pedestrian life.

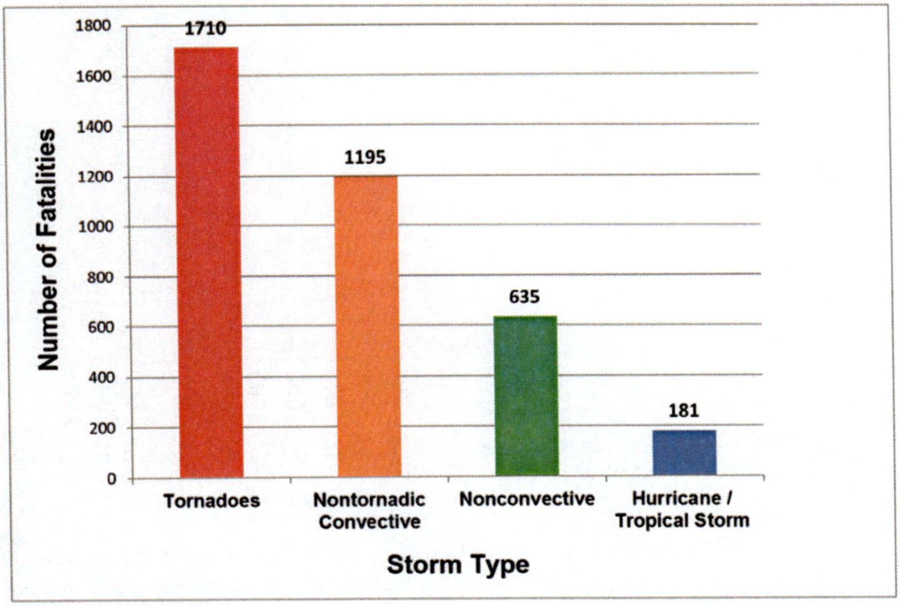


**Fig. 6.** Wind-related fatalities by storm type, between 1977–2007 in United States (after A. W. Black, 2009 [9])

On the other hand, high speed winds (with significant increase in maximum wind gusts) and severe windstorms (tropical storms, hurricanes, tornadoes (Fig. 7) or nontornadic convective winds) cause enormous loss to life and property worldwide (Fig. 6). Hurricane Andrew in 1992, with total economic losses estimated at $30 billion is labeled as one of the costliest natural disasters in the history of the United States of America.

Hurricane Georges in 1998 caused insured losses of about $3.5 billion in the USA and the Caribbean and Hurricane Floyd in 1999 caused insured losses of $1.96 billion in the USA. Thousands of people lost their lives in the "Super Cyclone" 05B that struck coastal India in October 1999.

The insured losses in the December 1999 European windstorms Lothar and Martin have been estimated at $5.8 billion and $2.4 billion, respectively, Hurricane Katrina in August of 2005 killed over 2,000 individuals, destroyed or damaged nearly 300,000 homes, and caused roughly $96 billion in damages, Hurricane Sandy in October of 2012 wreaked havoc along the East Coast killing 147 individuals and damaging or destroying over 650,000 homes [69]. These figures show the serious implications of wind-related catastrophes to both life and property [38].

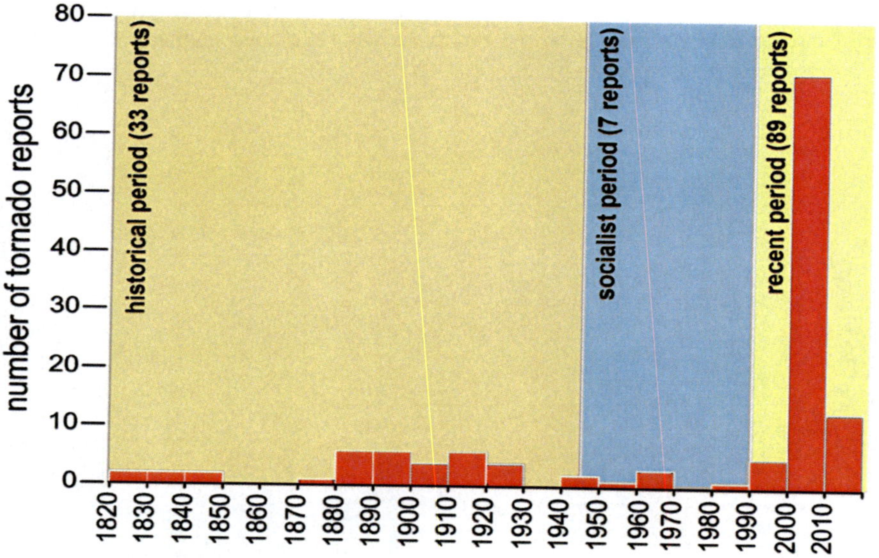


**Fig. 7.** The distribution of tornado reports per decade between 1822 and 2013. The first decade includes 1822–29 and the last decade includes 2010–13.

In Romania with a more temperate continental climate, the strong storms or tornadoes are less violent. There are studies summarizing the tornado climatology of Romania between 1822 and 2013 based on a dataset comprising 129 tornadoes reported on 112 days. The spatial distribution of tornado reports shows that tornadoes are more frequently reported over eastern Romania, with a maximum over south-eastern Romania [approximately 1.5–2.25 $(10^5\ \text{km}^2)^{-1}$ every 5 yr]. We speculate that a large number of tornadoes over south-eastern Romania can be attributed to the mesoscale environments over this region that are more favourable for tornadoes compared with other regions of the country, but further studies are necessary to confirm this hypothesis [4].

**Storms and Tornadoes Risk Mitigation Actions.** The reduction of the hazard risk due to the dynamic action of the common winds involve, as in the case of earthquakes, dynamic structural analysis in order to provide adequate strengthening, stiffness and dynamic behaviour to the construction by design methods. In many cases of high rise or aeroelastic structures, a more precise analysis can be carried out using studies on physical models of the building and surrounding aria tested in boundary layer wind tunnels. The hazard risk derived from the indirect wind effect (wind born debris) is more damaging with relation to humans, because the level of action is around ground where pedestrians can be always present.

"Objects raised and detached by wind can be carried out and, depending on the intensity of the wind, can cause significant impact on humans and on the building skin. Topics that should be taken into consideration in design are: impact resistant materials and elastic binds of façade and roof elements, but in the same time a more secure and homogenous site planning with green barriers and irregular relief and green roofs.

Benefits of this are, beside a friendly urban landscape, a wind filter that can attenuate winds and the amount of flying objects that can harm" [47].

Regarding the vulnerability of structures to extreme wind events, the studies are focused on three directions (a) damage assessment, (b) field examinations of wind-structure interaction and (c) hurricane risk assessment from the insurance perspective.

a) Damage assessment - Research mostly deals with observed damage from extreme wind events in order to correlate building damage intensity to measured wind speeds and to examine building performance when wind speeds were close to building code values. Buildings were grouped in various categories:

   – Fully engineered buildings, which performed well, even for wind speeds above the code specified values. The damage was limited on roofing material and façade.
   – Pre-engineered buildings suffered from structural framing damage for wind speeds close to, or over, the code-suggested values. Weak links (e.g. overhead door) were identified in such structures and held responsible for progressive damage.
   – Marginally engineered buildings, which were affected significantly at all wind speed regimes.
   – Non-engineered buildings, which were severely damaged when wind speed reached the code specified values.

Also, wind-induced damage can be classified to structural (lack of uplift load path, roof sheathing loss at corners and gable end wall loss) and non-structural (loss of roof shingles and vinyl siding, vulnerability of soffits, breach through attic vents and better performance of hip roofs over gable roofs).

b) Field studies of wind-structure interaction - This set deals with the effect of wind-structure interaction, which can be very important for groups of buildings, where the wind force acting on each building is heavily influenced by the nearby structures that may be shielding it or channelling the wind.
c) Hurricane risk assessment – Especially for the insurance and risk modelling, it has become important to estimate the risk faced by structures, especially in the catastrophic event of a hurricane. Risk information related to hurricane events in the

context of defining risk assessment models to be used for insurance purposes. The effort was mainly focused on refining the "general" vulnerability models by specific and detailed models related to factors such as different building types, occupancy, construction material, height etc.

# 3 Sustainable Rehabilitation of the Damaged Built Environment - Risk Mitigation Actions

## 3.1 Temporary Measures

In a world characterized by rapid change, disaster management must adapt rapidly and effectively to global and local tendencies in order to make the process as reliable as possible. Every day, people all around the world are exposed to devastating natural disasters, so that global statistics talk about a dramatic increase, by 93% in the last 40 years, of hazard risk.

The elements at risk in case of a hazard manifestation are: • Buildings and facilities; • Population; • Environment; • Production; • Economic activities [40].

It is highly important for certain facilities to be maintained functional after a hazard in order to provide the necessary services to the community. These facilities are: • Police stations; • Hospitals; • Fire stations; • Shelters (ex. Schools, churches).

The probability of an event to take place can be estimated based on a risk matrix, which takes into consideration in the evaluation process the degree of severity and the frequency of that potential event. The Risk Matrix provides three levels of risk: low risk (considered acceptable), moderate (may or may not be acceptable) and high (unacceptable). Manifestation of a risk comes associated with two types of losses, primary and secondary (Table 1).

In the context of six major natural hazards (earthquakes, volcanoes, tsunamis, tropical cyclone storm surge, tropical cyclone wind and floods), USA, Japan and European Union, restructured their policies and strategies in hazard management, investing time, effort and money to reduce the consequences. Characteristic for EU are joint strategies and action plans like Hyogo Framework for Action, "Managing risks to achieve resilience" [7, 8, 45] or a ten years plan on "Building the resilience of nations and communities to disasters" [24, 30, 44]. International Federation of Red Cross/Red Crescent Societies (IFRC) has developed a complete set of International Disaster Response Laws (IDRL). In November 2007, IFRC presented "Guidelines for the domestic facilitation and regulation of international disaster relief and initial recovery assistance" ("IDRL Guidelines"), unanimously adopted by all High Contracting Parties to Geneva Conventions.

Following the same trend as EU, Romania implemented several hazard management decisions and laws, resilience strategies [11, 37, 43], like Decision no. 762/2008 for National Strategy for Emergency Situations Prevention, and became a contributory member of International Strategy for Disaster Reduction (ISDR) in order to put into practice Hyogo Framework for Action. Even more challenging is to create a joint strategy and policy at NE Romanian border with Republic of Moldavia, since the legislation and intervention process is different.

Since 2004, Romania has a National Management System associated with Emergency Situations, IGSU – General Inspectorate for Emergency Situations [31], a medical

**Table 1.** Primary an secondary loses

| | Human - social | Physical | Economical |
|---|---|---|---|
| Primary effect | • Fatalities<br>• Injuries<br>• Loss of income or employment opportunities<br>• Homelessness | • Ground deformation or loss of ground quality<br>• Structural damage or collapse to buildings and infrastructure<br>• Non-structural damage and damage to contents | • Interruption of business due to damage to buildings and infrastructure<br>• Loss of productive workforce through fatalities, injuries and relief efforts<br>• Capital costs of response and relief |
| Secondary effect | Disease<br>Permanent disability<br>Psychological impact<br>Loss of social cohesion due to disruption of community<br>Political unrest (govt. response is perceived as inadequate) | Progressive deterioration of damaged buildings and infrastructure which are not repaired | • Losses borne by the insurance industry weakening the insurance market and increasing premiums<br>• Loss of markets and trade opportunities through short term business interruption<br>• Loss of confidence by investors, withdrawal of investment<br>• Capital costs of repair |

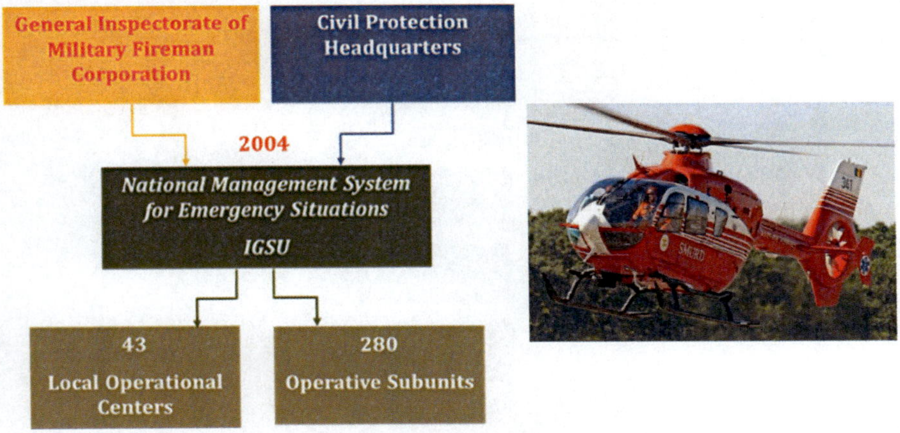


**Fig. 8.** Romanian National Management System for Emergency Situations

emergency unit SMURD – Mobile Emergency Service for Resuscitation and Extrication and an emergency phone number 112 (Fig. 8).

### 3.2 Restoration of Services

An integrated design approach shall provide opportunities for energy efficiency [35, 80]. The issues to be reviewed shall include site, plan form, orientation, passive ventilation and passive solar strategies and daylighting (Fig.9).

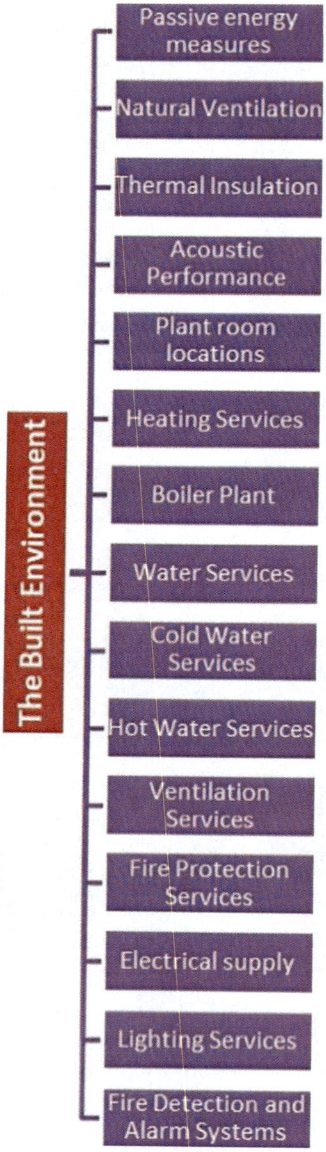


**Fig. 9.** Restoration of building services for energy efficiency

The use of passive energy measures to achieve a comfortable internal environment shall be employed where possible [67]. The form of the building shall be developed to take account of the need to minimise energy consumption with particular emphasis on maximising the use of natural ventilation, daylighting, useful solar gain and minimising heat losses and unwanted heat gains [64]. Ventilation where possible shall be natural ventilation by means of permanent wall vents and windows with opening sections.

Thermal insulation standards shall meet or exceed the prevailing standards but shall also be considered in the context of the balance of heat loss and gain to minimise the running costs and maintain comfort conditions.

The heating services shall comprise fuel installation, the heating centre plant room installation, the space heating and distribution services and controls.

Adequate protection services to enable the building occupants to evacuate safely against the risk of fire shall be provided in the form of handheld type fire extinguishers and fire blankets.

**Energy system**

The energy provided to the built environment need to adopt new strategies based on the impact category, on energy supply, and energy demand.

**Energy Supply Strategies:**

- Construct additional or redundant transmission or distribution line capacity to offset anticipated efficiency losses.
- Establish new coastal power plant siting rules to minimize flood risk.
- Install solar PV technology to reduce the effects of peak demand.
- Upgrade local transmission and distribution network to handle increased load associated with higher temperatures.
- Expand hazard preparedness programs.
- Require utilities to develop storm hardening plans regularly.
- Retrofit power plants so they use less cooling water.
- Automate restoration procedures to bring the energy system back online faster after weather-related service interruption.
- Provide additional support for distributed generation systems to spread climate risk over a larger area.

**Energy Demand Strategies:**

- Install steam-powered chillers to reduce the burden on local power system on hot days
- Establish or expand demand-response programs which encourage consumers to voluntarily reduce power consumption during peak demand events
- Establish public education programs to promote lifestyles that are less energy-dependent
- Employ passive building design strategies to maintain minimum comfort or lighting levels even in situations where energy system losses occur.

### 3.3 Scenario-Based Mitigation Measures

Each hazard manifestation has a certain level, impact and recovery level. A disaster arises when an extreme natural event (ex. storm, earthquake, flooding) strikes a vulnerable society. Whether a natural event becomes a disaster depends on the social, economic, ecological and political characteristics of the society in question.

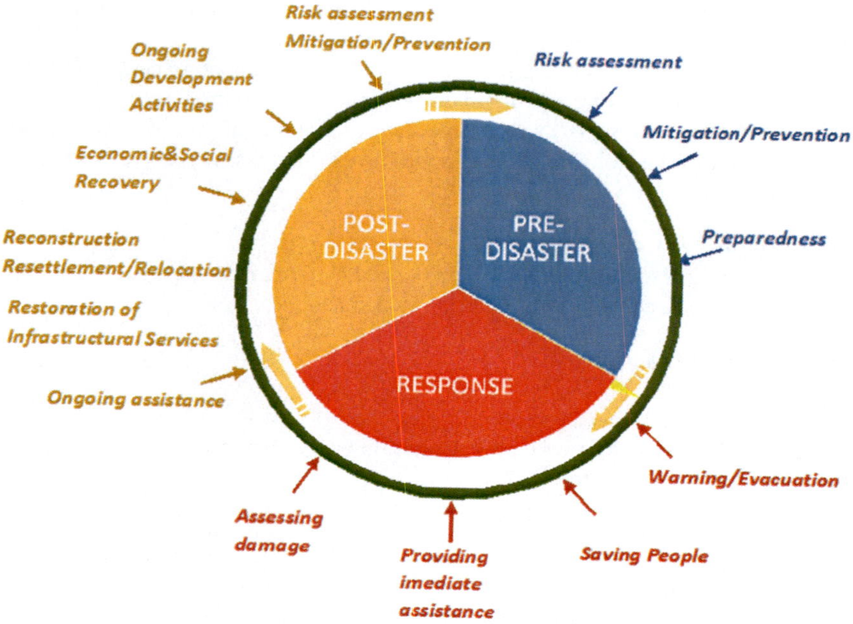


**Fig. 10.** Disaster risk management diagram [27]

The question is: are we, as individuals, and the authorities, as managers of the situation, prepared to face a worst-case scenario?

In a potential example of worst case scenario for the City of Iași was created an open list of main disruptions, like: • Building collapse; • Transport infrastructure destruction; • Human casualties: deceased or wounded; • Destruction of connections with other cities and with main emergency units; • Lack of food and water; • Lack of salvage units and shelters; • Economic breakdown.

The process of planning, implementing, evaluating and adapting strategies, procedures and measures relates to the analysis, reduction and transfer of disaster risks, with the aim of reducing hazards and vulnerability and strengthening the adaptation capacities of individuals, households, communities and state structures. Disaster risk management is a continuous process that involves physical and non-physical measures and takes into account the underlying risk factors within a society (Fig. 10) [46]. Disaster risk management [85] aims to avoid the generation of new risks, improves resilience to the effects of natural events and contributes to sustainable development [10, 12, 22].

Worst Case Scenario for Natural Hazards associated with Road Network in Iaşi County is based on the information received from Regional Direction of Roads and Bridges. For the scenario, the main components that can represent "critical points" in case of a natural hazard have been considered, points that might be part of a worst-case scenario for traffic communication sector.

Worst case scenario is correlated with destruction of main communication systems, bridges and roads. These constructive systems have the role to connect different areas. In their absence, a certain area remains blocked, potentially without possibility to provide food and water. According to DRDP (Road and Bridges Administration) representatives, in case of a natural hazard, they are capable of intervening with different types of temporary components for bridges and roads, in order to give into usage the network as soon as possible.

Statistics mapping the hazard risk at national level situate Romania on the 82nd position, in a total of 173 countries, with high risk of natural disasters. Romania position in European ranking of exposure to hazards is confirmed by annual hazards manifestation, but mostly by several dark historic events of natural disasters, like May 1970 floods resulting in over 50 deaths and significant economic loses or March 1977 Vrancea Earthquake, with over 1,500 deaths and huge material losses [34].

In order to underline the importance of a strategic risk management, worst case scenario focuses on an specific area from Iaşi city centre, namely Podu Roş Intersection and Bahlui River Bay (Fig. 11).

The city of Iaşi still has a large number of buildings built previously to 1977 based on old standards, buildings that were affected by the earthquake, that are unstable, that have not been properly rehabilitated. A significant number of buildings of this type are situated close to the most important traffic node, Podu Roş Bridge. These buildings, marked with"a red bullet" represent public danger and a significant risk of crashing in case of an earthquake, producing, according to experts, a large number of victims.

Podu Roş Intersection has been considered for the case study since it is the most significant road intersection in the city, connecting the city centre with 5 large neighbourhoods leading to city exists. More than this, a fail in accessing this intersection would mean breaking the connection with the city health and safety institutions, since all the hospitals are in the city centre.

Another characteristic of the area, from strategic point of view, is that Bahlui River crosses a large area of Iaşi, breaking the City in half. All connections between the two parts are made by bridges. In the last years the City developed greatly outside the old City area. Still, main emergency services, hospitals and police are situated in the old part of the City, leaving the only few private hospitals with reduced capacity, low level of traffic facilities (Arcadia Hospital, Providenţa Center) and no emergency services to the newly developed part of the city.

The two parts of Iaşi City are connected by several bridges, along Bahlui River, some of them only with pedestrian traffic, others cover also vehicle traffic. Of these, most significant, from capacity and traffic point of view, are the ones in Podu Roş and Tudor Vladimirescu (Fig. 12, Fig. 13). It is vital to preserve the functionality of these bridges in case of a hazard, since they provide access to medical assistance and emergency units.

**Fig. 11.** Podu Roş Intersection and Bahlui River Bay in Iaşi city, Romania

In the worst case scenario, failure of these bridges would mean isolation for half of City population, limited access to emergency units and delay in salvage operations.

## 3.4 Case Studies

**Earthquakes**
Following the 1940 earthquake which has been considered a major one, the 4th March 1977 Vrancea earthquake proved that seismic load is not yet correctly assessed in Romania.

At 21:22, Bucharest was hit by an earthquake of 7.4 on Richter scale (Fig. 14). It lasted around one minute and killed more than 1,500 people and injured around 11,300 in Romania and also in the neighbouring countries. 90.2% of the casualties occurred in Bucharest due to the collapse of 19 high-rise buildings, which were constructed as reinforced concrete frame structures, considering only the gravitational loads, according to regulation from the time. Two buildings constructed in 1962 and 1974 of 10 and 11 floors partially collapsed [6]. The main failure mechanism was similar to those observed after the 1940 earthquake, pancake collapse pattern [83, 84, 87]. Three public buildings collapsed, but considering the late hour the number of casualties was significantly reduced.

Other casualties occurred across Romania in Prahova (Plopeni), Ploiesti, Iasi and Craiova where unreinforced masonry buildings collapsed killing and injuring several people [6]. Some data recovered from Rosiorii de Vede, Turnu Magurele and Alexandria showed that gable walls of old masonry house have been the major source of injuries. Zimnicea city, in the south of Romania, suffered an 80% buildings destruction.

The earthquake was felt across the entire Balkan Peninsula. The most affected city was Svistov from Bulgaria where three block of flats collapsed killing 100 people [83].

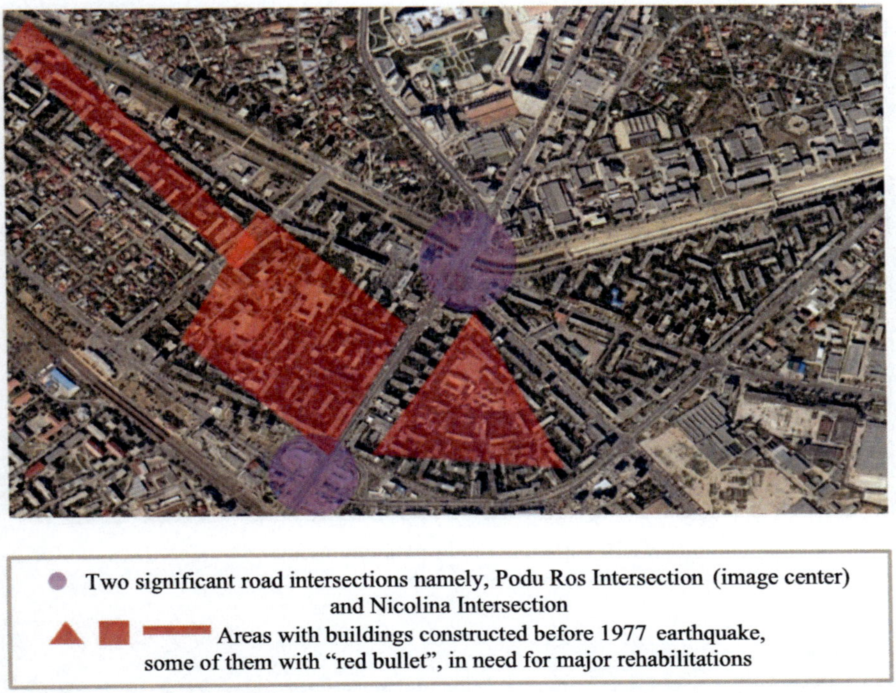


**Fig. 12.** The worst-case scenario – Podu Roş Intersection, Iaşi City, Romania

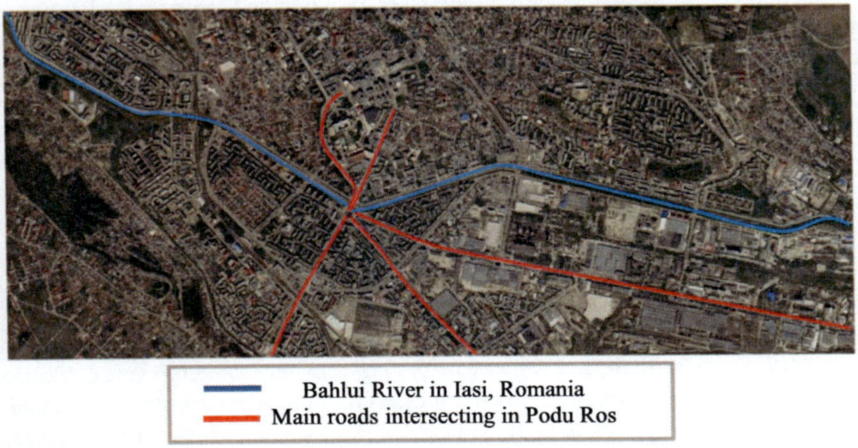


**Fig. 13.** The worst-case scenario –Podu Roş Intersection, Iaşi City, Romania

The problems arising from this earthquake highlighted the need for better conformity to earthquake of buildings, as well as the need for new seismic prescriptions, more accurate regulation on site, namely the urgent need for some draft laws regarding the responsibility of designers and executors. Following this event the 1978 seismic code appeared. Currently we design according to Eurocode, but the national norm has the spectrum defined based on the recordings of 1977 Vrancea earthquake.

**Fig. 14.** 4th March 1977 earthquake, Romania

**Landslides**

Pârcovaci locality, Iaşi county, Romania, is located west of the town of Hârlău. It spreads about 7 km along the Bahlui River [52]. The hills to the right of the Bahlui River are prone to landslides. The regularization of the flood rates occurs under the management of the Department of Water Resources of the Prut River, through the Storage Lake Pârcovaci, a hydropower work installed at approx. 10 km upstream of the town of Hârlău. The landslide occurred in Pârcovaci village in early December 1996 (Fig. 15). The landslide was profound, induced on a wide-spread area in continuous expansion, determining the damage of 97 peasant households out of the 186 located in the area.

a) Characteristics of the landslide: The affected area is located on the right side of the Bahlui River, downstream of the Pârcovaci accumulation dam (2–3 km). b) Characteristics of the slide: the length measured on the steepest slope gradient up to the axis of the Bahlui River is about 2.5 km with the width of 1 km, with extension tendency [51]. The area strongly affected is about 250 ha. Geomorphologically, the area presents a very pronounced relief energy, there being a level difference between the Bahlui River bed and the ridge slope over 250 m. The slip surface comes into the site in the upstream area as very rough, with vertical cracks in the order of tens of centimetres. Vertical displacements attributed to zonal collisions turned up, some with shallow overlaid compressed layers, others through stretching due to the layer bending. In the order of several meters, visible slides through horizontal movements go out between the slip surfaces and seemingly stable areas. The general deep slip surface rounded with a partial sliding mass directed to the valleys of the torrents facing the Bahlui River. The area had high humidity as a result of the existence of the groundwater at shallow depths highlighted by a series of springs and a very high level of water in the local wells (1–1.5 m from the ground elevation).

c) On the land affected by the landslide (250 ha), 186 peasant households (houses and annexes) settled down, and the rest availed for agricultural use and grazing. The peasant houses covered with tile, tin or asbestos cement held local materials (adobe, shallow foundations). The landslide affected 97 households (400 people) and destroyed a large number of houses (cracks in the walls and foundations of about 10–15 cm, Fig. 16).

**Fig. 15.** Lanslide – Pârcovaci village, 1996

**Fig. 16.** Cracks in the walls, Pârcovaci village, 1996

The landslide changed the hydrostatic regime of the area, completely changing the directions of water movement, also having effects on the well water [57]. By raising the slope of the Bahlui River about 40 cm, the drainage slope reduced, the riverbed narrowed, which resulted in an accumulation of water upstream with the continuous rise of the level, leading to the danger of flooding 300 peasant households (1,300 individuals). The decrease of the flow section of the Bahlui River by sliding mass obstruction could have affected the normal functioning of the Pârcovaci storage lake (2.75–4.02 million cubic meters of water) preventing the taking-over of the floods from the winter-spring period.

Measures: Deterioration of buildings made them unfit to habitation, endangering the lives of people, necessitating emergency evacuation, finding acceptable housing solutions in hygienic-sanitary conditions. The slip area being strongly fragmented and with sliding tendencies, it was necessary to allocate plots for new farms. To avoid the flooding of more or less than 300 households as well as the disturbance of the normal functioning of Pârcovaci storage lake by raising the Bahlui River thalweg and reducing its flow section, the riverbed of the water storage upstream of the slipped area necessitated urgent corrective measures. Technical improvements to limit the landslides and their effects in Pârcovaci village were: 1. Interception-collection-evacuation works of the groundwater from the plateau area carried out by the rapid capture and evacuation of the surface stagnant waters originated from heavy rains [63]. They employ interception drains, absorption drains, or spring water catchment chambers. 2. Interception-collection-evacuation works of the water within the slip surface originated from heavy rains through a network of open drainage collection channels, systematic drainage networks through absorbent drains, spring water catchment chambers, land modelling works, consolidation of slopes and the riverbed of the existing cloughs. 3. Anti-erosive roads and water-related works to consolidate the technological road platform through compaction and ballast addition. 4. 2.85 km of marginal channels along to protect the road platform; 5. Nine platforms at the junction of roads with canals of which a paved bridge; 6. Consolidation of the river bank with a pitching wall on a concrete foundation.

**Storms and tornadoes.** A building that shelters the archaeological vestiges of the ancient Goto-Dacical civilization discovered on Catalina hill, near Cotnari village, Iaşi County, Romania is a cable supported structure with curved hyperbolic paraboloid (or saddle shaped) roof. It is situated on the ridge of a steep hill (at 100 m over the rest of the site) near the crest. The structure is submitted to strong winds because the aerodynamic roughness parameters of the ground surface are specific to a category 3 open area - zo $\cong$ 0.1 m -on a distance of some km around. In addition, the particular position on the top of the hill emphasizes the wind action on the structure.

The thin covering made of fibber reinforced plastics was destroyed by the fluttering of the supporting cables and because the distance between the fixing points was too large comparing to the thickness of the covering.

The rigid model (Fig. 17) represents the saddle shaped structure at the length scale of 1/100. The boundary layer developed in the tunnel simulates open country exposure with a vertical velocity profile represented by a power-low with exponent a $= 0.185$. The similarity of Reynolds number is provided by artificially roughening of the external surface of the curved model. The model was equipped with 40 pressure taps. The first 39 placed on the envelope measure the local normal components of the dynamic pressure from the outer surface and the last one measures the internal pressure inside the model.

The mean dynamic velocity pressure used as reference is measured with a Pitot probe placed in the upstream and at the middle high of the model roof.

After measuring, the internal and local external pressures are composed to determine the local normal resultant of wind action on the envelope. The results are expressed as local pressure coefficients related to the reference pressure.

**Fig. 17.** Aspects of the damaged structure in 1994 and the building scaled model placed in the boundary layer wind tunnel

Using a software package based on the finite element method a numerical model of the same structure was created (Fig. 18). The software package that was tested in the Department of Structural Mechanics of our University develops a non-linear geometrical analysis, which takes into consideration the deformation of the structure.

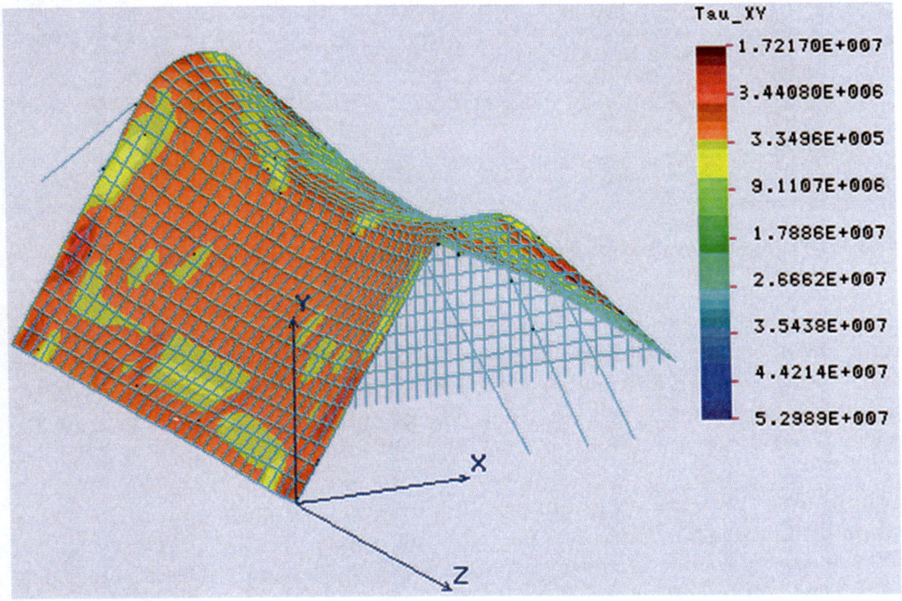


**Fig. 18.** Unitary shear stresses on envelope numerical model for the wind direction $\alpha = 0^{o}$.

The numerical model is made of 1072 finite elements and has 6120 degrees of freedom. As loading data, the values of the local normal pressure measured on the physical model in the boundary layer wind tunnel were used. We made the hypothesis that between the measuring points the pressure evolution is linear. The local pressure coefficients were multiplied by the basic dynamic pressure of the site and by a factor that reflects the influence of the hill where the structure is constructed. The basic dynamic pressure according EN 1991-1-4:2005 corresponds to a basic wind speed $\bar{U}10 = 30$ m/s (so gv = 550 N/m$^2$). This basic dynamic pressure of the site was multiplied with a factor that reflects the influence of the hill. This factor (c = 1.75 for the basic wind velocity) was determined in the wind tunnel on a 1:400 model where the hill and the structure has been reproduced.

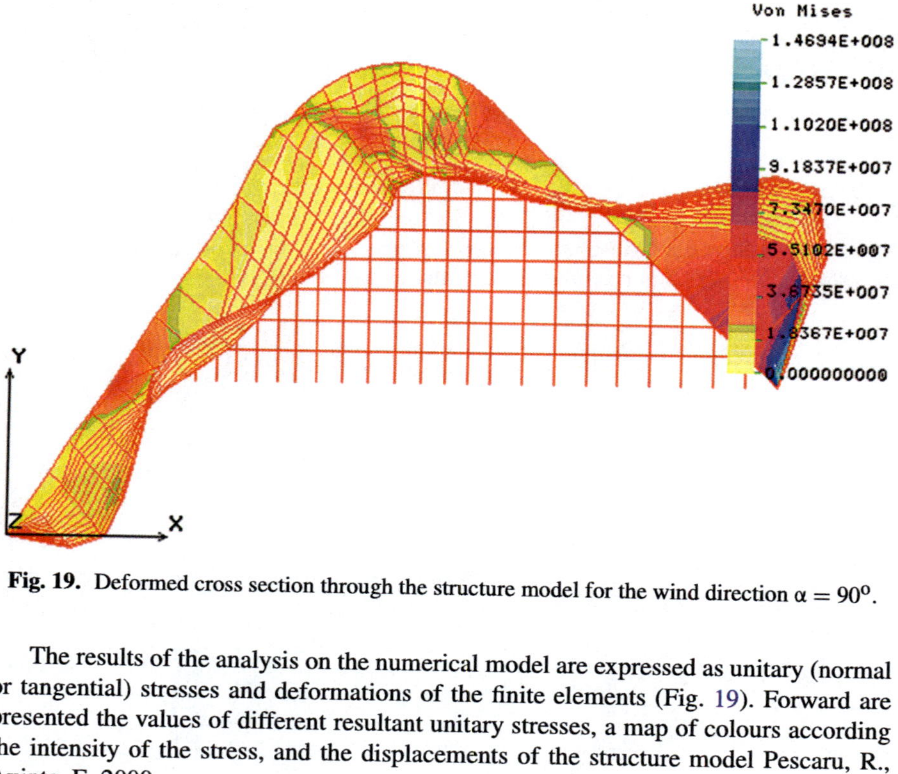


**Fig. 19.** Deformed cross section through the structure model for the wind direction $\alpha = 90^o$.

The results of the analysis on the numerical model are expressed as unitary (normal or tangential) stresses and deformations of the finite elements (Fig. 19). Forward are presented the values of different resultant unitary stresses, a map of colours according the intensity of the stress, and the displacements of the structure model Pescaru, R., Axinte, E. 2000.

- The results of the wind tunnel tests on the rigid model of the cable supported structure from the Catalina hill emphasises the following aspects:
- The maximum wind loads considered in the initial design of the structure were underestimated. An important aspect, which was neglected, is the particular position of the building which has a major influence especially on the negative pressure, more developed comparing with the usual conditions. Concerning these aspects, the revised

edition from 1990 of STAS-10101/20 regarding the wind loads on buildings, stipulate the necessity of wind tunnel tests for the design of lightweight structures with large size and unusual shape.

- The results of this program of measurements is a more accurate assessment of the wind loads on a hyperbolic paraboloid suspension structure, that has no analytical solution in STAS 10101/20-1990, and provide the appropriate data for the proposal of new solutions of reconstruction.

The membrane with an open area of 1.25%, showed that the magnitude of the surface pressure coefficients was generally considerably reduced and that the loading across the membrane material were alleviated in almost all conditions.

## 4 Conclusions: Hazard Risk Mitigation Opportunities

Mitigation opportunities are available to reduce losses from natural hazard events [86].

- Zoning as a form of land-use management and control can help regulate population and industrial development in hazard-prone areas. It can be used to control building density, adjust the timing of development plans, and better define "allowable" development. Maps that identify high-hazard areas should be adopted and used to guide, restrict or limit development. Models to identify hazard areas need to be developed or tested to verify accuracy.
- Control or protective structures may be useful in protecting life and property. Examples include dams to control floods and structures to mitigate de risk or control landslides.
- Building codes designed to improve construction, reinforcement, and anchoring of buildings. A nationwide hazard-based code may help to ensure implementation of standards appropriate for hazard- and damage-resistant structures. Examples of progress in this area are the seismic regulations for new and existing buildings that have been developed in Romania.
- Evacuation planning and preparedness programs are helpful in protecting residents in areas subject to imminent danger. In general, evacuation saves lives but does not result in significant damage reduction.
- Warnings and forecasts are useful for alerting communities and citizens to an impending hazard event. Both real-time, and longer range forecasts should be provided. Warnings and forecasts are issued in preparation of possible evacuations and to prompt property protection measures.
- Education and awareness efforts provide hazard information to the public to make them aware of the impacts of possible hazards. Information can include, but is not limited to, graphic depictions of hazard areas and evacuation routes, and simple, effective mitigation actions.
- Monitoring and data collection are necessary to support research, to provide affected communities and citizens with better warnings and forecasts, to understand hazards, and to develop loss reduction methodologies.
- Relocation of utilities and transportation routes out of extremely high risk areas can be beneficial.

- Research on hazard processes and model development are needed to understand hazards and their consequences. Dedicated hazard-specific research facilities could coordinate research efforts with academic institutions and international organizations.

## References


1. Alcántara-Ayala, I.: Geomorphology, natural hazards, vulnerability and prevention of natural disasters in developing countries. Geomorphology **47**, 107–124 (2002)
2. Alexander, D.: Principles of Emergency Planning and Management. Dunedin Academic Press, Terra Publishing (2002)
3. Anderson, J. E., Wulfhorst, G., & Lang, W.: Energy analysis of the built environment - a review and outlook. Renew. Sustain. Energy Rev. (2015). https://doi.org/10.1016/j.rser.2014.12.027
4. Antonescu, B., Bell, A.: Tornadoes in Romania, Monthly Weather Rev. (2015)
5. Ardizzone, F., Cardinali, M., Carrara, A., Guzzetti, F., Reichenbach, P.: Impact of mapping errors on the reliability of landslide hazard maps. Nat. Hazards Earth Syst. Sci. **2**, 3–14 (2002)
6. Balan, St., Cristescu, V., Cornea, I.: The Romania Earthquake of 4 March 1977. Editura Academiei, Bucharest (1982). (in Romanian)
7. Bene, C., Newsham, A., Davies, M., Ulrichs, M., Godfrey-Wood, R.: Review article: resilience, poverty and development. J. Int. Dev. **26**, 598–623 (2014)
8. Bene, C.: Towards a Quantifiable Measure of Resilience, Institute for Development Studies, working paper, vol. 2013, No. 434 (2013)
9. Black, A.W., Ashley, W.S.: Nontornadic convective wind fatalities in the United States, Nat Hazards (2009). https://doi.org/10.1007/s11069-009-9472-2
10. Boorman, J., Fajgenbaum, J., Ferhani, H., Bhaskharan, M., Arnold, D., Kohli, H.A.: The centennial resilience index: measuring countries' resilience to shock. Global J. Emerg. Mark. Econ. **5**(2), 57–98 (2013)
11. Briguglio, L., Cordina, G., Farrugia, N., Vella, S.: Economic Vulnerability and Resilience: Concepts and Measurements, The World Institute for Development Economics Research (WIDER), Research Paper No. 2008/55 (2008)
12. Bristow, G., Healy, A.: Regional resilience: an agency perspective. Reg. Stud. **48**(5), 923–935 (2014)
13. Broersma, S., Fremouw, M., Van Den Dobbelsteen, A.: Energy potential mapping: Visualising energy characteristics for the exergetic optimisation of the built environment. Entropy (2013). https://doi.org/10.3390/e15020490
14. Carter, W.N.: Disaster Management: A Disaster Managers Handbook. Asian Development Bank, Manila (1992)
15. Centre for Research on the Epidemiology of Disasters, Natural Disasters 2018 (2019)
16. Centrul de Cercetare pentru Evaluarea Riscului Seismic (2013). http://www.encipedia.org/articole/proiectare/resurse-utile/harti-de-zonare/harta-de-zonare-seismica-din-p100-1-2013.html. Accessed 30 Oct 2018
17. Cioca, M., Cioca, L. I., Cioca, A.: Disaster Management Spatial Information System": Proceedings of the 4th IASME/WSEAS International Conference on Energy, Environment, Ecosystems and Sustainable Development (EEESD'08), Algarve, Portugal, 11–13 June, Published by WSEAS Press, ISBN: 978-960-6766-71-8, ISSN: 1790-5095, pp. 338-342, (2008)

18. Coppola, L., Nardone, R., Rescio, P., Bromhead, E.: Reconstruction of the conditions that initiate landslide movement in weathered silty clay terrain: effects on the historic and architectural heritage of pietrapertosa, Basilicata, Italy. Landslides **3**(4), 349–359 (2006). Published by Springer Berlin / Heidelberg, ISSN: 1612-510X (Print) 1612-5118 (Online)
19. Corbeanu, H., Brikowski, T., Aiken, C., Enoiu, P.: Landslide Monitoring in Romania. GPS World, March (2000)
20. Crosta, G.B., Dal Negro, P., Frattini, P.: Soil slips and debris flows on terraced slopes. Nat. Hazards Earth Syst. Sci. **3**, 31–42 (2003)
21. Cruden, D.M.: A Simple Definition of a Landslide. Bull. Int. Assoc. Eng. Geol. **43**, 27–29 (1991)
22. Cutter, S., Barnes, L., Berry, M., Burton, C., Evans, E., Tate, E., Webb, J.: A place-based model for understanding community resilience to natural disasters. Glob. Environ. Change **18**, 598–606 (2008)
23. David, Al.: Confronting Catastrophe: New Perspectives on Natural Disasters. Oxford University Press, Oxford (2000)
24. Davoudi, S., Shaw, K., Haider, L.J., Quinlan, A.E., Peterson, G.D., Wilkinson, C., Fünfgeld, H., McEvoy, D., Porter, L.: Resilience: a bridging concept or a dead end? "Reframing" resilience: challenges for planning theory and practice interacting traps: resilience assessment of a pasture management system in Northern Afghanistan Urban resilience: what does it mean in planning practice? Resilience as a useful concept for climate change adaptation? Politics Resil. Plan. Cautionary Note, Plan. Theory Pract. **13**(2), 299–333 (2012)
25. De Dear, R., Brager, G.S.: The adaptive model of thermal comfort and energy conservation in the built environment. Int. J. Biometeorol. (2001). https://doi.org/10.1007/s004840100093
26. Ercanoglu, M.: An overview on the landslide susceptibility assessment techniques. In: Proceedings of the 1st WSEAS International Conference on Environmental and Geological Science and Engineering (EG 2008), Malta, September 11-13, Published by WSEAS Press, ISSN: 1790-5095, ISBN: 978-960-474-001-7, pp. 131–134 (2008)
27. Facts + Statistics: Global catastrophes. https://www.iii.org/fact-statistic/facts-statistics-global-catastrophes (2018)
28. Federal Ministry for Economic Development and Cooperation, Disaster Risk Management. Approach and Contributions of German Development Cooperation, Sprintout Digitaldruk GmbH Publisher. www.bmz.de. Accessed 09 2019
29. Flentje, P., Chowdhury R.: Quantitative landslide hazard assessment in an urban area. In: Proceedings of 8th Australia New Zealand Conference on Geomechanics: Consolidating Knowledge, Australian Geomechanics Society, pp. 115–120 (1999)
30. Folke, C.: Resilience: the emergence of a perspective for social-ecological systems analyses. Glob. Environ. Change **16**(3), 253–267 (2006)
31. General Inspectorate for Emergency Situations – IGSU. https://www.igsu.ro/, https://www.igsu.ro/index.php?pagina=adaposturi_prociv. Accessed 09 2019
32. Glicksman, L.R.: Energy efficiency in the built environment. Phys. Today (2008). https://doi.org/10.1063/1.2963035
33. Guzzetti, F.: Landslide fatalities and the evaluation of landslide risk in Italy. Eng. Geol. **58**(2), 89–107 (2000)
34. Haukkala, H.: The European Union as a regional normative hegemon: the case of European neighbourhood policy. In: Whitman, R.G. (ed.) Normative Power Europe: Empirical and Theoretical Perspectives, pp. 45–64. Palgrave Macmillan (2011)
35. Hernandez, P., Kenny, P.: From net energy to zero energy buildings: Defining life cycle zero energy buildings (LC-ZEB). Energy Buildings (2010). https://doi.org/10.1016/j.enbuild.2009.12.001
36. Hewitt, K.: Regions of risk: a geographical introduction to disasters. Addison Wesley Longman, Reading (1997). https://reliefweb.int/report/world/natural-disasters-2018

37. Human Development Report 2014: Sustaining Human Progress: Reducing Vulnerabilities and Building Resilience (2014). http://hdr.undp.org/en/content/human-development-report-2014
38. Khanduri, A.C., Morrow, G.C.: Vulnerability of buildings to windstorms and insurance loss estimation. J. Wind Eng. Ind. Aerodyn. **91**, 455–467 (2003)
39. Kohler, N., Moffatt, S.: Life-cycle analysis of the built environment. Ind. Environ. (2003)
40. Law, S.H., Lim, T.C., Ismail, N.W.: Institutions and economic development: a granger causality analysis of panel data evidence. Econ. Syst. **37**(4), 610–624 (2013)
41. Lotteau, M., Loubet, P., Pousse, M., Dufrasnes, E., Sonnemann, G.: Critical review of life cycle assessment (LCA) for the built environment at the neighborhood scale. Buil. Environ. (2015). https://doi.org/10.1016/j.buildenv.2015.06.029
42. Lungu, D., Arion, C., Aldea, A., Văcăreanu R.: Seismic hazard, vulnerability and risk for Vrancea events. In: International Symposium on Strong Vrancea Earthquakes and Risk Mitigation, Bucharest, Romania, 4–6 October 2007
43. Manyena, S.B., O'Brien, G., O'Keefe, P., Rose, J.: Disaster Resilience: a bounce back or bounce forward ability? Local Environ. **16**(5), 417–424 (2011)
44. Martin, R., Sunley, P.: On the notion of regional economic resilience: conceptualization and explanation. J. Econ. Geogr. **15**(1), 1–42 (2014)
45. Mitchell, T., Harris, K.: Resilience: a risk management approach, overseas development institute (2012). https://www.odi.org/sites/odi.org.uk/files/odi-assets/ publications-opinion-files/7552.pdf
46. Nicuță, A.M., Țigănașu, R.: Risk management analysis in a worst case scenario: an urban example. Bulletin of the Transilvania University of Braşov, vol. 11 (60) No. 1, Series I: Engineering Sciences (2018)
47. Oprița, R.: Risks influences on the built environment. In: Proceedings of the International Conference on Risk Management, Assessment and Mitigation, Romania (2010)
48. P100-1/2013: Earthquake Design Code. www.mdrap.ro/userfiles/reglementari/Domeniul_I/I_22_P100_1_2013.pdf. Accessed Sept 2019
49. Pleşcan, C., Rotaru A.: Asspects concerning the improvement of soils against liquefaction. Bull. Polytech. Inst. Jassy, Constr. Architecture. Sec. **LVI (LX)**(3), (2010)
50. Rotaru, A., Kolev, C.: Addressing issues of geoenvironmental risks in Dobruja. Romania/Bulgaria, Environ. Eng. Manage. J. **9**(7), 961–969 (2010)
51. Rotaru, A.: Causes and mechanisms of landslides triggered on foundation soil areas, Geophysical Research Abstracts, 12, EGU2010-13952, 2010EGU General Assembly (2010)
52. Rotaru, A., Răileanu, P.: Landslide in Pârcovaci, Iași County, International PIARC Seminar on "Managing Operational Risks on Roads", Iaşi, 5–7 November 2009
53. Rotaru, A., Oajdea, D., Răileanu, P.: Analysis of the landslide movements. Int. J. Geol. **1**(3) (2007)
54. Rotaru, A., Oajdea, D., Răileanu, P.: Dynamics of a Landslide Surface, Environmental Problems and Development, Environmental Problems and Development – Energy and Environmental Engineering Series, pp. 22–27, (2008)
55. Rotaru, A., Nicuță, A.: Some aspects of landslide risk evaluation taking into account their distribution and properties. J. Mater. Meth. Technol. Int. Sci. Publ. **2**(1), 47–57 (2008)
56. Rotaru, A.: Landslides triggered in hard soils and soft rocks in Romania. In: Proceedings of the 15th European Conference on Soil Mechanics and Geotechnical Engineering - Geotechnics of Hard Soils – Weak Rocks, Athens, 12–15 Septembrie 2011
57. Rotaru, A.: Landslides in Romania. In: Proceedings of the International Conference on Geotechnical Engineering, Lahore, Pakistan, pp. 229–236, 5–6 November 2010
58. Rotaru, A.: Knowing the complex of consequences produced by the earthquakes in order to increase the safety of buildings. In: International Symposium New Solutions for Essential Requirements in Buildings, pp. 160–165, Iaşi, 7–8 October 2004

59. Rotaru, A., Răileanu, A.: Seismic waves propagation in soil deposits taking into account soil stratification. In: Proceedings of International Conference VSU 2004, 20–22 May 2004, Sofia, Bulgaria, vol.1, pp. I-90–I-99 (2004)
60. Rotaru, A., Babor, T.D.: Some criteria for liquefaction and the influence of the geologic time in liquefaction process. In: Proceedings of International Conference VSU 2004, 20-22 May 2004, Sofia, Bulgaria, vol. 2, pp. IV-67–IV-71 (2004)
61. Rotaru, A.: Some consideration upon landslide repair and correction. In: International PIARC Seminar "Adapting Road Earthworks to the Local Environment", Iaşi, 31 May–2 June (2007)
62. Rotaru, A.: Some consideration upon lanslide risk mitigation, In: Proceedings of International Seminar on Managing Operational Risk on Roads, Iaşi, Romania, 5–7 November 2009
63. Rotaru, A., Răileanu, P.: The importance of Hydrogeolgical Analyses of Groundwater Behaviour in the Slope Stability Analyses, Analele Universităţii "Ovidius" Constanţa, Seria Construcţii, Anul VI (2004), Proceedings of the International Symposium "Civil Engineering 2004", section VIII "Water Resurces Management, Environmental Engineering", Constanţa (2004)
64. Rupp, R.F., Vásquez, N.G., Lamberts, R.: A review of human thermal comfort in the built environment. Energy Build. (2015). https://doi.org/10.1016/j.enbuild.2015.07.047
65. Shukuya, M.: Exergy: Theory and Applications in the Built Environment. Green Energy Technol. (2013). https://doi.org/10.1007/978-1-4471-4573-8
66. Smith, K.S.: Environmental Hazards: Assessing Risk and Reducing Disaster, 3rd edn. Routledge, London (2001)
67. Soares, N., Bastos, J., Pereira, L.D., Soares, A., Amaral, A.R., Asadi, E., Gaspar, A.R.: A review on current advances in the energy and environmental performance of buildings towards a more sustainable built environment. Renew. Sustain. Energy Rev. (2017). https://doi.org/10.1016/j.rser.2017.04.027
68. Sokolov, V., Bonjer, K.P., Wenzel, F., Grecu, B., Radulian, M.: Ground-motion prediction equations for the intermediate depth Vrancea (Romania) earthquakes. Bull. Earthquake Eng. **6**, 367–388 (2008)
69. Sullivan, K., Uccellini, L.: Service Assessment: Hurricane/ post-tropical cyclone sandy. National oceanic and atmospheric administration. National Weather Service (2013)
70. Swan, W., Brown, P.: Retrofitting the Built Environment. In: Retrofitting the Built Environment (2013). https://doi.org/10.1002/9781118273463
71. Takewaki, I., Moustafa, A., Fujita, K.: Improving the Earthquake Resilience of Buildings. The worst case approach, Springer Series in Reliability Engineering (2013)
72. Takewaki, I.: Preliminary report of the 2011 off the Pacific coast of Tohoku Earthquake. J. Zhejiang Univ.-Sci. A **12**(5), 327–334 (2011). https://doi.org/10.1631/jzus.A1100111
73. Tsang, H.-H., Lam, N.T.K., Lo, S.H.: The importance of crustal shear wave velocity profile for ground motion modeling. Int. J. Geol. **1**(1), 1–5 (2007)
74. UNISDR, United Nations Office for Disaster Risk Reduction: Global Assessment Report 2013, Geneva (2013). www.unisdr.org/we/inform/publications/33013
75. United Nations Development Programme (UNDP), Time for Global Action (2014). http://www.undp.org/content/undp/en/home/librarypage/corporate/annual-report-2014-2015.html
76. Vamvatsikos, D., Kouris, L.A., Panagopoulos, G., Kappos, A.J., Nigro, E., Rossetto, T., Lloyd, T.O., Stathopoulos, T.: Structural Vulnerability Assessment under Natural Hazards: A review - Urban Habitat Constructions under Catastrophic Events (Proceedings) – Mazzolani (Ed). © Taylor & Francis Group, London (2010). ISBN 978-0-415-60685-1
77. Van Beek, L.P.H., Van Asch, T.W.J.: Regional assessment of the effects of land-use change and landslide hazard by means of physically based modeling. Nat. Hazards **30**(3), 289–304 (2004)
78. Wang, G., Sassa, K.: Pore-pressure generation and movement of rainfall-induced landslides: effects of grain-size and fine-particle content. Eng. Geol. **69**(1–2), 109–125 (2003)

79. Waugh Jr., W.L.: Living with hazards, dealing with disasters: an introduction to emergency management. M.E.Sharpe, Armonk, New York (2000)
80. Wilkinson, P., Smith, K.R., Beevers, S., Tonne, C., Oreszczyn, T.: Energy, energy efficiency, and the built environment. Lancet. https://doi.org/10.1016/S0140-6736(07)61255-0
81. World Bank annual report 2014: Main report (2014). http://documents.worldbank.org/curated/en/111781468170952958/Main-report
82. Wu, J.H., Chen, C.H.: Evaluating the Shear Strengths of the Tsaoling Landslide by Combining Accelerograph and GPS Data. In: Proceedings of the 2nd IASME/WSEAS International Conference on Geology and Seismology (GES'08), University of Cambridge, Cambridge, UK, February 23-25, Published by WSEAS Press, ISSN: 1790-2769, ISBN: 978-960-6766-39-8, pp. 113–119 (2008)
83. https://ro.wikipedia.org/wiki/Cutremurul_din_1977_(Romania)
84. http://infp.infp.ro/bigsees/New/Bigsees_Catalogue.html
85. SSC Report 99-04, Proposition 122: Seismic retrofit Practices Improvement Program, Earthquake Risk Management: A toolkit for Decision-Makers. California Seismic Safety Commission (1999)
86. FEMA, Mitigation Ideas – A Resource for Reducing Risk to Natural Hazards (2013)
87. https://ro.wikipedia.org/wiki/Cutremurul_din_1977_(Romania)

# The Impact of Environmental Degradation: Atmospheric and Geological Issues in Built Areas

Ancuţa Rotaru[1](✉) and Gupinath Bhandari[2]

[1] "Gheorghe Asachi" Technical University of Iasi, 67 Prof.dr.doc. D.Mangeron Bvd, Iaşi 700050, Romania
ancuta.rotaru@tuiasi.ro
[2] Jadavpur University, Kolkata, India

**Abstract.** The paper aims to present in the form of case studies chosen from Sundarban (near Kolkata, India) and Oltenia Plain (South-west Romania), areas located thousands of kilometers apart, how the loss of vegetation can cause distinct environmental degradation through depletion of resources such as air (global warming), water (changes of groundwater level), and soil (landslides), with significant impact on the built-up areas such as Kolkata (India), Craiova, Drobeta Turnu Severin and Turnu Magurele (Romania).

The morpho-dynamic of the forest beauty "Sundarbans" is part of the Ganga-Brahmaputra-Meghna delta built at the confluence of the Bay of Bengal [13]. It is the largest contiguous block of mangroves in the world spanning about 350 km in width over Southern Bangladesh and the state of West Bengal in India, the Indian Sundarbans delta covering about 9,630 sq.km [10]. Mangroves protect the coast from erosion, surge storms (especially during hurricanes), tsunamis, also protecting the city of Kolkata, West Bengal [12]. However, the Sundarbans Delta is being threatened by the heavy loss of vegetation coverage that is proving to be fatal to Kolkata City in terms of climate drowning it increasingly towards global warming [9]. The paper's research methodology on how the climate is changing the city is carried out based on Remote Sensing and GIS technology, Secondary Data, State forest reports and reports available from the Central Government of India.

The Southern part of the Oltenia Plain is one of the regions with the most reduced forestry surface in Romania, the area being highly exposed to drought and aridity. The study aim is to present how the groundwater level changes and land slides due to deforestation and heavy seasonal rains with huge impact on the stability of soil in areas like Craiova City, Drobeta Turnu Severin and Turnu Măgurele.

**Keywords:** Remote sensing · GIS · Vegetation coverage · Climate change · Landslides


A. Rotaru (Ed.): CRIT-RE-BUILT 2019, SSGG, pp. 35–46, 2021.
https://doi.org/10.1007/978-3-030-61118-7_2

# 1 Introduction

Spanning across the two neighboring countries of India and Bangladesh, Sundarbans covers an area of about 25,500 sq.km [8]. Placed at the confluence of Bay of Bengal, the region is a part of the world's largest Ganga-Brahmaputra Delta. Sundarbans, with an endless diversity of vibrating nature, is a suitable habitat of world famous luxuriant mangroves and Royal Bengal tiger [14]. The present study is conducted on Indian Sundarban Region covering an area of about 9,630 sq.km. The larger part of Indian Sundarban is with the South 24 Parganas. District is bordering the Bay of Bengal and is bounded by 21°32′–22°40′N latitude and 88°05′–89°00′E longitude [16].

With 64 species of mangroves and 1586 species of fauna, the region depicts a rich biodiversity image that helps in maintaining the ecological balance of the region, even through the whole state of West Bengal [6]. Tropical oceanic type of climate prevails in the zone [21]. Apart from these, the region is contributing to various ecosystem services like soil formation, soil protection and regulation of hydrological cycle, moisture contents, evaporation, climate and protection of the area from the natural calamities [2].

As a matter of fact, it is acting as a protective shield to the city of Kolkata. Such ecologically enriched region is being threatened by the loss of vegetation coverage which in turn is triggering the climate change scenarios [15, 30].

Kolkata being a congested overpopulated zone is already at the alarming stage of its atmospheric conditions. Moreover to that, the degrading ecological condition of the Indian Sundarban is acting as a catalyst to it [17]. Changing climatic patterns of the city is impacting the health issues of the people and also the groundwater levels [18].

In the past few decades, global climate change has accentuated the intensification of aridization in the South-western part of Romania, with direct consequences on the quality of forest ecosystems (deforestation) and indirect consequences on landslide triggering. These severe effects meant, first of all, a significant decline in the quality of the soil as a foundation support in built areas such as Craiova City, Drobeta Turnu Severin and Turnu Magurele, the most important urban places in the studied area.

The present study explores how the vegetation coverage is impacting the climatic conditions and the trend of the impact as will be identified in the present study 2010, 2011, 2012, 2013, and 2014 will show the direct relationship of vegetation coverage increase and decrease with the major climatic factors.

# 2 Methodologies and Data Analysis

## 2.1 Sundarbans Region, India

Vegetation coverage change is being depicted from the remote sensing analysis of the cloud-free images of Landsat 4–5 TM (Thematic Mapper) and Landsat L7ETM + SLC off –sensors – (Path 138/ Row 45) of the month of February of years 2010, 2011, 2012, and 2013 [3].

For the meteorological data, the study relied on the Customized Rainfall Information System (CRIS) - Hydro met Division - India Meteorological Department - Ministry of Earth Sciences and on data by Weather station: 428090 (VECC) (Online portal).

## 2.2 Remote Sensing Data Analysis

Georeferenced satellite images as mentioned earlier are being classified in Erdas Imagine 9.2 software (Unsupervised classification) up to 50 classes [5, 7]. The vegetation coverage classes are grouped at the penultimate stage and the actual area of the vegetation is being ascertained [19]. Such a typical classification is represented in Fig. 1. Identified and calculated vegetation coverage is then used for time series analysis using MS-Excel platform (Fig. 2).

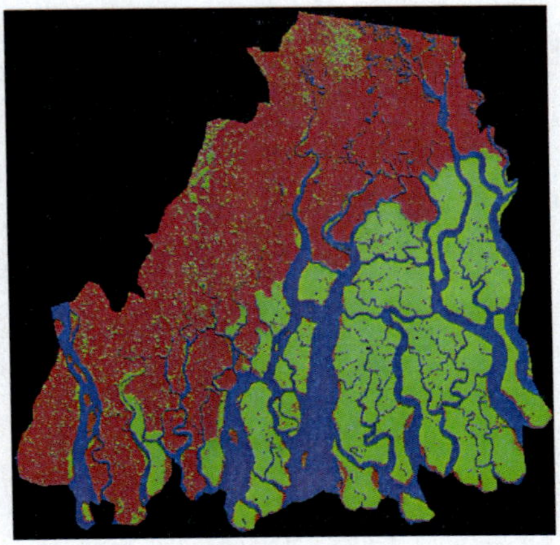

**Fig. 1.** A typical classified image of ISD

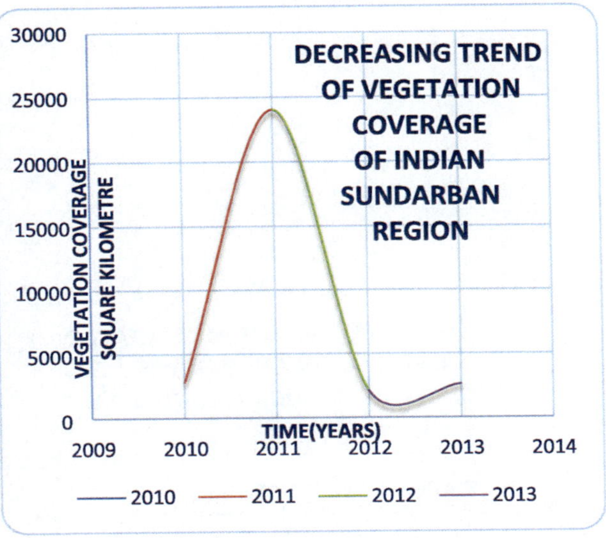


**Fig. 2.** Vegetation coverage change with the time

### 2.3 Correlation Between Vegetation Coverage - Annual Average Rainfall of Indian Sundarban Region

Rainfall pattern has been analysed on the basis of secondary data and hence have been graphed to visualize the changing pattern with time (Fig. 3). Vegetation coverage account as depicted in Fig. 2 is then correlated with the rainfall pattern in Fig. 3 to see the relation between vegetation coverage and rainfall in Fig. 4 [11].

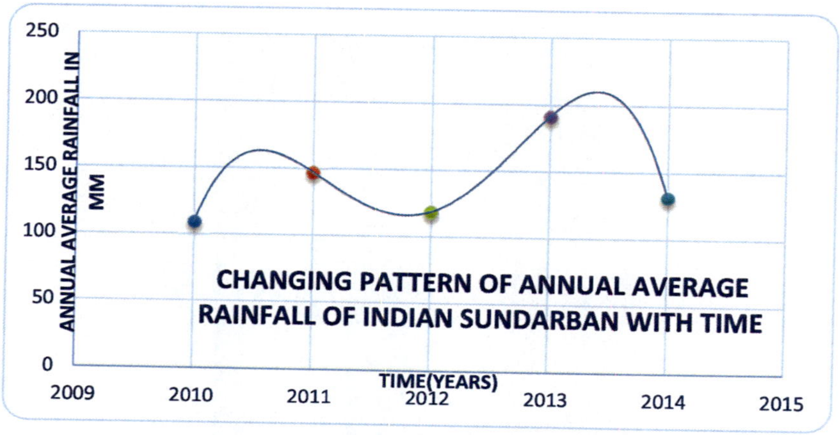


**Fig. 3.** Changing rainfall (Annual average)

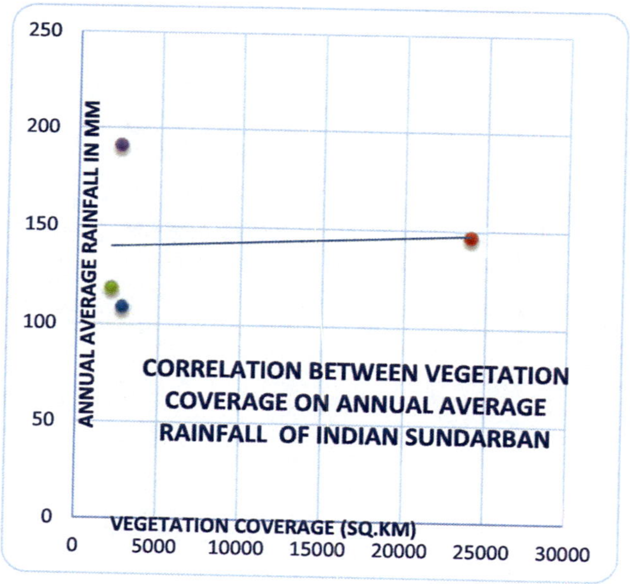


**Fig. 4.** Vegetation pattern and rainfall (correlation) of ISD

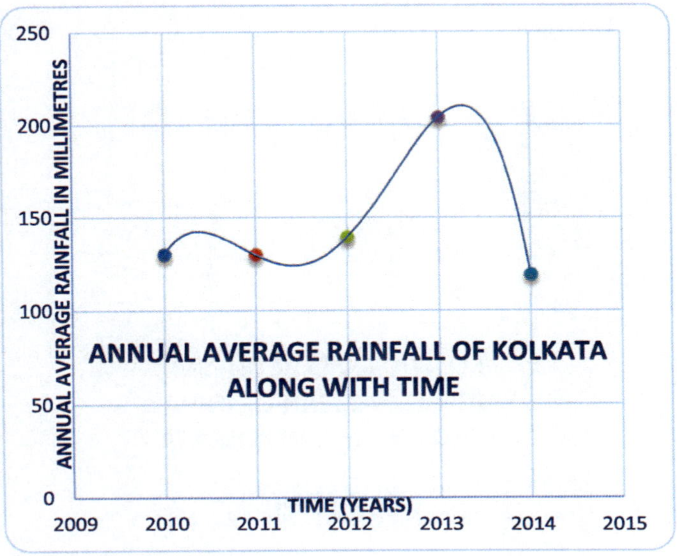


**Fig. 5.** Changing rainfall pattern in Kolkata City with time

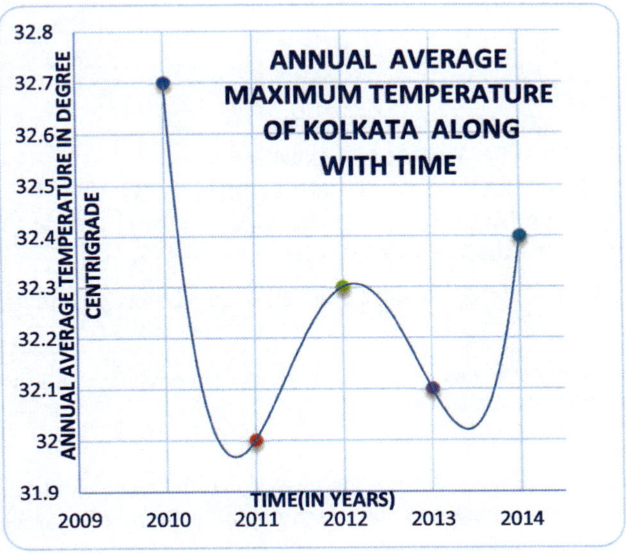


**Fig. 6.** Maximum temperature pattern of Kolkata

Figure 5 and Fig. 6 show the changing rainfall pattern and the maximum temperature in Kolkata City with time. Figure 7 shows the relationship between raifall patterns of

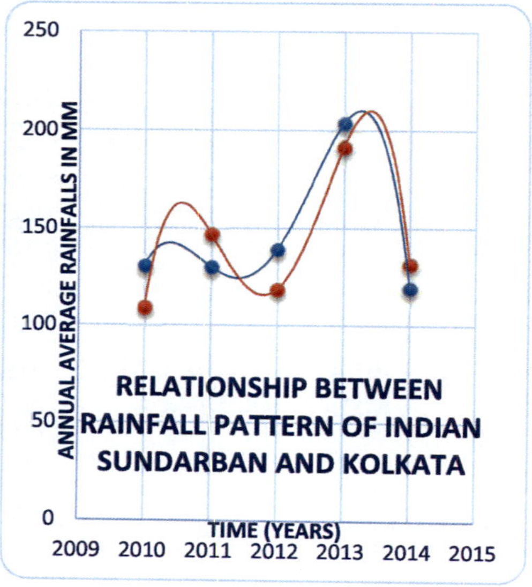


**Fig. 7.** Relationship between rainfall patterns

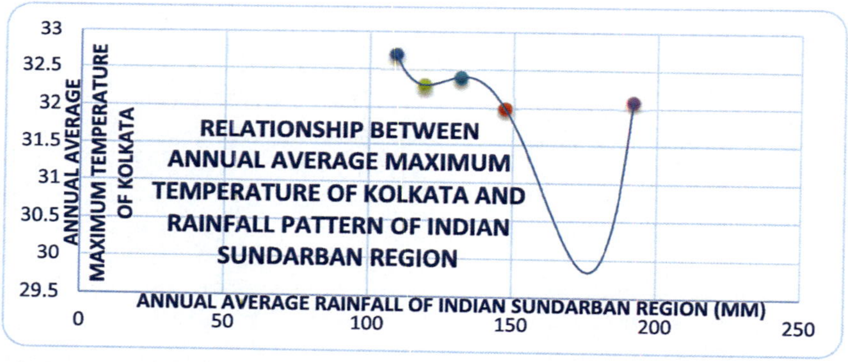


**Fig. 8.** Relationship between average annual maximum temperature of Kolkata and average annual rainfall of Indian Sundarban Region

Sundarban Region and Kolkata climate. Figure 8 depicts the relationship between annual average maximum temperature of Kolkata and rainfall pattern of Sundarban Region.

## 2.4 Landslides in the Southern Oltenia Plain, Romania

The Southern part of the Oltenia Plain, one of the regions with the most reduced forestry surfaces in Romania, is highly exposed to drought and aridity, deforestation and heavy seasonal rains changing the groundwater level and triggering landslides on large surfaces. Such phenomena occur either at the contact between clayey formations and sand deposits that slip on the clayey layer moistened by rainwater, thus creating scope for loosening and slipping in excessive wet conditions of some layers of thin rocks, either because of some steeper slopes that drain and leak out the water from rainfall resulting in their slip deposits of soil and debris. Strong ruptures, usually in the form of an arch, stand within the perimeter besides these sometimes stabilised but sometimes young landslides, developed along 200 m length and up to depths of 7–8 m located mainly transverse to the level curves [4].

In the framework of the actual climate changes, these restrictive climatic phenomena represent the main natural hazards which are affecting the study-area [1], but climatic issue is not the only responsible factor for triggering lanslides [22]. Geological issues (high seismicity, difficult rocks for building foundations), geomorphological issues (high susceptibility to landslides, slope processes, high altitude over 1200, relief fragmentation, steep slopes) and hydrological issues (high incidence of floods, insufficient water resources, low groundwater depth, under 2 m) increase the incidence of landslide occurring [25].

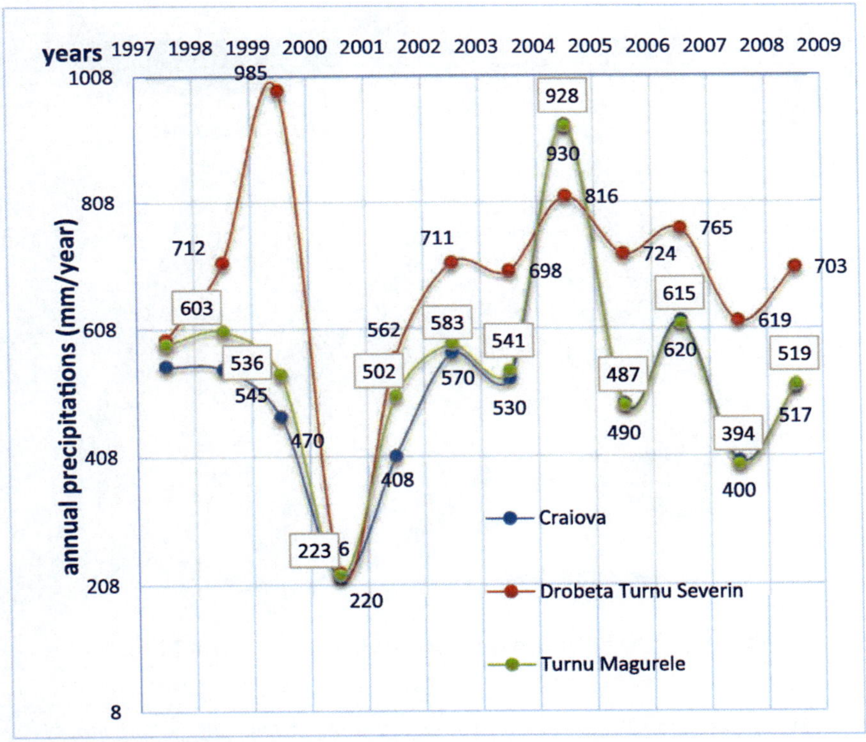


**Fig. 9.** Average annual precipitations in Oltenia Plain (1997–2009)

The analysis of the evolution of the amount of precipitations at the three weather stations located in the most important cities of the Oltenia Plain (Craiova, Drobeta-Turnu Severin and Turnu Măgurele) indicates a general trend of decline of the amount of precipitations (Fig. 9) that leads to aridization [20].

Aridization contributes to deforestation and triggers landslide processes [28].

During 1990–2006, the surface of the traditional woodland schrub shrank by up to 2.6 thousand ha. The climate change impact on deforestation is particularly seen in sandy soil areas (Sadova, Bechet, Corabia, Apele Vii, etc.) through intense aridization phenomena liable to enhancing desertification over larger surfaces. The uncontrolled deforestation of protection belts and the absence of irrigation accelerated the extension of desertification-affected surfaces and landslides [23].

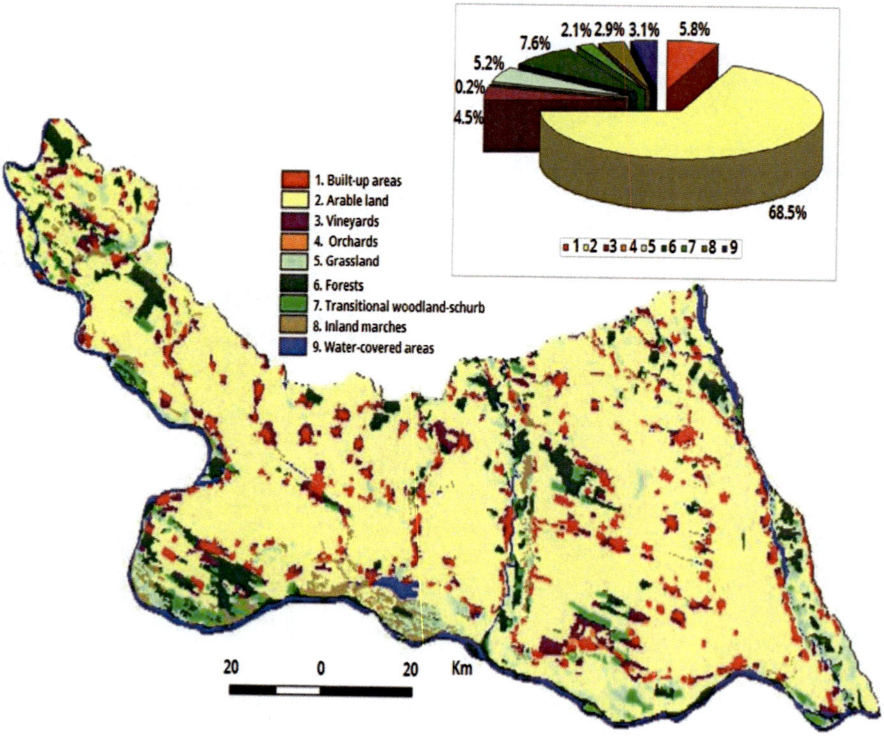


**Fig. 10.** Oltenia Plain: land use map (after Corne Land Cover, 2006)

Changes occur most often with transitions from the dense-vegetation category to the average-density category indicating the deterioration of the forest in those areas. Quite vast in size, were registered in the sandy soil grouped into small and compact areas [24], particularly East of the Jiu River and inside the Danube River bend. The soil, much more fragile because of deforestation, becomes much more exposed to landslide processes [27].

About 10% of the studied area is covered by very sparse vegetation or areas without any vegetation, bodies of water included (Fig. 10). By 2011, the situation had changed radically. The land with sparse vegetation had risen by one third to the detriment of dense vegetation. The most spectacular changes encompassed the inner lands West of the Olt River and the lands between the Jiu River and Ohrincea [1].

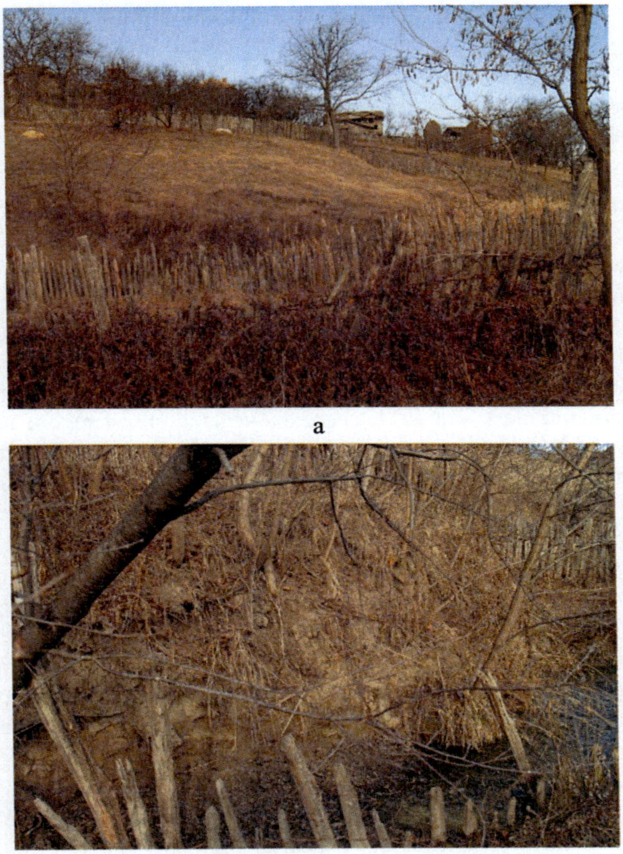

a

b

**Fig. 11.** Landslides in Oltenia Plain, Romania, triggered by deforestation and heavy rainfall

The clay deposits in the area are frequently affected by cracks and friction plans that constitute preferential sliding plans. The land area shows wavy and frequent uneven and during the sunny season develops a network of cracks that often vary between 5 and 10 cm. Cracks formed due to the presence of contractile clays promote the accumulation of rainfall in the permeable layers of the slopes producing increased weight and therefore moving land under the slope [26]. Thus, new landslides can trigger and reactivate during periods of heavy rainfall (Fig. 11) [29]. Of the three distinct areas of the investigated perimeter corresponding to dense built environment due to the proximity of the cities (Craiova, Drobeta-Turnu Severin and Turnu Măgurele), only one wooded area has a relative stability, the other zones being temporarily stabilised, but triggering active landslides. Even so, the area with relative stability has bumps and fractures of meters due to very old and stabilised landslides.

Slippery landslide area depends on the highly developed hydrographic network of the region that fully or partially activates the landslide in the rainy season [26]. The large cracks favour the rainfall accumulation in the landslide mass. Meadows cover the affected area. The landslide has a detachment of about 3 m thick and a deposit of 2–4 m thickness on the top. Active landslide zone formed due to erosion processes of the valley since the spring of 1987.

## 3 Results and Discussions

Vegetation pattern both in Sundarbans (India) and Oltenia Plain (Romania) areas is clearly indicating the massive deterioration (Fig. 2, Fig. 10). Rainfall pattern change in case of Indian Sundarban and Romanian Oltenia area is quite dynamic (Fig. 3, Fig. 9). Figure 2 and Fig. 3 depict that with the changing of vegetation coverage the rainfall also changes in the Sundarbans region. As such in 2010 where the vegetation coverage is less, then the rainfall is less – in 2011 where the vegetation coverage is more, rainfall is more – in 2012 where the vegetation coverage is again less, then the rainfall is less, and in 2013 where there is again a bit increase in vegetation coverage, then the rainfall also increased. Again, the correlation between rainfall and vegetation coverage of ISD (Fig. 4) indicates the positive impact of vegetation coverage on rainfall.

The decrease of vegetation coverage in turn, which decreases rainfall contents of Indian Sundarban Region, indicates the environmental and atmospheric degradation of the region. Such degradation is impacting Kolkata climate a lot being also depicted in the present study. Figure 7 indicated that where there is a decrease in the rainfall pattern of ISD the rainfall degeneration is also noticed in the case of Kolkata City. Similarly, the rainfall decrease of ISD is increasing the maximum temperature of Kolkata City (Fig. 8), which clearly indicates the deteriorating atmospheric conditions of Kolkata and the direct impact of ISD atmosphere on Kolkata climate.

Figure 9 shows the variation of the annual precipitations in Oltenia Plain (Romania) with time. After a period of big variations concerning rainfalls (1997–2001), the rainfall quantity decreases (2001–2009) affecting the vegetation coverage (Fig. 10).

The decrease of rainfall contents in Oltenia Plain, Romania in the last three decades, which decreases the vegetation coverage leading to the deforestation of the Southwestern region of Romania, indicates the geomorphological degradation of the area, triggering landslides with negative impact on the built environment.

## 4 Conclusions

The environmental condition of Indian Sundarban on the one hand and Oltenia Plain in Romania on the other hand is greatly impacted by the vegetation coverage of the region as well as the climatic condition. As a matter of fact, the deteriorating climatic condition with time will lead more and more both regions to the adverse global warming impacts. Hence, this impacts differently the built environment, mainly the urban places located near both remote areas, i.e. Kolkata City in India and Craiova City, Drobeta Turnu Severin and Magurele in Romania through either geomorphological changes (landslides) affecting the stability of buildings or atmospheric changes affecting the health issues of the population. Hence, immediate action is expected from the joint venture of researchers, educators, government, NGOs and other institutions to save the fragile "Indian Sundarban" ecosystem in India and also Oltenia Plain in Romania from being destructed for the sake of the built environment.

In this study, the impact of reduced vegetation coverage is limited to atmospheric and geomorphologic factors affecting the built environment, and there is a need to develop this knowledge at wider scales to address the threats and challenges posed by climatic change. The study raises several areas for future research. Of particular interest is to investigate in the near future how environmental natural factors may change in response to medium and long term climatic change affecting the buit environment. Second, to explore how the severity and distribution of atmospheric and geological impacts varies when considering other hazard phenomena (e.g. flooding or groundwater changes) with negative impact on the built environment.

## References


1. Bălteanu, D., Dragotă, C.S., Popovici, A., Dumitraşcu, M., Kucsicsa, Gh, Grigorescu, I.: Land Use and Crop Dynamics Related to Climate Change Signals During the Post-Communist Period in the South Oltenia. The Publishing House of the Romanian Acad, Romania (2013)
2. Barua, P., Chowdhury, N.S.M., Sarkar, S.: Climate change and its risk reduction by mangrove ecosystem of Bangladesh. Bangladesh Res Publication J **4**(3), 208–225 (2010)
3. Blasco, F., Gauquelin, T., Rasolofoharinoro, M., Denis, T., Aizpuru, M., Caldairou, V.: Recent advances in mangrove studies using remote sensing data. Mar Freshwater Res, 49(98), 298–296
4. Cernea, P.: Amenajare teren şi lucrări de construcţii pe traseul benzilor de steril spre Valea Şoimului – Cariera Bustuchin – geotechnical study
5. Das, S., Singh, T.P.: Mapping vegetation and forest types using Landsat TM in the western ghat region of Maharashtra, India. Int. J. Comput. Appl. **76**(1), 33–37 (2013)
6. Feller, C.I., Sitnik, M.: Mangrove Ecology: A Manual for a field course (A field manual focused on the bio-complexity on Mangrove Ecosystems)
7. Green, P.E., Mumby, J.P., Edwards, J.A., Clark, D.C.: A Review of remote sensing for the assessment and management of tropical coastal resources. Coastal Manage. **24**, 1–40 (1996)
8. Ghosh, A., Schmidt, S., Fickert, T., Nusser, M.: The Indian Sundarban mangrove forests: history, utilization, conservation strategies and local perception. Diversity **7**, 149–169 (2015)
9. Ghosh, A.: Kolkata and Climate Change, Climate Change Policy, Paper IV
10. Hossain, M.S., Sam, W., Shamsuddoha, M.: Care Mangrove Forest Care Coastal People, Institute of Marine Sciences and Fisheries, vol. 25. University of Chittagong, Bangladesh (2008)
11. Hong-Shao, W.U., Ta, P., Feng-Shan, H.E.: Climate Change Risk Research: A Case Study on Flood Disaster Risk in China. Adv. Clim. Change Res. **3**(2) (2012)
12. Jegillos, S.R., Lunde, G., Kawate, H., Dzung, T.V.: Vietnam Red Cross mangrove and disaster preparedness in the Red River Delta and northern coastal Vietnam (1994–2005). Final Evaluation Report for the Danish Red Cross. Available from the Danish Red Cross (2005)

13. Lacambra, C., Spencer, T., Moeller: tropical coastal ecosystems as coastal defences, Literature Review (Report), 1–61, (2008)
14. Lee, L., et al.: Ecological role and services of tropical mangrove ecosystems: a reassessment, Global Ecology and Biogeography. Global Ecol. Biogeogr. **23**, 726–743 (2014)
15. Mahapatra, M., Ramakrishnan, R., Rajawat, A.S.: Mapping and monitoring of land use and land cover changes using Remote Sensing and GIS techniques. Int. J. Geomatics Geosci. **4**(1), 242–248 (2013)
16. Manna, S., Mondal, S.S., Mukhopadhayay, A., Akhand, A., Hazra, S., Mitra, D.: Vegetation cover change analysis from multi temporal satellite data in Jharkhali Island, Sundarbans, India. Indian J. Geo-Marine Sci. **42**(3), 331–342 (2013)
17. Nayak, S.: Role of remote sensing to integrate coastal management, Sp Appl Centre (2000)
18. Nayak, S., et al.: Manual for mapping of coastal wetlands/land forms and shoreline changes using satellite data. Technical Note, Space Applications Centre, Ahmedabad. IRSUP/SAC/MCE/SN/32/91.63 (1991)
19. Nayak, S., Bahuguna, A.: Application of remote sensing data to monitor mangroves and other coastal vegetation of India. Indian J. Marine Sci. **30**(4), 195–213 (2001)
20. Peptenatu, D., Sîrodoev, I, Pravalie, R.: Quantification of the aridity process in South-Western Romania. Journal of Environ. Health Sci. and Eng. **11**(5) (2013)
21. Rahman, R.M., Asaduzzaman, M.: Ecology of Sundarban, Bangladesh. J. Sci. Found. **8**(1&2), 35–47 (2010)
22. Rotaru, A., Kolev, C.: Addressing issues of geoenvironmental risks in Dobruja, Romania/Bulgaria. Environ. Eng. Manage. J. **9**(7), 961–969 (2010)
23. Rotaru, A.: Causes and mechanisms of landslides triggered on foundation soil areas, Geophysical Research Abstracts, 12, EGU2010-13952, General Assembly (2010)
24. Rotaru, A, Răileanu, P.: Landslide in Pârcovaci, Iaşi County. In: International PIARC Seminar on "Managing Operational Risks on Roads", Iaşi, 5–7 November 2009
25. Rotaru, A., Oajdea, D., Răileanu, P.: Analysis of the landslide movements. Int. J. Geol. **1**(3) (2007)
26. Rotaru, A., Oajdea, D., Răileanu, P.: Dynamics of a Landslide Surface, Environmental Problems and Development – Energy and Environmental Engineering Series, pp. 22–27 (2008)
27. Rotaru, A., Nicuță, A.: Some aspects of landslide risk evaluation taking into account their distribution and properties. J. Mat. Meth. Technol. Int. Sci. Publ. **2**(1), 47–57 (2008)
28. Rotaru, A.: Landslides triggered in hard soils and soft rocks in Romania. In: Proceedings of the 15th European Conference on Soil Mechanics and Geotechnical Engineering - Geotechnics of Hard Soils – Weak Rocks, Athens, 12–15 Septembrie 2011
29. Rotaru, A.: Landslides in Romania. In: Proceedings of the International Conference on Geotechnical Engineering, Lahore, Pakistan, pp. 229–236, 5–6 November 2010
30. Roy, A: Vulnerability of the Sundarban Ecosystem. J. Coastal Environ. **1**(2) (2010)

# Mexico's Earthquake: Description, Effects and Reconstruction. A Particular Case

Manuel E. Ruiz-Sandoval(✉), Emmanuel Rivera, and Luciano R. Fernandez-Sola

Universidad Autónoma Metropolitana, 02200 Azcapotzalco, CDMX, Mexico
{mrh,lrfs}@azc.uam.mx, emmanuel.2706@hotmail.com

**Abstract.** On September 19, 2017 an earthquake of magnitude of 7.1 occurred at 120 km south of Mexico City. In that region where it was not considered to happened earthquakes of that scale. The effects in the city affected a zone called transition, between rock and lake zones. The maximum accelerations reached 780 gals. Affecting mainly buildings of 7 to 10 floors. In this paper it is study in detailed a building located in the transition zone. The building is of habitational use, structured with reinforced concrete columns and beams and planar slabs. The main damage was located in the base columns, since that floor was used as parking space (weak floor). An ambient vibration test was done in the roof in 6 different locations. The first three frequencies were identified by the Eigensystem realization algorithm (ERA) along with stabilization diagrams. A mathematical model was constructed with a commercial software. Results shown that a match with an error less of 7% with respect of experimental values.

**Keywords:** Structural health monitoring · Earthquake engineering · System identification

## 1 Introduction

### 1.1 Mexico's Earthquake Description

On September 19, 2017 an earthquake of magnitude of 7.1 occurred at 120 km south of Mexico City. In that region, it was not considered to happened earthquakes of that scale. The effects in the city affected a region called transition zone, situated between the rock and lake zones. The maximum accelerations reached 780 gals [1] (see Fig. 1). Affecting mainly buildings of 7 to 10 floors (around one second of period). A map of the seismic intensity based on accelerations is presented in Fig. 2, this map was elaborated by Instituto de Ingenieria UNAM [2].

The seismic rupture was located inside of the Coco's plate, therefore was considered as intraplate earthquake. Depth was estimated to be at 57 km [3]. The local characteristics of Mexico City subsoil makes seismic amplification to be a serious dynamic problem.

Apparently, recently designed buildings did not suffer damage. However, buildings constructed before 1985, and did not sustained damage, collapsed. This may imply that some of them had a hindered structural problem or entered in resonance with the waves of the earthquake.


A. Rotaru (Ed.): CRIT-RE-BUILT 2019, SSGG, pp. 47–57, 2021.
https://doi.org/10.1007/978-3-030-61118-7_3

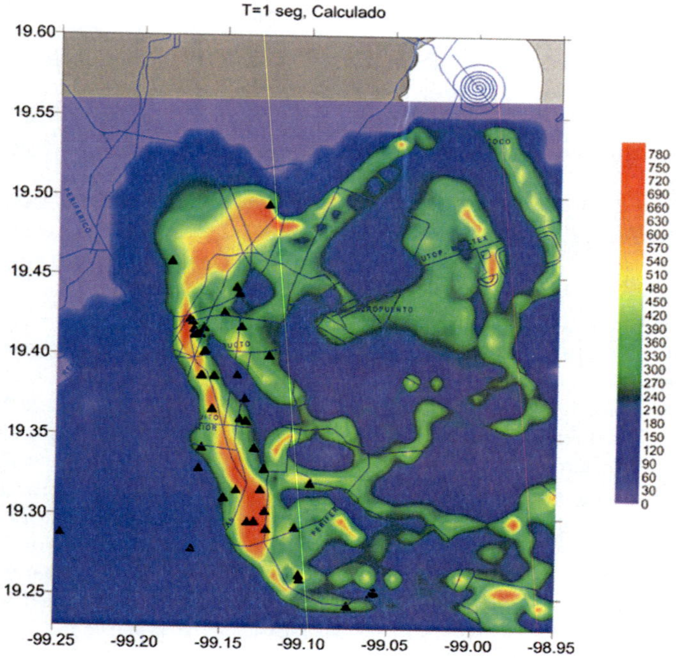


**Fig. 1.** Mexico City spectral acceleration map (T = 1 s). The black triangles represent severely damage or collapses [1].

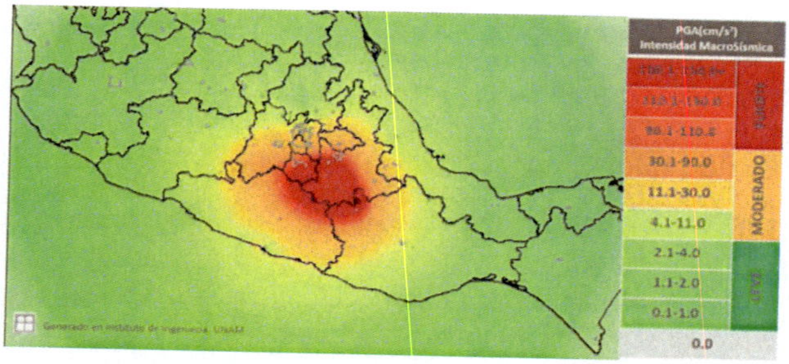


**Fig. 2.** Seismic intensity of 2017's earthquake [2].

## 2 Census of Building Damage or Collapse

After the earthquake, the government, along with public institutions organized brigades determine the status of buildings [4]. Professors, students and professional engineers walked the streets, taking pictures, and pointed out the type of damages encounter. The total of 210 buildings were found to have a type of deterioration, and in some cases collapses. In Fig. 3 can be seen the description of the types of failures encounter.

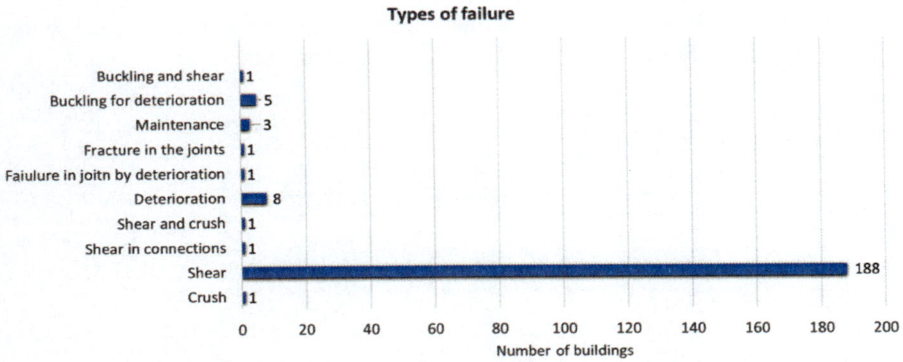


**Fig. 3.** Description of the type of failures encounter.

As can be seen in Fig. 3, most of the failures encounter were due to shear. That means that either the seismic forces sustained in the buildings were bigger than the allowed in their design, or that a deficiency in the construction was present.

In their report, the brigadiers also include the characteristics of damage. These range from non-structural elements and week floors, among others. In Fig. 4 can be seen this information.

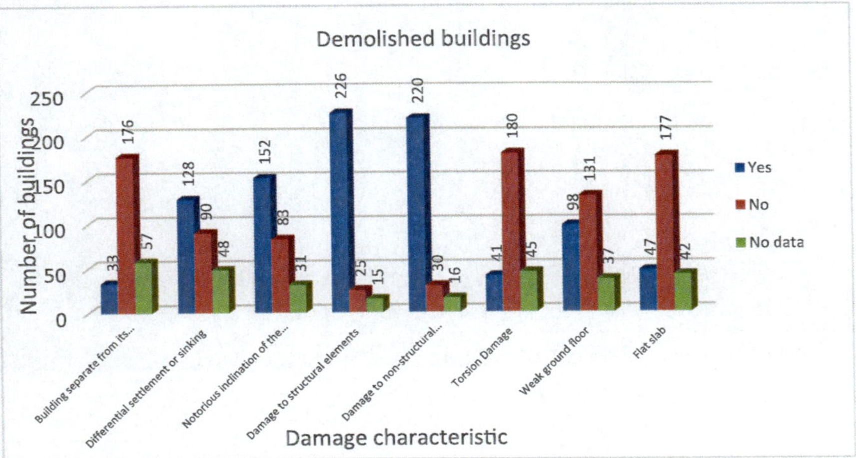


**Fig. 4.** Damage characteristics of demolished buildings.

## 3 Description of the Building

The property is located within a housing complex of 1,764.47 m$^2$ of surface, with a construction of 5,030 m$^2$, located in the south of Mexico City. The front of the property overlooks the street Pirineos in an approximate length of 25.28 linear meters, and on the east side, to Odessa Street in a length of 53.34 m (Fig. 5 and Fig. 6).

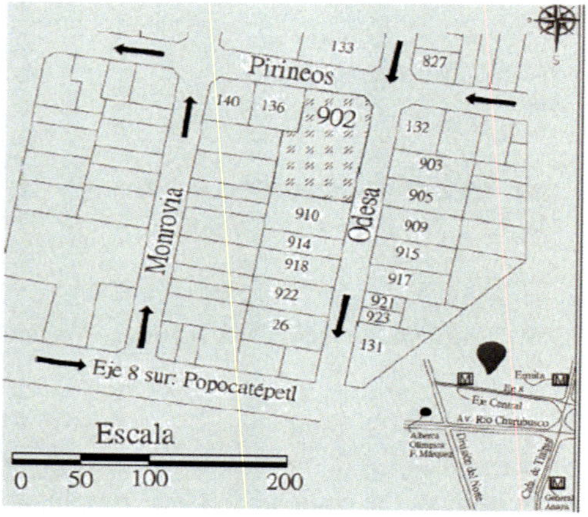


**Fig. 5.** Location of the building.

**Fig. 6.** Street view from Odesa.

The structure has a ground floor and three levels. The ground floor is used as a parking space (weak floor). The rest of the levels are for habitational purposes. Each has two apartments of approximately 95 m$^2$ of area. In the roof exists six water tanks.

The structural system consists of reinforce concrete structure with an irregular distribution of columns (see Fig. 7). The red dots in the figure represent the columns that have damage.

Damage in the ground floor is attributed to the presence of short columns. This situation is because the street level and the ground floor have a 1.25 m difference. The damage can be seen in Fig. 8.

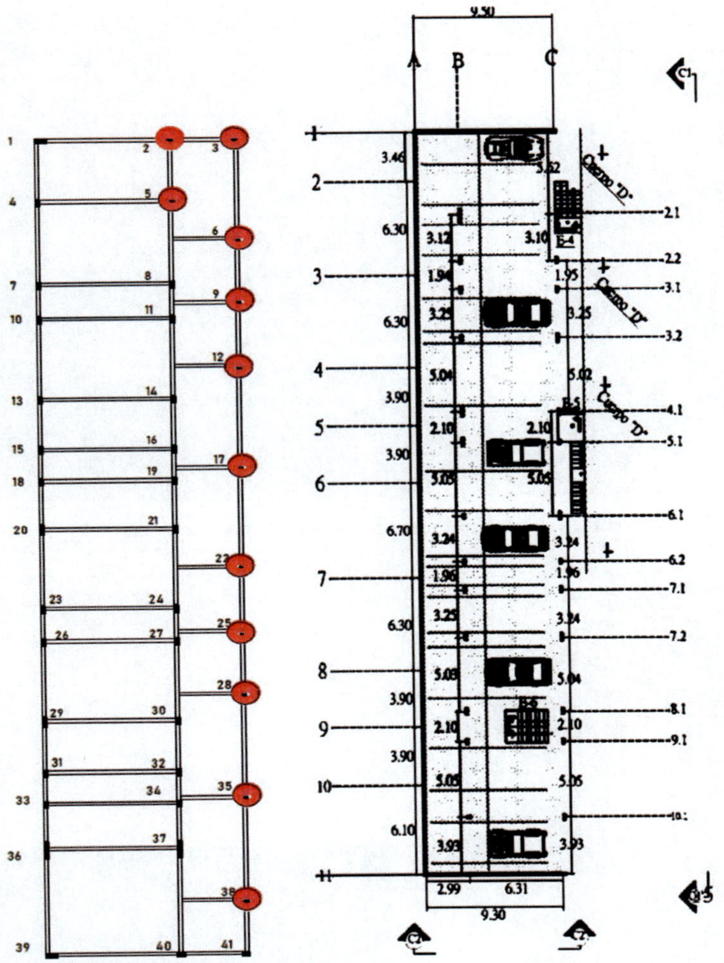

**Fig. 7.** Plant of the building. Red dots are where damage is present in the ground floor.

A non-destructive test was conducted to characterize the properties of the buildings. A Hardness Drop Tester was used to determine the elastic modulus of the concrete. The Mexican norm NMX-C-192-2008-ONNCCE was followed. After 200 tests in columns, concrete walls and beams the elastic modulus of elasticity was found to be 9,799.78 MPa.

A Bartracker was used to determine the position and number of steel bars inside of each of the concrete columns (Fig. 9).

**Fig. 8.** Damage present in the ground floor columns.

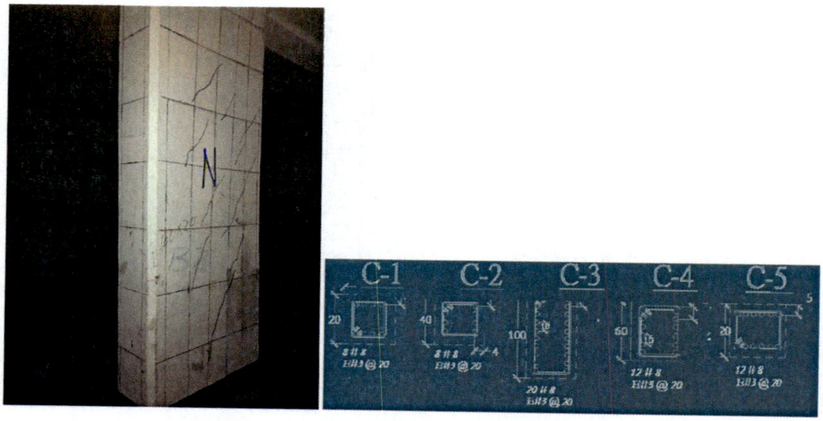


**Fig. 9.** Distribution of reinforce steel after using a bartracker.

## 4 Mathematical Model of the Structure

A Mathematical model of the structure was constructed using the program SAP2000 [5]. All structural elements: columns, beams, slabs and wall were considered. Also, nonstructural elements, such as stairs and roof elements were modeled. The measured physical dimensions of the structure, as well as the mechanical properties, were used. A 3D view is presented in Fig. 10.

The first two natural frequencies of the structure were: 2.14 and 2.48 Hz. The first one is a torsional mode due to the eccentricity between torsional and mass center. The second mode is predominant in the longitudinal direction (Figs. 11 and 12).

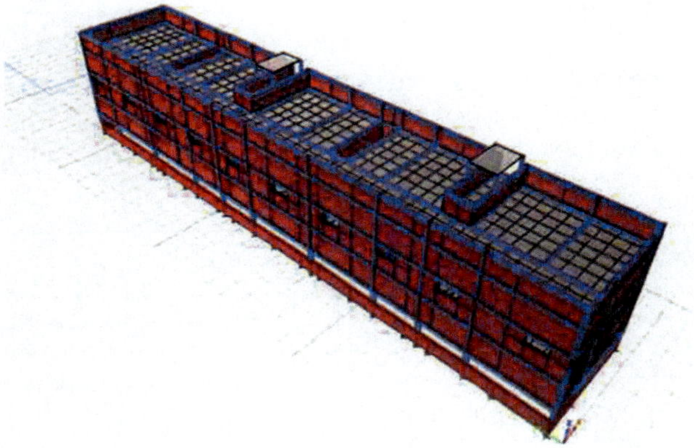

**Fig. 10.** 3D mathematical model of the structure (SAP2000).

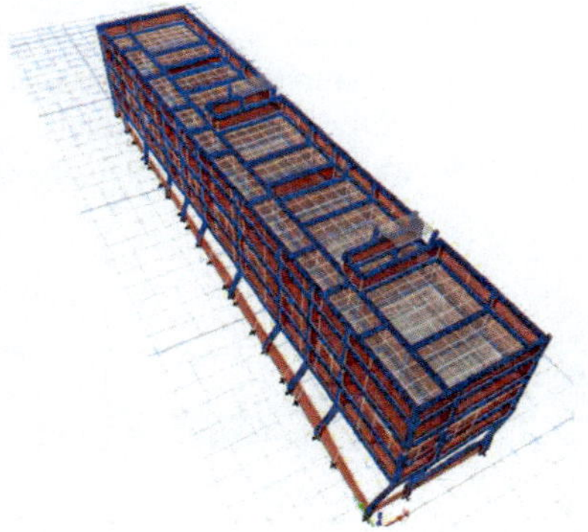

**Fig. 11.** First natural frequency of the building (2.14 Hz).

**Fig. 12.** Second natural frequency of the building (2.48 Hz).

## 5 Ambient Vibration Test

An ambient vibration test was conducted to determine the dynamic characteristics of the building. Two sets of sensor placement were employed (see Fig. 13). The objective was to able to determine at least the first two natural frequencies of the building.

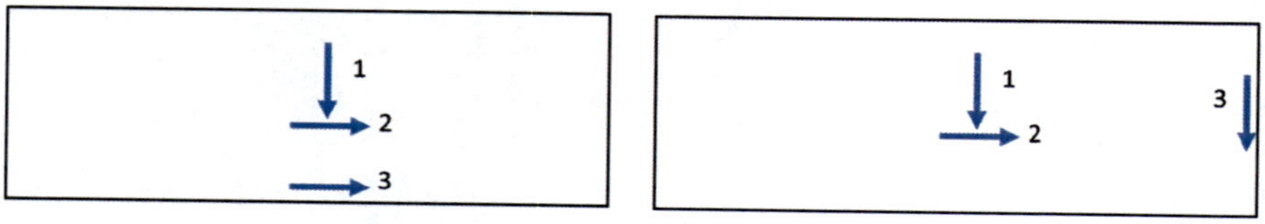


**Fig. 13.** Disposition of sensors in the roof level.

The data acquisition system used was the SigLab 20–42, that has 20 bits of A/D conversion with an anti-alias filter of a delta-sigma type. In conjunction with this DAQ four sensors PCB model 393B31 were used.

It was recorded six sets of data of 5 min each. The records were base line corrected and low frequencies were removed. An example of signals recorded in original state, after their corrections, as well as their power spectral density spectrum are presented in Fig. 14.

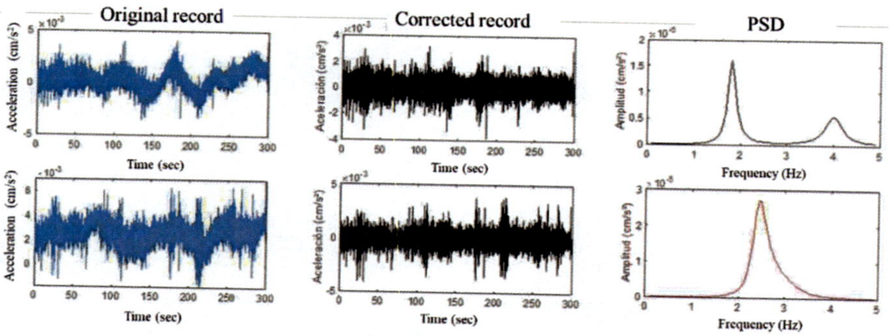


**Fig. 14.** Typical set of data recorded over the structure. Original, corrected and PSD

The method NeXT-ERA [6, 7] was used to determine the frequencies of the signals. A Modal Assurance Criterium (MAC) of above 95% was set. A typical Eigen Realization Algorithm, along with its stabilization plot, is presented in Fig. 15.

After analyzing all the data, it was found that the fist experimental frequency calculated was 1.594 Hz, that compared with the 2.14 Hz from the mathematical model it has a 25.53% of difference. The damage present in the ground floor columns it is affecting the natural frequency of the structure.

In the case of the second frequency, the value obtained experimentally was 2.536 Hz. This has a difference of 2.24% with respect of the mathematical model.

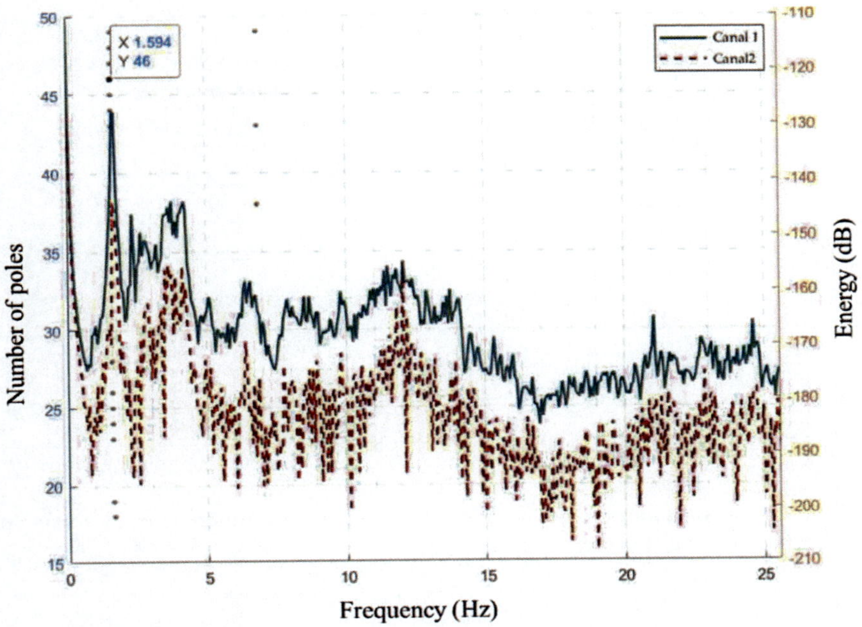


**Fig. 15.** A typical NeXT-ERA plot calculated from recorded data.

## 6 Calibrated Analytical Model

In order to calibrate the mathematical model, the authors focus in the zone where damage is located. That is, the columns of the first floor. Different approaches can be used. However, the main goal is to reduce their stiffness. A reduction in the area of some of the columns was conducted. A typical area reduction of the cross section of about 55% was necessary to achieve the match of the model (see Table 1).

**Table 1.** Original and damage dimensions of the columns of ground floor.

| Column | Axis | Original dimension | | Damaged dimension | |
|---|---|---|---|---|---|
| | | Base (cm) | Height (cm) | Base (cm) | Height (cm) |
| C-1 | 1-A | 60 | 20 | 20 | 20 |
| C-3 | 2-A | 40 | 20 | 40 | 20 |
| C-3 | 3-A | 40 | 20 | 40 | 20 |
| C-2 | 3-B | 20 | 40 | 20 | 30 |
| C-2 | 3-C | 20 | 40 | 20 | 30 |

*(continued)*

**Table 1.** (*continued*)

| Column | Axis | Original dimension | | Damaged dimension | |
|---|---|---|---|---|---|
| | | Base (cm) | Height (cm) | Base (cm) | Height (cm) |
| C-2 | 3-D | 20 | 40 | 20 | 30 |
| C-2 | 3-E | 20 | 40 | 20 | 30 |
| C-2 | 3-F | 20 | 40 | 20 | 30 |
| C-2 | 3-G | 20 | 40 | 20 | 30 |
| C-2 | 3-H | 20 | 40 | 20 | 30 |
| C-2 | 3-I | 20 | 40 | 20 | 30 |
| C-2 | 3-J | 20 | 40 | 20 | 30 |
| C-3 | 3-K | 40 | 20 | 40 | 20 |

With the above changes in the columns the mathematical model reaches a value of 1.60 Hz, therefore a difference of 0.3%.

## 7 Conclusions

The assessment of damage in real structures is a growing area of interest for various research groups around the world. However, characterizing a real structure requires specialized studies, countless inspection visits, and numerous and diverse material and human resources.

The work presented herein involves the study of a building affected by the recent earthquake of September 19, 2017, which damaged hundreds of structures in several states of the Mexican Republic. It was required to carry out exhaustive research in the field regarding the structure. The data collected was the characterization of materials, type of structuring, location of reinforce steel and description and extension of damage elements. The above allowed to elaborate approximate computational models without damage of the building. In the case of characterization with damage, environmental vibration tests were performed, as well as tests using robust mathematical models recently developed. The applied technique offers computational advantages, since it is fast in execution, and easily modifiable.

Based on both fronts of attack of the investigation, it was possible to discern between the approximate degree of damage existing in the perimeter columns of the body, by calibrating the original models, and thus finally suggesting a slightly more rational design proposal, to the one that would have come from only having preliminary visual inspection opinions.

## References

1. ERN Ingenieros Consultores: ERNTerate, Nota de interes al respecto del sismo del 19 de septiembre de 2017

2. Servicio Sismologico Nacional de Mexico ssn.unam.mx
3. Miranda, E., Galvis, F., Heresi, P., Davalos, H., Silos, J.R.: Preliminary statistics of collapsed buildings in Mexico City in the September 19, 2017 Puebla-Morelos Earthquake (2017)
4. https://www.sismosmexico.org/informes
5. CSI (Computers and Structures Inc.): SAP2000 v10 integrated finite element analysis and design of structures. CSI, Berkeley (2004)
6. James, G.H., Carne, T.G., Lauffer, J.P.: The natural excitation technique (NExT) for modal parameter extraction from operating wind turbines, Sandia national laboratories, Albuquerque, NM and Livermore, CA, SAND92-1666 (1993)
7. Juang, J.-N., Pappa, R.S.: An eigensystem realization algorithm for modal parameter identification and model reduction. J. Guid. **8**, 620–627 (1985)

# Data Analysis on the Environmental Characterization of Groundwater in Fundão - Portugal

Paulo Eduardo Maia de Carvalho(✉) and Luís José Andrade Pais

Department of Civil Engineering and Architecture, Beira Interior University, Covilhã, Portugal
pemc@ubi.pt

**Abstract.** The area of implementation of the Cova da Beira's irrigation system, due to its geographical situation and its sustainable socio-economic development requires the execution of geo-environmental studies, focusing on the behaviour of a set of variables with important ecological significance. This study is aimed to characterize the distribution patterns of variables in groundwater samples and their characterization in terms of chemical quality. Hence, an exploratory data analysis was performed in order to characterize the overall sample, followed by a stage in which a multivariate analysis was carried out through a dimensionality reduction technique (analysis of main components). The study of the dynamics of natural processes, anthropic and its influence on the dispersion, fixation and (re)mobilization of chemical elements, has contributed decisively to the optimization of the diagnostic strategy and the environmental management as well.

**Keywords:** Groundwater · Statistical

## 1 Introduction

This study aims to answer some questions about groundwater characterization in the Fundão area, starting with a set of 20 groundwater analyzes collected specifically for this work, focusing on the behavior of a set of chemical elements included in the Environmental Protection Agency's "List of Priority Pollutants" [1]. The quantitative and qualitative importance of groundwater is a factor of extreme importance for the socio-economic development of the region in which this study is inserted. Water for the purposes of agronomy is closely linked to processes of infiltration, storage and circulation both superficial and underground. The contamination of the environment by chemical elements of multiple origins implies the occurrence of these elements in form assimilable or easy immobilization in concentrations above the advisable or admissible levels. Among the chemical elements of concern are the pesticides, fertilizers, sewage sludge, irrigation water, etc. The areas where this type of contamination usually appears are areas of high industrial deployment and high agricultural development (Ross 1994) [2]. The fundamental objective is the geoenvironmental characterization of groundwater, at


A. Rotaru (Ed.): CRIT-RE-BUILT 2019, SSGG, pp. 58–66, 2021.
https://doi.org/10.1007/978-3-030-61118-7_4

a local scale, in the area of the Cova da Beira Irrigation Project. For the accomplishment of this work were also defined some objectives namely:

Calculate, analyze and interpret the sample characteristics;
Describe the variability of the data with a smaller number of uncorrelated variables;
Explain through a smaller number of factors the correlations between variables.

## 2 Characterization of the Study Area

The municipality of Fundão is located in the District of Castelo Branco, in the central region and in the sub-region of Beira Baixa called Cova da Beira. It is in this area that the city of Fundão is implanted, at the northern foot of the Mountain of Gardunha, leaning against the Mount of S. Brás at an altitude of 497 m.

Of the municipalities that constitute the Cova da Beira area, the municipality of Fundão is the one with the lowest environmental performance value (Vieira 1999) [3]. Thus, the municipality of Fundão constitutes, an excellent case of geoenvironmental study.

Sustained use of river systems, soils and aquifers requires the identification of problems and their assessment, with a view to developing scientific foundations to support the proper management of the environmental impacts of human interventions.

For the detailed characterization of the study area, it is essential to obtain data on geology, climatology, hydrology and possible anthropic influences, in order to estimate the natural variability and to differentiate natural concentrations of contamination/pollution levels (Dassenakis et al. 1998) [4].

Water for the purposes of agronomy and forestry, is closely linked to processes of infiltration, storage and circulation both superficial and underground. Thus, its quantitative and qualitative importance is a factor of extreme importance for the socioeconomic development of the region in which this study is inserted.

### 2.1 Geographical Location

The study area is located in the central region of Portugal. It is inserted between the Mountain of Estrela in the northwest and Mountain of Gardunha in the south.

Topographically it is a slightly undulating plain, with heights varying between 400 and 500 m (Ribeiro 1949) [5].

### 2.2 Geology

According to the geotectonic division, the area is located in "Zona Centro Ibérica". It is characterized by the existence of Fundão's hercinic plutonite among the most recent hercinic batholiths of Castelo Branco and Guarda.

The plutonite of the Fundão is intruded in a strip of the Schist-Gravel Complex, constituted by alternating shales and gravels. The metamorphic halo surrounds the plutonite, so it is discordant to the schists.

### 2.3 Climate

The current climate of the study area presents characteristics of a temperate climate with mediterranean characteristics, being named, according to the classification of Koppen (1930) [6], by Csa (C - temperate climate, s - temperature of the hottest month >22 °C). The classification fits perfectly into the two major climatic domains of the national territory, calculated by Mendes and Bettencourt (1980) [7].

The months of higher precipitation and relative humidity are those of November, December, January and February, while the months of July and August record the lowest precipitation.

The lowest average monthly temperatures occur in January and the highest in July and August.

### 2.4 Hydrographic Network

In the whole area, the hydrographic network is well developed, dendritic type and with a high drainage density. From the geological point of view, the substrate is essentially constituted by granite schist and gravel rocks, which are assigned low infiltration values and consequently high surface runoff rates (Carvalho 2003) [8].

### 2.5 Underground Flow Network

According to Carvalho (2003) [8], the piezometric surface is of the radial hyperbolic type (Castany 1975) [9]. Since the flux is a perpendicular vector at each point to the equipotential surface, the flow network presents a fan distribution, appearing as nuclear zones of dispersion the topographically higher points. The water table follows the topography of the site.

### 2.6 Hydrochemistry

According to Carvalho (2003) [8], from the hydrochemical point of view, the groundwater of the study area can be classified into two groups, one with predominantly calcium chlorosulfate intermediates and the other with predominantly sodium/potassium bicarbonate intermediates.

## 3 Sampling

In a geoenvironmental study, methodologies for identification and quantification of contaminants/pollutants are compulsory, examples of which are sampling methods, sample preparation, as well as instrumental methods of chemical analysis. Twenty samples of groundwater were collected during the months of September and October. Sampling was performed in large diameter wells (Galofré and Batista 1988) [10], scattered throughout the study area (Fig. 1).

The samples were analyzed for the following parameters: Copper, Lead, Zinc, Nickel, Manganese, Iron, Chromium, Magnesium, Potassium, Sodium, Phosphorus, Calcium, Nitrites, Ammonium, Bicarbonates, Nitrates, and Sulphates.

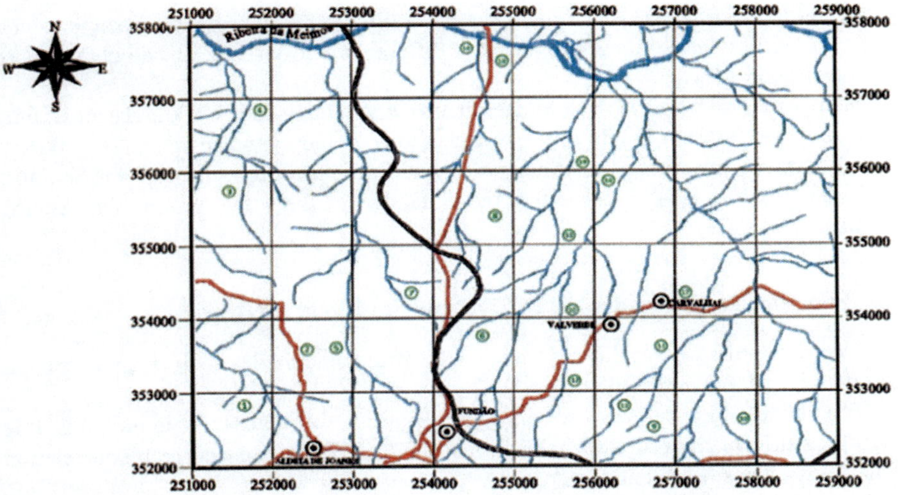


**Fig. 1.** Study area with the collection site of the 20 samples

# 4 Exploratory Data Analysis

## 4.1 Sample Characteristics Using Univariate Statistical Analysis

Copper, Lead, Zinc, Nickel and Chromium do not have any valid sample characteristics, since the values that resulted from the chemical analyzes are all below the limit of quantification.

Concentrations of Manganese vary between 0.07 and 0.44. This variable presents an average of 0.1830 higher than the median (0.1550). Asymmetry shows a positive value.

For the Iron variable, concentrations range from 0.01 to 0.03 and show a small variability around the mean (0.0115).

Magnesium presents a slightly positive asymmetry with a range of values of 31.88.

Potassium concentrations in the analyzed samples presented a minimum of 0.02 and a maximum of 0.36, presenting an average of 0.1795 lower than the median (0.1850).

Sodium is the only analyzed variable that presents a negative asymmetry. The mean (11.7110) is slightly higher than the median (11.2800).

Phosphorus has an amplitude of values of 2.50 with a small variability of values in relation to the mean.

For the variable Calcium an average value of 7.6070 is observed, presenting a standard deviation of 5.75202. It presents a slightly positive asymmetry.

Nitrites present the same values for the 1st quartile, 2nd quartile and 3rd quartile (0.03).

In the case of the Ammonium variable concentrations vary between 0.03 and 0.95. It presents a positive asymmetry (1,950) and a slight variability of values around the mean value.

Bicarbonates present an average value of 30.2 and a positive asymmetry (1,688).

The Nitrate variable shows a positive asymmetry (2,888). It presents the same values in the 3 quartiles since only 2 samples have different values of 22,200, namely samples number 12 and 14.

With respect to the variable Sulphates, it has a minimum value of 2.00 and a maximum value of 41.00. The variability of the values is low around the mean value (16.6500).

All variables present a positive asymmetry with the exception of the variable Sodium;

In all variables it is observed that the variability of the values around the mean is small.

## 4.2 Sample Characteristics Using Bivariate Statistical Analysis

In order to identify the correlations between the different variables analyzed, the Pearson correlation method was used.

The variable Mn presents low correlation values, lower than 0.664 in modulus. It is slightly positively correlated with the variables $K^+$ and P and presents a null correlation with the variables $Mg^{2+}$ and $Ca^{2+}$.

The Fe variable is slightly positively correlated with the variables $K^+$, $Ca^{2+}$ and $NH_4{}^+$.

The variable $Mg^{2+}$ is positively correlated with the variables $Na^+$, $Ca^{2+}$ and $SO_4{}^{2-}$ and has a null correlation with the variable Mn.

The variable $K^+$ is slightly positively correlated with the variable $HCO_3{}^-$.

The $Na^+$ variable is correlated with the variables $Mg^{2+}$ and $SO_4{}^{2-}$.

In the case of the variable P, it presents correlation values greater than 0.5 for the variables Mn and $HCO_3{}^-$.

The variable $Ca^{2+}$ is correlated with the variable $Mg^{2+}$ and presents a null correlation with the variable Mn.

The variable $NO_2{}^-$ is slightly positively correlated with the variable $HCO_3{}^-$.

The variable $NH_4{}^+$ is not correlated with any variable, however presenting the value of 0.586 with the variable Fe.

The variable $HCO_3{}^-$ is slightly positively correlated with the variables $Mg^{2+}$, $K^+$, P, $NO_2{}^-$ and $NO_3{}^-$.

The variable $NO_3{}^-$ is slightly positively correlated with the variable $HCO_3{}^-$.

The variable $SO_4{}^{2-}$ presents a positive correlation with the variables $Mg^{2+}$ and $Na^+$.

In the global sample the correlations have low values, lower than 0.826.

## 4.3 Sample Characteristics Using Multivariate Analysis

In order to explore the data using multivariate analysis, the principal component analysis (PCA) method was used.

This study is based on 20 samples and 12 variables. The Cu, Pb, Zn, Ni and Cr variables were excluded because they had a null variability around the mean.

In order to try to clarify the relationships between variables, we created a matrix generated from the binomial total number of samples versus variables ($20 \times 12$).

It starts from a set of 12 variables and it is intended to find a new set of uncorrelated variables, the main components. These are a linear combination of the original variables.

Thus, a smaller number of these variables may explain a significant percentage of the variability of the data. If the explanation were total, we would have to use the 12 major components and the study would be unnecessary.

The determination of the principal components to be retained generally follows an empirical criterion on the basis of the eigenvalues greater than 1 (Davis 1986) [11].

Tables 1 and 2 show, respectively, the coordinates of the active variables in the factorial axes, the eigenvalues and percentage of the explained variance, for each of the factorial axes resulting from the analysis of the main components of the 20 groundwater samples.

**Table 1.** Coordinates of the active variables in the factorial axes

| | 1 | 2 | 3 | 4 |
|---|---|---|---|---|
| Mn | 0,382 | 0,828 | −0,059 | −0,222 |
| Fe | 0,498 | 0,572 | −0,600 | 0,563 |
| $Mg^{2+}$ | 0,855 | −0,421 | −0,156 | 0,113 |
| $K^{+}$ | 0,616 | 0,308 | −0,650 | −0,017 |
| $Na^{+}$ | 0,623 | −0,643 | −0,132 | 0,038 |
| P | 0,606 | 0,350 | −0,191 | −0,511 |
| $Ca^{2+}$ | 0,738 | −0,189 | −0,246 | 0,452 |
| $NO_2^{-}$ | 0,752 | 0,018 | 0,058 | −0,233 |
| $NH_4^{+}$ | 0,363 | 0,240 | 0,464 | 0,651 |
| $HCO_3^{-}$ | 0,856 | 0,208 | −0,061 | −0,169 |
| $NO_3^{-}$ | 0,752 | 0,018 | 0,583 | −0,233 |
| $SO_4^{2-}$ | 0,556 | −0,711 | −0,047 | −0,127 |

**Table 2.** Own values and percentage of variance explained for each of the factorial axes

| Axis | Own value | % variance explained | % cumulative variance |
|---|---|---|---|
| 1 | 5,109 | 42,574 | 42,574 |
| 2 | 2,464 | 20,531 | 63,105 |
| 3 | 1,471 | 12,257 | 75,362 |
| 4 | 1,424 | 11,868 | 87,229 |
| 5 | 0,520 | 4,330 | 91,560 |
| 6 | 0,358 | 2,982 | 94,541 |
| 7 | 0,721 | 2,262 | 96,803 |

*(continued)*

**Table 2.** (*continued*)

| Axis | Own value | % variance explained | % cumulative variance |
|---|---|---|---|
| 8 | 0,197 | 1,638 | 98,441 |
| 9 | 0,123 | 1,027 | 99,469 |
| 10 | 0,044 | 0,366 | 99,835 |
| 11 | 0,020 | 0,165 | 100,00 |
| 12 | 2,082E−16 | 1,735E−15 | 100,00 |

The retention of the first four factorial axes explaining 87.229% of the total variance. (Using this method, the "dimensionality" of the problem, 12 original variables, was reduced to 4 "latent" variables).

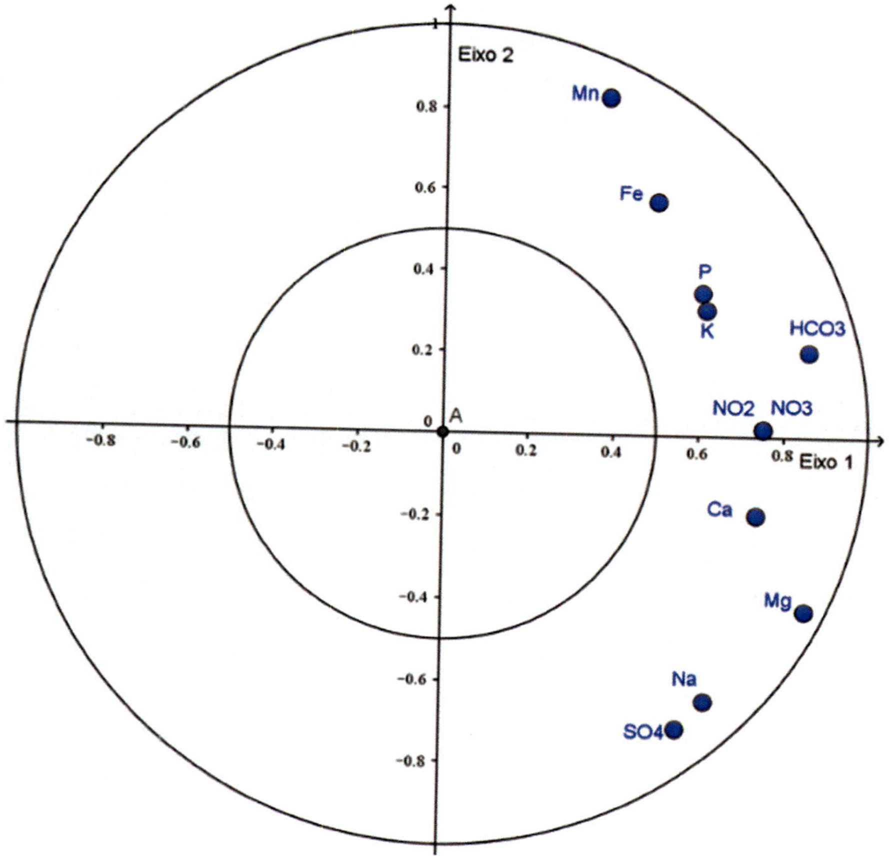


**Fig. 2.** Projection of variables in the first main plane

Factorial axis 1 explains the variables $Mg^{2+}$, $K^+$, $Na^+$, P, $Ca^{2+}$, $NO_2^-$, $HCO_3^-$, $NO_3^-$ and $SO_4^{2-}$.

The factorial axis 2 explains the variables Mn, Fe, as opposed to the variables $Na^+$ and $SO_4^{2-}$.

The factorial axis 3 explains the variable $K^+$.

Factorial axis 4 explains the variable $NH_4^+$.

Figure 2 shows the projection of the different variables in the first main plane, not having projected the variables that have a correlation coefficient with less than 0.5.

The first main plane consisting of the main axes 1 and 2 contains 63.105% of the information contained in the correlation matrix.

Axis 1 explains the association (P and $K^+$), whose origin is to be attributed to the agricultural activity of the region and the association ($NO_2^-$ and $NO_3^-$), whose origin is due to the farming activity practiced in the region. This axis also explains the variables $SO_4^{2-}$, $Na^+$, $Ca^{2+}$ and $Mg^{2+}$ that shows the natural quartz of the groundwater of the region.

Axis 2 separates the agricultural and farming component of the component that shows the natural chemical matrix of the groundwater of the studied area.

Axis 2 explains the association (Mn and Fe), whose origin is to be found in the geological complexity of the region.

## 5 Conclusions

The application of univariate statistics allowed us to characterize the "behavior" of the variables under study. In all the variables it is verified that the variability of the values around the mean is small and that with the exception of the variable Sodium, all the variables present a positive asymmetry.

The Pearson correlation method allowed us to define the correlations between the different variables. In the overall sample, these correlations are generally low.

The application of the Statistical Analysis of Major Components technique to the overall data allowed us to study the intercorrelation between the variables analyzed and to establish the distribution structure of these variables. The projection of the variables in the first factorial plan, allowed to test the connection of the variables and to separate those of essentially lithological character (Mn and Fe), those of essentially hydrochemical character ($SO_4^{2-}$, $Na^+$, $Ca^{2+}$ and $Mg^{2+}$) and those of essentially anthropogenic character (P, $K^+$, $NO_2^-$ and $NO_3^-$). The application of the PCA also allowed the identification of the hydrogeochemical image of the agricultural component (P and $K^+$) and of the farming component ($NO_2^-$ and $NO_3^-$) practiced in the study area.

The low concentration of the variables in the analyzed samples could be attributed to several factors, namely at the time of sampling at the end of the dry period (September and October), the low infiltration values characteristic of the study area, the non-use of groundwater for irrigation, which may increase the dilution of concentrations, and the environmentally sustainable practices of agricultural and livestock farming in the studied area.

**Acknowledgment.** This work is financially supported by national funds through the FCT - Foundation for Science and Technology, I.P., under the GEOBIOTEC project - UID/GEO/04035/2019.

## References

1. EPA-Environmental Protection Agency: Handbook for Sampling and Simple Preservation of Water and Wastewater. (EPA/604/4-82/029). United Sates Environmental Agency, p. 402 (1982)
2. Ross, S.M.: Toxic Metals in Soil - Plant Systems, p. 469. Wiley, Hoboken (1994)
3. Vieira, J.M.: A Economia do Turismo em Portugal, p. 89. Publicações Dom Quixote, Lisboa (1999)
4. Dassenakis, M., Scoullos, M., Foufa, E., Krasakopoulou, E., Pavlidou, A., Kloukiniotou, M.: Effects of multiple source pollution on a small mediterranean river. Appl. Geochem. **13**, 197–211 (1998)
5. Ribeiro, O.: A Cova da Beia, Controvérsia de Geomorfologia. Com. Ser. Geol., Lisboa, T.XXX, pp. 23–41 (1949)
6. Koppen, W., Gerger, R.: Handbuck der Klimatologie. Gebruder Borntraeger, Berlin, Véare vol I, parte (c) (1936) para um análisio general del sistema de climatologia de Koppen, vol. 5 (1930)
7. Mendes, J.C., Bettencourt, M. L.: Contribuição para o estudo do balanço climatológico de água no solo e classificação climática de Portugal continental. Facs. XXIV. Lisboa INMG (1980)
8. Carvalho, P.E.M.: Contribuição para a Caracterização Geo-ambiental da Zona Sul da Cova da Beira. Tese de Doutoramento, Universidade da Beira Interior, p. 195 (2003)
9. Castany, G.: Prospección y Explotación de las Aguas Subterraneas. Ediciones Omega, Barcelona (1975)
10. Galofré, A., Batista, E.: Medição y Registro de Nivels de Água em Pozos e Sondagens, Águas Subterrâneas, Instrumentação, Medida e Toma de Muestras. Pressa XXI, S. A., pp. 23–55 (1988)
11. Davis, J.C.: Statistics and Data Analysis in Geology. 2ª edição, Wiley, New York, p. 646(1986)

# Damage Due to Earthquake and Reconstruction of Residential Houses After the 2006 Yogyakarta Earthquake

Yoyong Arfiadi[1(✉)], Ade Lisantono[1], and Lucida[2]

[1] Universitas Atma Jaya Yogyakarta, Yogyakarta 55281, Indonesia
yoyong.ar@uajy.ac.id
[2] CV Tri Matra, Yogyakarta, Indonesia

**Abstract.** Residential houses are generally vulnerable to earthquakes. On May 27, 2006, an earthquake of magnitude M6.3 hit the Special Province of Yogyakarta, Indonesia. Due to the 2006 Yogyakarta earthquake, thousands of houses were damaged. More than 200000 houses were estimated to be heavily damaged and collapse. Those damages are mainly because of the residential houses in Indonesia were usually built by the workers who have no or limited capability on earthquake-resistant design. After the earthquake, the government intended to start the reconstruction quickly. The government implemented the policy that the community groups should do the majority of rehabilitation and construction processes to speed up the recovery time. This policy has speeded up the recovery time after the earthquake. However, due to massive construction activities and less construction experience, the results of construction have to be assured to the requirement of earthquake-resistant buildings standard. The survey has been conducted to the built house in Special Province of Yogyakarta and Central Java, to ensure the compliance of the houses to the guidelines for earthquake resistant structures. A data collection form was prepared to evaluate the reconstructed houses. The survey showed that the houses in the Special Province of Yogyakarta have better fulfillment compared to the ones in Central Java.

**Keywords:** Damage · Earthquake · Reconstruction · Yogyakarta earthquake · Houses · Residential buildings

## 1 Indonesia Seismicity

Indonesia is an archipelago country located in the earthquake-prone area. Some areas of the country are located on the boundaries of tectonic plates, i.e., Indo-Australian plate, Pacific plate, and the Eurasian plate. The Indo-Australian plate moves approximately from south to north and collides with the Indonesian region of the Eurasian plate. Pacific plate moves roughly from the east to the west, along the boundary of the Indo-Australian plate. Two tectonic plates with different directions and velocities approach the more stable Eurasian plate. The junction of those plates is approximately near the Maluku


A. Rotaru (Ed.): CRIT-RE-BUILT 2019, SSGG, pp. 67–83, 2021.
https://doi.org/10.1007/978-3-030-61118-7_5

islands. In addition to those seismicity sources, some potential seismic faults are also present within Indonesia regions. Those conditions result in many earthquakes occur in Indonesia. Therefore, the built infrastructure has to be designed according to the building codes to prevent structural collapse. Figure 1 shows Indonesian seismicity from 1991–2019.

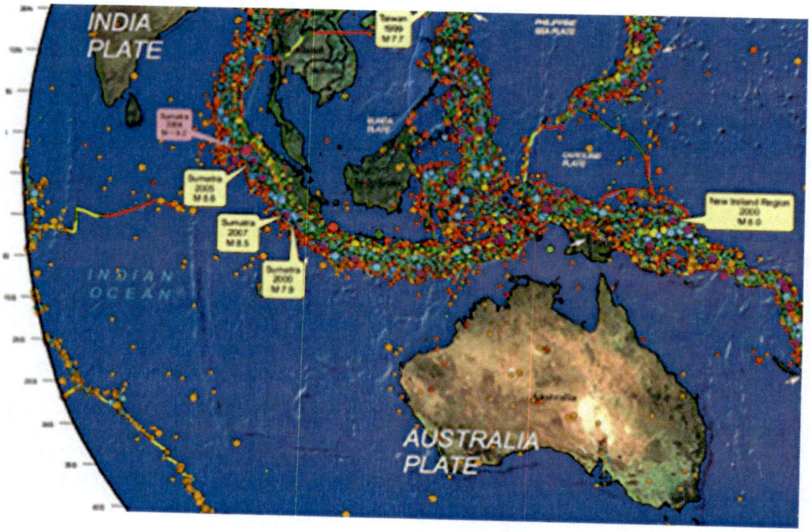


**Fig. 1.** Indonesia seismicity 1998–2009 [1]

## 2 Damage Due to Yogyakarta Earthquake

On May 27, 2006, an earthquake of M6.3 struck the Yogyakarta region, causing a large number of damages to buildings and houses. The shock caused by the subduction of the Indo-Australian plate north-northeastward beneath the Eurasian plate (or precisely Sunda plate). According to USGS [2], the coordinate of the earthquake was at 7.961°S–110.446°E, with a focal depth of 12.5 km (Fig. 2). The Yogyakarta earthquake was one of the deadliest and damaging earthquakes in Indonesia. More than 5000 people were killed [3], and many buildings and houses collapse. Tables 1 and 2 show the fatality of the 2006 Yogyakarta earthquake.

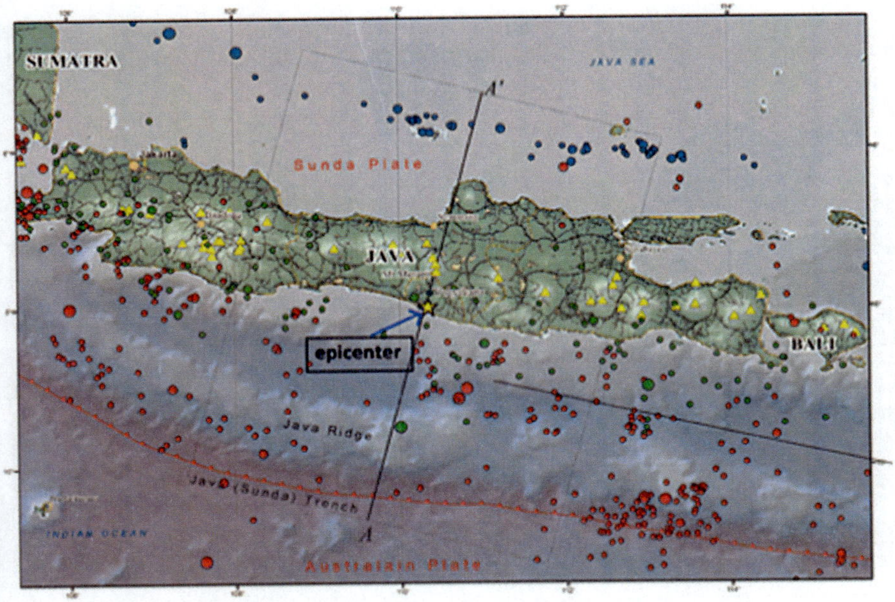


**Fig. 2.** The epicenter of 2006 Yogyakarta earthquake

**Table 1.** The causalities of the 2006 Yogyakarta earthquake in Special Province of Yogyakarta

| District | Death toll | Number injured |
|---|---|---|
| Bantul | 4121 | 12026 |
| Sleman | 240 | 3792 |
| Yogyakarta downtown | 195 | 318 |
| Kulon Progo | 22 | 2179 |
| Gunung Kidul | 81 | 1086 |

**Table 2.** The causalities of the 2006 Yogyakarta earthquake in Central Java Province

| District | Death toll | Number injured |
|---|---|---|
| Klaten | 1041 | 18127 |
| Magelang | 10 | 24 |
| Boyolali | 4 | 300 |
| Sukoharjo | 1 | 67 |
| Wonogiri | – | 4 |
| Purworejo | 1 | 4 |

As a result of the Yogyakarta earthquake, many buildings and houses collapsed. These include commercial buildings, educational facilities, and residential houses. Figures 3, 4, 5 and 6 show some examples of damaged buildings and houses. BAPENAS [3] estimated 156,664 houses collapsed, and 202,032 houses were damaged, as shown in Table 3.

**Fig. 3.** The collapse of the Indonesia Institute of Art building

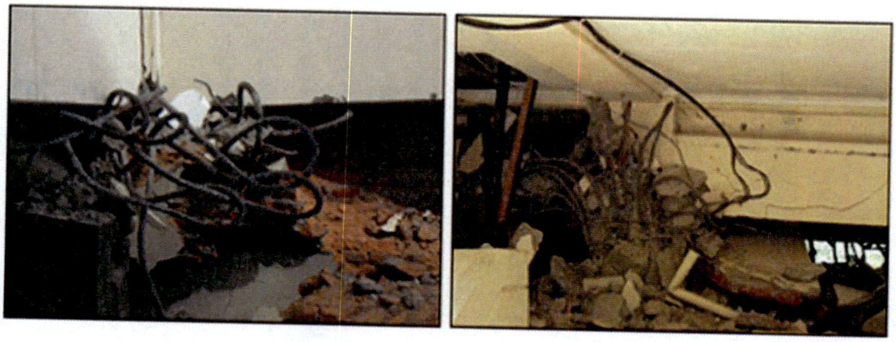

**Fig. 4.** First story column damage

## 3 Damage of Residential Houses

As shown in Table 3, the 2006 Yogyakarta earthquake caused thousands of houses to be destroyed and collapsed. The massive damage is due to the poor quality of materials and the skill of the workers. Most of the houses built in the past contained masonry without reinforcement. Although the thickness of the brick is about 30 cm, the bonding between brick masonry is not good enough. In this case, the bonding in each layer or between masonry has limited or no mortar cement. In modern houses in Indonesia, the mortar cement consists of cement, sand and water are used to lay and bond within the brick. Most of the old houses have limestone and water without or with a small amount of cement. Although the wall might resist vertical gravity loads, it is vulnerable under lateral ground

**Fig. 5.** Financial and development audit agency building

**Fig. 6.** Sang Timur primary school

**Table 3.** Distribution of damage houses in Yogyakarta and Central Java

| District | Totally collapse | Damage |
|---|---|---|
| Special Province of Yogyakarta: | | |
| Bantul | 46,753 | 33,137 |
| Sleman | 14,801 | 34,231 |
| Yogyakarta downtown | 4,831 | 3,591 |
| Kulon Progo | 6,793 | 9,417 |
| Gunung Kidul | 15,071 | 17,967 |
| Central Java: | | |
| Klaten | 65,849 | 100,817 |
| Magelang | 499 | 488 |
| Boyolali | 715 | 825 |
| Wonogiri | 23 | 70 |

shaking. This quality of material for bonding the brick might be considered as one of the sources why so many houses collapsed during the Mw 6.3 Yogyakarta earthquake. Some other causes of damage to the old masonry houses are:

(a) no columns between masonry wall corner joint,
(b) no ring beam enclosing the masonry wall,
(c) no connection between roof and masonry, and
(d) missing reinforcement.

Figures 7 and 8 show the damage of the houses due to the 2006 Yogyakarta earthquake.

**Fig. 7.** Houses damage 1

**Fig. 8.** Houses damage 2

## 4 Rehabilitation and Reconstruction

After the earthquake, the government of Indonesia had the initiative to reconstruct and rebuild residential housing immediately. The main objective is that the people do not stay too long in the temporary shelter. However, this is not too easy to build a large number of houses in a limited time frame. First, sufficient quality materials must be available in hand. At that time, cement and reinforcing steel shops were in demand. Besides, brick makers are also in great need. Other difficulties for massive construction are providing good contractors, workers, and inspectors that understand the vital requirement for earthquake-resistant buildings. At that time, an effort had been made by several agencies and universities to provide manual and guidance for earthquake-resistant buildings. The community also did some training activities to build safer houses. Note also that in carrying out housing rehabilitations, some residents constructed their own by their funding or family and relatives, some of them received government assistance, and some received international donor funding. Those difficulties for rehabilitation and reconstruction after the earthquake have also been identified in [4] and [5].

The government of Indonesia implemented the policy that the community groups should do the majority of rehabilitation and construction processes to speed up the recovery time. The government initiated a community group as a basis for the reconstruction. This community group is named as Residential Self-Help Community, as shown in Table 4.

Usually, in each Residential Self-Help Community Group, an inspector as supervisor was available to help the community to build houses according to the standard with the hope that the building will comply with the building standard. However, some constructions have taken place without assistance from the inspector of the Residential Self-Help Community Group. They made an effort altogether to build the house according to their understanding based on the manual provided by the government, universities, and other non-profit agencies. Figure 9 shows the help from the community to clean up of residential house.

**Fig. 9.** Community help to clean up the ruin of damage house

## 5 Quality Assurance in Rehabilitation and Reconstruction

The rehabilitation and reconstruction work of the post-earthquake houses has to be known to assure a safer building against earthquakes. The rehabilitation has taken place with and without the supervision and guidance from the inspector or supervisor. After reconstruction, a survey had been performed to 42,052 houses to ensure the fulfillment of the reconstructed housed. A data collection form was prepared to evaluate the compliance of buildings to the guidelines and standards. The assessment was done in the year 2007 [6] to the houses that were built by the community. Table 4 shows the distributions of the houses to be surveyed.

**Table 4.** Distribution of sample in the Residential Self-Help Community Group

| Cluster | No of Residential Self-Help Community Group | Number of samples |
|---|---|---|
| Special Province of Yogyakarta: | | |
| Yogyakarta | 590 | 1080 |
| Bantul-1 | 1998 | 3996 |
| Bantul-2 | 1864 | 3728 |
| Bantul-3 | 1516 | 3032 |
| Bantul-4 | 1865 | 3730 |
| Bantul-5 | 1450 | 2900 |
| Kulon Progo | 206 | 412 |
| Gunung Kidul | 400 | 800 |
| Central Java Province: | | |
| Klaten-1 | 2649 | 3296 |
| Klaten-2 | 2734 | 3468 |
| Klaten-3 | 2618 | 5236 |
| Boyolali | 81 | 162 |
| Sukoharjo | 111 | 222 |
| Karanganyar | 12 | 24 |
| Wonogiri | 17 | 34 |
| Magelang | 72 | 144 |
| Purworejo | 33 | 66 |
| Temanggung | 7 | 14 |
| Kebumen | 1 | 2 |
| Total | 21,026 | 42,052 |

Each house was evaluated based on the list of evaluation decided to fulfill the requirement as the earthquake-resistant buildings. The surveyors were trained on how to fill in the collection data form before they began to get involved.

## 5.1 Structural Drawing

One of the evaluations of the survey of the built house is the presence of the structural drawing. The surveyor had to fill in whether or not the house was constructed based on structural drawings.

## 5.2 Plan

To evaluate whether the building has a symmetrical plan and no irregularities are important for the earthquake-resistant buildings. There are two evaluations in the data collection form for the plan of the buildings. The first evaluation is whether or not the plan of the buildings is symmetric. The second is whether the houses have irregularities in the horizontal direction. The maximum size for re-entrant corner irregularities in the floor plane is 25%.

## 5.3 Foundation

In the foundation part, there are five questions to be answered as follows: (a) whether the depth of the foundation is minimum of 600 mm, (b) whether the minimum width of the base is 600 mm or not, (c) whether or not the embedded length of the column reinforcement to the foundation is longer than the 40-time diameter of steel bars, (d) whether or not the foundation used river stones or hard white stones, (e) whether the foundation has excellent bonding between the river stones or not by evaluating the mortar to bound the river stone. In this case, the ratio of cement: sand should be no larger than 1: 4.

## 5.4 Tie Beam

A tie beam is necessary for the house to assure the integrity of the house. It connects columns and also connects the upper structure to the foundation. The investigation for the tie beam had six criteria. The fulfillment criteria for the tie-beam are as follows: (a) the dimension of the tie beam 150 mm × 200 mm, (b) available longitudinal bar: 4D12, (c) available stirrups: D8 at 150 mm or D6 at 125 mm, (d) available steel anchorage to the foundation, (e) good concrete pouring quality (no segregation in concrete, (f) composition cement: sand: gravel ratio of 1: 2: 3.

## 5.5 Column

The function of the column here is to join between brick walls, especially at the corner of the masonry wall. The column has to have the requirements: (a) minimum dimension 150 mm × 150 mm, (b) the longitudinal bars minimum 4D12, (c) minimum stirrups D8 at 140 mm or D6 at 125 mm, (d) good concrete pouring quality (no segregation in concrete, (e) minimum composition cement: sand: gravel ratio of 1: 2: 3.

### 5.6 Masonry Wall

There are three objects in the requirement for the masonry walls. The requirements for the unreinforced masonry wall were: (a) maximum area of the wall that confined by columns and tie-beam was 9 $m^2$, (b) available anchorage to columns, (c) mixture of mortar for wall bonding 1 part of cement: 4 part of sand.

### 5.7 Ring Beams

The presence of the ring beams is to confine the masonry wall and to ensure the integrity of the structure. The requirements for the ring beams were as follows: (a) minimum dimension 120 mm × 150 mm, (b) longitudinal reinforcement minimum 4D12, (c) minimum stirrups: D8 at 150 mm or D6 at 125 mm, (d) good concrete pouring quality (no segregation in concrete), (e) minimum composition cement: sand: gravel = 1: 2: 3.

### 5.8 Beam-Column Joint

Beam column joint has to have good detailing and be in good construction results. The main concern on this column joint is the anchorage of the reinforcement either from beams or columns. The development length of the steel reinforcement, either from columns or beams, must be larger than 40 × the diameter of the steel reinforcement. If a hook is available, the development length was taken as 30 × the diameter of the steel reinforcement.

### 5.9 Reinforcement Lap Splices

The requirement for the lap splices of reinforcement was that the splices were of minimum 40-time diameter of the steel reinforcement.

### 5.10 Gable Wall

In many residential houses, gable walls exist at the end of the house's plan, as the boundary of the houses. The observation for gable walls were: (a) the availability of the ring beam at the top the roof, (b) good concrete pouring quality (no segregation in concrete), (c) minimum dimension of the ring beam 120 mm × 150 mm, (d) minimum reinforcement of ring beam 4D12, (e) minimum stirrups D8 at 150 mm or D6 at 125 mm, (f) availability of the bracing between gable walls or between gable walls and roof trusses.

### 5.11 Roof Truss

Roof trusses of the residential houses are typically made of timber trusses. The requirements for the roof trusses are as follows: (a) minimum dimension of the timber is 60 mm × 120 mm, (b) all joins are strengthened by steel plate, (c) the availability of bracing between truss and or gable wall, (d) the availability of anchorage to the base of truss, (d) the color is dark.

## 5.12 Result

The results of the survey showed that the constructed houses in the Special Province of Yogyakarta have higher conformity to the earthquake resistant structures compared to the ones in Central Java Province. The results of the survey are summarized as follows.

There are 74.6% of houses in the Special Province of Yogyakarta built based on construction drawings. In comparison, only 48.4% of the houses in Central Java were built according to the construction drawing.

Figure 10 shows the compliance of the constructed houses for the floor plan. From this figure, the constructed houses in the Special Province of Yogyakarta have better performance compared to the ones in Central Java Province.

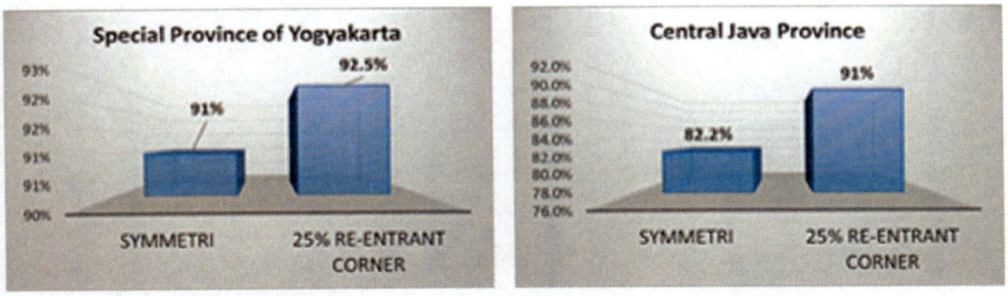


**Fig. 10.** Conformity diagram for the floor plan

For the foundation, the results from the survey showed that the conformity of the built houses in the Special Province of Yogyakarta and Central Java Province has good results, as shown in Fig. 11. Of the five criteria, four criteria have scored above 90%, i.e., for the depth, width, anchorage, type of stone, while for mixture criterion, the results are above 70% for both provinces.

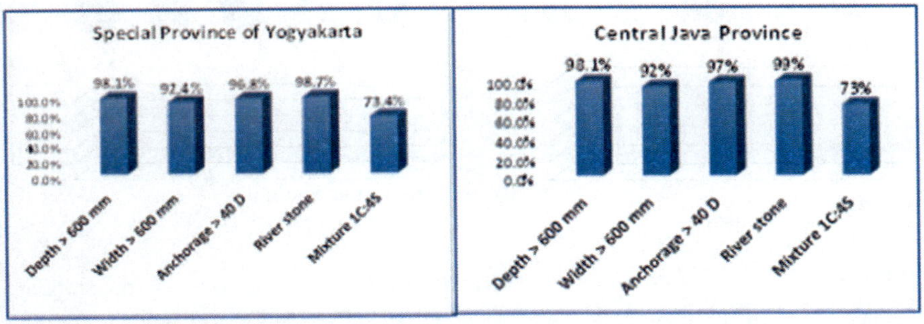


**Fig. 11.** Conformity diagram for the foundation

Figure 12 shows the conformity of the built houses for the tie beam. The compliance of the built houses is above 80% for both provinces, except for the criterion of the reinforcement. For the size of the steel reinforcement, the conformity is only 42.5% for the Special Province of Yogyakarta and 41.6% for Central Java Province.

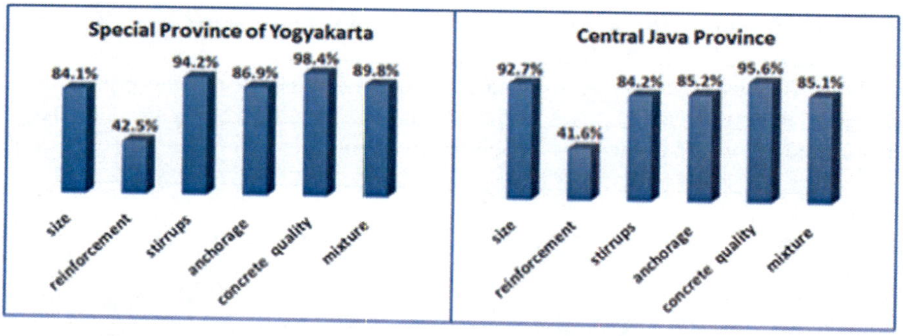


**Fig. 12.** Conformity diagram for the tie beam

For the columns, the conformity of the built houses is above 85% for both provinces, except for the reinforcement. In this case, the compliance of the Special Province of Yogyakarta is 46.5%, and Central Java is 39.3%. Figure 13 shows the conformity diagram of the columns of the constructed houses.

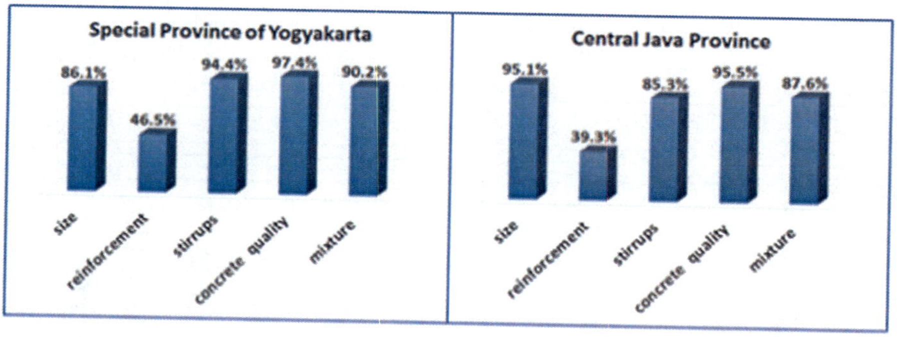


**Fig. 13.** Conformity diagram for the column

Figure 14 show the compliance for a masonry wall of the built houses to the standard and guidelines. From Fig. 14 it can be seen that the agreement of the wall to the maximum area and the availability anchorage are good, where the score is higher than 80% for both provinces. But, for the use of a good mortar for bonding are 70% and 67.5% for Special Province of Yogyakarta and Central Java Province, respectively.

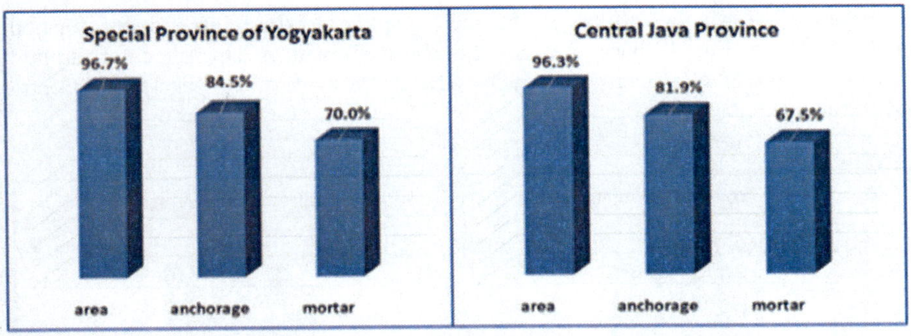


**Fig. 14.** Conformity diagram for the wall

Figure 15 shows the fulfillment of ring beams to the guidelines. From Fig. 15 it can be seen that four criteria have good conformity where the scores are above 80 for both provinces. However, for the reinforcement, the fulfillment score is small, where 38.3% for the Special Province of Yogyakarta and 31.9% for Central Java Province.

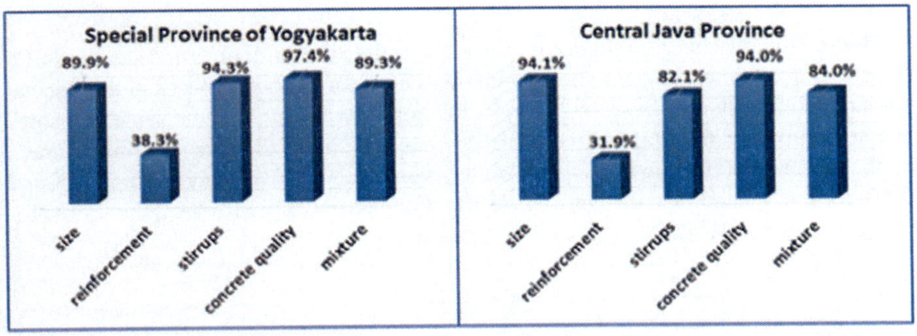


**Fig. 15.** Conformity diagram for the ring beam

Figure 16 shows the fulfillment of the detailing of reinforcement at beam-column joints of the built houses. In this case, the fulfillments are 89.7% and 81.4% for the Special Province of Yogyakarta and Central Java, respectively.

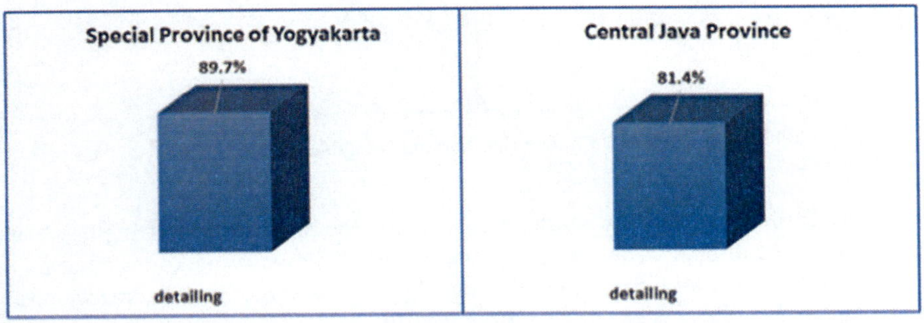


**Fig. 16.** Conformity diagram for beam-column joint

Figure 17 shows the fulfillments for the lap splices of the reinforcement, for both provinces. From Fig. 17 it can be seen that the fulfillment of lap splices of the built houses is 71.4% and 74%, respectively, for Special Province of Yogyakarta and Central Java Province.

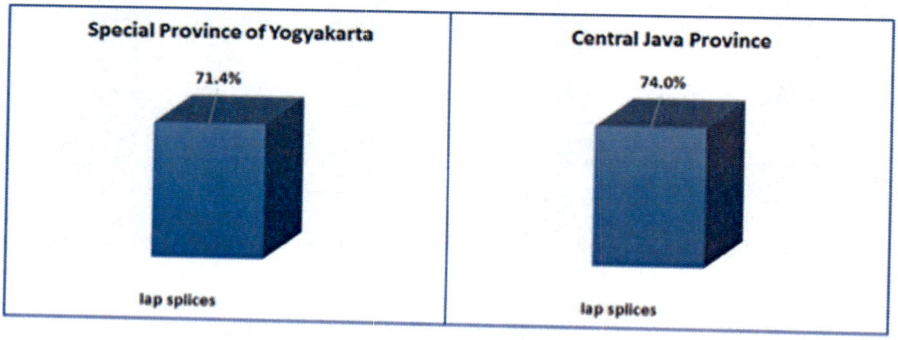


**Fig. 17.** Conformity diagram of lap splices

Figure 18 shows the compliance for the gable wall to the guidelines and standards. Of six criteria, four criteria have scored higher than 80% for the built houses in the Special Province of Yogyakarta. In comparison, for Central Java Province, the same measures have scored higher than 60%. Two other criteria, which are the size of reinforcement and the presence of bracing in the gable wall has small scored. The scores are 37.4% and 50.6% for the Special Province of Yogyakarta. For Central Java Province, the scores for those two criteria are 40.3% and 43.2%, respectively.

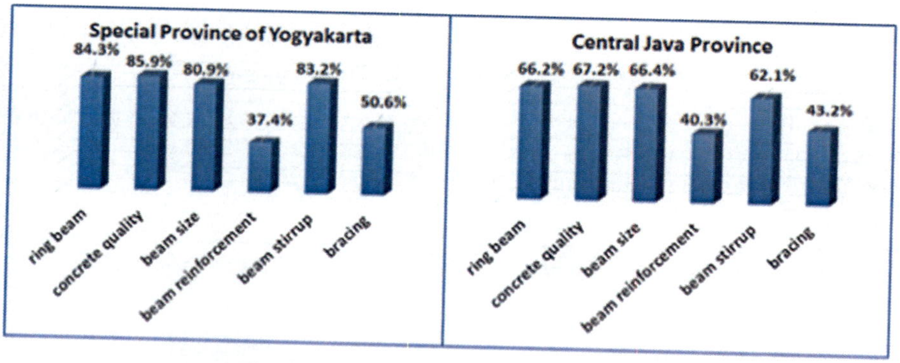


**Fig. 18.** Conformity diagram for gable wall

For roof trusses, the survey results indicated that the presence of bracing has low scores compared to other requirements. For the Special Province of Yogyakarta, this score is 55.7%, while for Central Java Province 53.5%. Other roof trusses criteria have scored higher than 85%, as can be seen from Fig. 19.

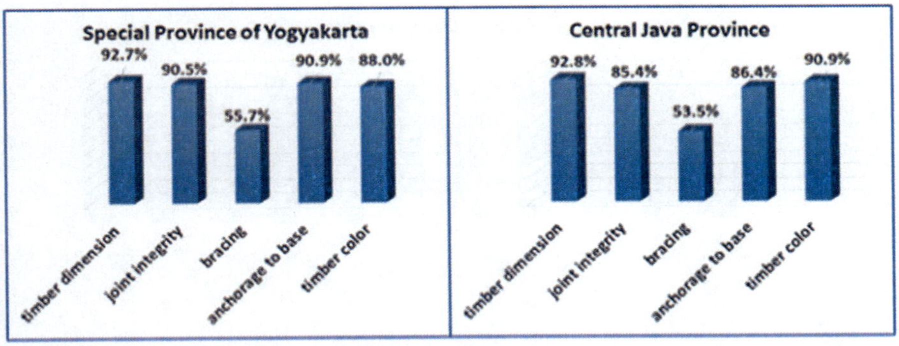


**Fig. 19.** Conformity diagram for roof trusses.

To summarize, the fulfillment of the rehabilitation and reconstruction for the Special Province of Yogyakarta is shown in Fig. 20, while for Central Java Province, is shown in Fig. 21. For the Special Province of Yogyakarta, the lowest score is the criterion for the gable wall, where the score is 70.4%. Very similar results were observed for the lap splices and structural drawings. Other measures are higher than 80%. For the Central Java region, the availability of the structural drawings before the construction has the lowest score. On average, the fulfillment of the built houses in Yogyakarta is better than the ones in Central Java Province, as can be seen in Figs. 20 and 21.

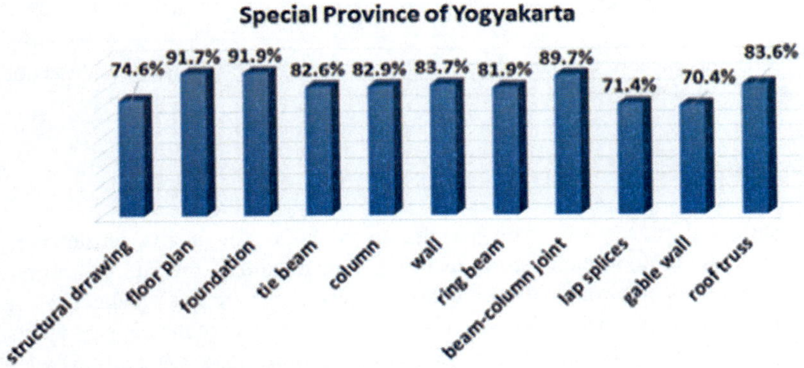


**Fig. 20.** The fulfillment to each component for Special Province of Yogyakarta

Figure 22 shows the fulfillment of the built houses after the 2006 Yogyakarta earthquake in general. From Fig. 22 it can be seen that not all houses have fulfilled the requirement for earthquake resistant structures. The houses in the Special Province of Yogyakarta have better compliance with the guidelines for the earthquake resistant structures than the ones in Central Java Province. The percentage of houses that fulfill more than 80% of the requirements is 66.4% in the Special Province of Yogyakarta and 55.1% in Central Java Province, respectively.

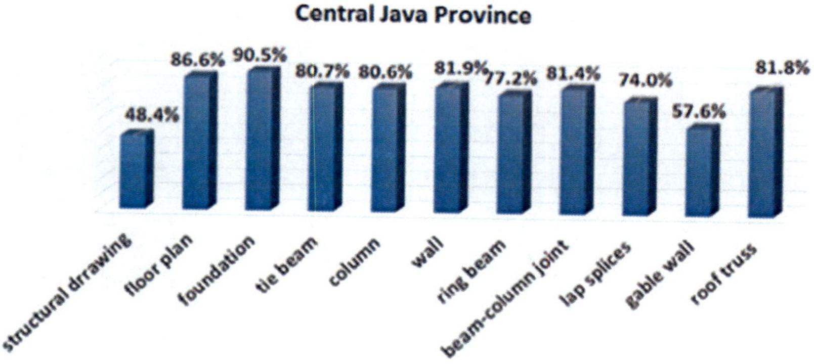


**Fig. 21.** The fulfillment to each component for Central Java Province

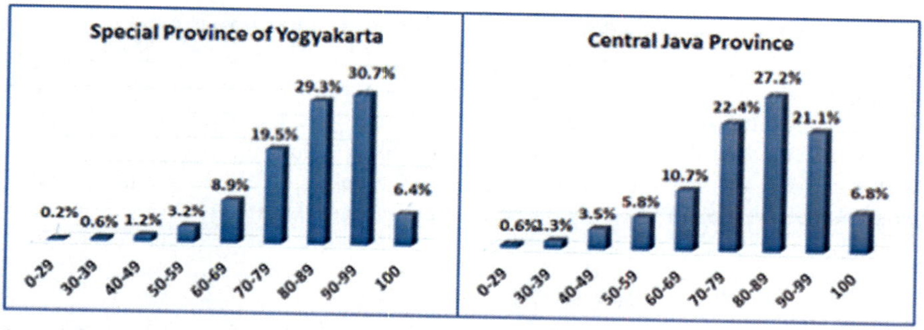


**Fig. 22.** The fulfillment of the built houses in Special Province of Yogyakarta and Central Java Province to the guidelines

## 6 Conclusions

Many residential houses were damaged due to the 2006 Yogyakarta earthquake. The damage was due to poor construction material, poor detailing, and lack of compliance with the guidelines of earthquake-resistant structures. After the earthquake, the government intended to speed up the rehabilitation and reconstruction of the damaged house. In this case, the government took the policy for rehabilitation and reconstruction by gathering as much community participation as the community-based construction. The survey has been conducted to the built house in Special Province of Yogyakarta and Central Java, to assure the fulfillment of the houses to the guidelines for earthquake resistant structures. The survey showed that not all houses followed the guidelines and standards. The houses in the Special Province of Yogyakarta have better fulfillment compared to the ones in Central Java.

## References


1. Global Background Seismicity 1998–2009. http://www.tectonics.caltech.edu/images/maps/seismicity.pdf. Accessed 25 Oct 2019
2. USGS Earthquake Hazards Program. https://earthquake.usgs.gov/earthquakes/eventpage/usp000ej1c/executive. Accessed 25 Oct 2019
3. BAPPENAS: Preliminary damage and loss assessment Yogyakarta and Central Java Natural disaster, Jakarta (2006)
4. Pribadi, K.S., Kusumastuti, D., Rildova.: Learning from recent Indonesian earthquakes: an overview to improve structural performance. In: The 14th World Conference on Earthquake Engineering, China (2008)
5. Boen, T.: Reconstruction of house in Aceh, three years after the December 26, 2004 tsunami. In: International Conference on Earthquake Engineering and Disaster Mitigation, Jakarta (2008)
6. PT: Multi area conindo: final report: quality assurance and quality control for post-earthquake rehabilitation and reconstruction work in Yogyakarta and Central Java (in Bahasa Indonesia), Yogyakarta (2007)

# Landslides – Effects on Built Environment and Solutions

Oana Elena Colţ(✉)

"Gheorghe Asachi" Technical University of Iaşi, Iaşi, Romania
Oana.colt@tuiasi.ro

**Abstract.** The development of the built environment (transport infrastructure and modern inhabited zones) is strongly influenced by the extension of the areas with sliding potential, that is why it is necessary to have a clear knowledge of the slipping potential, of the triggering factors that could influence the landslides development, in order to ensure a harmonious and continuously interaction between the inhabited and the natural space.

For a sustainable built environment, besides knowing all the factors that influence a landslide, it is needed to apply the right solution to assure the stability (retaining walls, changing the soil profile, drains, piles, geosynthetics etc.).

**Keywords:** Landslide · Build environment · Effects

## 1 Landslides

Landslides are natural disasters (destructive geological or meteorological natural phenomena) with a complex nature and a multitude of factors that are influencing the development and advancement. On a global view (Fig. 1), [1] of the landslides susceptibility it could be said that this type of natural disaster is an issue on international level.

### 1.1 Factors and Types

**Factors**
The factors that can influence and develop a landslide are coming from multiple sources including both natural events and human-related activities:

- overloaded slopes;
- deforestation and changes made to the flow of groundwater;
- quarry operations excavations;
- underground excavations (shallow mining);
- various dynamic actions (different technologies used on the construction area for soil improvement);


A. Rotaru (Ed.): CRIT-RE-BUILT 2019, SSGG, pp. 84–89, 2021.
https://doi.org/10.1007/978-3-030-61118-7_6

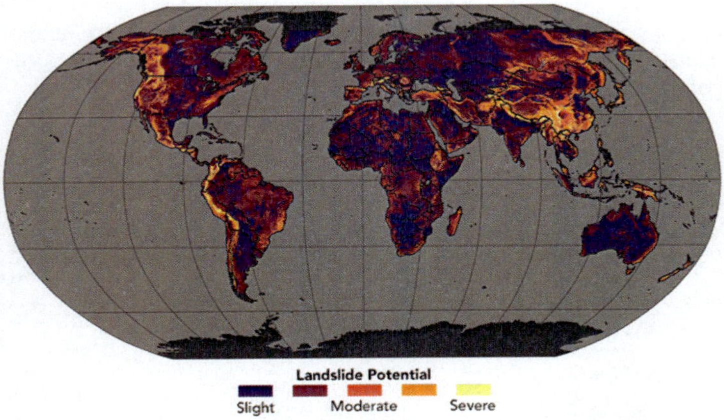


**Fig. 1.** A global view of landslide Susceptibility

- changes in the built environment such as excavation for new transport infrastructure systems and modern inhabited;
- destruction of vegetation by droughts, fires, and logging;
- erosion of the natural or artificial slopes caused by streams, rivers, especially during floods;
- variation of the water level due to high level of precipitations that causes changes of the soil characteristics;
- earthquakes;
- torrential erosions caused by gaps created by karstic phenomena.

These factors/changes, which can occur simultaneously or successively, influences either the state of tension or the shear resistance, or both parameters. From this point of view, an example, the water changes the pressure within the slope (example the pore pressure) that will affect the forces of gravity.

**Table 1.** Types of landslides

| Type of movement | Type of material | | |
|---|---|---|---|
| | Bedrock | Engineering soil | |
| | | Predominantly coarse | Predominantly fine |
| Falls | Rock fall | Debris fall | Earth fall |
| Topples | Rock topple | Debris topple | Earth topple |
| Rotational and translational slides | Rock slide | Debris slide | Earth slide |
| Lateral spreads | Rock spread | Debris spread | Earth spread |
| Flows | Rock flow | Debris flow | Earth flow |
| Complex | Combination of two or more principal types of movement | | |

All these types of factors, natural or human – related, may determine the loose of the shear resistance of the constituent rocks in certain points and the appearance and formation of the failure surface [2, 3].

**Types of Landslides**

Starting from the multitude of factors that influence the landslides phenomena there are different types of landslides that can be differentiated by the type of material involved and the way of movement. In the Table 1 and Fig. 2 is presented the classification of the landslides according to Varnes' classification criteria [4]: type of material involved in sliding, mode of movement.

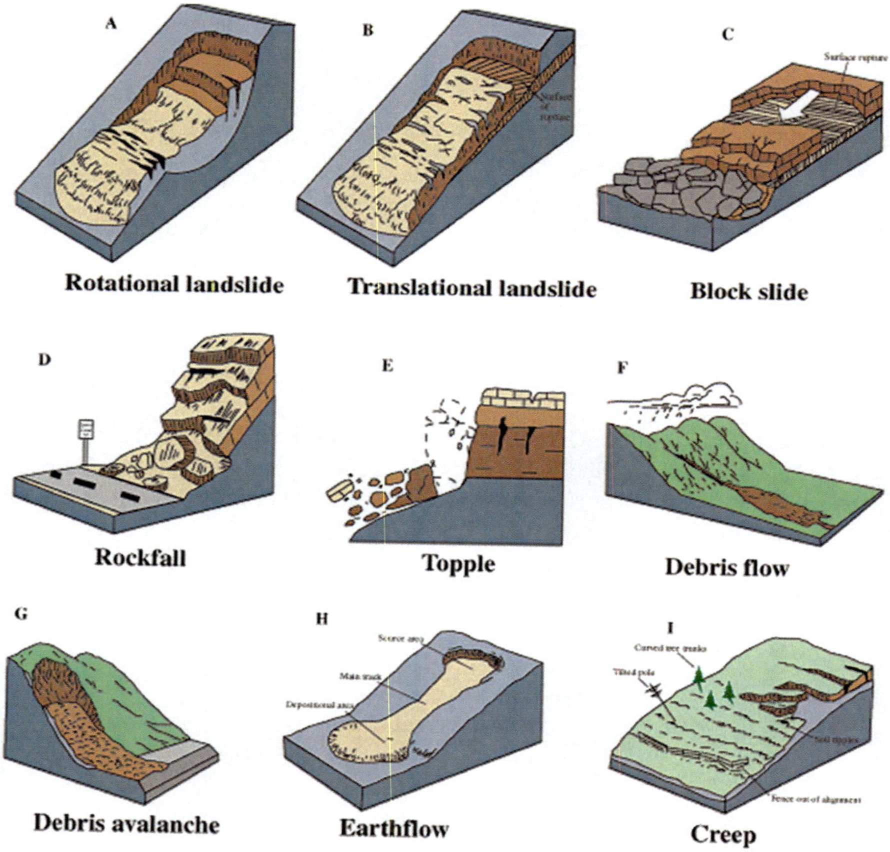


**Fig. 2.** Types of landslides

## 2 Environment and the Landslides

Landslides are destructive natural hazard that can have long-lasting effects on the environment (built and natural). Landslides can affect streams and rivers, can wipe out large tracts of forest, destroy wildlife habitat and remove productive soils from slopes; can cause tsunami, seiches, or outburst floods.

Landslides cause social disruptions when humans place parts of the built environment in their path. Slope instability along transportation routes or near communities poses dangers and is a source of considerable economic losses. From this point of view there is a continuum between the socioeconomic costs of landslides and the environmental costs of landslides. In situations when a landslide causes the destruction of the built environment, or have impact on natural environmental the socioeconomic dimension is affected.

Knowing all the factors that can influence a landslide it is possible to develop long-term strategies in terms of monitoring and control of the landslides, that may help to prevent, or, if triggered landslide, to reduce the negative effects of these natural disasters, primarily aimed to save lives, protecting construction works, buildings, preventing traffic disruptions or, where landslides are near rivers, preventing the flooding.

Reducing the global impact of landslides on risk-prone populations and critical infrastructure is an economic, social, and environmental necessity [5].

## 3 Landslides Control

When we talk about landslides we should take in consideration two aspects: preventions and solutions to apply to reduce the effects.

Landslides prevention may be divided into four different directions: economic, social, environment and infrastructure [7].

The economic aspects regard the government that should inform the public about the risks and the causes of this natural phenomenon. Also, the government has to monitor and to implement solutions, such as retaining structure, to prevent the development of the landslide in the affected areas.

In term of social aspects after being informed about the risk of landsliding the locals living in areas with sliding potential should take precautions about their property and to respect the solutions that the geotechnical engineers suggest.

For the environmental aspects, to prevent from landslide to occur, the geotechnical expertise on the properties that are going to be constructed is needed. Another way to prevent the landslide environmentally is to reduce deforestation and plant vegetations on slopes.

In term of infrastructure we could talk about the technology needed for monitoring the landslides hazard in susceptible areas and for warning the population in case of a landslide.

### 3.1 Monitoring/Data Acquisition

Monitoring is essential to predict the behaviour of the landslides and forecasting the causes that can trigger large numbers of landslides. Nowadays, for monitoring the unstable slopes, different types of equipment and techniques are used: open piezometers, extensometers, drilling inclinometers, terrestrial or space geodesy, electro-optic equipment for measuring distances, terrestrial and aerial photogrammetry, etc. [6].

Having all the information's from the monitoring systems, a data acquisition, storage, could help us to collect all the data and to perform virtual modelling studies of areas with sliding potential, in order to evaluate the risk and to recommend the adequate solutions for each situation. As example in a case of an active landslide, the analysis of the slide mass movements accompanied by virtual modelling analysis can provide information on the onset of these slide and one can take the right decision to reduce the effects of these phenomena.

### 3.2 Solutions to Prevent/Reduce Sliding

Having all the needed data the decision may vary depending the risk value [8]: to accept the risk; to avoid the risk when is possible to change a certain solution or a emplacement for a construction; to reduce the risk by using a consolidation/stabilization method; to reduce consequences by setting defensive measures of stabilization; to introduce monitoring and warning systems in the area of risk on a permanent or temporary basis that can be regarded also as measures to reduce consequences as well; to transfer, compensate, or share the risk, involving insurance companies.

Between the consolidation/stabilization methods that could be apply in order to reduce the risk we may adopt solutions as:

- Soil mass moving;
- Reducing natural slope inclination;
- Improve the stability by using geosynthetics;
- Slope vegetalization;
- Retaining wall (classical or geosynthetics retaining walls);
- Soil nailing;
- Drainage systems;
- Anti-slide piles;

This consolidation/stabilization methods may work alone or we may think a combination of them, in order to reduce the landslide risk.

## 4 Conclusions

For a sustainable built environment, besides knowing all the factors that influence a landslide, it is needed to apply the right solution to assure the stability. All the above-mentioned solutions should be regarded, or combinations of them may also be appropriate to consider, especially since large reductions of risk may be achieved. Depending the landslide risk, the geotechnical characteristics of the site, the economic interest and the related costs for implement a certain stabilization method an adequate solution will be taking for each particular situation in order to prevent and to preserve the natural or the built environment.

## References

1. Nasa Earth Observatory Homepage. https://earthobservatory.nasa.gov/images/89937/a-global-view-of-landslide-susceptibility. Accessed 30 Oct 2019
2. Stanciu, A., Lungu, I.: Fundatii – volumul I, Editura Tehnică (2006)
3. Colţ, O.: Landslides monitoring – hazard and risk management case study. In: Special Issue: XVI DECGE 2018 Proceedings of the 16th Danube - European Conference on Geotechnical Engineering, vol. 2, no. 2–3, pp. 287–292. Wiley (2018)
4. Varnes, D. J.: Slope movement types and processes. In: Schuster, R.L., Krizek, R.J. (eds.) Special Report 176: Landslides: Analysis and Control Transportation and Road Research Board, National Academy of Science, Washington D.C., pp. 11–33 (1978)
5. Turner, A.K.: Social and environmental impacts of landslides. Innovative Infrastruct. Solutions, **3**(1), (2018)
6. Cruden, D.M.: Investigation and mitigation. Special report - Transportation Research board National Academy of Science (1993)
7. Mase-landslide blog. http://mase-landslide.blogspot.com/2009/05/solutions.html. Accessed 30 Oct 2019
8. Lungu, I., Boţi, N., Stanciu, A., Donciu, O.: Present trends in landsliding control. Acta Technica Napocensis, section Civil Engineering-Architecture, **51**(III), pp. 197–203. Cluj (2008)
9. Cardona, O.C., et al.: The need for rethinking the concepts of vulnerability and risk from a holistic perspective: A necessary review and criticism for effective risk management. In: Bankoff, G., Frerks, G., Hilhorst, D. (eds.) Maping Vulnerability, Disasters, Development And People. Earthscann Publications, London (2004)
10. Villagran de Leon, J.C.: Vulnerability – A conceptual and methodological review, SOURCE, Publications Series of UNU-EHS, nr. 4 (2006)
11. Manea, S.: Evaluarea riscului de alunecare a versantilor. Editura Conspress, Bucuresti (1998)
12. Homepage. http://www.springer.com/lncs. Accessed 21 Nov 2016

# The Role of Groundwater and Land Subsidence Analysis for Sustainable Development of Infrastructure in Some SE Asian Cities

Pham Huy Giao(✉), Krit Saowiang, and Nguyen Thi Hai Anh

Asian Institute of Technology, Pathum Thani, Thailand
hgiao@ait.asia

**Abstract.** Many cities in SE Asia are founded on soft deltaic soil and have faced land subsidence due to groundwater extraction, which can cause a number of problems to sustainable development of their infrastructure such as damages to the buildings, roads, highways, aggravated flooding, increase in the construction cost etc. Groundwater and settlement monitoring is needed to assess the subsidence situation of the concerned cities. However, the monitoring cannot follow promptly and cover enough the continuously expanding urban areas. But when monitoring cannot the numerical analyses can help. To highlight the useful role of numerical analysis in assessing land subsidence two case studies from two major cities in the SE Asia, Hanoi and Bangkok, are presented. For Hanoi, land subsidence along an urban mass rapid transit line is predicted using an one-dimensional FEM consolidation model. For Bangkok, after a long period from 1950 to 1997 of increasing extraction the groundwater levels have started to recover in all pumped aquifers since 1997, 3D groundwater modelling and coupled consolidation analysis were successfully conducted to help understand better on groundwater recovery and its effects on stability of substructure.

**Keywords:** Land subsidence analysis · Sustainable development · SE Asia

## 1 Introduction

### 1.1 A Subsection Sample

Land subsidence due to groundwater extraction is a persistent problem facing infrastructure development in major cities in the SE Asia such as Bangkok (Thailand), Hanoi and Ho Chi Minh City (Vietnam), Jakarta (Indonesia), Yangon (Myanmar), Kuala Lumpur (Malaysia) etc. [1–3] and it continues to be for many years ahead with the continuing groundwater extraction and the increasing urbanization in this dynamic region. Land subsidence is common for many countries all over Asia and beyond, and for a city development it can have significant impacts on groundwater management, flooding protection and control, foundation design and stability etc. Distribution of a number of cities that are located in the deltaic plains, underlain by compressible soft soils, is shown in Fig. 1 as slightly modified from [2].


A. Rotaru (Ed.): CRIT-RE-BUILT 2019, SSGG, pp. 90–100, 2021.
https://doi.org/10.1007/978-3-030-61118-7_7

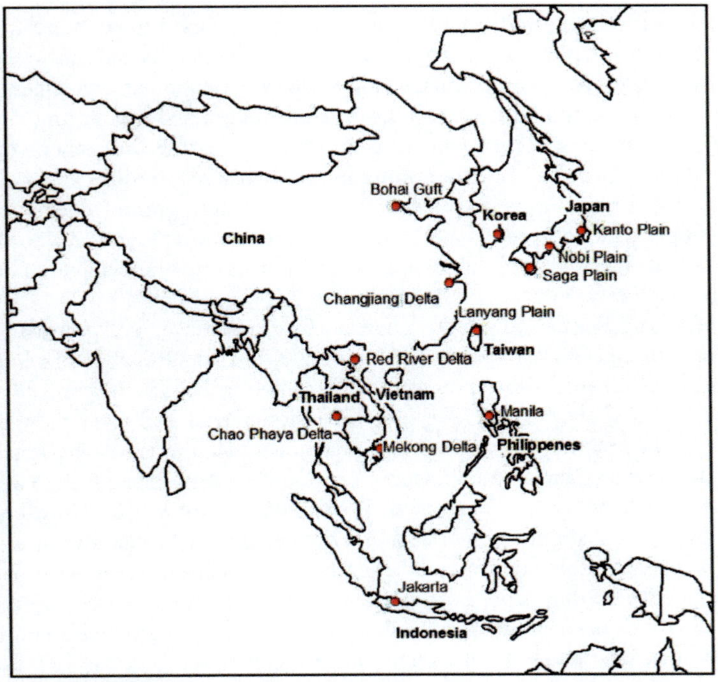


**Fig. 1.** Some major cities Asia, located on deltaic plains and experiencing land subsidence due to groundwater extraction [2].

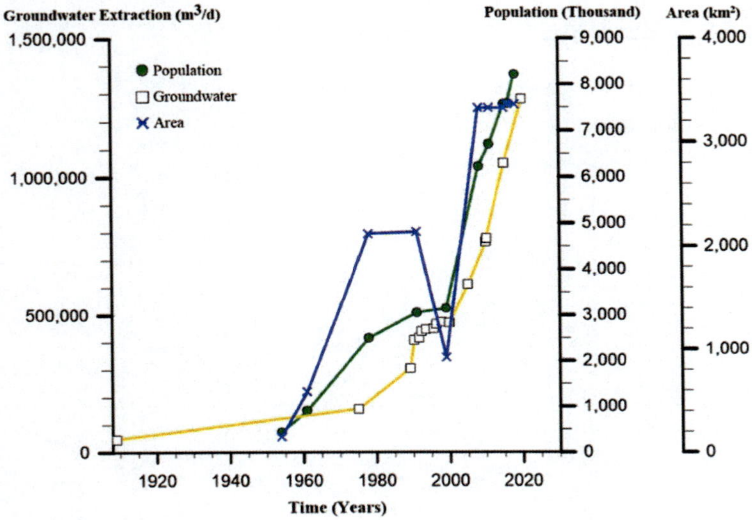


**Fig. 2.** Groundwater extraction in Hanoi versus urban expansion and growing population [4]

For a land subsidence study there are a few things one used to be interested in, and namely: (i) how to assess that a city is going to have or it has already had subsidence?; ii) if a city shows the signs of land subsidence, how to do investigation and what geotechnical parameters need to be monitored: iii) after instrumentation and monitoring it will take some time before the monitoring data can be made use of, is there any other way around to assess and predict land subsidence? Some answers to these questions are briefly given as follows: i) if a city that is founded on soft soils, where groundwater is extracted for water supply and the groundwater pumping increases time together with the city's expansion and its increasing population then land subsidence is inevitable to occur. An example of Hanoi city, Vietnam, is shown in Fig. 2.

The land subsidence can be visually observed by groundwater well protrusion, differential settlements and cracks of pavement, building etc.; ii) identification of geotechnical parameters of the subsoil profile, monitoring of piezometric head in the pumped aquifers, pore pressure in the adjacent clay layers, compression of each soil layer and the surface settlement are the key things one need to do for understanding land subsidence; iii) in reality the monitoring data are often insufficient or not existent due to lack of awareness on the issue, limited budget and manpower, in particular for the less developed countries like many in the SE Asia. In such cases numerical analysis of groundwater and subsidence can be very helpful to understand better the subsidence situation of the concerned city. The objective of this paper is to present a few case studies of groundwater and subsidence analyses in some major cities in the SE Asia to highlight their important role for sustainable infrastructure development of urban centers.

## 2 Prediction of Land Subsidence Along a MRT Line in Hanoi

### 2.1 Land Subsidence Monitoring in Hanoi

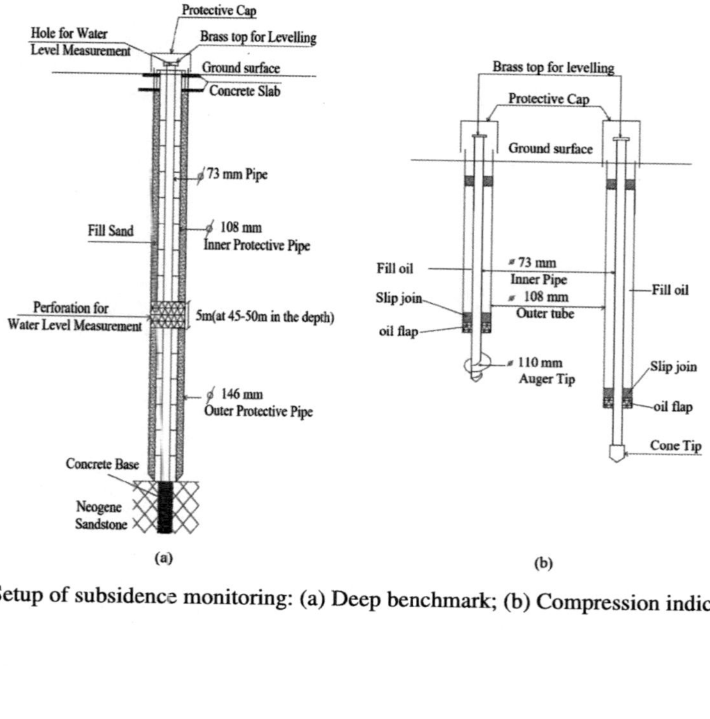


**Fig. 3.** Setup of subsidence monitoring: (a) Deep benchmark; (b) Compression indicators [5]

The aquifer system underlying Hanoi city is up to 80 m deep and consists of an upper Holocene aquifer, underlain by an Pleistocene aquitard, followed by the Pleistocene aquifer with Upper and Lower Pleistocene units, and the Neogene sandstone bedrock. Subsidence monitoring in Hanoi started in the early 1990s by Hydrogeological Division No.2 (K2) [5] with 30 surface settlement points. However, the monitoring based on levelling conducted by K2 has ceased since 1995 due to the lack of funding. The Hanoi Institute of Building Technology (HIBT) has taken over with construction of a number of subsidence monitoring stations, whose number is however quite limited. Instrumentation at a land subsidence monitoring station is shown in Fig. 3.

The land subsidence monitoring data recorded by now are far from being satisfactory, being few and insufficient. In particular, the pore pressure in the clay layers, which is one of the most important indicators on whether or not land subsidence has occurred, has not been well adequately recorded. As seen in Fig. 4 the locations of land subsidence monitoring stations in Hanoi most follow the locations of the well fields and groundwater

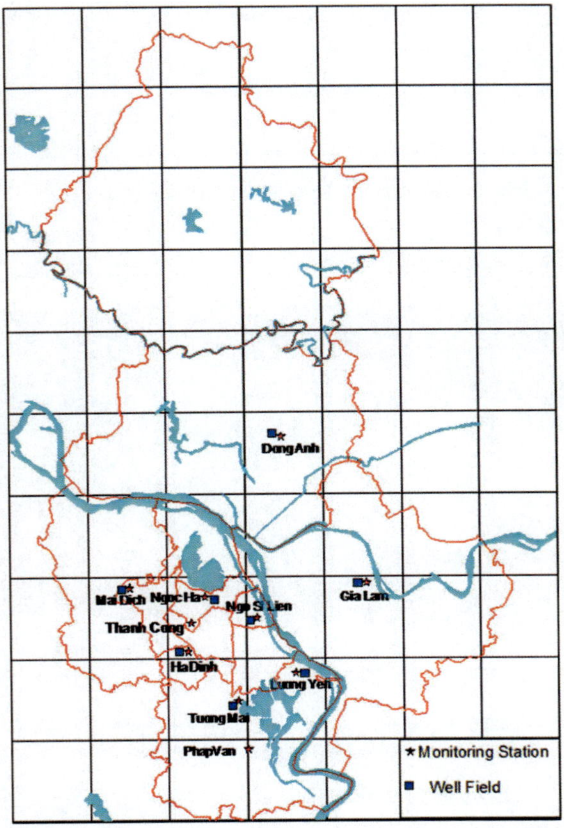


**Fig. 4.** Location of well fields and land subsidence monitoring stations in Hanoi [5]

exploitation plants. Thus, for assessment of land subsidence over a large area one needs to have other tools beside the traditional monitoring approach.

## 2.2 Prediction of Land Subsidence for an Urban Mass Rapid Transit (UMRT) Line in Hanoi

**Fig. 5.** Location of the UMRT Line 3 in Hanoi [6]

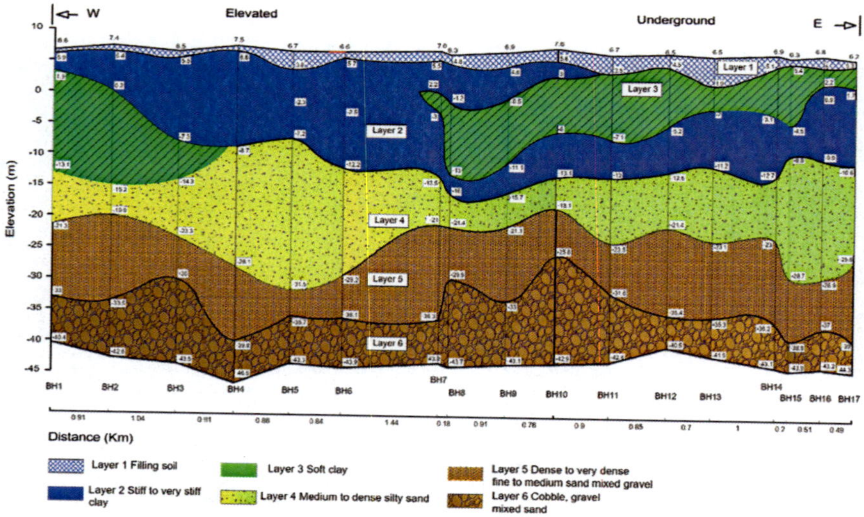


**Fig. 6.** Soil profile along UMRT line no.3

Hanoi has a total area of 3.329 $km^2$ and it is home to about 8 million people by 2020. To meet the fast growing needs of public transportation an Urban Mass Rapid Transit

(UMRT) system consisting of 8 lines in both elevated and underground sections with a total length of 318 km is planned, including the UMRT line no. 3, whose first segment is 21.5 km long, running from south to north through the city center as shown in Fig. 5. The subsoil profile along this line no. 3 is plotted in Fig. 6. The geotechnical parameters of the soil layers used in a 1D FEM analysis are presented in Table 1 below, where UW, e0, Cv, Cs, CR, RR, OCR and E are unit weight, initial void ratio, consolidation coefficient, compression ratio, recompression ratio, over-consolidation ratio and Young's modules, respectively.

**Table 1.** Soil parameters

| Layer | | UW ($kN/m^3$) | $C_v$ ($m^2/y$) | CR | RR | OCR | E (MPa) |
|---|---|---|---|---|---|---|---|
| 1 | Soft clay | 17 | 2.92 | 0.71 | 0.013 | 2.51 | |
| 2 | Stiff clay | 19.9 | 4.91 | 0.284 | 0.018 | 1.5 | |
| 3 | Sand | | | | | | 19 |

A FEM program for subsidence analysis was developed by Giao [7] and was applied in this study. By applying Green's Lemma to Terzaghi's one-dimensional consolidation equation Eq. (1) one can obtain the weak form for a generic linear element as shown in Eq. (1) below:

$$\int_{\Omega^e} c_v \frac{\partial N_i}{\partial z}\frac{\partial N_j}{\partial z} u_j dup\Omega^e + c_v N_i \frac{\partial N_j}{\partial z} u_j \Big|_{R^e} - \int_{\Omega^e} N_i N_j \frac{\partial u_j}{\partial t} d\Omega^e = 0 \tag{1}$$

$$\int_{\Omega^e} c_v \frac{\partial N_i}{\partial z}\frac{\partial N_j}{\partial z} u_j d\Omega^e - \int_{\Omega^e} N_i N_j \frac{\partial u_j}{\partial t} d\Omega^e = 0 \tag{2}$$

$$dS_c(z,t) = b_i.[RR(z).\log\frac{P_c^{'}(z)}{\sigma_{v0}^{'}(z)} + CR(z).\log\frac{\sigma_{v0}^{'}(z) + \Delta u(z,t)}{P_c^{'}(z)} \tag{3}$$

$$S_c(t) = \sum_{z=1}^{n} dS_c(z,t) \tag{4}$$

$$S_a = \frac{1}{E}.b.\gamma_w.dh \tag{5}$$

Based on the pore pressure change calculated by FEM code Eq. (2) the compression of each sub layer can be calculated by Eq. (3), and the total primary consolidation of the a clay layer made of n sub-layers can be calculated by Eq. (4). Where: dSc(z,t) is the consolidation settlement of the sub-layer at z depth, time t; Sc(t) is the total consolidation settlement of the whole clay layer at time t; b: is the thickness of the considered sub-layer; $P'_c(z)$ is the pre-consolidation pressure of the considered sub-layer; $\sigma' v0(z)$ is the effective stress of the considered sub-layer; $\Delta u(z,t)$ is the dissipation of pore pressure of the considered sub-layer at time t; CR(z) is the compression ratio of the considered sub-layer; RR(z) is the recompression ratio of the considered sub-layer. For a sand layer, its compression is rather instantaneous with groundwater extraction and can be calculated by Eq. (5), where b is the thickness of the sand layer (m), E is modulus of compression (kPa), $\gamma_w$ is the unit weight of water, dh is the groundwater drawdown or head drop (m).

An integrated approach, combining geotechnical characterization with a 1D consolidation analysis was applied to predict land subsidence the UMRT line no. 3. The input data for consolidation analysis are shown in Table 1. Land subsidence for clay layers 2 and 3 (see Fig. 6) is calculated by using a FEM code developed by Giao [7]. The observed groundwater drawdowns in some wells along the UMRT line no. 3 were collected and used as boundary conditions. The subsidence results are plotted in Fig. 7 for a drawdown of 20 m, and a period of 25 years starting from 2010 to 2035. The subsidence is not uniform along this UMRT line, at some locations such as BH1 where the soft clay is thick the subsidence can reach 0.75 m after 25 years.

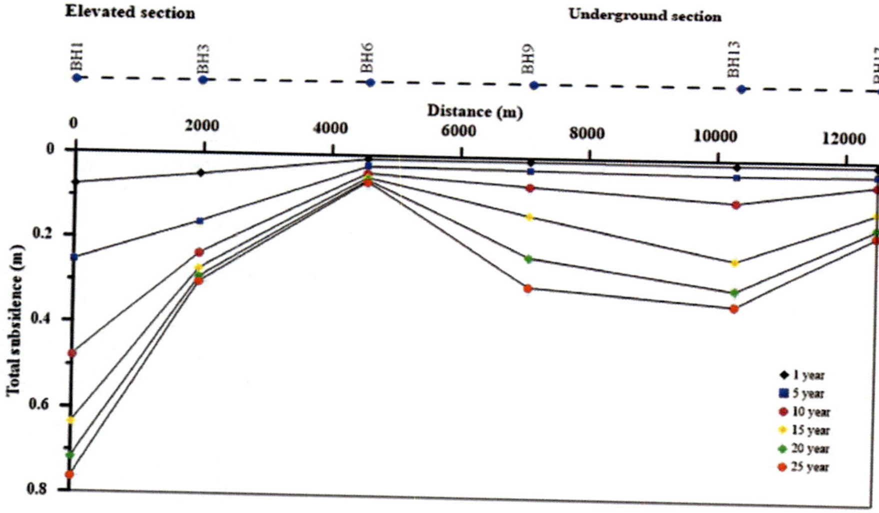


**Fig. 7.** Prediction of land subsidence along the UMRT line No. 3 for a drawdown of 20 m

# 3 Groundwater and Land Subsidence Analysis for Bangkok

## 3.1 Bangkok Aquifer System and Groundwater Extraction

The multi-aquifer system in Bangkok consist of nine aquifers up to more than 500 m deep (Fig. 8). Groundwater has been extensively extracted from the second, third and fourth aquifers since 1950s, causing significant compression of the Bangkok soft clay layer on top (see Fig. 9). Due to a good groundwater pumping control and management since 1997 the groundwater levels in all the aquifers has recovered (see Fig. 9. This groundwater recovery created another type of concern to infrastructure development in Bangkok [9].

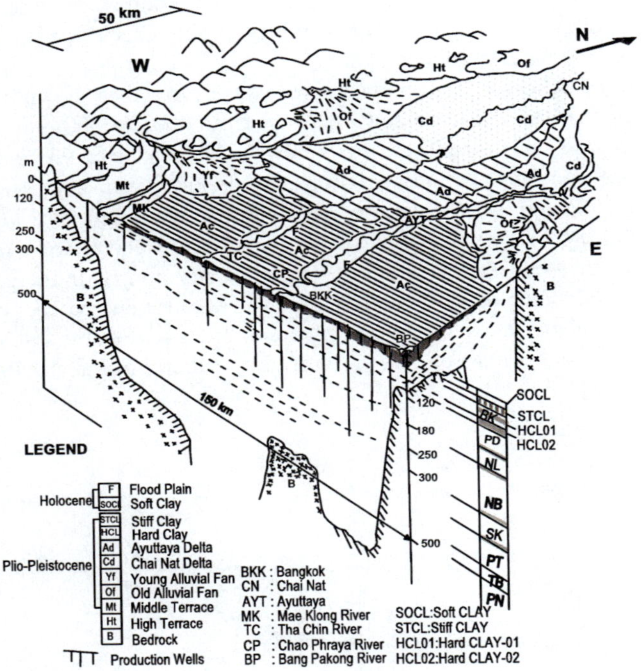


**Fig. 8.** Bangkok aquifer system and distribution of Bangkok soft clay [8]

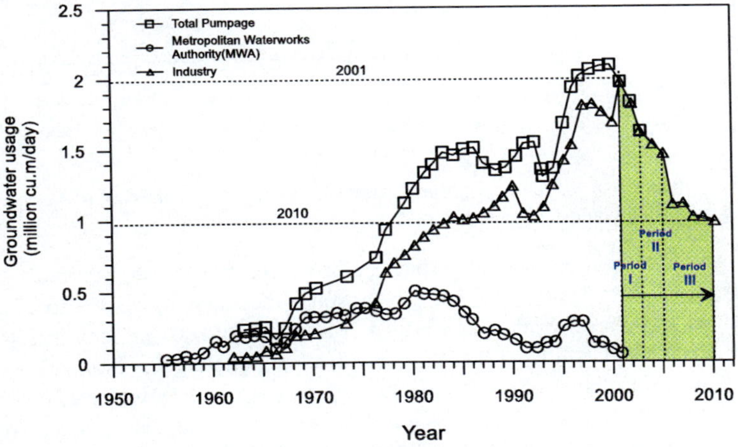


**Fig. 9.** Groundwater extraction in Bangkok for the last 60 years

The 3D groundwater model was conducted to explain the behavior of groundwater recovery in the Bangkok aquifer system for short periods of decreasing groundwater pumping, i.e., from 2 million $m^3$/day in 2001 to 1 million $m^3$/day in 2010 (see periods I, II and III in Fig. 9). The groundwater model is of $122 \times 140$ $km^2$ and 385 m in depth, covering seven provinces including Bangkok, Samut Prakan, Samut Sakhon, Nonthaburi, Pathum Thani, Nakhon Pathom, and Ayutthaya. The model consists of 5 aquifers namely Bangkok (BK), Phrapradaeng (PD), Nakhonluang (NL), Nonthaburi (NB), and Samkok (SK). All the aquifers are assigned as the confined aquifer except the Bangkok aquifer that is considered as the unconfined aquifer.

As a result the behavior of groundwater recovery is found as shown in Fig. 10 and explained below: i) Period I (drawdown period): from 2001 to 2003, the discharge is higher than recharge, which implies that groundwater level in the aquifer is still reducing due influence of pumping; ii) Period II (balance period): from 2003 to 2005, when more water is recharged into the aquifer. Although the pumping rate is decreased, the discharge is still higher than recharge; iii) Period III (recovery period): from 2005 to 2010, the discharge is constant as pumping has practically stopped in Bangkok area and the rate of pumping in the outskirt areas of Bangkok has also decreased. Consequently, the recharge-related groundwater volume increases and it crosses the discharge-related groundwater volume in 2008 (see Fig. 9), indicating that groundwater in the aquifers is recovered.

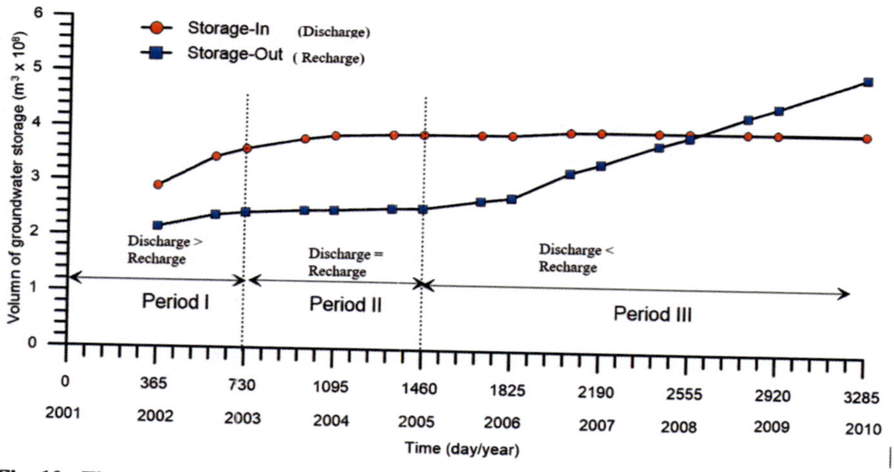


**Fig. 10.** The changes in volume of groundwater storage volume behavior of storage-in and storage-out in Bangkok aquifer system

The numerical analyses of subsurface deformation due to long-terms of groundwater drawdown from 1960–1997 and groundwater recovery from 1997 to 2016 were conducted for three testing sites, i.e., NBTA, MBTA and SBTA as seen in Fig. 11. A consolidation analysis followed the 3D groundwater modeling for Bangkok soft, stiff and hard clays up to the depth of 80 m. Subsurface deformation was calculated using the FEM transient coupled pore pressure/effective stress analysis with the Modified Cam-Clay model available in Abaqus/CAE software, indicating that the groundwater drawdown from 1906 to 1997 caused a subsidence of 51, 62 and 76 cm, and the following groundwater recovery from 1997 to 2016 could pull back a rebound of 9.8, 9.7 and 5.3% at the three study sites as shown in Fig. 12, respectively [10].

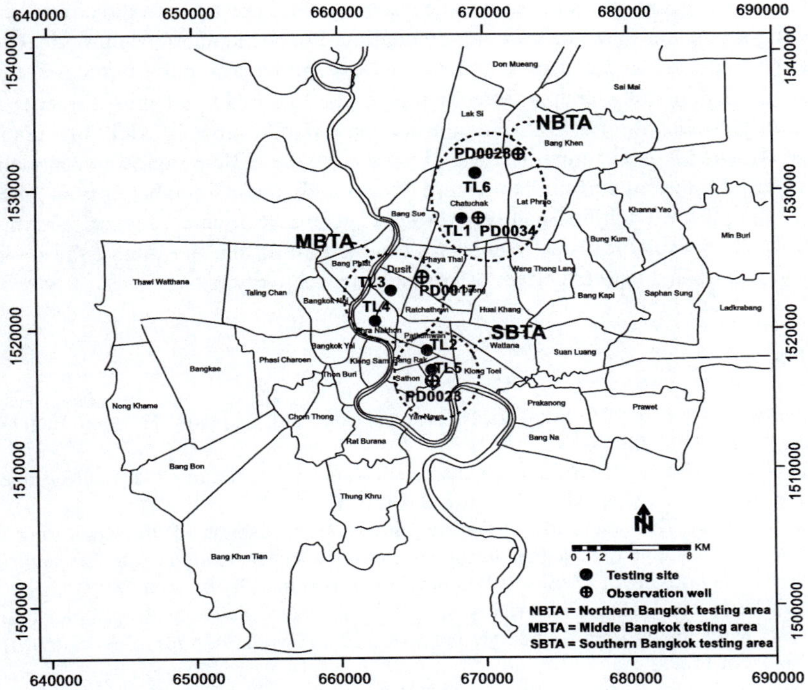


**Fig. 11.** Locations of testing sites and observation wells

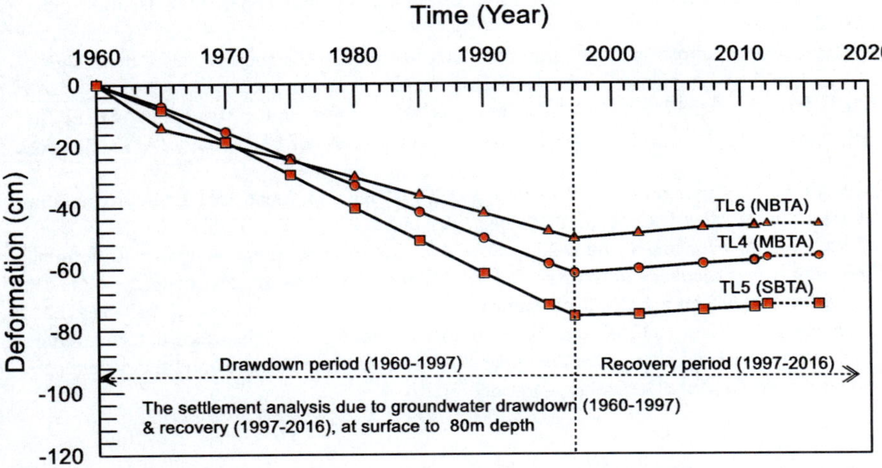


**Fig. 12.** The results of surface deformation analysis for three sites of NBTA, MBTA and SBTA in Bangkok after [10]

## 4 Conclusions

Monitoring is very important for land subsidence investigation and assessment. But for major SE Asian cities located on large area of thousands square kilometer, the number of monitoring stations is never sufficient and cannot keep up with the continuously expanding urban areas and the growing population. For such situations numerical analyses of groundwater and consolidation can help. To prove this point two case studies for Hanoi and Bangkok were presented in this research. For Hanoi land subsidence was successfully predicted for a 21-km long urban mass rapid transit (UMRT) line planned to be constructed in near future. For Bangkok, a large scale 3D groundwater modelling and a coupled transient consolidation were done to help understand better how groundwater in the pumped aquifers can recover after more than four decades of continuing drawdown. The results form these analyses can be used further to evaluate the stability of pile foundations due to the effect of groundwater recovery.

## References

1. Phien-wej, N., Giao, P.H., Nutalaya, P.: Land subsidence in Bangkok, Thailand. Eng. Geol. **82**(4), 187–201 (2006)
2. Thoang, T.T., Giao, P.H.: Subsurface characterization and prediction of land subsidence for HCM city, Vietnam. Eng. Geol. **199**, 107–214 (2015)
3. van der Horst, T., Rutten, M.M., van de Giesen, N.C., Hanssen, R.F.: Monitoring land subsidence in Yangon, Myanmar using Sentinel-1 persistent scatterer interferometry and assessment of driving mechanisms. Remote Sens. Environ. **217**, 101–110 (2018)
4. Giao P.H., Hue, V.T., Han, T.D., Hai Anh, N.T., Minh, N.N.: Land Subsidence Prediction for a new UMRT line in Hanoi, Underground Space, Elsevier (2019). https://doi.org/10.1016/j.undsp.2018.11.002
5. Giao, P.H., Minh Son, N.M., Hai Anh, N.T., Minh, N.N.: Role of monitoring data in subsidence prediction for the MRT lines and stations in Hanoi. In: Proceedings of International Conference on Sustainability in Civil Engineering, 24—25 November, 2018, Hanoi, Vietnam (2018)
6. Urbanist Hanoi: https://urbanisthanoi.com/hanoi-news/13692-hanoi-reveals-train-design-of-metro-line-3,-seeks-public-feedback (2018)
7. Giao, P.H.: Artificial Recharge of the Bangkok Aquifer System for Mitigation of Land Subsidence, D.Eng. Dissertation, No. GE/96-2, Asian Institute of Technology (AIT), Bangkok, p. 319 pp (1997)
8. Takaya, Y., Thiramongkol, N.: Chao Phraya Delta of Thailand, Asian Rice-Land Investigation, A Description Atlas 1. Center for Southeast Asian Studies, Japan (1982)
9. Giao, P.H., Paveechana, T., Saowiang, K.: Consolidation settlement analysis with reference to groundwater recovery in the Bangkok multi-aquifer system. In: Proceedings of the 18th Southeast Asian Geotechnical Conference, 29–31 May 2013, Singapore, pp. 567–573 (2013)
10. Saowiang, K., Giao, P.H.: Abaqus-based numerical analysis of subsurface deformation due to groundwater level changes in the upper 100-m zone of the Bangkok aquifer system paper under review for Acta Geotechnica journal (2020)

# Sustainable Infrastructure Design and Maintenance

# Sustainable Infrastructure Design and Maintenance

Wei Wu(✉)

Institute of Geotechnical Engineering, University of Natural Resources and Life Sciences, Feistmantelstrasse 4, 1180 Vienna, Austria
wei.wu@boku.ac.at

**Abstract.** Infrastructure are long-lasting products which have huge impacts on the environment during their whole lives. The design and subsequent maintenance of infrastructure should take into consideration long-term environmental and economic benefits. Through a systematic literature review this report presents the concept of sustainable infrastructure design and maintenance (SIDM), identifies the key factors during the application of SIDM, and introduces corresponding sustainability assessment tools. Particularly, a life cycle assessment approach is introduced in detail. Finally, an existing case related to the life cycle assessment method is demonstrated in order to assist the decision makers.

**Keywords:** Sustainable infrastructure · Environmental and economic benefits · Life-cycle assessment

## 1 Introduction

Nowadays, many developing countries in the world are experiencing a process of urbanization at an unprecedented scale and speed. This trend is expected to result in, by 2050, an additional 1.2 billion people who are projected to live in cities in Asia [1], as shown in Fig. 1. As urban population grows, not only will new cities and new urban districts need to be developed, but existing cities will also need to be retooled to improve livability.

Since infrastructure is central to development in developing countries, massive infrastructure construction is urgently required for ensuring the connectivity, productivity, efficiency and overall competitiveness. For instance, there are about 40 billion square meters of buildings in China and this figure will reach 70 billion square meters by 2020 [2]. However, the challenge of meeting massive infrastructure construction needs in the face of its significant impacts on environment, and dwindling resources [3]. It is estimated that approximately 80% of the total greenhouse gas (GHG) emissions come from cities, with 25% attributed to urban transportation, 32% to build environment, and an additional 5% to municipal solid waste. Urban transportation accounts approximately 20% for the global energy consumption, and building construction and operation responds to an additional 25% [4]. Therefore, design for sustainable infrastructure is an urgent task for the construction industry.


A. Rotaru (Ed.): CRIT-RE-BUILT 2019, SSGG, pp. 103–115, 2021.
https://doi.org/10.1007/978-3-030-61118-7_8

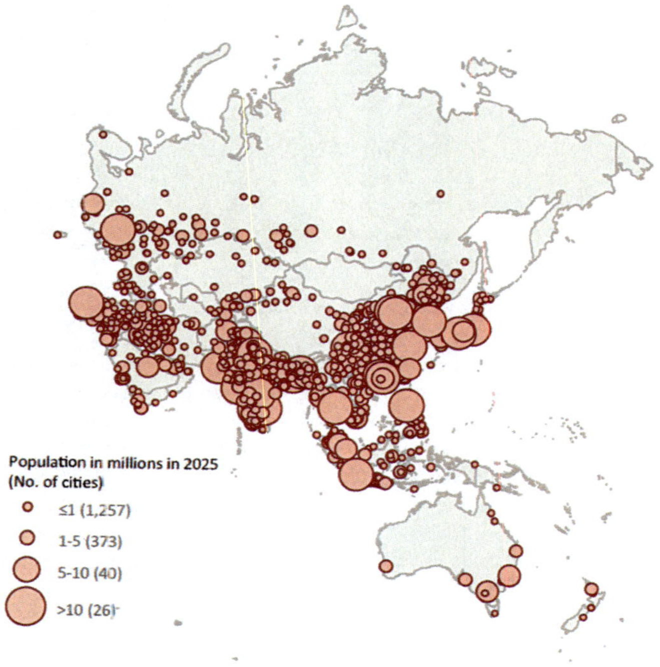


**Fig. 1.** An expected distribution of cities in Asia by population size in 2025 [1].

Sustainability is a concept of meeting the current need without jeopardizing the future generations' capability of meeting their own needs. It is an approach to meeting the current need with the future in mind. While this is a generally acceptable concept, there remain significant gaps on the approaches to operationalizing in many sectors of the economy especially in some developing countries where majority are struggling to make the ends meet. Although the approach to design is essentially traditional whereby the design focus is basically on functionality, cost and aesthetics [5]. The world bank statement that "As of 2000, about 80% of all government-owned water systems in small towns were nonoperational" is an eye opener and a good pointer to the need for a different approach. A new design approach that enables incorporation of technical, economic, institutional, environmental and socio-cultural factors into design is sustainable design. However, the current sustainability assessment system focuses mainly on the design stage. As a critical set of systems, infrastructure forms much of the foundation for quality of life and enables national development and progress. However, it also consumes vast material resources and energy [6]. Sustainable infrastructure design and maintenance (SIDM) considers the environmental influences of a building through its whole life, that is, from the initial construction process to the future operation stage of the building [7]. Thus, it is essential that the infrastructure be designed and maintained using the SIDM approach that consider social, environmental, and economic impacts over buildings' whole lives.

## 2 What Is Sustainable Infrastructure [8]

The **Global Commission on the Economy and Climate** defined Infrastructure as: structures and facilities that underpin power and other energy systems (including upstream infrastructure, such as the fuel production sector), transport, telecommunications, water, and waste management. It includes investments in systems that improve resource efficiency and demand-side management, such as energy and water efficiency measures. Infrastructure includes both traditional types of infrastructure (including energy to public transport, buildings, water supply and sanitation) and, critically, also natural infrastructure (such as forest landscapes, wetlands and watershed protection).

SIDM refers to infrastructure projects that are planned, designed, constructed, operated, and decommissioned in a manner to ensure economic and financial, social, environmental (including climate resilience), and institutional sustainability over the entire life cycle of the project. The guiding principles for each of the dimensions of sustainability is shown in Fig. 2.

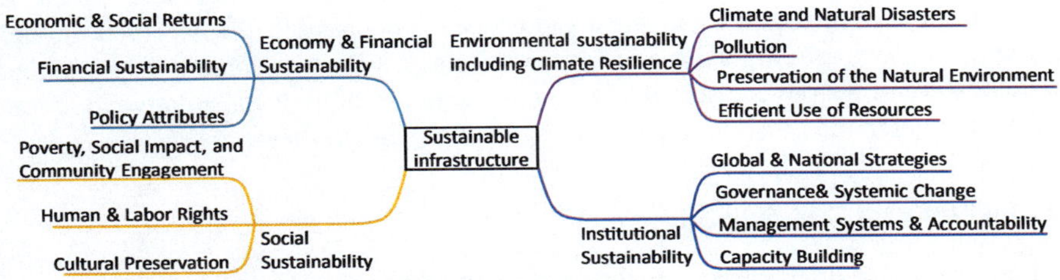


**Fig. 2.** The Four Dimensions of Infrastructure Sustainability [8].

### 2.1 Economic and Financial Sustainability

Infrastructure is economically sustainable if it generates a positive net economic return, considering all benefits and costs over the project life cycle, including positive and negative externalities and spillovers. In addition, the infrastructure must generate an adequate risk adjusted rate of return for project investors. Sustainable infrastructure projects must therefore generate a sound revenue stream based on adequate cost recovery and be supported, where necessary, by well-targeted subsidies (to address affordability) or availability payments (when users cannot be identified), or where there are large spillover effects. Sustainable infrastructure must be designed to support inclusive and sustainable growth and boost productivity and to deliver high-quality and affordable services. Risks must be fairly and transparently distributed to the entities most able to control the risk or to absorb its impact on the investment outcomes over the life cycle of the project.

### 2.2 Environmental Sustainability, Including Climate Resilience

Sustainable infrastructure preserves, restores, and integrates the natural environment, including biodiversity and ecosystems. It supports the sustainable and efficient use of natural resources, including energy, water, and materials. It also limits all types of pollution over the life cycle of the project and contributes to a low-carbon, resilient, and resource-efficient economy. Sustainable infrastructure projects are (or should be) sited and designed to ensure resilience to climate and natural disaster risks. Sustainable infrastructure often depends on national circumstances, where the overall performance will need to be measured compared to what could have been built or developed instead.

### 2.3 Social Sustainability

Sustainable infrastructure is inclusive and should have the broad support of affected communities - it serves all stakeholders, including the poor—and contributes to enhanced livelihoods and social well-being over the life cycle of the project. Projects must be constructed according to good labor, health, and safety standards. Benefits generated by sustainable infrastructure services should be shared equitably and transparently. Services provided by such projects should promote gender equity, health, safety, and diversity while complying with human and labor rights. Involuntary resettlement should be avoided to the extent possible and when avoidance is not possible, displacement should be minimized by exploring alternative project designs. Where economic displacement and relocation of people is unavoidable, it must be managed in a consultative, fair, and equitable manner and must integrate cultural and heritage preservation.

### 2.4 Institutional Sustainability

Institutionally, sustainable infrastructure is aligned with national and international commitments, including the Paris Agreement, and is based on transparent and consistent governance systems over the project cycle. Robust institutional capacity and clearly defined procedures for project planning, procurement, and operation are enablers for institutional sustainability. The development of local capacity—including mechanisms of knowledge transfer, promotion of innovative thinking, and project management—is critical to enhance sustainability and promote systemic change. Sustainable infrastructure must develop technical and engineering capacities as well as systems for data collection, monitoring, and evaluation, to generate empirical evidence and quantify impacts or benefits.

## 3 Leading Factors and Assessment Tools for SIDM

Infrastructure provision has considerable impacts on resource utilization, quality of the environment and overall quality of life. Infrastructure provision process should meet performance requirements in terms of economic, ecological and social aspects and that the design approach has to be in a way to harness technologies and meet human needs by working with nature, instead of solving problems at nature's expense. Thus, key design elements for sustainable infrastructure design should be identified for the application of SIDM in developing countries.

### 3.1 Leading Factors for SIDM [7]

To identify the key design stage factors that significantly influence the delivery of sustainable infrastructure, a systematic literature review was conducted by Nuramo and Haupt [9]. The results were found to be important toward delivery of sustainable infrastructure in the context of developing countries. Table 1 outlines refined 50 key elements and corresponding factors for sustainable infrastructure design.

### 3.2 Sustainability Assessment Toolkits [7]

The incorporation of the aforementioned factors for SIDM has brought sustainability issues to the forefront of infrastructure and building design, construction, and operations practice in the different countries. For example, the BRE Environmental Assessment Method (BREEAM) system is an integrated sustainability assessment tool used in the UK, which covers waste, water and energy, as well as transport, pollution, community engagement, the health and wellbeing of building occupants, the choice of material, enhancing biodiversity and building management. The BREEAM series can assess several types of buildings' design, including BREEAM Courts, Healthcare, Industrial, International, Multi-residential, Prison, Offices, Retail, Education, Communities and Bespoke.

The other sustainability assessment tool for buildings is The Code for Sustainable Homes [10] to be used in residential building designs in the UK covers the following areas:

(1) Energy efficiency/CO2.
(2) Water efficiency.
(3) Surface water management.
(4) Site waste management.
(5) Household waste management.
(6) Use of materials.
(7) Lifetime homes (applies to Code Level 6 only).

Another well-known sustainability assessment tool is Leadership in Energy & Environmental Design (LEED) [11] used mainly in the US. It covers new construction, existing buildings, commercial interiors, core and shell, schools, retail, healthcare, homes, and neighbourhood development. The checklist scoring system includes the following main aspects:

(1) Sustainable sites.
(2) Water efficiency.
(3) Energy and atmosphere.
(4) Material and resources.
(5) Indoor environmental quality.
(6) Innovation and design process.

**Table 1.** The main factors in sustainable infrastructure design and maintenance.

| |
|---|
| **Factor 1: Material Selection** |
| 1. Prescribing low energy materials<br>2. Use of locally available materials<br>3. Use of durable/high-performance materials<br>4. Use of materials with low health risk and pollution<br>5. Material reuse |
| **Factor 2: Economic Considerations** |
| 1. Cost/benefit analysis<br>2. Life cycle analysis<br>3. Cost efficiency<br>4. Bankability |
| **Factor 3: Policy and Regulations** |
| 1. Presence of design sustainability regulatory requirements<br>2. Presence of sustainability rating systems<br>3. Inclusion of sustainability requirements in public project briefs |
| **Factor 4: Social Considerations** |
| 1. Public/beneficiaries participation<br>2. Client participation<br>3. Accessibility of the infrastructure to the public including people with specific needs<br>4. Health and safety consideration for construction workers and the public during construction and operation stages<br>5. Security consideration during construction and use<br>6. Satisfaction of the public<br>7. Protection of cultural heritage<br>8. Protection of landscape, historical areas and archaeological sites<br>9. Risk analysis and disaster mitigation |
| **Factor 5: Design and Project Management** |
| 1. Early contractors' involvement at design stage<br>2. Early suppliers' involvement at design stage<br>3. Selection of appropriate contract/project delivery type<br>4. Inclusion of sustainability related clauses in contract documents<br>5. Proper construction quality control procedure |
| **Factor 6: Technical Considerations** |
| 1. Exhaustive site survey and ground investigation<br>2. Considering alternatives prior to proposing a solution<br>3. Multi-disciplinary integrated design team beginning from feasibility study stage<br>4. Meeting functional requirements and users' comfort<br>5. Robustness and less maintenance products<br>6. Completeness and clarity of design documents<br>7. Value engineering 8. Harmony with the surrounding environment |

*(continued)*

**Table 1.** (*continued*)

| **Factor 7: Environmental Considerations** |
|---|
| 1. Mitigating effects of natural disasters and climate change |
| 2. Climate resiliency (resistant to climate change) |
| 3. Ensuring efficient energy utilization both during construction and operation phases |
| 4. Optimizing uses of natural resources |
| 5. Optimizing site potentials (land use) |
| 6. Uses of less energy during construction and operation |
| 7. Waste minimization/design optimization |
| **Factor 8: Design Professionals and the Design Process** |
| 1. Awareness of clients about sustainability |
| 2. Awareness of designers about sustainability |
| 3. Knowledge and experience of designers |
| 4. Skill of designers |
| 5. Appropriating adequate time for design |
| 6. Reasonable financial compensation for design work |
| 7. Presence of design guidelines/procedures for sustainable infrastructure |
| 8. Proper coordination among designers from different disciplines |
| 9. Willingness of designers to implement sustainability design concept and practices in their designs |

The Waste and Resources Action Programme (WRAP) recycle content toolkit is designed by the WRAP [12] in order to increase the amount of recycled content incorporated into buildings. It includes some 3000 building components recycle content that estimates for new build and existing buildings. In the IPCC fourth assessment report Metz [13] suggested key mitigation technologies and practices by seven sectors.

(1) **Energy supply** - improved supply and distribution efficiency; fuel switching from coal to gas; nuclear power; renewable heat and power.
(2) **Transport-more fuel** - efficient vehicles; hybrid vehicles; cleaner diesel vehicles; biofuels; modal shifts from road transport to rail and public transport systems; non-motorized transport; land-use and transport panning.
(3) **Buildings** - efficient lighting and day lighting; more efficient electrical appliances and heating and cooling devices; improved cook stoves, improved insulation; passive and active solar design for heating and cooling; alternative refrigeration fluids, recovery and recycle of flu orientated gases.
(4) **Industry** - more efficient end-use electrical equipment; heat and power recovery; material recycling and substitution; control of non-CO2 gas emissions and a wide array of process specific technologies.
(5) **Agriculture** - improved crop and grazing land management to increase soil carbon storage; restoration of cultivated peaty soils and degraded lands; improved rice cultivation techniques and livestock and manure management to reduce CH4emissions; improved nitrogen fertilizer application techniques to reduce $N_2O$ emissions; dedicated energy crops to replace fossil fuel use; improved energy efficiency.

(6) **Forestry** - forestation; reforestation; forest management; reduced deforestation; harvested wood product management; use of forestry products for bioenergy to replace fossil fuel use.
(7) **Waste** - landfill methane recovery; waste incineration with energy recovery composting of organic waste-controlled wastewater treatment; recycling and waste minimization.

The advantages and disadvantages of those sustainable design tools are listed in Table 2. Other resources for generating sustainable design solutions include BRC [14], Constructing Excellence [15], SEEDA [16], The Carbon Trust [17], UK Government Office of Government Commerce [18]. There are an enormous number of literature and sustainable design tools providing numerous sustainable design options for designers to choose; however not all of them can be implemented in practice. In the current industrial environment, the application of sustainable design solutions in a project is limited by the affordability and risks the investors willing to take in practice. The opinions of the practitioners are crucial to the final decision. Therefore, the key factor that influence the practice of the sustainable infrastructure design and maintenance should be clarified.

**Table 2.** The advantages and disadvantages of some main sustainable assessment tools [7].

| Tool name | Advantages | Disadvantages |
|---|---|---|
| BREEAM | Covers a wide range of design aspects | Serves the UK public non-commercial buildings. Limited sustainable design options |
| Code for sustainable home | Higher standard requirements on energy efficiency. Request zero carbon for new houses on level 6 | Limits to residential projects only. Limited sustainable design options |
| LEEDs | Covers various types of buildings | Exclude waste management. Limited sustainable design options |
| WRAP recycle content tool | Building breaking down into components in order to select the high recycle content building materials for projects. Suitable for all types of buildings | Focuses on only the recycle content aspect of sustainable design. Limited sustainable design options |
| IPCC | Includes new considerations such as industry, agriculture and forestry, etc. | Just a general guidance for sustainable design. Limited sustainable design options |

# 4 Life Cycle Assessment

The existing sustainable design tools are designed for the general guidance of building design. These sustainable design tools for buildings mostly focus on the environmental impact of the buildings and the capital cost of the design options. However, the life cycle costs, and the practitioners' opinion have been neglected, even they are just as equally important as the environmental impact. Buildings are long-lasting products; the design of buildings should practically reduce the life cycle costs and improve the sustainability of the buildings.

In this aspect, life cycle assessment is the best tool to combine both the long term environmental and the economical evaluations of building designs. Life cycle assessment also includes economic and risk evaluations in analysis. The key economic analysis is life cycle costing, which is a technique to estimate the overall costs of design options during the economic life of the building. In order to compare the long-term economic performance of the design options, life cycle costing should be included in strategic designs because the economic concerns drive the decision makers may be sometimes more than the other concerns. There are many cost models that have been developed to estimate life cycle costs such as NHS hospital buildings [19] running cost for building element [20], noise barrier wall selection [21, 22], and water supply system [5, 23].

## 4.1 Procedure for SIDM Using Life Cycle Approach

Generally, life cycle design is a multi-stakeholder's collaborative approach. It may involve the qualitative evaluation from practitioners and the quantitative data from technical engineers and quantity surveyors. Various parties involved in the assessment process allow the tool taking into consideration not only the long-term environmental and economic impacts of the building design option but also the feasibility evaluation of the practitioners. Their data are finally integrated into a single measurement to rank the design options. The integrated tool includes the following steps:

(1) Generating sustainable building design options through literature review.
(2) Feasibility assessment by practitioners through educational workshops.
(3) Technical analysis on life cycle costs and risk levels of design options by technical experts.
(4) Multi-criteria decision-making process to combine multiple evaluations by direct weighting method.

Dunmade et al. [5] reported an approach to lifecycle design and sustainable management of Nigerian water infrastructure. Figure 3 is an illustration of various sustainable design concepts that need to be harmonized for a sustainable water supply infrastructure design. These design concepts include Design for Assembly and Disassembly, Design for simplicity and reliability, Design for manufacturing and remanufacturing, Design for Cost, and Design for Reuse and Upgrading. Furthermore, sustainable material selection for the system's hardware manufacturing, resource consumption minimization during system operational life as well as lifecycle extendibility are other things that would need to be put into consideration in the process of designing the water infrastructure for sustainability. This would ensure that sustainability is incorporated at the design, manufacturing, utilization and end-of-life management stages of the water supply facility's lifecycle. In designing a water supply system for sustainability, the energy requirement of the system and the source(s) of the energy need to be assessed and factored into the design. This is necessary in order to ensure availability and reliability of the requisite energy for the water system operation.

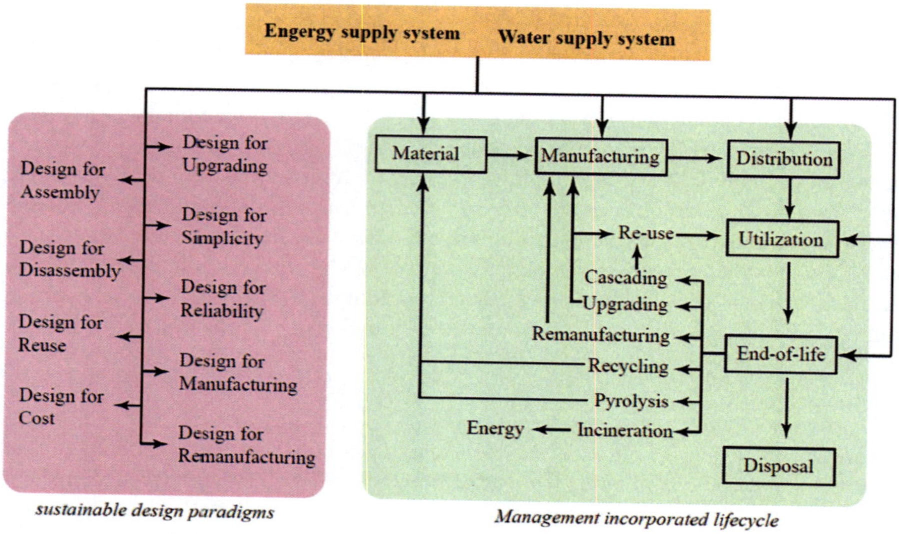


**Fig. 3.** An illustration of where sustainability can be incorporated [5].

Figure 4 is an illustration of the lifecycle design and sustainable management framework developed as a result of the study. It is a multi-stakeholder's collaborative approach. An application of the lifecycle design and sustainable management-based water supply infrastructure development model would start with the assemblage of coalition partners. Their first assignment would be to identify credible water needs of the community and to articulate their groups' interests in the project. The next step involves making arrangement for how the needs will be addressed. This would be achieved through an evaluation of existing policies suitability and possible reformulation of such policy to facilitate incorporation of innovative approach if/when necessary. This is expected to be followed by an assessment of water sources and their conditions in the locality, utilizable options, required energy and affordability. The next step would be generation of conceptual designs and applying various lifecycle design concepts (DFXs) on the conceptual designs. DFX refers to design for X where X could be material (DFMt), modularity (DFMd), assembly (DFMA), manufacturability (DFMf), disassemblability (DFD), maintainability (DFS), use and reuse (DFUR), upgradability (DFUG), remanufacturability (DFRm), recyclability (DFR), energy efficiency (DFEE), minimum residue (DFMR) and so on. Each conceptual design is then evaluated in turn for environmental friendliness by using E-LCA.

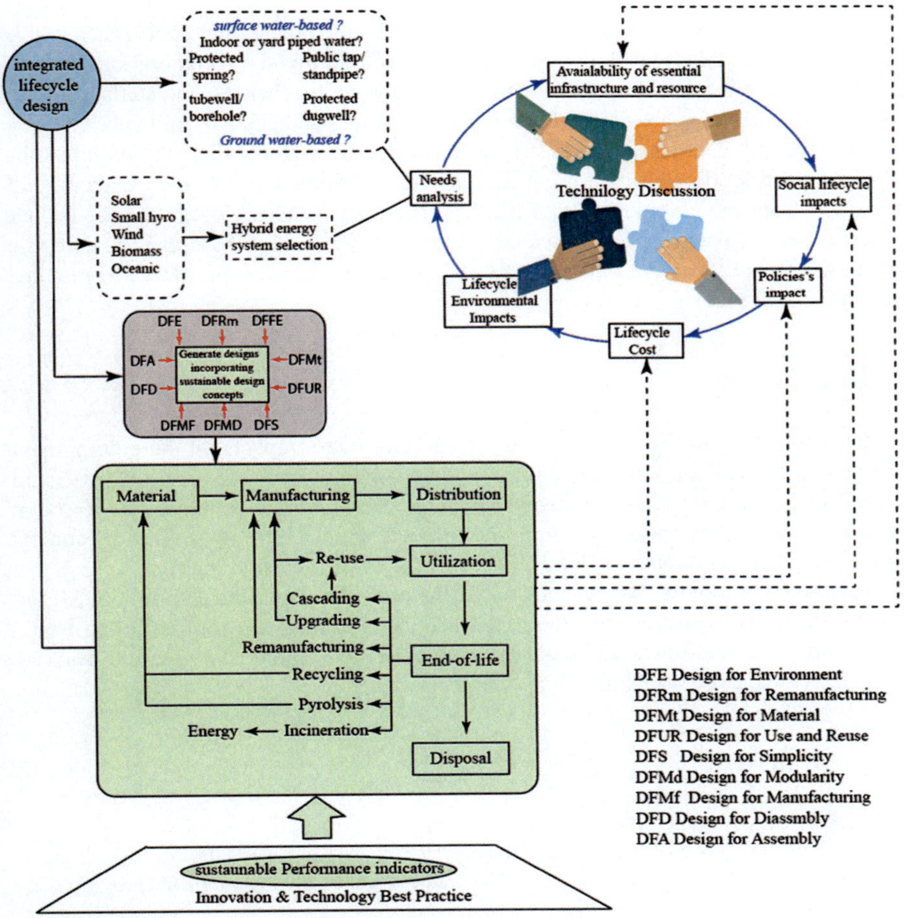


**Fig. 4.** An illustration of Lifecycle design and sustainable management based water supply system infrastructure [5].

## 4.2 Possible Challenges for Life Cycle Approach to SIDM

Without doubt, a comprehensive, life cycle approach to sustainability focused design will be difficult to implement. Significant challenges quickly come to mind; Some of the challenges include the current lack of sustainable design knowledge, and the degree of willingness to learn the new approach. Other challenges include the huge cost whenever new project would have to be undertaken in their region and how to wean them from personal aggrandizement and take ownership of the new system. Overcoming these challenges would require tact, consistency and patience in the process of educating the “unruly stakeholders” and managing incidences along the collaborative sustainable design and development process.

It should also bear in mind that the transition from conventional approaches to life cycle design may take very long time. And this shift required the collaboration of many academic disciplines, such as architecture, engineering, mechanics, and statistics, with contributions from economics (economic loss modeling and costing) and public policy (building codes). The transition to life cycle design for sustainability will require substantially more collaboration among academics, practitioners, and policymakers, and will draw from the diverse fields of biology, chemistry, and sociology for the proper establishment of ecosystem carry capacities and social norms. On a positive note, the rate at which collaborative thinking and research have moved from theory to practice has accelerated greatly in Europe.

## 5 Conclusions

Infrastructure are long-lasting products which have huge impacts on the environment during their whole lives. The design and subsequent maintenance of infrastructure should take into consideration long-term environmental and economic benefits. The selection of sustainable design options for building projects depends not only on the technical issues but also on economic and political concerns of the decision makers.

This report presents the concept of SIDM, and the assessment tools for SIDM. Among the existing assessment tools, the life cycle assessment tool is the best one, which combined feasibility study, life cycle costing, environmental evaluation, and risk analysis in the design decision.

## References

1. Bank, A.I.I.: Study on Sustainable Cities (2018)
2. CEIN. Buildings consume nearly half the energy each year in China (2007)
3. Sarte, S.B.: Sustainable Infrastructure: The Guide to Green Engineering and Design, Wiley, Hoboken (2010)
4. Martos, A., et al.: Towards successful environmental performance of sustainable cities: intervening sectors. A Rev. Renewable Sustain. Energy Rev. **57**, 479–495 (2016)
5. Dunmade, I., Onawumi, A., Ajayi, O.: An approach to lifecycle design and sustainable management of Nigerian water infrastructure. Sustain. Energy Build. Res. Adv. **5**(1), 11–20 (2016)
6. Moncarz, P.: The Future Design of Sustainable Infrastructure (2018)
7. Wang, N., Chang, Y.-C., Nunn, C.: Lifecycle assessment for sustainable design options of a commercial building in Shanghai. Build. Environ. **45**(6), 1415–1421 (2010)
8. Bank, I.-A.D.: What is Sustainable Infrastructure - A Framework_to Guide Sustainability Across the Project Cycle (2018)
9. Nuramo, D.A., Haupt, T.C.: Key Elements for Sustainable Infrastructure Design in Developing Countries 1Denamo A. Nuramo (2017)
10. BRE: Code for sustainable home, building research establishment, London (2007)
11. Council, U.G.B.: Leadership in Energy and Environmental Design (LEED) Green Building Rating System. US Green Building Council, US (2007)
12. WRAP, WRAP recycle content toolkit, Waste and Resources Action Programme. UK (2007)
13. Metz, B.: Mitigation of climate change. The IPCC fourth assessment report, G. WMO (2007)

14. BRC: BRC global standard – consumer products. British Retail Consortium (2007)
15. Excellence, C.: Sustainable construction fact sheet. Construction Excellence (2007)
16. SEEDA, Sustainability checklist. The South East England Development Agency (2007)
17. Trust, T.C.: Energy saving fact sheet: retail. The Carbon Trust (2007)
18. Commerce, U.G.O.o.G.: Achieving excellence in construction procurement guide – Sustainability (2007)
19. Kirkham, R.J., Boussabaine, A.H., Awwad, B.H.: Probability distributions of facilities management costs for whole life cycle costing in acute care NHS hospital buildings. Construction Manage. Econ. **20**(3), 251–261 (2002)
20. Wang, N., Horner, R.: Fuzzy rule-based system approach to assess cleaning costs of carpet flooring. J. Finan. Manage. Property Construction **12**(1), 25–36 (2007)
21. Morgan, S.M., Kay, D.H., Bodapati, S.N.: Study of noise barrier life-cycle costing. J. Transport. Eng. **127**(3), 230–236 (2001)
22. Oltean-Dumbrava, C., Watts, G., Miah, A.: Transport infrastructure: making more sustainable decisions for noise reduction. J. Cleaner Prod. **42**, 58–68 (2013)
23. Barbara Minsker, M.A.: Progress and recommendations for advancing performance-based sustainable and resilient infrastructure design. J. Water Res. Plann. Manage. **141**, 12 (2015)

# Experiment to Evaluate the Impulse Plate Compaction in a Test Embankment Section

Chavdar Kolev(✉)

Higher School of Transport "Todor Kableshkov", 158, Geo Milev Street, Sofia, Bulgaria
ch_kolev@abv.bg

**Abstract.** The compacting of the impulse plate prototype constructed was tested twice in test embankment sections. The plate operates at 12.5 Hz, with an eccentricity of 7.5 mm, and has a mass of 280 kg. The results of the three measurement cycles for the two test sites are presented in a comparative table. The results obtained indicate that the impulse plate has a much better compacting effect than commercially available vibrating plates and can replace a 6-ton vibration roller for embankment compaction. The considerable mass of the plate, when counterweighted, makes it difficult to push manually, but it can be coupled with one or two more in place of the bucket on the front loader and thus work instead of a roller.

**Keywords:** Compaction · Oedometer modulus · Dynamic modulus · Frequency · Granulometry

## 1 Introduction

Embankment compacting is an activity crucial for the high quality and durability of construction. On the other hand, compacting itself requires a lot of energy and time. Therefore, the aspiration of machine constructors is to create even more efficient mechanisms to achieve the goal at lower cost.

Mechanical compaction consists of two main types proven in practice: (1) impulse and heavy ram seals; (2) vibration damping. In the first type, the compaction is at the expense of mechanical work applied to each point of the embankment surface - multiplication of the falling force and the height of the gravity. In the second type, these two factors are not dominant at all. It relies on the resonance between the forced frequency of the vibrating mechanism and the natural frequency of the compacting grains, as well as on the humidity and friction of the compacting body. Pulse compaction machines operate with greater acceleration and less mass than heavy ram seals.

The compacting mechanisms are applied to large embankments (for railways, roads, bridges, under industrial buildings, etc.). Another group of them, which are lighter, serve to compact sidewalks, trenches, small playgrounds and more. The first group of machines is usually self-propelled or attached to tractor engines, while the second one is towed or dragged by hand.

The purpose of this paper is to evaluate the properties of a new pulse compaction machine compared to vibratory plates and rollers.


A. Rotaru (Ed.): CRIT-RE-BUILT 2019, SSGG, pp. 116–122, 2021.
https://doi.org/10.1007/978-3-030-61118-7_9

## 2 Description of Impulse Plate for Testing

Completely new impulse plate with a power pulse generator is constructed to combine the advantages of vibrating and heavy compaction machines. The plate is intended for embankment compaction of roads, railway lines, industrial halls, etc. The plate is moved by the help of a tractor. It is mounted or attached to the tow bar, independently or in a battery with two more other plates. The overall dimensions of the machine are 400/400/1500 mm, its own weight is 450 kN and its counter-weight is 1500 kN. The parameters of the electric motor are the following:

- Power – 3 kW;
- Engine frequency – 25 Hz;
- Crankshaft speed of the power pulse generator – 12,5 Hz;
- Crankshaft eccentricity – 7,5 mm;

Machine impulses have the following dynamic characteristics:

- Kinetic energy – 1.23 kJ;
- Inertia radius – 0.01575 kg.m$^2$;
- Impulse moment – 0.197 kg.m$^2$/s.

During the experiment, the machine moved by hand on the test path at a linear speed of 93.75 mm/s and with a linear acceleration – 1.17 m/s$^2$.

## 3 Standard Mechanical Work for Embankment Compaction

The standard work requires to compact the embankment through the well-known Proctor's method in the Soil Mechanics, which can be determined by the formula:

$$W = n.F.h/A.a \tag{1}$$

whereas:

n – number of strokes on one spot of the embankment;
F – stroke force of the vibrating plate eccentric;
h – flywheel eccentricity;
A – the contact area of the vibrating plate;
a – thickness of the compacted soil layer.

If in the formula (1) we replace the values of the parameters with those of the new machine, we will get that the ram has done the required work for about 2 700 kJ/m$^3$ and a time of about 30 s, while the lower value of about 600 kJ/m$^3$ has been achieved for a time of about 6 s.

## 4 Description of the First Test with Impulse Plate

The compacting work of the impulse plate prototype was tested twice in test embankment sections. The measurement of the achieved effect was according to Bulgarian State Standard (BSS) 8004: 1984 - “Soils construction. Determination of total deformation”.

The experiments were carried out on a construction site near Sofia, at the banks of the Iskar River. The bulk material is out of a standard crushed stone, fraction 0–63 mm, delivered from the quarry. The first attempt is to compact a 40 cm thick embankment in a *1* m/2 m trench, as shown in Fig. 1:

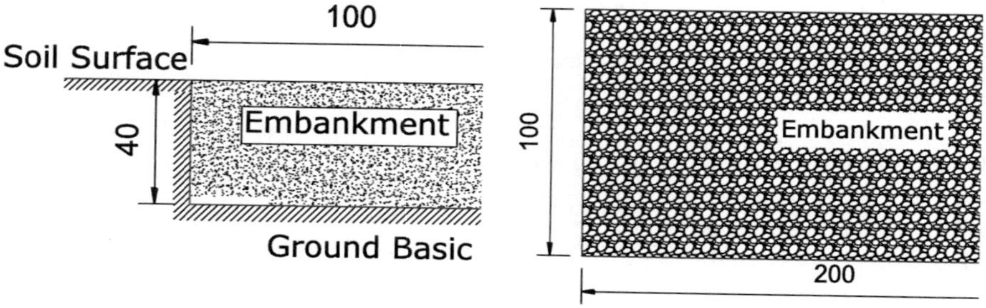


**Fig. 1.** Profile and embankment plan in the first tested section

The ground consist of an old mound of sandy gravelly clay, medium-plastic, very moist, but not homogeneous. The groundwater level in the area is high. The modulus of total deformation of the earth’s base is $E_0 = 20$ MPa. The embankment includes 2 layers of *20* cm.

The impulse plate worked to compact each of the layers for 2 min. The counter-weight on the plate was 1.5 kN and achieved a stroke force of 70 kN for each pulse. The static load test of the first test section gave very good results. The achieved *80* MPa modulus of elasticity was sufficient for a basic layer of first-class road or railway line, and the deformation modules of the first and second load cycles have had a very good ratio with value *2.1* at norm *2.2.*

### 4.1 First Test Results Analysis

It followed that at an impulse plate length of 40 cm and a layer thickness of 20 cm, the impulse plate should move at a speed of 3 cm/s (0.1 km/h). Thus, these compaction results was able to achieve. The lower density was reached with a moving speed of 13.65 cm/s (0.5 km/h).

As a whole, the average speed of such vibratory plates in the market is 40 cm/min. If the plate moves at such a speed, its mechanical performance shows to be much lower than the standard values (2 700 kJ/m$^3$ and 600 kJ/m$^3$). And when we compare the mechanical operation of the impulse plate with that of a roller weighing 6 t, we obtain the following results:

The roller has compacted a layer of crushed stone, 25 cm thick, with 9 passes through a point. The recommended average moving speed was 4 km/h [1]. Therefore, for the 9 passes, it was able to compact a 500 m stretch for 1 h. Relative to the impulse plate performance, the roller was 5 times more efficient to achieve a higher standard density (2 700 kJ/m$^3$). The ratio between the mass of the roller and that of the vibrating plate was about 21 times.

- The average static pressure exerted by the impulse plate on the ground without the motor being switched on was 17.5 kPa. The average roller tension was from 30 to 40 kPa, i.e. about 2 times more than the one of the plate.
- The average dynamic pressure exerted by the vibrating plate on the ground when the engine is running was 468 kPa.

All calculations have been made with the assumption that the impulse force of the impulse plate was 75 kN and the impact of such plates was from 10 kN to 38 kN.

In general, impulse plates are used to compact lightly loaded embankments, such as sidewalks, to backfill narrow trenches or pits, and more. The modulus of elasticity to achieve after such a compaction does not exceed 40 ÷ 50 MPa. For comparison, the normal stress created by the weight of 1 person stepping on a 40 × 40 cm sidewalk is only about 6 kPa.

The results thus obtained show that the performance of the impulse plate under study was completely satisfactory and does not concede to the results obtained from the vibrating plates being marketed.

## 5 Description of the Second Test with and Impulse Plate

The second test section was about 5 m away from the first one and has dimensions of *2.30* m/*1.00* m. The section was pre-excavated and aligned to a depth of 50 cm, as shown in Fig. 2. Half of the exposed and leveled ground was covered with a layer of *20* cm crushed stone with a fraction of *0–63* mm and has been compacted with an impulse plate, but this time with double the counter weight of 3.1 kN and an impulse force of 180 kN.

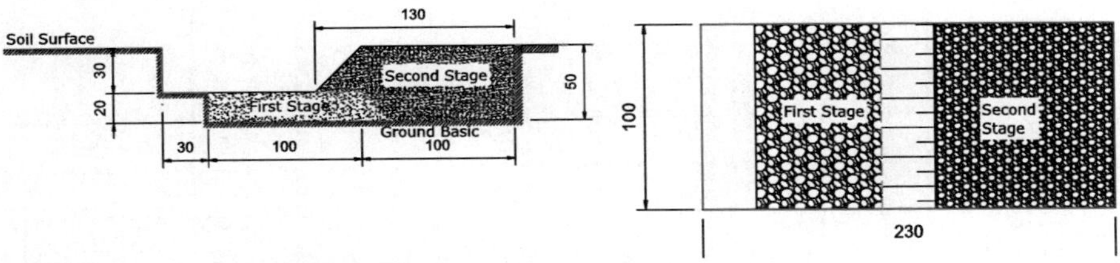


**Fig. 2.** Profile and plan of the second test section

A static load plate measurement has been done, showing a modulus of elasticity of *57* MPa and a ratio of deformation modules of the second and first load cycle $E_2/E_1 = 51/36 = \mathbf{1.4} \ll \mathbf{2.2}!$

A 50 cm layer of crushed stone was then laid on the second half of the trench and compacted at the same time with the same impulse plate with a counterweight of 31kN. Thus, the modulus of elasticity $E_e = 154$ MPa and the ratio of the deformation modules $E_2/E_1 = 132/63 = \mathbf{2.1 < 2.2!}$

Parallel to these measurements in the second test area, measurements with dynamic load plate have been performed, which still has followed the standards and has worked in comparison with the of static load plate.

The dynamic plate actually measured the modulus of elasticity according [2], but its values was calibrated with those of the static plate for each particular section.

The results of experiments from both sites have been presented in the comparative Table 1 below.

**Table 1.** Comparison of experiment results

| | | 1st test with static load plate - 40cm | | 2nd test with static load plate - 50 cm | | | |
|---|---|---|---|---|---|---|---|
| | | | | First phase 20 cm | | Second Phase 30 cm | |
| **Indicator** | **Symbol** | **Value** | **Measurement Unit** | **Value** | **Measurement Unit** | **Value** | **Measurement Unit** |
| Deformation modulus from the first loaded branch | $E_1$ | 26.8 | *MPa* | 36 | *Mpa* | 63 | *Mpa* |
| Deformation modulus from the second load branch | $E_2$ | 56.3 | *MPa* | 51 | *Mpa* | 132 | *Mpa* |
| Deformation modulus from the third load branch | $E_3$ | 57 | *MPa* | 38 | *Mpa* | 130 | *Mpa* |
| Ratio between the modules | $E_2/E_1$ | **2.10** | | **1.42** | | **2.10** | |
| Modulus of elasticity | $E_e$ | **80** | *MPa* | **57** | *Mpa* | **154** | *Mpa* |
| Ratio $E_e/E_2$ | $E_e/E_2$ | 1.42 | | 1.12 | | 1.17 | |
| Modulus of elasticity with compaction plate | $E_{v},d_1$ | | *MPa* | 29.68 | *Mpa* | 38.40 | *Mpa* |
| Ratio $E_v/E_e$ | | | | 0.52 | 1.92 | 0.25 | 4.01 |
| Ratio $E_v/E_1$ | | | | 0.82 | 1.21 | 0.61 | 1.64 |
| Ratio $E_v/E_2$ | | | | 0.58 | 1.72 | 0.29 | 3.44 |
| Maximum Load | $P_{max}$ | 0.60 | *MPa* | 0.20 | *Mpa* | 0.50 | *Mpa* |

**Second Test Results Analysis**

The most important conclusion from the above table is that the impulse plate has reached the required normative ratio among the deformation modules whose value is below 2.2 within both experiments. Also, the values of the modulus of elasticity were sufficient under the prerequisites of the studies. For layers tick 40 cm and 50 cm respectively, the values of 80 MPa and 154 MPa have been obtained at the first and second sections, respectively. The first value was required by the norms for first class road or rail, and the second exceeded the norm of 120 MPa for highway or railway. The achieved value of 57 MPa has exceeded the requirements for second category of railway lines and roads when having 20 cm thick layer.

It is noteworthy that the deformation modules for the second test section were higher than those for the first test one by about 2.3 times. This was explained both by the greater thickness of the embankment (50 cm by 40 cm) and by twice the mass of the counterweight in the second section. The impulse stroke was even 2.5 times greater. Similar was the ratio of the modulus of elasticity Ee for the two sections - 1.93. The ratio of the modulus of elasticity obtained from the static and the dynamic plate was 1.92, which stayed within the expected values - about 2. The static plate settlement with 0.5 MPa for the first section was 4.38 mm, and for the second phase of the second section it was only 2.72 mm, i.e. 62% compared to the former ones (Table 2). This fact was again explained by twice the counterweight (3.1 kN > 1.5 kN) at the second section, where the higher density and stiffness of the embankment come from.

**Table 2.** Settlement under the load of 0.2 MPa

| | | 1st test with static load plate - 40cm | | 2nd test with static load plate - 50 cm | | | |
|---|---|---|---|---|---|---|---|
| | | | | Lower layer 20 cm | | Upper layer 30 cm | |
| Loading of the plate | 0.2 MPa | 1.87 | mm | 1.03 | mm | 1.33 | mm |
| | 0.5 MPa | 4.38 | mm | | mm | 2.72 | mm |
| | 0.6 MPa | 5.22 | mm | | mm | | mm |

The technical capability of this impulse plate was to withstand a maximum counter-weight with 700 kg of mass and an impulse force of 400 kN. This was limited by the impulse generator power with an area of 150 $cm^2$ segments.

If such a large counter-weight is realized, the density does not increase so much. This dependency has long been proven by Proctor and Modified Proctor methods. In any case, it is certain that such a plate can reach the maximum soil compaction capacity.

From the presented results it is evident that the pulse plate has high efficiency and can be applied in practice. The considerable mass of the plate, when the ram weight is attached, makes it difficult to push manually, but it can be hung in a battery with one or two more plates. The plates are hooked in place of the bucket on the front loader and thus work instead of a roller.

## 6 Conclusion

The results show that the impulse plate has much better compacting effect than commercially available vibrating plates. It can also replace a vibrating roller with a mass of 6t for compacting embankments.

The performance of the impulse plate is better than the one of the corresponding vibrating rollers and thus better technical and economic efficiency in construction of embankments can be achieved.

## References

1. Pistol, J., et al.: Ambient vibration of oscillating and vibrating rollers. In: Vienna Congress on Recent Advances in Earthquake Engineering and Structural Dynamics, Paper No. 167, Vienna, Austria (2013)
2. CEN CWA 15846: Measuring method for Dynamic Compactness & Bearing Capacity with SP-LFWD (2008)

# Sustainable and Durable Road Pavements Through Fundamental and Thorough Laboratory Evaluations

Wynand JvdM Steyn(✉)

University of Pretoria, Pretoria, South Africa
Wynand.steyn@up.ac.za

**Abstract.** A sustainable and durable road pavement network is essential to support the economy of a country. Traffic and the environment places a demand on the combination of materials in a pavement structure, and pavement design and management requires balancing this demand and supply problem. As road construction materials (bitumen, asphalt, aggregate, etc.) are variable natural materials, it is essential to ensure that the materials used in road pavement construction are well quantified in terms of their properties, homogeneity, variability etc. Various types of laboratories can be used for assaying the material properties, including field, commercial and research laboratories. Each of these laboratories plays a specific role in ensuring that the materials that are used on the site has specific properties, compared to a design material. In this paper the role of a fourth, the reference laboratory, is discussed indicating its specific role and functions, and the opportunities to develop databases of materials properties from which inferences can be developed around the materials using various data science techniques. A proper reference laboratory can support the ongoing improvement in fundamental materials understanding through it effects on technician capabilities, comparative data studies and development of improved understanding of material properties as evaluated under different environments.

Most relevant topic: Sustainable Infrastructure Design and Maintenance/The Durability of Sustainable Materials and Structures.

**Keywords:** Reference laboratory · Proficiency testing · Road materials

## 1 Introduction

The South African road network consists of almost 750 000 km of road, of which at least 20% is paved (Table 1). This is currently the 11$^{th}$ longest road network on the world. Continuous maintenance and rehabilitation of this network is essential to support the economy and social network in the country, as a sustainable and durable road pavement network is essential to support the economy of any country. A pavement typically deteriorates through the demand placed on it by a combination of traffic and the environment. Pavement design and management requires balancing this demand and supply (pavement bearing capacity) problem.


A. Rotaru (Ed.): CRIT-RE-BUILT 2019, SSGG, pp. 123–129, 2021.
https://doi.org/10.1007/978-3-030-61118-7_10

**Table 1.** South African road network details.

| Paved | Gravel | Total |
|---|---|---|
| 153 719 km | 593 259 km | 746 978 km |
| 20.6% | 79.4% | |

Road construction materials (bitumen, asphalt, aggregate, etc.) are variable natural materials, and it is essential to ensure that these materials are well quantified in terms of their properties, homogeneity, variability etc. Various types of laboratories can be used for assaying the material properties, including field, commercial and research laboratories. Each of these plays a specific role in ensuring that the materials used in the pavement structure has specific properties, compared to the design material.

This paper focuses on the role of a fourth type of laboratory - reference laboratory - indicating its specific role and functions. It discusses the principles of such reference laboratories and proficiency testing schemes, as well as the opportunities to develop databases of materials properties from which inferences can be developed around the materials using various data science techniques.

A proper reference laboratory should support ongoing improvements in fundamental materials understanding through it effects on technician capabilities, comparative data studies and development of improved understanding of material properties as evaluated under different environments.

### 1.1 South African Pavement Engineering Background

The South African pavement engineering history stretches back since the original tracks and routes made to travel into and within the southern part of the continent by local inhabitants. These tracks were upgraded through continuous use, and many became the alignments of modern roads due to the locations that have been optimized through trial and error. Modern pavement engineering started in the 1950s with investments into the improved understanding of local road building materials and incorporation of various international pavement engineering and evaluation developments into local practice [1].

Through the development of institutions like the Council for Scientific and Industrial Research (CSIR) [2] and the related development of civil engineering departments at the local universities (mainly the University of Pretoria (UP) and University of Stellenbosch (US)), experience was built up into the appropriate and applicable use of local materials in the local environment and under local traffic demands. This is visible in the large number of publications, design guidelines and manuals, software packages and the South African Road Design Method and the South African Pavement Engineering Manual [3, 4]. Many of these guidelines are used in sub-Saharan African countries, while technologies such as the Heavy Vehicle Simulator and Mobile Load Simulator (both Accelerated Pavement Testing (APT) devices) have been developed and exported from South Africa since the 1970s [5].

Over the last few decades, constraints in the local economy caused a decrease in the number of civil engineers in the country, compared with other countries. The South

African Institution of Civil Engineers (SAICE) indicates in their Infrastructure Report Card that, compared to the 130 to 450 people per engineer in Europe, North America, India and China, South Africa's ratio is 3 200 to 1 [6]. This also applies to technologists, technicians and artisans resulting in a severe lack of qualified personnel throughout the transportation sector. Apart from the staff challenge, South Africa also currently do not have access to a national reference materials testing laboratory that can conduct duplicate testing of project laboratories and operate a national proficiency scheme to perform quality control and quality assurance in transportation infrastructure construction. Such a national reference laboratory should ideally also participate in international proficiency schemes ran by relevant authorities in the US [7] and Europe (Forum of European National Highway Research Laboratories (FEHRL)) to ensure international accreditation of material testing performed in the roads industry in South Africa.

These challenges has partly lead to the development of an Integrated Education, Reference and Research Laboratory that can provide an internationally renowned platform for academic and vocational training support in transportation infrastructure materials testing; high quality research facilities; and a transportation materials reference testing platform [8].

## 2 Basic Pavement Materials Engineering

Any infrastructure, including road pavement networks, requires regular maintenance and rehabilitation as traffic uses it and the environment affects the condition of the materials [9]. These constant changes in material properties, together with the inherent variability in the source material properties, requires a system where a database of typical material properties, as well as levels of variability (e.g. Coefficient of Variation of property) are kept for a country.

In order to ensure that the data in the database are valid, and that day-to-day laboratory test data are valid, a proficiency scheme is required. This is a scheme where comparative testing is conducted on standard materials in-between various laboratories and technicians, to ensure that the bias due to different equipment and technicians are minimized. It should be appreciated that all pavement design, maintenance and rehabilitation decisions are ultimately based on the results of laboratory tests of materials. If this test data are flawed or biased, it will affect the quality and performance of the road pavement directly.

Pavement material properties can be divided into three levels, being performance properties, engineering properties and fundamental properties. Performance properties provide an understanding of the way the specific material will perform under very specific conditions (e.g. surfacing permanent deformation). Engineering properties of a specific material provide a more detailed understanding of the material and its potential responses to loading conditions or changes (e.g. stiffness modulus). Fundamental properties of a material are those properties that are not determined by external conditions but always remain the same and directly influence engineering and performance properties of the material (e.g. material grading) [10].

# 3 Road Material Laboratories

## 3.1 Laboratory Types

As indicated, there are essentially four types of road material laboratories. These are field or site, commercial, research, and reference laboratories. Each of these has a specific role to play in the provision and maintenance of road pavement infrastructure. Field or site laboratories are used to evaluate the properties of materials as they are constructed into the pavement structure, and thus provides a running record of actual properties. This becomes as-built data.

Commercial laboratories are typically sued to determine properties of materials before design commences, with the purpose of determining the general properties of the available materials in the area. Research laboratories are used to conduct research on existing or novel materials, typically under non-standard conditions, to determine the expected behavior and performance of such materials when used under various conditions.

Reference laboratories are finally used to conduct duplicate testing of materials that were submitted to the field or site laboratory. The objective of this is to act as a third-party evaluation of the outcomes of the field or site laboratory, and thus to ensure that the as-built data are not biased due to issues with the site or field laboratory, it staff or equipment.

## 3.2 SANRAL Reference Laboratory Role and Function

As an example of such a reference laboratory approach, the SANRAL reference laboratory at the University of Pretoria is used. This reference laboratory is a fully accredited laboratory (equipment, procedures and staff) where reference testing are conducted for various roads authorities. All tests are conducted as per standard test protocols, and no deviations from these protocols (as for typical research laboratories) are allowed.

The reference testing procedure essentially consists of a dual set of materials being evaluated - one at the field or site laboratory and one in the reference laboratory (Fig. 1). Comparison of the results should indicate similarity within a certain level of variation, depending on the material and test method. As the reference laboratory is fully accredited, its results are typically seen as the standard that the field or site laboratory data must resemble.

Apart from the reference testing role, the laboratory also participate in international proficiency schemes ran by relevant authorities in the US (American Association of State Highway and Transportation Officials (AASHTO)) and Europe (Forum of European National Highway Research Laboratories (FEHRL)). This participation provides international accreditation for the roads industry in South Africa, which is important to ensure that local practices are on par with international practices.

The proficiency scheme also allows for accreditation of competent laboratories and staff in the country and the region, to ensure formal recognition of such laboratories, and staff are competent to carry out specific tasks in terms of the Accreditation for Conformity Assessment, Calibration and Good Laboratory Practice Act (Act 19 of 2006) (Fig. 2).

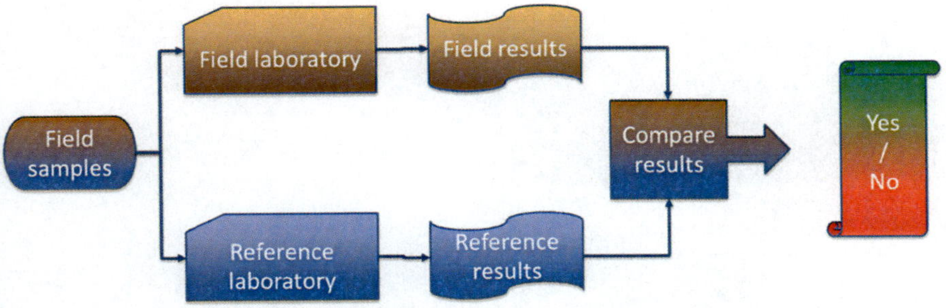


**Fig. 1.** Schematic depiction of duplicate testing of road material samples process.

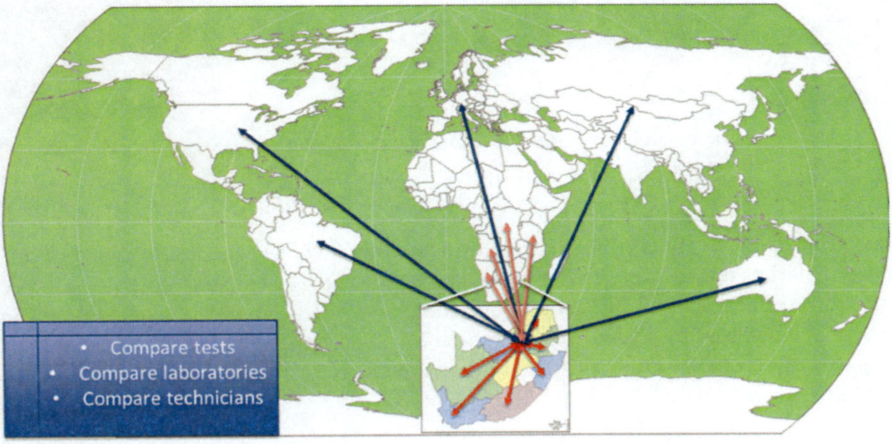


**Fig. 2.** Proficiency scheme comparison between local, regional and international laboratories.

This facility also incorporates an Accelerated Pavement Testing (APT) track and a real traffic test lane (Fig. 3). This allows for a hierarchy of testing to be conducted, starting with individual materials properties, followed by the properties and behavior of the full pavement structure (combination of materials in APT test) and finally the performance of the pavement structure under real traffic and environmental conditions.

### 3.3 Materials Database

The centralized process of collecting road pavement materials laboratory test data provides for the option where all such data are collected and housed in one database. This database can be used as the national reference center for material properties in the country. Although it is standard practice to evaluate each material for a specific roads project as a unique material, and determine its specific properties, the availability of a database with indications of typical values as well as variability of the various materials properties provides engineers with a valuable resource when conducting initial designs. Such a database is also of primary importance when the actual variability in different material

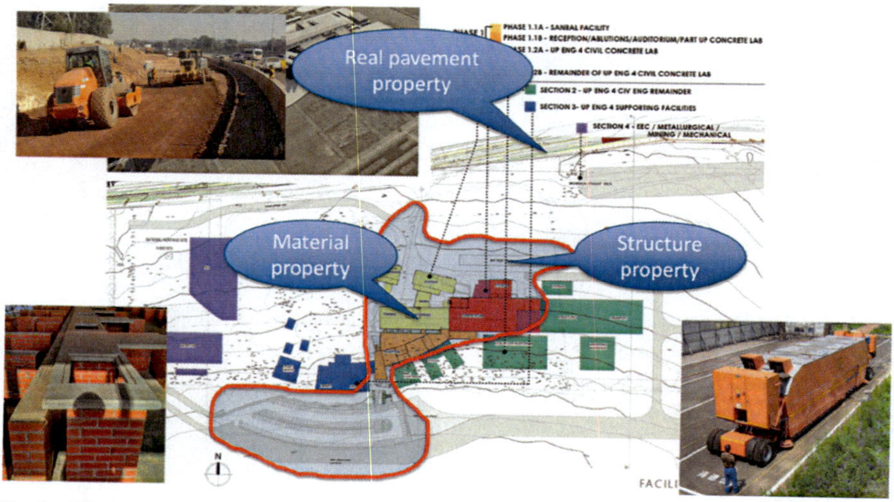


**Fig. 3.** Hierarchy of materials, pavement structure and real pavement (incorporating environment and traffic) evaluation of pavements.

properties for different test methods are evaluated. It is known that the accuracy of different materials tests differs. This is caused by factors such as inherent material property variations, effects caused by testing equipment, test methods and test operators. The materials database provides a sounding board against which the results of laboratory tests can be compared to assist in the determination of the relative accuracy of the test results.

### 3.4 International Proficiency Testing

The AASHTO Accreditation Program is an international program that formally recognizes competency of all participating testing laboratories to perform specific tests on construction materials. Participating in such a program provides the opportunity to demonstrate conformance to specific national and international standards, enhance confidence in test results and allows for trends in international testing procedures to be observed and identified [7].

### 3.5 Big Data/Data Science Evaluation of Database for Trends, Variability, etc.

The science of big data provides for the analysis of large datasets to identify trends and patterns that may not have been visible when only focusing on smaller sets of the same data. It also allows for developing relationships between various parameters that are part of the database.

In the case of the reference and proficiency testing discussed in this paper, the big data science options allows the databases of materials properties that are built up over the years to be interrogated to identify such relationships between the various materials properties that may affect its behavior in the field, but that may not be apparent from individual test results.

It further allows for the development of for example Artificial Neural Network (ANN) type analyses on test data to optimize the road pavement design procedure. One such an example is the application of an ANN to optimize asphalt surfacing design, allowing a much quicker and more robust route to finding the optimal bitumen content and aggregate grading for specific road surfacings [11]. Such ANN design options can be applied effectively in obtaining initial designs for various engineering materials, thereby limiting the testing required on the final materials.

## 4 Conclusions

In this paper the roles and possibilities of a reference laboratory, and the related issues such as databases of material test results are discussed. In a country such as South Africa, the vast road network requires focuses attempts to ensure that the maintenance and rehabilitation of this network is conducted in a sustainable manner, providing durable road networks.

The focused use of a national reference laboratory together with national and international proficiency schemes supports the provision of such a sustainable road network. Access to the type of materials test database described in the paper, and initiated in the reference laboratory, supports the ongoing development of transportation infrastructure, as well as the maintenance and rehabilitation of such infrastructure in the country.

Through appropriate research on the test results available in the materials testing database, generation of improved materials designs and test methods becomes a reality based on actual data.

## References

1. Weinert, H.H.: Natural Road Construction Materials of Southern Africa. CSIR, Pretoria (1980)
2. Basson, N.: Passage to Progress. The CSIR's Journey of Change, 1945–1995. Jonathan Ball Publishers, Johannesburg (1995). ISBN 1868420345
3. Technical Recommendations for Highways. https://www.nra.co.za/live/content.php?Item_ID=246. Accessed 22 Feb 2020
4. South African Pavement Engineering Manual (SAPEM). https://www.nra.co.za/live/content.php?Item_ID=232. Accessed 22 Feb 2020
5. Steyn, W.J.vdM.: Significant findings from full-scale accelerated pavement testing. A synthesis of highway practice. NCHRP Synthesis 433. Transportation Research Board, Washington D.C. (2012). ISBN 978-309-22366-9
6. SAICE 2017 Infrastructure Report Card for South Africa. SAICE, South Africa (2017)
7. AASHTO Proficiency Scheme. http://www.aashtoresource.org/aap. Accessed 22 Feb 2020
8. Engineering 4.0. http://www.up.ac.za/eng4. Accessed 27 Jan 2020
9. Hudson, W.R., Haas, R., Uddin, W.: Infrastructure Management: Integrating Design, Construction, Maintenance, Rehabilitation and Renovation. McGraw-Hill Professional Publishing, New York (1997)
10. Steyn, W.J.vdM.: Applications of observational techniques in pavement engineering. In: Proceedings of Southern Africa Transportation Conference (SATC), Pretoria, RSA, July 2007 (2007)
11. Sebaaly, H., Varma, S., Maina, J.W.: Optimizing asphalt mix design process using artificial neural network and genetic algorithm. Constr. Build. Mater. **168**, 660–670 (2018)

# Stabilization of Roads Located on Banks of Mountain Flowing Waters

Ancuța Rotaru(✉) and Dana-Mădălina Pohrib

Gheorghe Asachi Technical University, 700050 Iași, Romania
ancuta.rotaru@tuiasi.ro

**Abstract.** The location of a road embankment near a flowing mountain water contributes to its degradation. The ceding of the embankment or retaining wall occurred due to erosion and scouring phenomena. The scouring phenomena have led to the destruction of the existing support structures and the loss of the stability of the river bank, and the erosion phenomena have led to the increase of the land mass displaced below the road structure. The article presents the causes of the failure analyzing and proposing solutions for the stabilization of two roads located on the banks of mountain flowing waters in the Romanian Oriental Carpathian mountains.

**Keywords:** Earthwork · Scouring phenomena · Erosion · Embankment stabilization · Retaining wall stabilization

## 1 Introduction

The continuous development of human settlements has led to the arrangement of different constructions on difficult foundation soils, on areas with high slip potential or in the minor riverbed or major riverbed of stream waters. In most practical cases, the geotechnical engineer finds areas with combinations of factors mentioned above [8].

The present paper addresses the problem of a landslides located in the minor riverbed of mountain running waters by determining the causes which triggered the slip, analyzing and proposing intervention solutions.

The landslides are gravitational displacements of the masses of rocks, soils or combinations of these, on a slope. The generation of these phenomena represents a threat to buildings and implicit to humans [9]. In most cases, the importance given to these phenomena is low, especially in the case of landslides which do not occur with high speeds.

In Romania, the area subjected to landslides computes about 900,000 ha [12]. The most important and known landslide is the displacement of Suhardel Peak, which led to the obstruction of the Bicaz River and the formation of the Red Lake.

Large landslides occurred at Malul cu Flori and Vârfuri in February 1980, both of them located in Dâmbovița county, Zemeș in Bacău County (in 1992) and Izvoarele in Galați County (August 1993).


A. Rotaru (Ed.): CRIT-RE-BUILT 2019, SSGG, pp. 130–141, 2021.
https://doi.org/10.1007/978-3-030-61118-7_11

The process of landslides includes three phases [10]:

- preparatory phase, slow slip, incipient phase (ante-step processes);
- the actual slip (crossing of the geomorphological step);
- natural stabilization (balancing, post-step processes).

Two categories of factors, natural and anthropic [11], control the occurrence/reactivation of landslides. Natural factors include natural water level modification, torrential rains, seismic movement, erosion and scouring of soil. The anthropic factors involve uncontrolled completion of building on the slopes with sliding potential and deforestation/uncovering the vegetation on the same slopes.

## 2 Case Studies

### 2.1 Case Study 1 – Zemeş, Bacău County, Romania

The site is an area nestled in Zemeș village, Bacău County, in the minor riverbed of the Tazlăul Sărat River, an area known because of its high landslide potential [5].

On the occasion of the on-site visit, the followings were detected [6]:

- the riverbed of the Tazlăul Sărat River was almost 50–60% blocked by large boulders as well as tree trunks and vegetation debris (Fig. 1.B);

**Fig. 1.** Tazlăul Sărat River. a) remains of old concrete walls (left); b) large boulders along the riverbed (right)

- remains of some ruined concrete walls and gabions (Fig. 1.A);
- a landslide like a bank collapse impacted the road embankment [3]; the landslide occurred due to the water erosion of the river Tazlăul Sărat that undermined the foot of the slope over which the DC 130 road locates;
- downstream the sliding area, in the riverbed of the Tazlăul Sărat River, there is a gravity retaining wall that maintains the road embankment made by plain concrete; the visual analysis displays that its section next to the studied area underwent a GEO-type transfer following a horizontal translation (Fig. 2.A); the other sectors show

structural degradation as well as tracks of a rotational GEO motion as a result of the sub-erosion of the foundation soil; currently, foundations appear in the console for approximative 30–40% of their area (Fig. 2.B).

**Fig. 2.** Weathered gravity retaining wall. A) horizontal translation; B) rotational motion

Geotechnical prospecting works reveal the following geological stratification as from the upper zone of the borehole to its base: 60 cm of boulders supported by 230 cm of weak clay bounded in a clay-gravel-boulder mix underlain by 90 cm of a gravel-bolder mix supported by 100 cm grey marl clay and the bedding composed of compact grey sandstone.

## 2.2 Case Study 2 – Topliţa Municipality, Harghita County, Romania

The case study refers to a 275 m length retaining wall of Măgura Street, Topliţa municipality, Harghita County, Romania, executed in 1940–1950 epoch, at which crest an underpinning has been made in 2006. The retaining wall is located in the Mureș River Valley.

From a geographical point of view, the retaining wall is located in the northeastern part of Romania, in the central area of the Eastern Carpathians, in the Toplița Depression, in the town of Toplița, Harghita County [7].

The area has heights of 1500–1700 m in Gurghiului Mountains.

Măgura Street is a coastal mountainous road 6.0–7.0 m width only. On Măgura Street, there is a stone masonry retaining wall with a length of 275 m (Fig. 3 a, b). On certain parts on the right side of the road, there is another stone masonry retaining wall that stabilizes the area of individual houses [4].

The area settles on volcanic strata of sandstones, conglomerates, micro-conglomerates, sandstones and andesitic sands.

On both sides of the Mureș River, starting with the northern extremity of the Gheorghieni basin, there are remains of the lower and upper terrace. The lithological composition of terraces includes sand and rolled gravel with schists, covered by diluvial-proluvial

(a) (b)

**Fig. 3.** (a, b) The site view. Toplița municipality, Harghita County, Romania

deposits of sandy clays or clayey sands. The thickness of the terrace deposits is between 5–15 m in Toplița municipality.

The retaining wall has a length of 275 m, with an approximate height of 5 m, stabilized with buttresses. The retaining wall serves throughout its operation (so far) as an element of stability and strength to support the road (its left side) but incorrect operated and maintained. There are areas where parts of the retaining wall are missing or damaged (Fig. 4 a, b), significant cracks are detected, which influence the stability of the massif. Rainwater created channels damaging the retaining wall and transporting the granular material. Consequently, there were landslides, entrainment of vegetation, weak areas in terms of strength and stability of the slope. Structural degradations of the wall, as well as collapses of the granular material due to the washing of the aggregates by floods during the rainy periods were reported. The systems for taking over the meteoric waters are unmaintained and clogged, and the vegetation has infiltrated its roots in the granular material of the retaining wall.

(a) (b)

**Fig. 4.** (a, b) Existing retaining wall: missing or damaged parts

The retaining wall is the stability and strength element that supports the road, but it was not properly operated and maintained. There are significant cracks affecting the slope stability which can be observed. Stormwaters created ruts and transporting granular material, damaged the retaining wall. Consequently, collapses, displacements of vegetation, or weak areas in terms of strength and stability of the slope can be noticed.

Building a sidewalk on the left of Măgura Street, the City Hall of Topliţa, Harghita County, Romania requested a survey of the retaining wall because pedestrian access route is unsafe and uncomfortable at present, with no sidewalk set up. Investigations revealed that the retaining wall structural damage and collapse appeared in the rainy seasons due to washing aggregates of granular material by floods creating ruts. Systems for retrieving rainwater are poorly maintained and clogged, and vegetation has infiltrated roots in the granular material of the retaining wall.

Both structural and nonstructural elements of the retaining wall suffered significant degradation during operation. The beam where the metal safety railing belongs has discontinuities causing rainwater to flow in an unorganised course (Fig. 5 a, b, c). There is not a system for collecting the rainwater along the street. In the rainy seasons, the rainwater torrents coming from the high slope spreading on the right side reach the road and descend as floods washing the embankment of natural vegetation covering the retaining wall. The runoff had washed in this way, repeated over time, the soft, unconsolidated fabric of the slope creating gullies and infiltrating into the retaining wall what grows negative action on its stability and admissible settlements. The rainfall is not picked up by any system of drains being left to infiltrate chaotically water through the plant roots and gullies created in the granular material, thereby influencing the stability and settlements of the retaining wall.

(a) (b) (c)

**Fig. 5.** (a, b, c) Discontinuities of the beam where the metal safety railing belongs

Here and there, collapses develop at the contact between the road and the beam placed in the metal safety railing, the material beneath the hard covering of the road being washed by the rain.

# 3 Intervention Proposals and Methods

## 3.1 Intervention Proposals - Zemeş, Bacău County, Romania

Analyzing the physical-mechanical characteristics of the soil layers and the topography of the site, it results that the activation of landslides is due to the following factors:

- in time, the waters of the Tazlăul Sărat River eroded the foot of the slope supporting the DC 130 road, leading to its failure by downfall;
- the erosion induced by the river at the foot of the slope triggered the failure of the existing retaining walls; retaining walls downstream of the study area near the collapsed areas give stability to the area at the moment;
- a low percentage of foresting concerning adjacent areas;
- counter slopes upstream of the road, allowing the surface water stagnation over long periods, influencing the local physical-mechanical characteristics of the soil layers in site;

Considering the above, the current study analyses intervention proposals concerning the road embankment:

- Intervention without supporting elements disposal;
- The minimal intervention with supporting elements disposal (retaining walls);
- The maximal intervention with supporting elements disposal (piles);

**I. Intervention without supporting elements disposal**

The intervention propose a restructuring of the road embankment to a gentle slope, according to EUROCODE 7 [1] requirements.

It will be performed to unclog the riverbed in the analyzed area.

**II. The minimal intervention with supports elements disposal**

For the minimal intervention, the study proposes support for the road embankment through an elastic reinforced concrete retaining wall placed at the base of the road embankment slope towards the Tazlăul Sărat riverbed and dimensioned according to valid technical regulations.

The foundation depth of the retaining wall takes into account the maximum scouring depth. Works performed to unclog the riverbed in the analyzed area arrange the slope in front of the retaining wall. To that effect, the boulders from the riverbed placed downstream of the support wall create protection against foundation erosion.

**III. The maximal intervention with supports elements disposal**

The maximal intervention proposes support of the road embankment through a bored interspaced piles retaining wall dimensioned according to valid technical regulations, placed on the edge of the road towards the Tazlăul Sărat River.

Works performed to unclog the riverbed in the analyzed area will arrange the slope in front of the support wall. For that purpose, the boulders from the riverbed disposed downstream of the support wall will create protection against ground erosion between piles.

**Method: Stability Modelling for Intervention Proposals**
For the three intervention proposals, stability calculations were performed using the Plaxis 2D programme with the Mohr-Coulomb constitutive model considered as a first order approximation of the real soil behaviour [2] and the "Phi-c reduction" algorithm [3]. In the Safety approach, Plaxis successively and monotonically reduces the soil strength parameters (the friction angle and cohesion) until convergence of the boundary-value issue is no longer achieved. Failure of the structure occurs (the soil collapses) introducing deformations. The lowest reduction factor producing non-convergence is reported as the "factor of safety" of the boundary-value issue.

Table 1 presents the input values of ground characteristics used by the linear perfectly-plastic model considered by the programme, namely Young's modulus E, designed angle of internal friction $\phi_d$, designed cohesion $c_d$, natural unit weight $\gamma_{nat}$, and Poisson's ratio, $\nu$.

**Table 1.** Geotechnical input data for the on-site soil layers

| Layer | Design approach 1 – Combination 2 | | | | |
|---|---|---|---|---|---|
| | E | $\varphi_d$ | $c_d$ | $\gamma_{nat}$ | $\nu$ |
| - | kN/m$^2$ | ° | kPa | kN/m$^3$ | - |
| 1. Fillings | 1.7E+4 | 19.65 | 14.34 | 18.50 | 0.30 |
| 2. Boulders | 5.0E+3 | 10.45 | 8.85 | 19.20 | 0.27 |
| 3. Sandy clay | 2.5E+4 | 19.75 | 19.46 | 19.82 | 0.30 |
| 4. Sandstone | 4.0E+4 | 50.00 | 200.00 | 22.88 | 0.27 |

Following the design concepts of the EUROCODE 7 [1], to ensure stability and adequate strength in the structure and ground, the analysis is carried out for the Design Approach 1, Combination 2 for the GEO ultimate state. The partial factors are applied to material parameters of soil and to variable load actions, and the remain coefficients are set equal to 1.0.

Table 2 presents the values of stability factors for the stability analysis performed.

Table 3 presents the values of the theoretical total displacements for performed stability analyses.

## 3.2 Intervention Proposals – Topliţa Municipality, Harghita County, Romania

Two intervention solutions consolidate the retaining wall of Măgura Street, Topliţa municipality, Harghita County, Romania.

**Table 2.** Stability factors for the stability analysis performed

| Intervention | Calculus | Stability factor |
|---|---|---|
| **I. Embankment restructuring** | 1. Without loading | 1.48 |
| | 2. With loading | 1.29 |
| **II. Minimal intervention** | 1. Support wall without loading | 1.30 |
| | 2. Support wall with loading | 1.19 |
| | 3. Erosion without loading | 1.27 |
| | 4. Erosion with loading | 1.16 |
| **III. Maximal intervention** | 1. Pile wall without loading | 1.75 |
| | 2. Pile wall with loading | 1.72 |
| | 3. Erosion without loading | 2.02 |
| | 4. Erosion with loading | 2.15 |

**Table 3.** Values of the theoretical total displacements for performed stability analyses.

| Proposed intervention | Calculus | Total displace- place- ments [cm] | |
|---|---|---|---|
| **I. Embankment restructuring** | 1. Without loading | 57.95 | 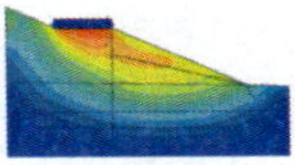 |
| | 2. With loading | 66.70 | |
| **II. Minimal intervention** | 1. Support wall without loading | 22.73 | 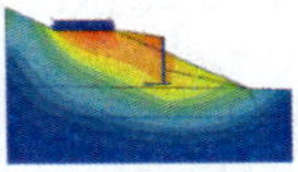 |
| | 2. Support wall with loading | 28.92 | |
| | 3. Erosion without loading | 31.37 | |
| | 4. Erosion with loading | 39.10 | |
| **III. Maximal intervention** | 1. Pile wall without loading | 5.14 | 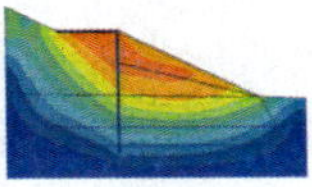 |
| | 2. Pile wall with loading | 7.98 | |
| | 3. Erosion without loading | 6.77 | |
| | 4. Erosion with loading | 8.56 | |

The first solution consists of the following steps:

1. To uncover the current slope in its integrity to the parament of the existing retaining wall.
2. Local repairs and strengthening of the existing retaining wall, where they are proper.

3. Carrying out a continuous vertical drain of river rocks no more than 1 m thick, followed by a continuous horizontal drain not exceeding 1 m thick, with a slope at most 4% to remove the collected water from the base of the retaining wall.
4. Filling of ballast.
5. A 1.50 m width sidewalk protected by a pedestrian guardrail at the canopy of the existing retaining wall.
6. Building an embankment of concrete stone pitching whose thickness could vary from 30 to 50 cm, between the end of the sidewalk and the top of the new retaining wall.
7. Making a new retaining wall to take over the displacements of the soil mass that the existing retaining wall is no longer able to take over.
8. The new retaining wall will be provided with barbicans at not less than 2.50 m and waterproofed with fillerized bitumen at the contact with the slope and drains.
9. A drainage culvert in the outer area of the retaining wall footing.

The second solution focused on the following goals:

1. To uncover the current slope in its integrity up to the parament of an existing retaining wall.
2. Local renovation and strengthening of the existing retaining wall, where they are proper.
3. Clay fill, clay and gravel, ballast, at a maximum slope angle not exceeding 33° for safety reasons. One metre high at the base of the filler, 10–50 cm width steps on the slope made of existing material at an angle of no more than 3%.
4. The fill protected by geocomposites.
5. A 5–10 cm layer of humus spreads out over geocomposites.
6. Making a 1.50 m width sidewalk protected by a pedestrian guardrail at the canopy of the existing retaining wall.
7. Making a new retaining wall to take over the displacements of the soil mass that the existing retaining wall is no longer able to take over.
8. Carry out, behind the retaining wall, of a vertical stone drain at least 1 m height, on which to place a reverse filter from granular material.
9. The new retaining wall will be provided with barbicans at not less than 2.50 m and waterproofed with fillerized bitumen at the contact with the slope and drains.
10. A drainage culvert in the outer area of the retaining wall footing.

## 4 Results and Discussions

### 4.1 Results Based on the Stability Calculations - Zemeş, Bacău County

The following results arose (Table 4):

- in the case of the slope in an embankment without load, the failure occurs at the foot, by hydraulic erosion; in the case of the loaded slope, the yielding surface has a circular cylinder shape while failure affects the entire road embankment. It is worth mentioning that the stability factors are only valid if the waters of the Tazlăul Sărat

**Table 4.** The variation of efforts values for maximal intervention calculus

| Maximal intervention | Deformed mesh | Bending moment [kNm] | Diagram allure | Shear force [kN] | Diagram allure |
|---|---|---|---|---|---|
| **1. Pile wall without loading** | 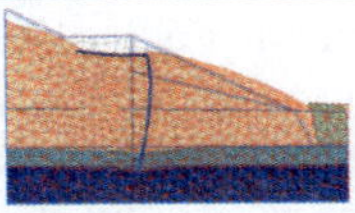 | 285 |  | 51 |  |
| **2. Pile wall with loading** | 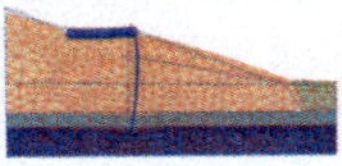 | 531 |  | 87 |  |
| **3. Erosion without loading** | 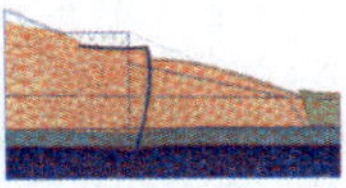 | 453 |  | 118 |  |
| **4. Erosion with loading** | 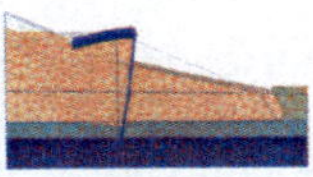 | 895 |  | 241 |  |

River do not erode the foot of the slope; if they eroded the foot of the slope, the stability factors have subunit values, which is the case on the site at the moment;

- in the case of the slope of the reinforced concrete retaining wall with a corner, a cylindrical circular yielding surface forms and extends in-depth covering the entire supported area; the erosion of the foot of the slope that would trigger the subsidence of the embankment in front of the retaining wall about 2.00 m is a critical situation; irrespective of the current study, the formation of a yielding surface downstream of the retaining wall is noted, indicating the possibility of losing the stability and bearing capacity of the wall by GEO-type yielding when rotating;
- in case of the retaining wall from bored piles, interspaced, supporting the embankment, the yield surface has a circular cylindrical shape that comes out just downstream of the retaining wall so that the road embankment shows stability; the erosion of the embankment foot, which could lead to about 3.00 m high subsidence of the roadbed in front of the retaining wall, is critical; in this case, an increase of the stability factor due to the yield surface formed in-depth comprising the whole supported area is observed.

## Discussions: Intervention Analysis

All the three intervention proposed above invest stability on the site but analyzing the building technologies and conditions of service in time, the following have resulted:

- remediating the slope of the road without requiring other protection works has applicability for a short term because of its erosion; on this account, the slope collapses again;
- open excavations with depths of about 2,00 m (settlement depth + embedding depth) leading to 3.00 m high vertical excavation are necessary for the completion of the retaining wall as to apply the minimum intervention proposal;
- the slope support system carried off at 5.00 m from the road embankment rest stable through earthworks; works for preventing the foundation soil scouring phenomenon performed to unclog the riverbed are necessary, with the disposal of the boulders recovered from the riverbed at the base of the support wall; as observed on the site, the discharge of the river leads to a notable landslide beneath the existing wall foundations (Fig. 3);
- completing boring piles to support the retaining wall is the maximum intervention solution regarding the road edge that assures stability to the embankment even though subsidence occurred 3.00 m downstream of the retaining wall. It is necessary to safely carry out the piles, unclog the riverbank, work out a slope on the embankment edge, and arrange boulders along the riverbank. Those supplementary measures confer slope stability during completion and service of boring piles playing anti-erosion function.

### 4.2 Results Based on Observations at Topliţa Municipality, Harghita County

Several results can be drawn:

- The relief conditions of the studied perimeter on the Mureş Valley promote surface and intermediate drainage depending on the season.
- The rainfall level determines the local hydrological conditions.
- The specific weather in the area is favourable to carry out construction ditches, culverts for drainage, retaining walls, culverts and rainwater collection system.

Sustainable rehabilitation of the retaining wall in Măgura Street, Topliţa is a priority, but both methods take into account the necessity of systems for retrieving rainwater to remove the collected water from the base of the retaining wall. The first solution proposes the collection of the rainwater through a continuous vertical drain of river rocks no more than 1.00 m thick followed by a continuous horizontal drain not exceeding 1.00 m thick. Unlike that, the second method proposes a 1.00 m high clay fill at the base of the filler and 10–50 cm width steps made of existing material at an angle of no more than 3%. First method proposes a 30 to 50 cm thick embankment of concrete stone pitching between the end of the sidewalk and the top of the new retaining wall, while the second method advises protection by geocomposites.

## 5 Conclusions

All situations which demand the design of support structures undertake relevant analysis of several intervention proposals to choose the most technically-economically efficient solution.

The article proposes solutions for stabilization of two roads near Tazlăul Sărat River and Mureş River, respectively, two mountain flowing waters in Romanian Oriental Carpathian ridges.

In the first case, the mountain water flow is responsible for the deterioration of the road embankment. The intervention considering a retaining wall completion is difficult to achieve. The pile intervention gives a high degree of stability to the ground, and it is easy to execute. Although the solution is more expensive, it reduces the volume of work and the execution risks and increases the degree of safety in service. Therefore, it is the most recommended intervention in this instance.

In the second case, the heavy rains directing the torrents towards the river and washing the soil are responsible for the degradation of the retaining wall. Consolidation develops either using drains or clay fill at the base of the filler.

Both cases show that the presence of water near earthworks influences their exploitation. The scouring phenomena have led to the destruction of both existing support structures and the loss of the stability of the river banks, and the erosion phenomena have led to the increase of the land masses displaced below the road structures.

## References


1. *.*.EUROCODE 7: Geotechnical Design, EN 1997-1:2004
2. *.*.Plaxis Version 8. Materials models manual
3. *.*.Plaxis Version 8. Reference manual
4. *.*.Technical expertise on strengthening and widening the access road, wall support and embankments, Măgura St., Topliţa City, Harghita County (2014)
5. Pohrib, D.M, Plătică, D., Ţurcanu, P., Pleşcan, C.: Road embankment for bridge accessing rural area. In: CIBv Conference, Braşov, Romania (2019)
6. Pohrib, D.M., Ciobanu, I.E.: Solutions to stabilise the embankment of a road located in the streambed of a flowing mountain water. Bull. Transilvania Univ. Braşov Series I: Eng. Sci. **11**(60) Special Issue No. 1, 279–284 (2018)
7. Rotaru, A., Bhandari, G., Bridging new solutions for sustainable rehabilitation of structures damaged due to difficult soils or foundation design. Adv. Eng. Forum **21**, 346–351 (2017)
8. Rotaru, A., Nicuţă, A.: Some aspects of landslide risk evaluation taking into account their distribution and properties. J. Mater. Methods Technol. Int. Sci. Publ. **2**(1), 47–57 (2008)
9. Rotaru, A., Oajdea, D., Răileanu, P.: Analysis of the landslide movements. Int. J. Geol. **1**(3), 70–79 (2007)
10. Rotaru, A., Oajdea, D., Răileanu, P.: Dynamics of a Landslide Surface, Environmental Problems and Development, Environmental Problems and Development – Energy and Environmental Engineering Series, pp. 22–27 (2008)
11. Rotaru, A.: Causes and mechanisms of landslides triggered on foundation soil areas, Geophysical Research Abstracts, 12, EGU2010-13952. EGU General Assembly (2010)
12. Rotaru, A.: Landslides in Romania. In: Proceedings of the International Conference on Geotechnical Engineering, Lahore, Pakistan, 5–6 November 2010, pp. 229–236 (2010)

# Innovative Application of IoT Technologies to Improve Geotechnical Monitoring Tools and Early Warning Performances

Andrea Segalini[1(✉)], Roberto Savi[1], Edoardo Cavalca[1], Alessandro Valletta[1], and Andrea Carri[2]

[1] University of Parma, Parco Area delle Scienze 181/a, 43123 Parma, Italy
andrea.segalini@unipr.it
[2] ASE – Advanced Slope Engineering Ltd., Parco Area delle Scienze 181/a, 43123 Parma, Italy

**Abstract.** The integration of Internet-of-Things principles, giving the possibility to connect different physical sensors through a communication network, is becoming increasingly important in several fields of application. Within the geotechnical framework, one of most productive application of the IoT approach is related to monitoring systems, which can significantly benefit from the possibility to exploit more proficient and reliable connections between sensors, control units and visualization devices. In particular, these advantages allow for the development of advanced Early Warning Systems (EWS) able to communicate with high efficiency the occurrence of a potentially critical event, getting even closer to a *real-time* approach. An innovative application of IoT technologies to geotechnical monitoring is presented, involving the integration of modern devices with data loggers that can be controlled from remote, dedicated software implementing machine-learning algorithms, and web-based visualization platforms. The result of these components integration is a fully automated and network-connected system, defined Internet of Natural Hazards (IoNH). One application deals with the installation of a mechanical trigger on a rockfall barrier, in order to detect an impact on the structure and consequently read all installed sensors, alerting those responsible of the monitoring activity. Another case study includes the monitoring of a building with tilt-based sensors, automatically identifying the overcoming of predefined alert thresholds. The software is able to disseminate automatically a message in case of an overcoming of these levels and to activate different alarm devices (e.g. sirens, traffic lights, etc.).

**Keywords:** Monitoring · Internet of Things · Early Warning Systems

## 1 Introduction

Geotechnical monitoring has experienced significant improvements in the last few decades, thanks to technological advancements and the introduction of innovative devices.


A. Rotaru (Ed.): CRIT-RE-BUILT 2019, SSGG, pp. 142–146, 2021.
https://doi.org/10.1007/978-3-030-61118-7_12

One of the driving reasons that prompted these researches relates to the possibility to provide new systems able to operate with a real-time approach. In fact, even if there is no chance to prevent an impending failure, a device able to communicate adequately the occurrence of a critical event could be vital to avoid damages to work equipment and losses of human lives [1]. In this context, the integration of Internet-of-Things (IoT) principles could be a major component in order to provide more efficient and reliable Early Warning Systems (EWS). Typical examples of this approach can be found in literature referred to landslides [2–4], mines [5] and structural health monitoring [6].

## 2 Materials and Methods

The system presented in this paper, defined Internet of Natural Hazards (IoNH), represents a new approach to the integration of IoT technologies into the geotechnical monitoring field. It has been developed in order to provide a fully automated system able to manage all operations related to data acquisition, transmission, storage, elaboration and visualization (Fig. 1).

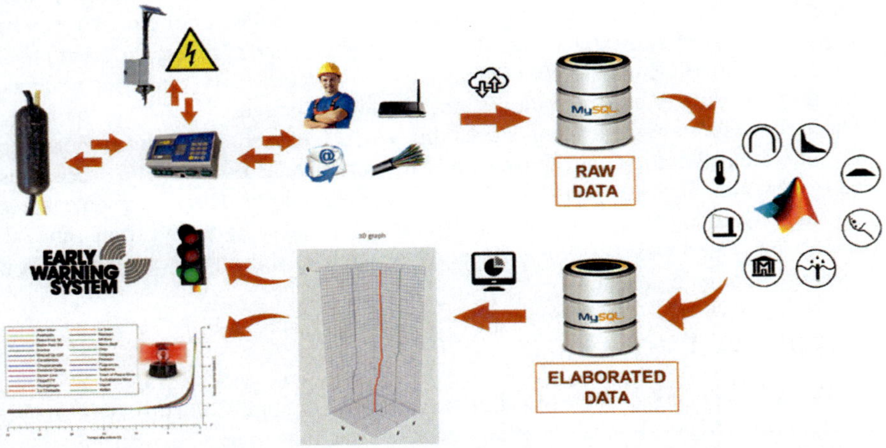


**Fig. 1.** Schematic representation of the Internet of Natural Hazards (IoNH) approach

In particular, the proposed methodology is intended to manage devices based on MUMS (Modular Underground Monitoring System) technology [7, 8], while its functioning principles could be expanded to other monitoring tools. MUMS instrumentation relies on a series of sensors, embedded in specifically molded epoxy resin nodes (called Links) and connected in order to form an arbitrary long array. The system is customizable with a wide range of different sensors, giving the possibility to record parameters such as displacement, pore pressure, inclination, temperature, etc.

The first phase consists of sensors reading operations, performed by a dedicated data logger that interrogates each sensor and saves the results on a local memory (i.e. SD-card). Depending on the application, sensors connection to the control unit can be performed with a quadrupole electric cable, or by exploiting individual radio modules. The control unit, which can be controlled and configured from remote, sends all sampled data to the elaboration center through an internet connection. Here, they are stored in a dynamic MySQL database with secure access and daily backups.

The elaboration step relies on a proprietary software, specifically developed to manage data sampled in different settings (e.g. landslides, underground constructions, rockfall barriers, etc.). In particular, the software elaborates raw data sent by the control unit by coupling them with appropriate calibration coefficients previously computed. This automatic routine integrates several different algorithms and statistical processes, including procedures to identify outliers and to filter spike noises. The elaboration process provides a preliminary results validation and check the dataset quality. The final output is a set of elaborated data representing the physical quantities monitored on-site, which are then stored in a parallel section of the MySQL database. Additionally, the software is able to detect the overcoming of predefined alert threshold and to identify potentially critical trends.

The following phase includes the representation of the elaborated datasets on an interactive web-based platform with private access, which can be checked anytime by using different network-connected devices (PC, tablet, smartphone, etc.). The visualization platform has been developed to provide a dynamic collector of data that can be easily browsed in order to extract any relevant information and correlate different physical quantities. Monitoring outcomes are displayed with interactive graphs and widgets, giving the possibility to export results as text files and/or images.

During the last step, the system checks the presence of any critical event identified by the software elaboration, and consequently triggers a series of automatic procedures aimed to alert authorities responsible of the monitoring activity. These safety measures include the activation of acoustic and visual devices, like traffic lights and sirens, and the dissemination of alert messages via SMS and/or e-mail reporting all relevant information concerning the ongoing event.

## 3 Results and Discussion

One of the main advantages of the IoNH system relates to its applicability to a wide range of different case studies. In fact, the proposed approach aims to provide a methodology able to adapt to various scenarios, in order to improve the overall performances of monitoring systems developed for different purposes.

The first example presented in this paper regards the monitoring activity of two rockfall barriers, located in Northern Italy and designed to protect the industrial area and the highway situated at the slope base. The system developed for this specific application, called Rockfall Safety Network (RSN), aims to identify impacts on the barrier and assess the work rate of the structure, thanks to a multi-parameter and redundant approach (Fig. 2). In particular, the RSN system comprises two different tilt sensors encased within an IP68-certified metal box (defined BPM module), which is installed on the barrier steel post in order to measure its inclination. Additionally, a strain gauge load cell measures the force acting on the upstream steel wire ropes, providing the stress state related to a single steel post. Lastly, the system includes an electro-mechanical Trigger device for Early Warning purposes.

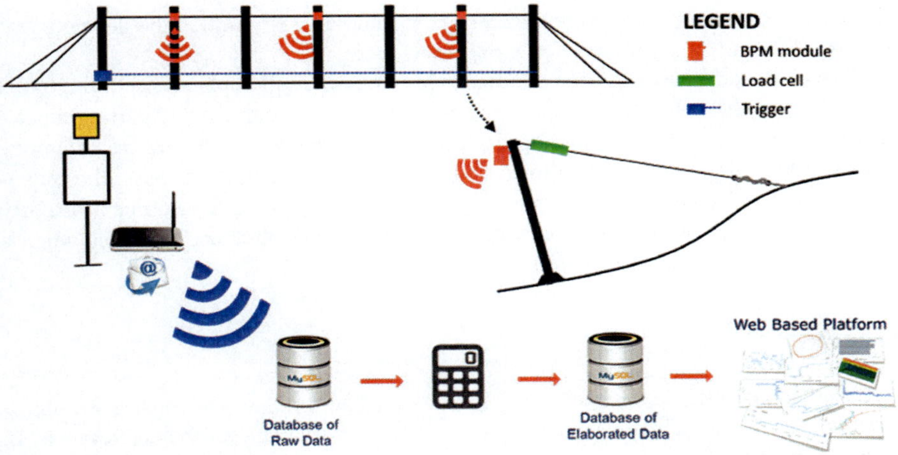


**Fig. 2.** Sensors composing the Rockfall Safety Network (RSN) system, installed for the monitoring of a rockfall barrier

When a specific event activates the Trigger component, the system automatically queries every sensor placed on the barrier and sends acquired data to the elaboration centre. Then, the software identifies the critical event and enables the dissemination of alert messages (SMS and/or e-mail) to predefined users and the activation of different alarm devices (traffic lights, variable message signs and acoustic alarms).

Another application of the proposed approach regards the monitoring of a building, specifically an industrial chimney, by exploiting both innovative and traditional instrumentation. In fact, IoNH technology features the possibility to convert analog signals into digital ones, thus giving the opportunity to integrate different devices within the same monitoring system. For this case study, the installed traditional instrumentation comprehends four crack meters and four strain gauge extensometers, providing a measure of the structure deformations. The inclination of the chimney is monitored by a MUMS-based clinometer (named Klino Array), which integrates a MEMS (Micro Electro-Mechanical System) sensor and an electrolytic tilt sensor enabling redundant measures and increasing the system reliability.

The monitoring activity involves also the definition of alert levels referred to specific inclination values. The elaboration software is able to detect the overcoming of one of these thresholds, automatically disseminating alert messages to authorities responsible of the monitoring activity with a near-real time approach.

## 4 Conclusions

This paper presents a new approach to geotechnical monitoring, defined Internet of Natural Hazards, which exploits IoT technologies and innovative tools in order to provide an advanced automatic system for monitoring data management, from their on-site acquisition to their elaboration and representation. The improved communication between sensors and the near-real time monitoring approach play a key role in the development of Early Warning Systems based on monitoring data.

Two case studies are reported in order to provide different applications of the IoNH approach, including the monitoring of two rockfall barriers with a new multi-parameter instrumentation, and the sampling of displacement data related to an industrial chimney thanks to the integration of traditional and innovative tools. In both cases, the implementation of IoT-related technologies played a major role in the development of Early warning processes, allowing to activate different risk management activities when a potentially critical event is detected.

## References

1. Klappstein, B., Bonci, G., Maston, W.: Implementation of real time geotechnical monitoring at an open pit mountain coal mine in Western Canada. Revista Minelor/Mining Revue **20**(3), 2–14 (2014)
2. Aggarwal, S., Mishra, P.K., Sumakar, K.V.S., Chaturvedi, P.: Landslide monitoring system implementing IOT using video camera. In: 2018 3rd International Conference for Convergence in Technology (I2CT), pp. 1–4 (2018)
3. He, H., Zongling, Y., Xiaosong, Z., Qingchen, W., Guoren, Z., Yu, Z., Yangfan, H.: Study on dynamic monitoring system of high slope based on the internet of things. In: 2016 IEEE International Conference on Cloud Computing and Big Data Analysis (ICCCBDA), pp. 407–412 (2016)
4. Sofwan, A., Sumardi, R.M., Goni, A.N.: Wireless sensor network design for landslide warning system in IoT architecture. In: 2017 4th International Conference on Information Technology, Computer, and Electrical Engineering (ICITACEE), pp. 280–283 (2017)
5. Singh, A., Singh, U.K., Kumar, D.: IoT in mining for sensing, monitoring and prediction of underground mines roof support. In: 2018 4th International Conference on Recent Advances in Information Technology (RAIT), pp. 1–5 (2018)
6. Wang, J., Fu, Y., Yang, X.: An integrated system for building structural health monitoring and early warning based on an Internet of Things approach. Int. J. Distrib. Sens. Networks **13**(1) (2017)
7. Segalini, A., Carini, C.: Underground landslide displacement monitoring: a new MMES-based device. In: Margottini, C., Canuti, P., Sassa, K. (eds.) Landslide Science and Practice, pp. 87–93. Springer, Heidelberg (2013)
8. Carri, A., Chiapponi, L., Giovannelli, R., Spaggiari, L., Segalini, A.: Improving landslide displacement measurement through automatic recording and statistical analysis. Procedia Earth Planetary Sci. **15**, 536–541 (2015)

# Vertical Vibro-Isulation for Building Foundations – A Full Scale Experiment

Chavdar Kolev(✉)

Higher School of Transport "Todor Kableshkov", 158, Geo Milev Street, Sofia, Bulgaria
ch_kolev@abv.bg

**Abstract.** The report presents the results of a field experiment conducted in the summer of 2019, which aims to determine the actual effect of a certain type of vertical barrier to reduce accelerations and horizontal displacements in the ground caused by vibrations of construction machinery and vehicles. The reason for the experiment were the cracks obtained on a number of old masonry buildings, some of which - cultural monuments exposed to heavy vehicle vibrations from trains, trucks or vibrating rollers that repair the road structure. The experiment was conducted on a special site about 50 km north of Sofia. The site is outside a settlement, at the foot of the mountain and around it there are no industrial sources of any vibration.

The schemes for conducting the experiment are presented and the results obtained are analyzed. Diagrams of the most significant values are drawn up and various design situations are compared. It is found that acceleration and displacement reduction of 70% to 90% has been achieved, which is quite satisfactory for the purpose. The effect of wave's diffraction below the building foundations barrier at a distance of 10 m backwards was also measured. The parameters of the mechanization used, the insulation materials and the earth base are also presented.

**Keywords:** Ground waves · Acceleration · Displacement · Roller · Building frequency · Diffraction

## 1 Introduction

The protection of buildings against vibration is a task that should be most often solved within the phase of designing their foundations. Another case when buildings and/or architectural monuments need protection is the one with heavy traffic nearby. And this is exactly the situation addressed in this report. The results of a large experiment are presented to prove the effectiveness of the designed passive vibration barrier. A design has been tested in line with the knowledge gained so far, but new results have been achieved which prove that Styrofoam is an extremely suitable material for vibration protection of buildings. The diffraction of waves passing under and around the barrier is also a matter of study and evaluation. We are aware of the principles of wave propagation, but their values and soil dumping needed to be proved by experiment.


A. Rotaru (Ed.): CRIT-RE-BUILT 2019, SSGG, pp. 147–155, 2021.
https://doi.org/10.1007/978-3-030-61118-7_13

## 2 State of the Art

During the course of the study, about 25 publications [1÷8 etc.] were reviewed, mainly American ones, showing the results of similar experiments. We have thoroughly studied the effect of active and passive vertical vibro-protection for foundations under machines; the spectrum and propagation of the waves in the soil caused by passing trains; the propagation of vibrations from the rollers on the construction site; the effect of pile drilling upon the surrounding buildings; etc. There were also analyses of the damping effect of various lightweight waste materials to fill the protective trenches.

Some important identified dependencies pertaining to this project were identified from the literature reviewed:

- The adverse effects of vibration on buildings are caused by surface waves (Relay's waves). Their propagation depth does not exceed 60 cm when generated by the relatively high frequencies of machines.
- Vibration trenches are the most reliable means of active and passive protection. Their depth in experiments is usually 60 cm; their width does not exceed 50 cm.
- Empty trenches isolate and reduce accelerations by about 70%.
- The trenches retain their efficiency to a great extent when filled with light materials - bentonite, rice blubber and other (e.g.by 50%). Filling with concrete has an effect only at very high vibrational frequencies.

## 3 Study of Suitable Vibration Protective Materials Available in the Market

In accordance with our intention to fill the trench with a very lightweight but sufficiently stiff, cheap and independent on frost material, we have studied the market for styrofoam and similar products intended for thermal and noise insulation. We have established by technical parameters and in practice that styrofoam for floor insulation is most suitable because it is stiff enough to withstand the earth pressure and the load from the pavement. Thus, Austrotherm EPS® T1000 styrofoam (foam-polystyrene) was embedded in the experiment.

The activities during the experiment were performed with the following construction mechanization: Roller 11t, brand model Liugong CLG 612; Excavator - 99 hp, brand model KOMATSU WB935S - 5; MERCEDES-BENZ dump truck, brand model ACTROS 5044 K, with 4 axles; BMW Reihe car, 150 hp.

## 4 Engineering and Geological Data About the Site for the Experiment

The geological profile in which the measurements were made consists of three clay layers and a coating of loose material with very thin old asphalt pavement. The groundwater level (−3.50 m) flows practically below the bottom of the trench.

## 5 Key Mile Stones of the Experimental Plan

The logical sequence of the full-scale experiment, subordinated to the goal of measuring the horizontal displacements and accelerations caused by machine vibrations was the following:

- Vibrations measurement of the operation of a vibrating roller with a motor turned on (in a static position and when moving).

The vibration was measured from a pre-dug trench with a depth of 3.60 m, as shown in Fig. 1. This trench simulates the basement floor of the protected building.

There were 4 steel rods, 25 mm in diameter and 50 cm long horizontally, nailed in the trench walls. The rods were vertically arranged at a depth of 20 cm, 1 m and 3 m respectively. The fourth rod was driven vertically to the bottom of the pit at a depth of 3.50 m. Thus, the rods represented a single measured vertical profile. Magnetic sensors are mounted at the rod ends. The sensors were connected by cables to the measuring equipment. The experiment profile is shown in Fig. 2.

A second trench was also dug at the construction site, located 10 m behind the first one. Its purpose was to simulate the back of the basement of a protected building, to the extent that only the vibrations generated by the diffraction of the waves below the vibration barrier have been reached. Steel rods were strung along the wall of the second trench in the same scheme and sensors were also mounted on them. Given that the measuring equipment was able to work with only 4 channels (sensor cables) at a time, measurements in the two trenches were carried out sequentially and not simultaneously. The vibration measurements taken proved the results of simulation of the following design points:

- Effects of vibrations from the mechanization on the wall in the basement below the facade of the protected building without having an insulating system (Fig. 3);
- Vibration actions on the basement wall under the facade from available mechanization, with an empty excavated trench for insulation (Fig. 4);
- Vibration actions from the mechanization on the basement wall under the facade, with a trench for insulation, filled with styrofoam (Fig. 4);
- Vibration effects from the motors on the basement below the rear facade, with an empty excavated trench for insulation (Fig. 5);
- Vibration actions from mechanization on the basement below the rear facade, with a excavated trench for insulation filled with styrofoam (Fig. 6);

In the analyses, it was taken into consideration, the percentage reduction of accelerations and horizontal displacements resulting from the insulation system as well as the large excitation of these parameters below the rear facade of the building being caused by wave diffraction.

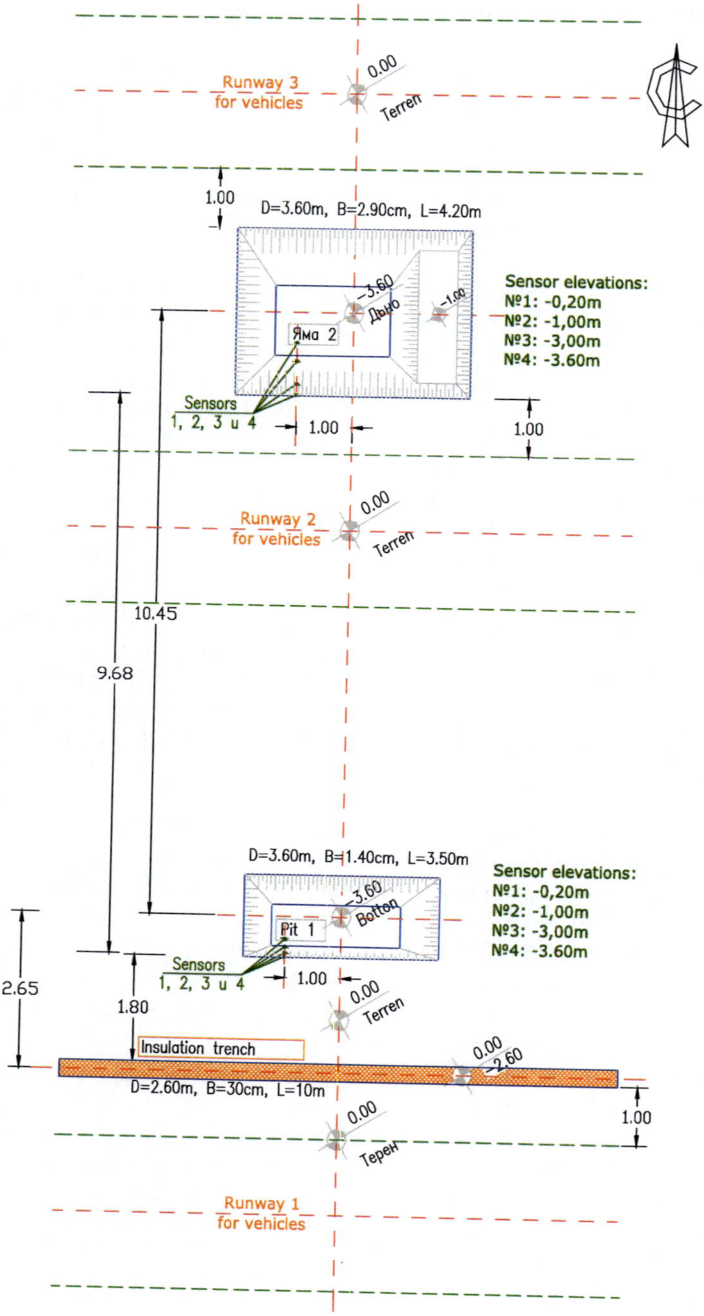


**Fig. 1.** Sketch of the experiment site

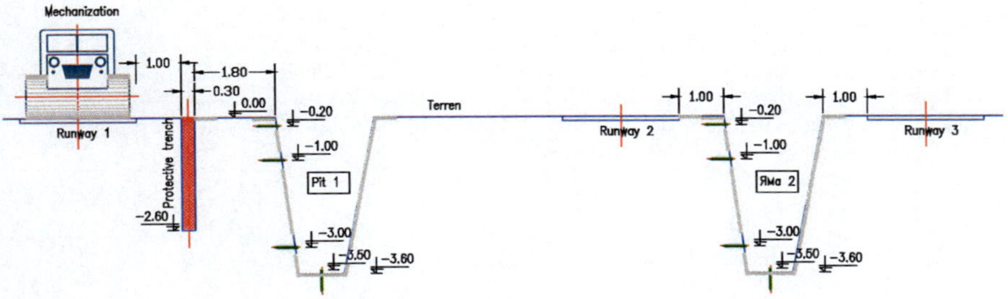


**Fig. 2.** Transverse profile of the experimental site

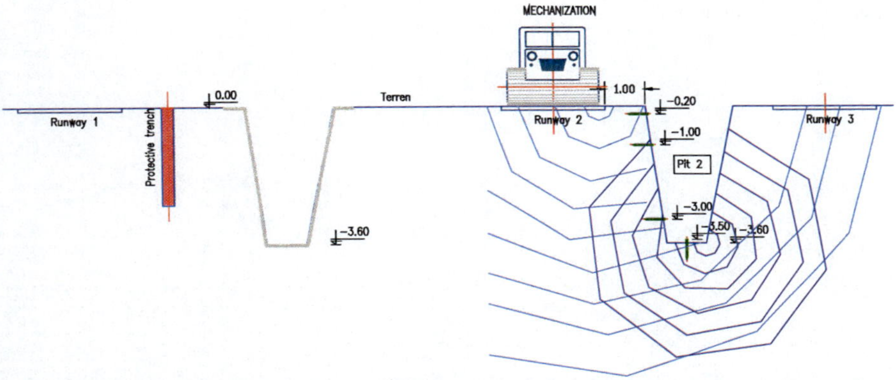


**Fig. 3.** Transverse profile of the site with simulating vibrations on the unprotected basement of the building

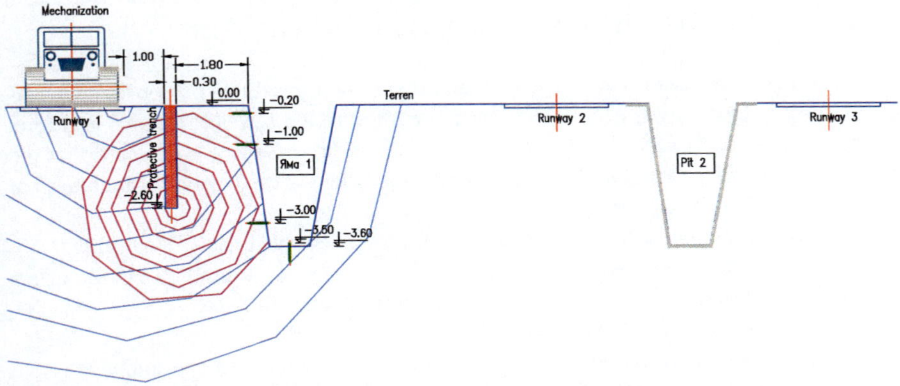


**Fig. 4.** Transverse profile of the site when simulating vibrations on the protected basement at the facade of the building

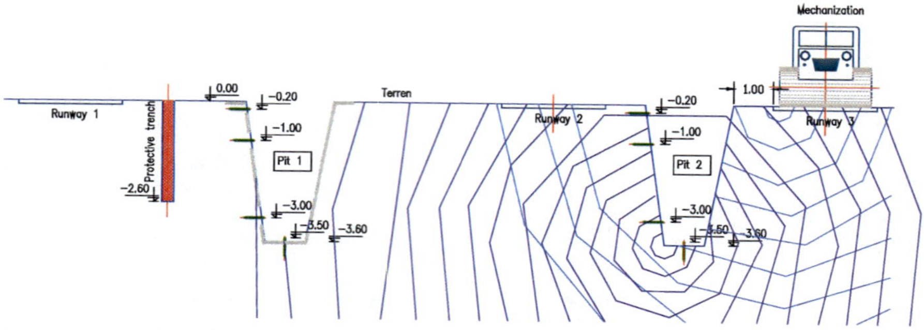


Fig. 5. Transverse profile of the site when simulating vibrations on the basement below the rear facade of the building

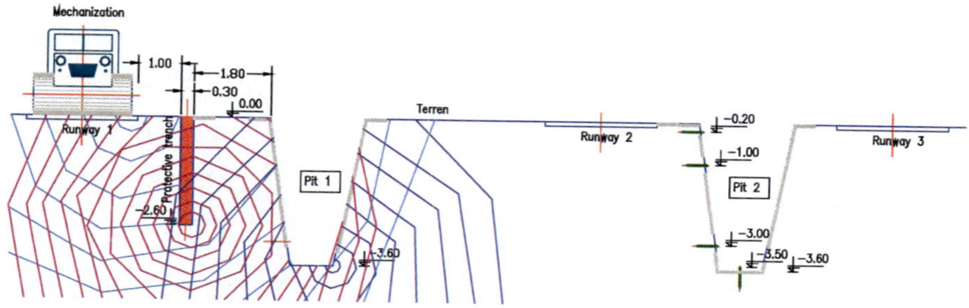


Fig. 6. Transverse profile of the site when simulating vibrations on the basement below the rear facade of the building in a filled with styrofoam trench

# 6 Analysis of Results Obtained and Comparison with the Published Information About Similar Experiments in the Past

The most interesting results of the experiment were the recordings of Sensor 1. It was located only 20 cm below the terrain and measures the surface waves of Relay. Therefore, the highest and worst values had to be obtained right there. Sensor 1 detected that the vibration barrier was reducing the vibrations of the roller (without vibrator included) at a frequency of 20 Hz by more than 82%.

The protection against the impact of the rolling tractor was over 77% at 30 Hz.

The truck's vibration protection was 33% at 30 Hz, but at around 32 Hz the effect rose to 75% and in the 14 ÷ 16 Hz range the reduction reached 85%.

The reported reduction for the car was almost 80% at 20 Hz.

In general, low frequencies create higher amplitudes of displacements and vice versa. Also, the largest displacements from oscillations in the soil occur on the surface, gradually decreasing to about 50–60 cm depth, and then continuing to decrease progressively. These universal dependencies were established, of course, there. Therefore, after an overview and comparison of results, it was naturally found that the machines on the move were giving less impact than the on-site engines, but boosted to a frequency of

2000 rpm. The most adverse effects were obtained from a stationary roller whose engine was running at low rpm - 1030 rpm and with the maximum impact of its vibrator. Diagrams of acceleration and displacement values in cases without and with a protective trench are shown in Fig. 7 and Fig. 8.

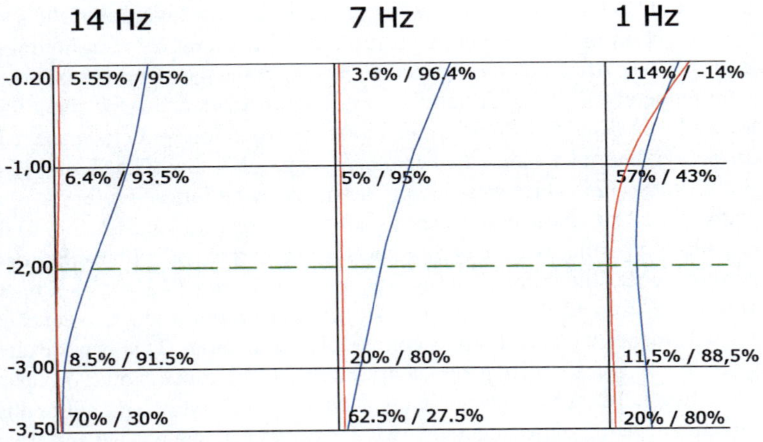


**Fig. 7.** Stationary roller – maximum acceleration diagram with 1030 rpm engine and maximum operating vibrator

Blue color in the diagrams is used for values without barrier and the red color is used for the ones in the presence of a styrofoam trench.

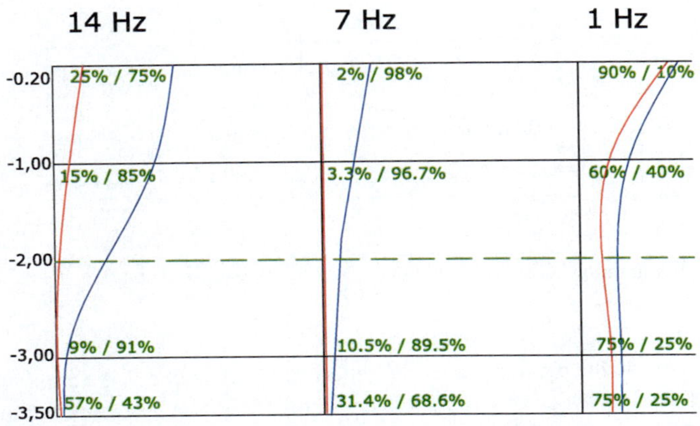


**Fig. 8.** Diagram of maximum displacements along the access X from a stationary roller with 1030 rpm engine and vibrator in maximum operating mode

The results were clear, and the percentages of reduction were also plotted. The relevant project values were obtained at a frequency of about 14 Hz, as well as at the octaves of this frequency. Extreme values were also obtained at the frequency of 1–2 Hz, but they were not relevant for the project, because the own building frequencies have been close to 14 Hz, and the low frequencies of 1–2 Hz haven't been practically released around the ordinary masonry buildings. Displacement reduction percentages from oscillations generally follow the course of the accelerations, although they were not exactly proportional. The explanation lies in the more complex picture of the spectrum from the wave action in the soil, the deposition of diffractions, the non-homogeneous soil base and other. After all, the important results from the diagrams show that at the relevant frequency of 14 Hz, surface wave accelerations decrease by about 95% and displacements by 75% to 85%. At a greater depth, the reductions only increase relatively due to diffraction, but the absolute values remain negligibly low. The picture is similar at 7 Hz, but the absolute values were lower and not relevant for the project.

It became clear from the measurements made in the rear trench, # 2 (Fig. 6) that for a stationary vibrating roller with an engine speed of 1030 rpm and with a vibrator turned on a maximum speed, the corresponding frequency had been 14 Hz with a maximum acceleration of 35 mm/s$^2$. Table 1 shows that the acceleration reduction reaches 96% and the displacement - 91%, and this is completely satisfactory. This is not exactly the result reported at 1 Hz. Low frequencies are known to be much slower in space and therefore the displacement remains unchanged, but the acceleration was still reduced by 28%. As noted above, this low frequency was not relevant to the project and the values of acceleration and displacement were much lower than those of the front facade of the building.

**Table 1.** Comparison of the results for the action on the rear trench # 2.

| Situation | 14 Hz | | 1 Hz | |
|---|---|---|---|---|
| | $a_{max}$. | DXmax. | $a_{max}$. | DXmax. |
| With barrier | 35 mm/s$^2$ | 0.010 mm | 0.5 mm/s$^2$ | 0.030 mm |
| Without barrier | 900 mm/s$^2$ | 0.120 mm | 0.7 mm/s$^2$ | 0.100 mm |
| Ratio | 3.9% | 8.3% | 71.5% | 30% |
| Difference | 96.1% | 91.7% | 28.5% | 70% |
| Evaluation | OK! | OK! | Satisfactory! | OK! |

The difference in the percentages of vibration damping between the empty trench and the filled with styrofoam one varied between 5 and 15%. The vibration damping effect of the empty trench reached 70% and corresponded entirely to the published results of similar experiments, which had been conducted worldwide in the past.

In a result, the trench was filled with material for the building to stay strong, but the mass of the filler was very small and the stiffness relatively high in order to preserve its damping effect.

## 7 Conclusion

The experiment thoroughly demonstrated the beneficial effect of making vibration isolation for buildings. All values were within the expected range, which had been achieved in previous experiments, according to scientific publications on the topic.

The reached effect of acceleration and displacement reductions was in the range of 70% for an empty trench and about 90% for a trench filled with styrofoam. All reduced acceleration and displacement values fell well below the critical requirements.

The dominant frequency was about 14 Hz. The surface waves (Relay) were strongly influenced. The diffraction of the waves beneath the building was not capable of causing alarmingly large values of accelerations and displacements.

The width of the protective trench was 30 cm and the depth was sufficient to be about 2 m. At a greater depth, the waves progressively decay and reduce the values of their accelerations and displacements. Higher frequency engines cause lower accelerations and displacements, respectively will be safer for the building.

Such protective trenches should be located at a distance of not more than 2 m from the walls of the buildings.

## References


1. Ahmad, S.: Simple model for active isolation of machine foundations by open trenches. In: 3rd International Conference on Recent Advances in Geotechnical Earthquake Engineering and Soil Dynamics, pp. 747–752. University of Missouri-Rolla, USA (1995)
2. Ahmad, S.: Experimental and numerical investigations on vibration screening by in-filled trenches. In: 3rd International Conference on Recent Advances in Geotechnical Earthquake Engineering and Soil Dynamics, pp. 757–762. University of Missouri-Rolla, USA (1995)
3. Conte E., Dente, G.: Screening of rayleigh waves by open trenches. In: 3rd International Conference on Recent Advances in Geotechnical Earthquake Engineering and Soil Dynamics, pp. 763–768. University of Missouri-Rolla, USA (1995)
4. Bencat, J.: Assesment of ground vibration from passing train. In: 3rd International Conference on Recent Advances in Geotechnical Earthquake Engineering and Soil Dynamics, pp. 753–756. University of Missouri-Rolla, USA (1995)
5. Tehranizadeh, M.: Sliding base isolation system for unreinforced buildings. In: 3rd International Conference on Recent Advances in Geotechnical Earthquake Engineering and Soil Dynamics, pp. 743–746. University of Missouri-Rolla, USA (1995)
6. Roesset, J., Huerta, C. Stokoe, K.: Effect of magnitude and type of damping on soil amplification. In: 3rd International Conference on Recent Advances in Geotechnical Earthquake Engineering and Soil Dynamics, pp. 725–732. University of Missouri-Rolla, USA (1995)
7. Jayawardana, W., De Silva, S., De Silva, Y.: Effectiveness of a trench filled with waste material in reducing the propagation of ground vibration induced by soil roller compaction. Engineer **XLIX**(02), 17–24 (2016). The Institution of Engineers, Sri Lanka
8. Orehov, V., Moghanlou, R., et al.: Investigation effects of trench barrier on the reducing energy of surface waves in soils. In: 15th World Conference on Earthquake Engineering, Lisbon (2012). ISBN 978-1-63439-651-6

# The Contribution of Satellite Radar Interferometry for Land Management Activities

Donato Infante[1], Diego Di Martire[2], Domenico Calcaterra[2], and Massimo Ramondini[1](✉)

[1] SINTEMA Engineering srl, Via Toledo 156, 80134 Naples, Italy
ramondin@unina.it
[2] Federico II University of Napoli, Naples, Italy

**Abstract.** Interferometric analysis of repeat pass radar images (InSAR) enables the monitoring and identification of changes in the Earth's surface. This new geodetic technique calculates the interference pattern caused by the difference in phase between two images acquired by a spaceborne synthetic aperture radar at two distinct times. The resulting interferogram is a contour map of the change in distance between the ground and the radar instrument. These data represent a useful tool for land management, from natural hazards mitigation to environmental alterations assessment related to anthropogenic activities.

In the last twenty years many applications have been carried out exploiting satellite technique: in this work, Satellite Radar Interferometry has been successfully applied as a remote-sensing tool to provide information both on spatial and temporal evolution of surface displacements and on interaction with existing structures/infrastructures in urban areas. The following case studies of remote sensing applications have been shown.

Firstly, a complementary approach to conventional techniques to provide integrated systems for a better investigation of ground instabilities, such as landslides, has been developed. To this regard, radar interferometry can offer a valuable contribution to characterize landslide kinematics, allowing to obtain the most precise information on slope failures mechanisms, including boundary, state of activity, triggering factors and temporal evolution.

Subsequently, the potentiality of interferometric products for structural health monitoring of man-made constructions, such as buildings, linear infrastructures (roads and bridges) and dams has shown. Historical DInSAR data have permitted to investigate structural performance of these exposed elements, as indicator of their vulnerability, representing a powerful tool for complementing classic forensic analysis.

**Keywords:** Interferometry · Monitoring · Vulnerability · Man-made constructions


A. Rotaru (Ed.): CRIT-RE-BUILT 2019, SSGG, pp. 156–164, 2021.
https://doi.org/10.1007/978-3-030-61118-7_14

# 1 Introduction

In Italy, mass movements represent an important cause of damage: in particular, the occurrence of ground instability phenomena can cause a serious threat to facilities, structures and infrastructures.

The cost of damage caused by natural events, such as landslides, is increasing globally, due to the intensification of their severity, on one hand, and to the higher exposure of socio-economic properties in prone areas, on the other.

Therefore, strategies for risk reduction are necessary, especially in highly urbanized areas, where buildings and infrastructures may often be destroyed or damaged. Existing buildings are sensitive to ground movements caused by slow-moving landslides that can generate rigid rotations or angular distortions as a result of differential displacements. Evaluation of structural performance, meant as assessment of damage conditions, plays a key role in risk prevention and mitigation activities. In fact, spatial and temporal analysis both of ground deformation and of building damage degree represents a useful tool to understand landslide evolution and structures behavior. This approach can be suitable to evaluate how the elements at risk in a landslide affected area may change with time, contributing to the risk assessment, which is very important for decision making by responsible authorities.

During the last decade, a significant number of researches have been carried out to investigate geotechnical and structural performance of exposed elements by using Synthetic Aperture Radar data [1–3].

Differential Interferometry SAR (DInSAR) technique may offer a viable source of independent information products to support infrastructures and building health assessments, thanks to the easy management of the data. It allows to obtain mean displacement rate maps and time series of deformations of wide areas with sub-centimetric accuracy, by using multitemporal stacks of images acquired at different dates.

The adoption of a new technology proposed for geotechnical asset management and monitoring in fact, requires the evaluation of potential technical benefits that can be obtained versus potential costs of implementing it. Such step is crucial to decide whether it is worth adopting it compared with other conventional alternatives. Traditional sensors based on mechanical and electrical transducers often require the intervention of specialized operators which increase complexity and costs, limiting the frequency of measurements. The feasibility and the cost optimization are the most important goals for the development of an adequate monitoring system. To this purpose, a considerable interest in innovative approaches has grown among the scientific community and land management institutions.

In this paper, we show some applications of DInSAR technique to different case studies. Firstly, the efficacy of remedial works planned in Pisciotta village to reduce the effects induced by the landslides has been investigated by analyzing displacement time-series corresponding to pre- and post-repair intervention configurations of a corner reflectors network. Subsequently, structural performance of a building affected by landslides in Moio della Civitella urban settlement has been assessed through the identification of part which can suffer a greater susceptibility to damage as consequence of deformation gradients. Finally, a geotechnical monitoring of Provincial Road "P.R.

264" in Salerno Province, which is affected along its linear development by several slope instabilities, has been provided.

## 2 Materials and Methods: DInSAR Approach

DInSAR approach is based on the analysis of phase difference in interferometric stacks of radar images: such technique operates at a full spatial resolution and identifies reliable scatterers by measuring their multitemporal coherence related to the phase stability.

Several improvements in the techniques and datasets quality led to more effective, near real-time assessment and response, and a greater ability to constrain dynamically changing physical processes.

Operating X-band Synthetic Aperture Radar sensors of new generation (COSMO-SkyMed satellite), in particular, can provide long historical archives of motion rates and displacement time series with a high spatial resolution (3 × 3 m), reduced revisiting time (4–16 days), non-invasiveness, and at reasonable costs.

SAR images have been processed by Coherent Pixels Technique (CPT) [4], a processing chain able to extract the deformation evolution during a long time-spans from a stack of differential interferograms. The interferograms, defined as the phase difference between two different acquisitions, are generated according to small temporal and spatial separation constraints in order to further reduce decorrelation effects associated with possible coherence losses typically due to topographic distortions, atmospheric patterns, and radiometric distortions.

CPT is divided into the following main steps:

- selection of interferograms, which consists in selecting a quality set of interferograms from all the possible combinations among available images;
- pixel selection, in which pixels that have a minimum phase quality will be selected. The aim of this selection method is to find the well-known permanent scatterers (PS) in the observed scene: a PS is a structure that behaves like an ideal reflector during all the time of study and independently of the point of view from which is observed;
- estimation of the linear term of deformation in order to define linear velocity, DEM error and azimuth position of targets;
- estimation of the non-linear deformation component and atmospheric artifacts to assess deformation evolution of selected pixels;
- geocoding of the results.

As results, mean yearly velocities and cumulative displacement time series provided by DInSAR products are measured along the satellite line of sight (LoS) direction on a discrete number of points (PS).

In the last twenty years many applications have been carried out exploiting DInSAR methods, which proved to be an efficient tool in the monitoring and the investigation of several types of natural hazards. Among the most significant DInSAR applications, the following can be mentioned:

- landslides and subsidence: ground instability monitoring with DInSAR methods gained a lot of attention in recent years by the scientific community, even though the detection of these phenomena is limited to slow and very slow movements [5, 6];
- structures and infrastructures: the monitoring of man-made infrastructure can be also conducted by means of DInSAR methods, for the estimation of the deformations of dams, buildings, roads, bridges, etc. [7, 8].

# 3 Case Studies: DInSAR Monitoring of Ground Instability Phenomena and Exposed Elements

In this work, interferometric products have been used to improve spatial and temporal characterization both of ground and interacting structures deformations, representing a useful tool to understand how instability phenomena kinematics and exposed elements performance may change with time.

## 3.1 Efficacy of Remedial Works in a Landslide-Affected Area in Pisciotta Village

The purpose of the presented application was to show the use of remote sensing data in the framework of remedial works monitoring activities within areas affected by slope instability phenomena. In this case study, we describe the approach adopted to control the efficacy of stabilization interventions carried out on unstable sectors of the Pisciotta village, which is located in the Southern part of Cilento, Vallo di Diano and Alburni National Park (Campania Region, Italy), and it is strongly affected by erosional and gravitational phenomena which cause severe damage to existing structures and infrastructures [9].

Remedial structural and hydraulic works have been designed by performing a back-analysis slope stability test, based on results of geotechnical monitoring in pre-intervention phase (boreholes, equipped with piezometers and inclinometers, laboratory testing, historical remote sensing data): hydraulic drainage system to reduce pore water pressure in the area, while three poles bulkheads, as well as retaining walls, to protect the area. Analysis of slope stability condition in post-intervention phase has been detected by performing a geotechnical monitoring of surface displacements, carried out by the processing of satellite radar images. The integrated assessment of displacement time-series corresponding to pre- and post-repair intervention configurations, allows to investigate the efficacy of remedial works planned to reduce the effects induced by the landslides (see Fig. 1). Sixty-nine high-resolution COSMO-SkyMed descending images spanning the period October 2011–October 2016 have been processed by means of CPT to assess landslide-induced movements in pre-intervention phase.

Displacement rate map showed a movement trend along the Line of Sight (LoS) characterized by velocities up to 15 mm/yr (Fig. 1a). To improve the satellite visibility of the vegetated areas and to detect more targets within the landslide boundaries, a network of 17 corner reflectors has been installed in September 2016 during the intervention phase (Fig. 1b). As to better infer spatial and temporal landslide evolution during and in post-intervention phase, additional 26 descending images have been taken in account in the period October 2016–January 2018. A decrease of "LoS" displacement rate due to the implementation of remedial works, has been observed in the analyzed area (Fig. 1c).

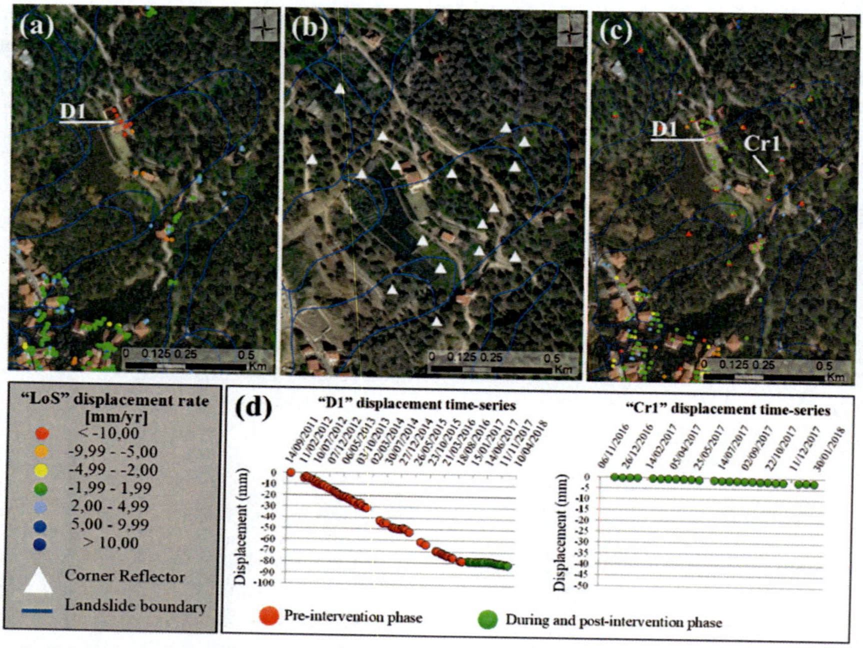


**Fig. 1.** DInSAR displacement rate map in pre-intervention phase (a), position of corner reflectors (b), displacement rate map in post-intervention phase (c) and displacement time-series corresponding to pre- and post- intervention configurations (d).

Displacement temporal evolution of a point (D1), located within the affected areas and identified both in pre- and post-intervention configurations, is characterized by a slowing trend from June-July 2017. Furthermore, the analysis of displacement time-series of target corresponding to corner reflector "Cr1" showed the occurrence of very low movement rates, in the order of some mm/yr, thus highlighting a progressive stabilization of slope instability after the intervention (Fig. 1d).

## 3.2 Structural Behavior of a Building in Moio Della Civitella Urban Settlement

In the following, the capability of DInSAR technique has been tested to investigate structural behavior of a reinforced concrete building in Moio della Civitella urban settlement (Salerno province) (see Fig. 2), whose territory is affected by several slow-moving landslides induced by differences in lithology and in the hydrogeological behavior of involved terrains. The availability of X-band COSMO-SkyMed data has permitted to assess cause-effect relationship between landslide kinematics and building damage susceptibility, thus obtaining a global interpretation of structural performance within a landslide affected areas [8]. Seventy-six ascending and sixty-six descending images acquired in the period 2012–2016 have been processed by means of CPT. Starting from discrete "ascending" and "descending" data (Fig. 2a), firstly, continuous maps of vertical and horizontal movements have been evaluated through an Inverse Distance Weighted (IDW) geostatistical interpolation method (Fig. 2b). Subsequently, the displacement gradient map, meant as the maximum rate of change in displacement value from each cell to its neighbors, has enabled to identify the direction along which maximum vertical and horizontal differential displacements have occurred (Fig. 2c). Such analysis showed a good agreement between the direction of the maximum distortion and real distribution of damage surveyed on building façades (Fig. 2d).

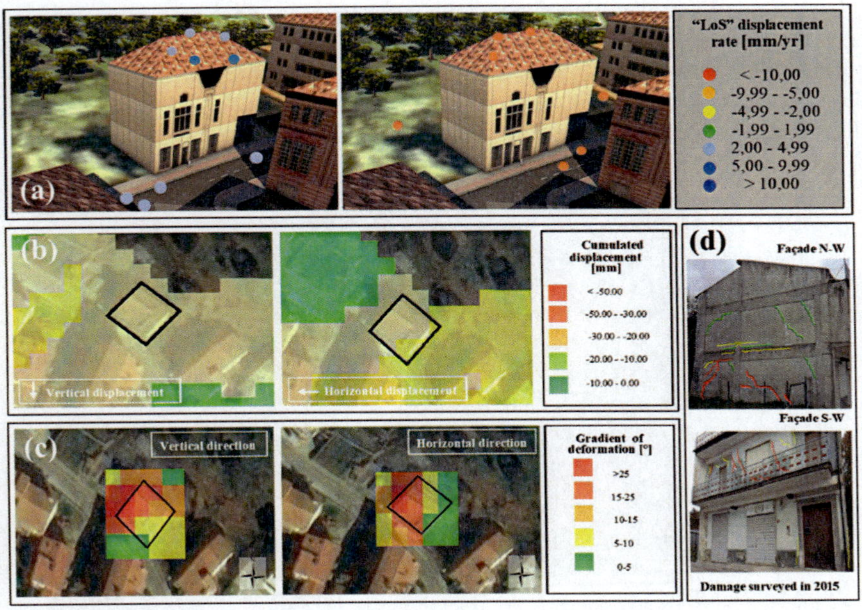


**Fig. 2.** "LoS" displacement velocity map of targets identified by ascending and descending COSMO-SkyMed images (a), cumulated displacement (b) and displacement gradient maps (c) along vertical and horizontal directions, and damage surveyed by in situ campaign (d).

The change in displacement-value describes a condition for which a part of a building foundation is moving at a different rate from other portions and, commonly, buildings affected by differential movements can suffer anisotropic stresses and, as consequence, a greater susceptibility to damage.

### 3.3 Stability Conditions of Provincial Road "P.R. 264" in Salerno Province

The main goal of such analysis was to investigate the capability of spaceborne data in advanced infrastructure monitoring activities, thanks to availability of a very high-resolution radar images characterized also by a shortened revisiting time. The study has been carried out on a part of Provincial Road "P.R. 264", located in Salerno Province, about 15 km long [10, 11]. The lithological and hydrogeological differences of involved terrains are among the main factors predisposing to instability: their lithologic and structural features result in a poor mechanical strength and, thus, in a high ratio between the mobilized and the available shear resistance, even in the case of low-angle slopes. The main product of remote sensing technologies applied for geotechnical monitoring of linear infrastructures is the information regarding stability conditions of pavement surface and surrounding areas. In detail, as a first result of DInSAR processing, historical surface displacement rate and time series of scatterers identified, are obtained.

Fifty-two COSMO-SkyMed ascending images, acquired in time-span January 2012-January 2015, have been processed by means of CPT.

Interferometric products highlighted the occurrence of very slow-moving phenomena, with displacement velocities in the order of some mm/yr and cumulated displacements in the considered time span up to 50 mm. A good correspondence among historical mapped phenomena, satellite measurements and results of field campaign evidences, as indicator of a preliminary cause-effect relationship, has been obtained (see Fig. 3).

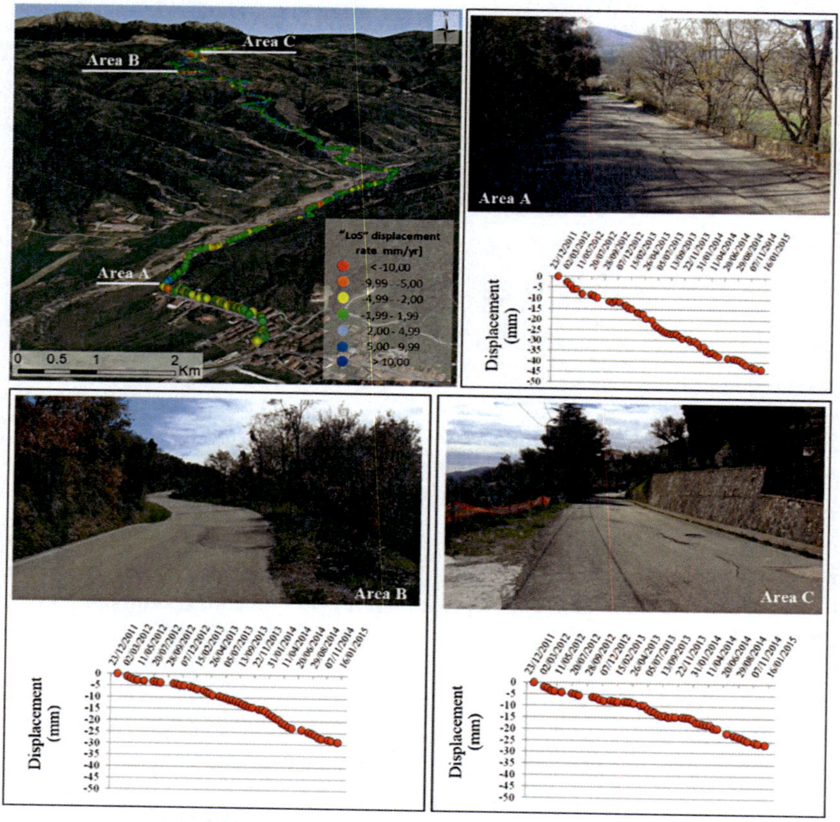


**Fig. 3.** "LoS" displacement rate map of target identified along "P.R. 264" and some examples of correspondence among historical DInSAR displacement time-series and damage occurred.

## 4 Conclusions

Prevention and mitigation of ground instability phenomena require effective technologies to reduce the vulnerability of existing facilities. Differential Interferometry SAR approaches have demonstrated to be powerful investigation tools for their high spatial and multi-temporal coverage, fast data acquisition, and overall reasonable costs.

The results of our study confirm that DInSAR technique can be conveniently used to support risk management activities.

Firstly, interferometric products can provide essential information to analyze the effectiveness of engineering remedial works within areas affected by slope instability phenomena. Furthermore, combined with constructive characteristics of exposed elements, they showed a helpful support for complementing classic forensic analysis aimed to investigate structural behaviour of buildings. Finally, the benefits of DInSAR application can be evaluated also in the context of geotechnical asset management, as an input for monitoring the performance of linear infrastructure, also where no in-situ measurements are available.

## References

1. Bianchini, S., Pratesi, F., Nolesini, T., Casagli, N.: Building deformation assessment by means of persistent scatterer interferometry analysis on a landslide-affected area: the Volterra (Italy) case study. Remote Sens. **7**(a), 4678–4701 (2015)
2. Peduto, D., Elia, F., Montuori, R.: Probabilistic analysis of settlement-induced damage to bridges in the city of Amsterdam (The Netherlands). Transp. Geotech. **14**, 169–182 (2018)
3. Milillo, P., Giardina, G., Perissin, D., Milillo, G., Coletta, A., Terranova, C.: Pre-collapse space geodetic observations of critical infrastructure: the Morandi Bridge, Genoa, Italy. Remote Sens. **11**, 1403 (2019)
4. Mora, O., Mallorquí, J.J., Broquetas, A.: Linear and non-linear terrain deformation maps from a reduced set of interferometric SAR images. IEEE Trans. Geosci. Remote Sens. **41**(10), 2243–2253 (2003)
5. Confuorto, P., Di Martire, D., Infante, D., Novellino, A., Papa, R., Calcaterra, D., Ramondini, M.: Monitoring of remedial works performance on landslide-affected areas through ground- and satellite-based techniques. CATENA **178**, 77–89 (2019)
6. Tessitore, S., Fernández-Merodo, J.A., Herrera, G., Tomás, R., Ramondini, M., Sanabria, M., Duro, J., Mulas, J., Calcaterra, D.: Comparison of water-level, extensometric, DInSAR and simulation data for quantification of subsidence in Murcia City (SE Spain). Hydrogeol. J. **24**(3), 727–747 (2016)
7. Pastor, J.L., et al.: Multi-Source data integration to investigate a deep-seated landslide affecting a bridge. Remote Sens. **11**(16), 1878 (2019)
8. Infante, D., Di Martire, D., Confuorto, P., Tessitore, S., Tòmas, R., Calcaterra, D., Ramodini, M.: Assessment of building behavior in slow-moving landslide-affected areas through DISAR data and structural analysis. Eng. Struct. **199**, 109638 (2019)
9. Infante, D., Di Martire, D., Confuorto, P., Ramondini, M.: Planning and monitoring of mitigation measures in a landslide-affected area through numerical modelling, conventional and satellite data. In: Geotechnical Research for Land Protection and Development, CNRIG 2019, Lecture Notes in Civil Engineering 40, Springer (2020)

10. Infante, D., Di Martire, D., Calcaterra, D., Miele, P., Scotto di Santolo, A., Ramondini, M.: Integrated procedure for monitoring and assessment of linear infrastructures safety (I-Pro MONALISA) affected by slope instability. Appl. Sci. **9**, 5535 (2019)
11. Infante, D., Di Martire, D., Di Santolo, A.S., Calcaterra, D., Ramondini, M.: The contribution of spaceborne SAR data in linear infrastructures monitoring activities. In: Geotechnical Research for Land Protection and Development, CNRIG 2019. Lecture Notes in Civil Engineering 40, Springer (2020)

# Evaluation of the Seismic Response of a Reinforced Concrete Footing to Increasing Peak Ground Acceleration Using Pseudo-dynamic Experimentation

Shane Hossell and Chris Roth(✉)

University of Pretoria, Pretoria, South Africa
chris.roth@up.ac.za

**Abstract.** The pseudo-dynamic experimentation technique was utilised to evaluate the damage occurring in a reinforced concrete footing due to the overall response of a linear elastic two-storey, two-bay moment resisting steel frame structure that is subjected to an earthquake excitation. Five pseudo-dynamic experiments were performed by scaling the El Centro ground motion record, which occurred in California on May 18, 1940, to produce peak ground accelerations that ranged between 0.34 g to 2 g. All the laboratory experiments were undertaken under a constant axial load for the duration of the applied earthquake excitation and utilised Rayleigh damping to model the energy loss within the overall structure. The pseudo-dynamic method provides a reliable method to relate damage suffered by the footing due to the overall structure's response to the applied earthquake excitation. The method enables the structural capacity and failure mechanisms of the reinforced concrete footing to be observed in relation to the seismic demand. The hysteretic response of the footings and energy dissipation characteristics were determined and was shown that the yield strength of the longitudinal reinforcement within the footing has a significant impact on the maximum shear capacity and damage incurred by the footing. The damage is more pronounced with an increase in the number of cycles of vibration, particularly at displacements that exceed the yield strength of the reinforcement. An increase in the hysteretic energy dissipated by the reinforced concrete footing results in a concomitant increase in the observed damage to the footing in the form of concrete cracking, reinforcement yielding and spalling of the concrete.

**Keywords:** Pseudo-dynamic experimentation · Reinforced concrete footing · Seismic performance evaluation · Hysteretic curves

## 1 Introduction

The pseudo-dynamic testing technique is a computer controlled experimental method whereby the dynamic behaviour of a structure is mathematically calculated on the computer and the resultant displacement is statically applied to a physical test specimen


A. Rotaru (Ed.): CRIT-RE-BUILT 2019, SSGG, pp. 165–178, 2021.
https://doi.org/10.1007/978-3-030-61118-7_15

of a portion of the structure at a common degree of freedom between the numerical model and the physical model using servo-controlled actuators in an on-line procedure [1]. The pseudo-dynamic testing technique uses the same equipment as conventional quasi-static tests; however, the analysis is controlled by a closed loop system comprising of computer software that is integrated and runs in tandem with the quasi-static experiment. The method utilises well-established step-by-step time integration methods and utilises the same numerical approach generally undertaken in nonlinear structural dynamics, however, the structural restoring force of apportion of the structure is based on experimental feedback from load cells as opposed to an idealised material model. The force obtained from the load cell is fed back into the computational model, which is used in successive iterations to determine the new deformation. Historically reinforced concrete footings have been experimentally tested using both quasi-static monotonic load tests and cyclic load tests to determine the performance of the structure, however, the experiments are not able to relate the progressive damage incurred to the reinforced concrete footing for the duration of a realistic earthquake excitation. Therefore, the pseudo-dynamic experimental method enables the cumulative damage to be related to earthquake intensity.

Pseudo-dynamic testing originated approximately forty years ago as an alternative to shake table testing with Takanashi et al. [2] seen as one of the pioneers of the testing method [3]. Pseudo-dynamic testing is especially efficient when having to test structures that are too heavy or too large to be practically tested on available shake tables [4].

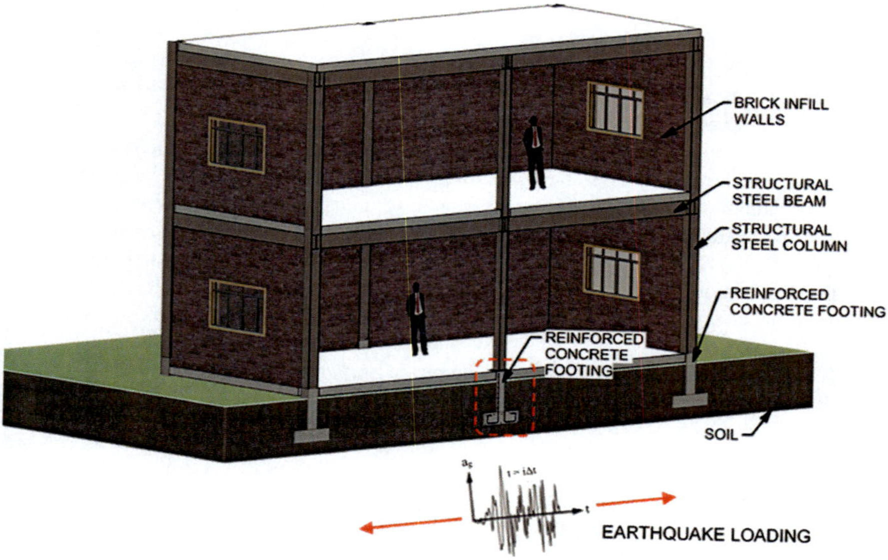


**Fig. 1.** Conceptual model of the overall structure

### 1.1 Objective of the Study

The primary objective of the study was to use the pseudo-dynamic experimental method to relate the resultant structural damage and hysteretic response capacity of a reinforced concrete footing under a constant axial load, that forms part of a two-bay two-storey moment resisting frame structure, to increasing peak ground acceleration (PGA). Figure 1 shows the structure that was considered in this research and the position of the reinforced concrete footing under investigation.

## 2 Application of the Pseudo-Dynamic Method

Figure 2 shows the discretised numerical model used in the pseudo-dynamic analysis. The frame structure was discretised into 28 degrees of freedom with the boundary conditions comprising of two pin supports at the external columns, and with the internal footing's lateral degree of freedom idealised as a single degree of freedom consisting of a constant lumped mass that was supported laterally by a massless non-linear spring of stiffness (). This degree of freedom coupled the calculation cycle with the laboratory test specimen. The horizontal displacement was taken at the top of the column where the steel column of the frame structure connects to the footing using a base plate and holding down bolts.

### 2.1 Test Specimen

The reinforced concrete footing that was used in the research is shown in Fig. 3. The reinforced concrete footing only required minimal reinforcement to resist the applied gravity and wind loads.

### 2.2 Test Setup

A conceptual model of the experimental test setup is shown in Fig. 4. The test setup is comprised of two actuators and two load cells with the vertical actuator being used to simulate the constant axial load applied by the structure on the footing. The horizontal actuator was used to simulate the varying horizontal shear load on the footing due to the applied earthquake excitation that was calculated from the computational model of the frame structure.

The displacement calculated in the computer model was applied directly to the footing using the horizontal actuator with the corresponding force measured using the horizontal load cell. The horizontal force was subsequently fed back into the numerical model at which time the numerical model computation continued.

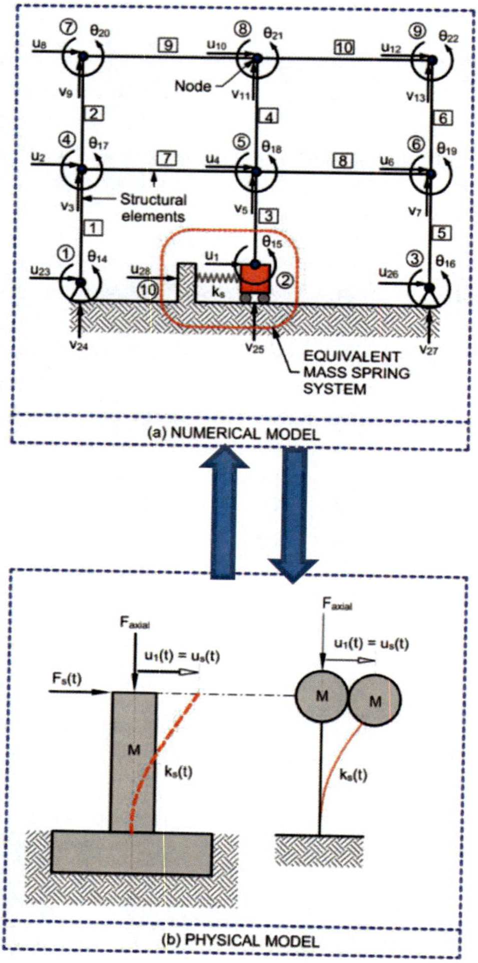


**Fig. 2.** Pseudo-dynamic numerical model and physical model

## 2.3 Earthquake Loading on the Structure

The El Centro, S00E of the event at Imperial Valley, California on May 18, 1940, was selected as the input ground motion for the pseudo-dynamic experiments. For each of the pseudo-dynamic tests, the peak ground acceleration was increased by scaling the amplitude of the El Centro earthquake.

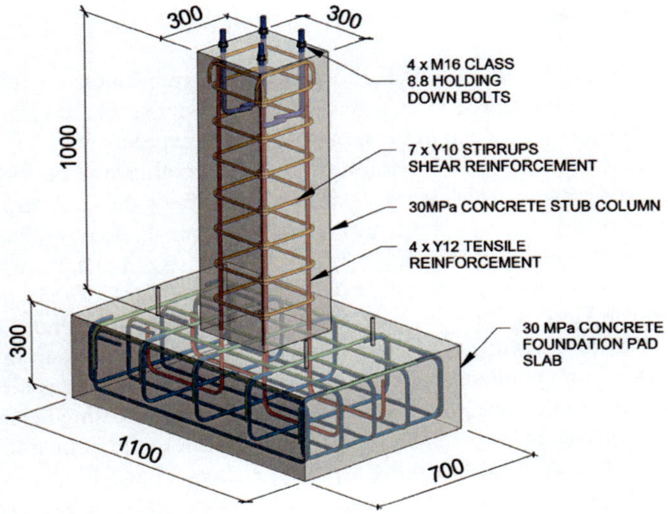


**Fig. 3.** Test specimen

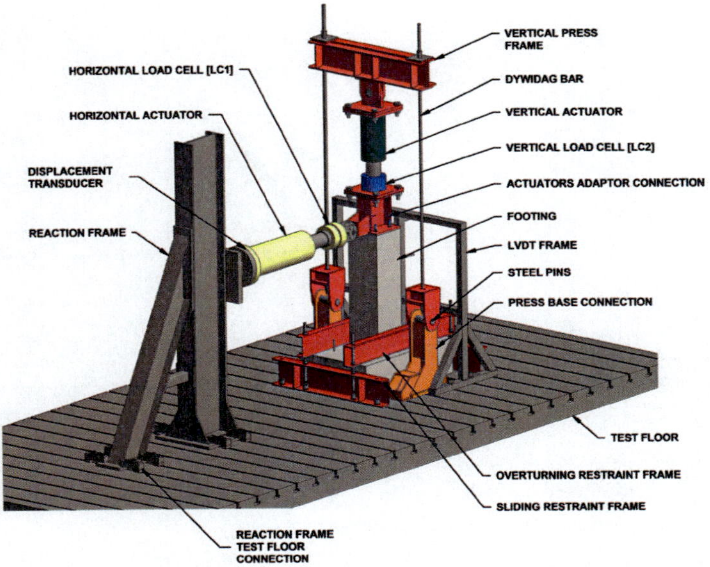


**Fig. 4.** Conceptual experimental test setup

## 3 Results of Pseudo-Dynamic Experiments

The hysteretic behaviour of each of the reinforced concrete footings is provided and the observed damage is discussed in terms of cracking, concrete spalling, reinforcement buckling, and fracturing for each of the pseudo-dynamic experiments.

Pseudo-dynamic analysis of a reinforced concrete footing was performed at five peak ground accelerations, which were obtained by amplifying the El Centro earthquake ground motion record. The hysteretic results and force versus time results for each of the experiments are shown in Figs. 5, 6, 7, 8, 9, 10, 11, 12, 13 and 14, with the peak ground accelerations (PGA) ranging from 0.34 g to 2 g. A PGA of 0.34 g resulted in only minimal cracking, while a PGA of 2 g resulted in complete failure. Figures 15, 16, 17 and 18 show the damage to the reinforced concrete footing during the 0.78 g PGA test. The column continued to carry the axial load until the cross-sectional area had reduced substantially, resulting in the axial capacity of the footing being exceeded. The structure subjected to the 0.78 g PGA test collapsed due to the formation of a plastic hinge and the loss of axial load carrying capacity. Figures 15 and 16 show the buckled and fractured reinforcement due to excessive lateral deformation of the footing during the 0.78 g PGA test, and Figs. 17 and 18 show the formation of the plastic hinge at collapse of the footing.

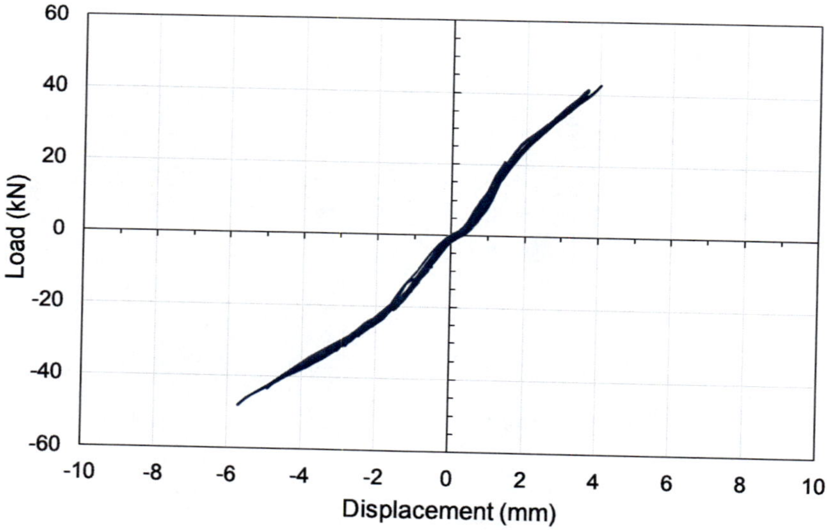


**Fig. 5.** Hysteretic response under El Centro earthquake scaled to 0.34 g

Table 1 summarises the results obtained during the pseudo-dynamic experiments for the following critical points: the minimum cracking force ($F_c$) and cracking deformation ($u_c$), the minimum reinforcement yielding force ($F_c$) and deformation ($u_y$), and the maximum achieved force ($F_m$) and deformation ($u_m$). The following observations were made during the pseudo-dynamic experiments:

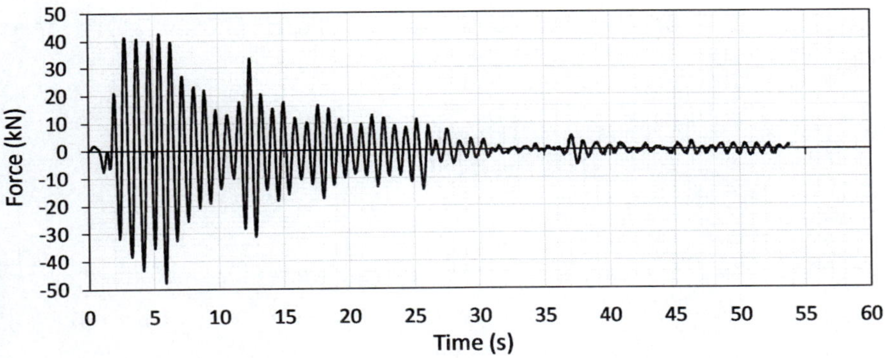


**Fig. 6.** Force vs time graph under El Centro earthquake scaled to 0.34 g

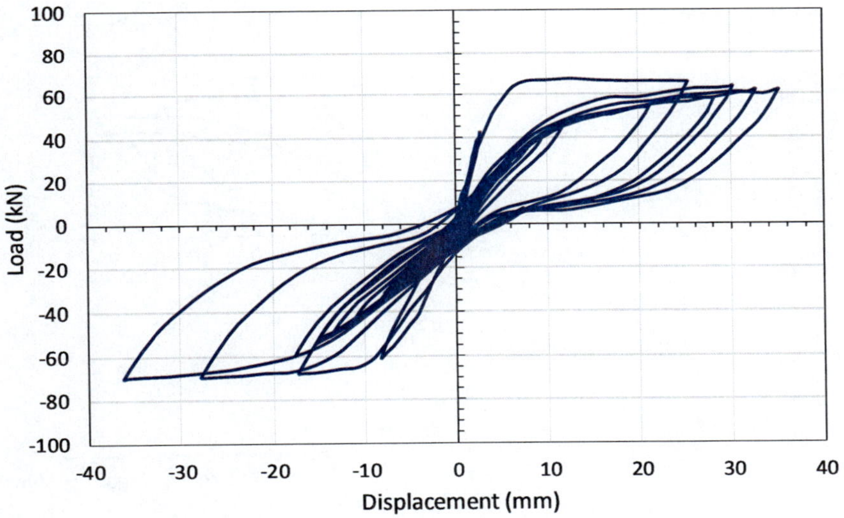


**Fig. 7.** Hysteretic response under El Centro earthquake scaled to 0.68 g

- With increasing deformation, the shear load carrying capacity tends to a constant value with a subsequent reduction in the lateral stiffness of the footing, and therefore, there is a maximum shear load that the structure is capable of transferring between the ground and the structure as deformation increases;
- An increase in the applied lateral deformation applied to the footing results in cracking and yielding of the reinforcement, which in turn results in a reduction in the lateral stiffness of the footing;
- The lateral capacity of the reinforced concrete footing and subsequent damage is predominately controlled by the yield strength and ductility of the reinforcement;
- The ductility of the reinforced concrete footing is controlled by the tensile reinforcement and the shear reinforcement;

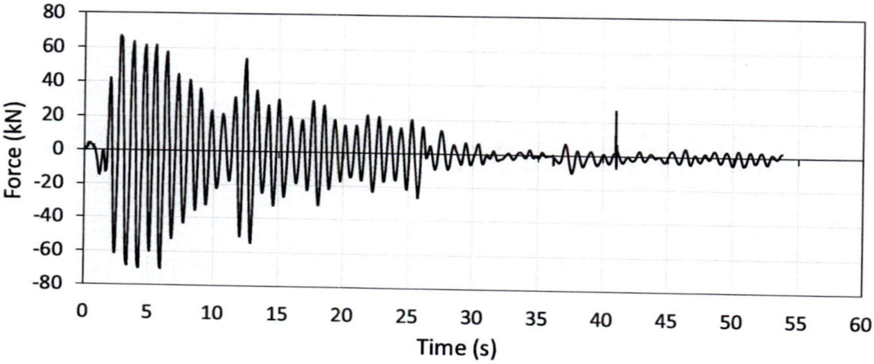


**Fig. 8.** Force vs time graph under El Centro earthquake scaled to 0.68 g

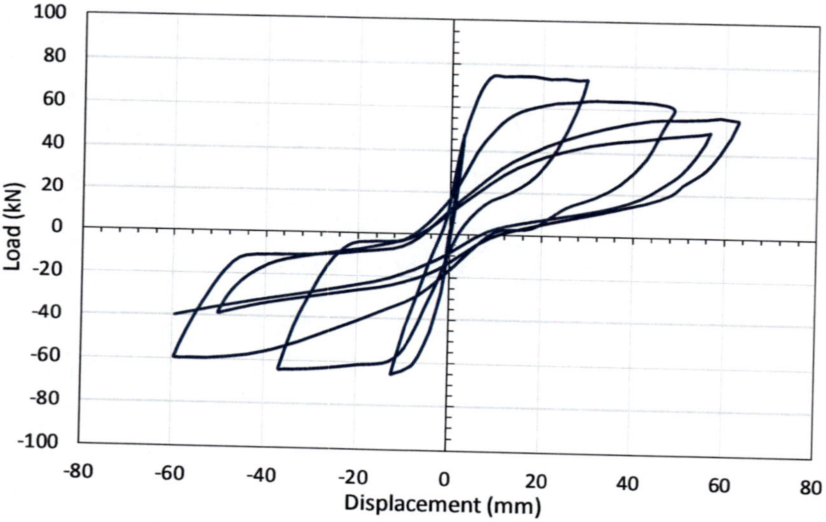


**Fig. 9.** Hysteretic response under El Centro earthquake scaled to 0.78 g

- Cracks occurred near the base of the column where the maximum moment was expected, however, the cracks do not always open at the interface between the base of the column and the top of the footing;
- The number of cycles of vibration increases the damage to the reinforced concrete footing, particularly when the load reverses after the reinforcement has yielded;
- Spalling of concrete occurs due to the buckling of reinforcement during load reversal from tensile loading to compression loading on either face of the concrete column in the direction of loading. This occurs due to the incompatibility between the brittle concrete material and the ductile reinforcement. Upon load reversal, the permanently elongated reinforcement is first mobilised in compression before the crack that has formed in the concrete can close and mobilise in compression. To overcome this

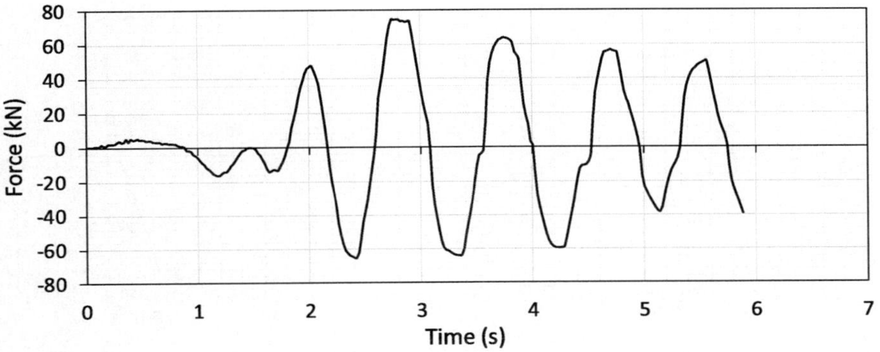


**Fig. 10.** Force vs time graph under El Centro earthquake scaled to 0.78 g

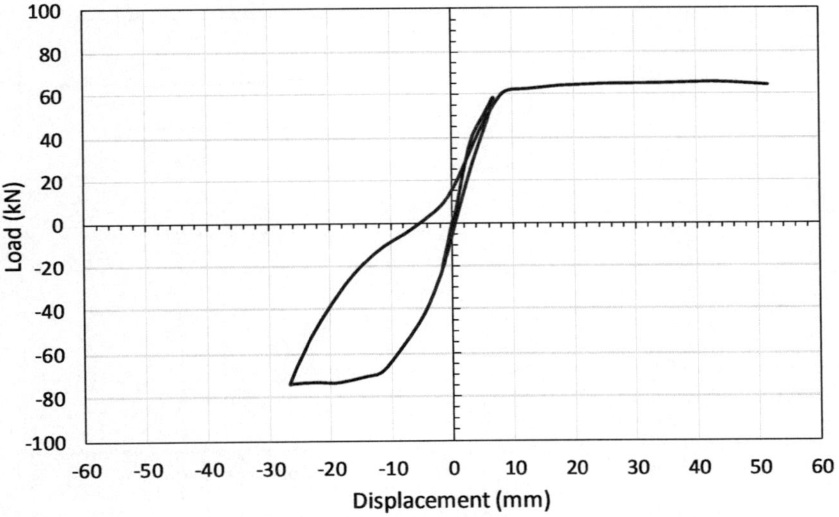


**Fig. 11.** Hysteretic response under El Centro earthquake scaled to 1 g

incompatibility, the reinforcement buckles, resulting in the spalling of the concrete. The spalling of the concrete concomitantly results in a reduction in the gross cross-sectional area of the column, which in turn results in a decrease in the axial and shear capacity of the footing;

- The unloading stiffness from the backbone curve of the footing is greater than the reloading stiffness into the backbone curve, which indicates that the structure is absorbing energy and incurring damage;
- Before the reinforcement yields, the response under cyclic loading remains predominantly perfectly plastic without any significant permanent deformation;

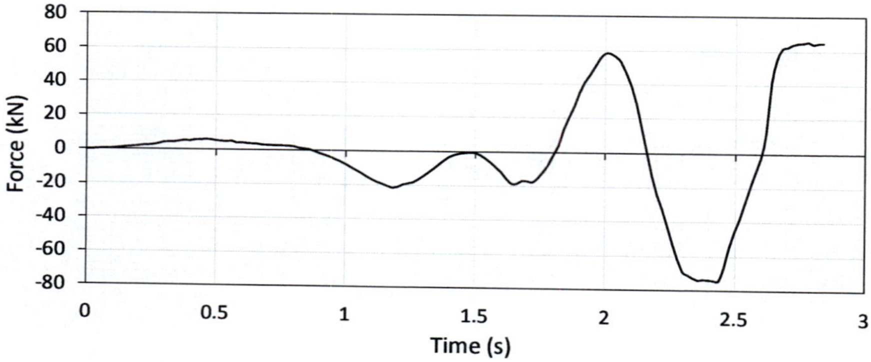


**Fig. 12.** Force vs time graph under El Centro earthquake scaled to 1 g

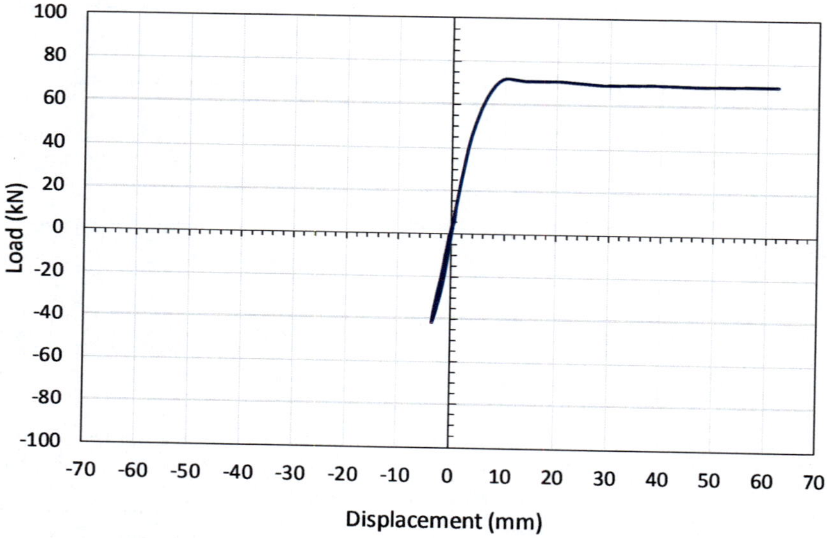


**Fig. 13.** Hysteretic response under El Centro earthquake scaled to 2 g

- The loss of moment capacity due to the formation of a plastic hinge is governed by the repeated cycling of the footing at a displacement less than the failure displacement of the footing and greater than the yielding displacement of the reinforcement; and

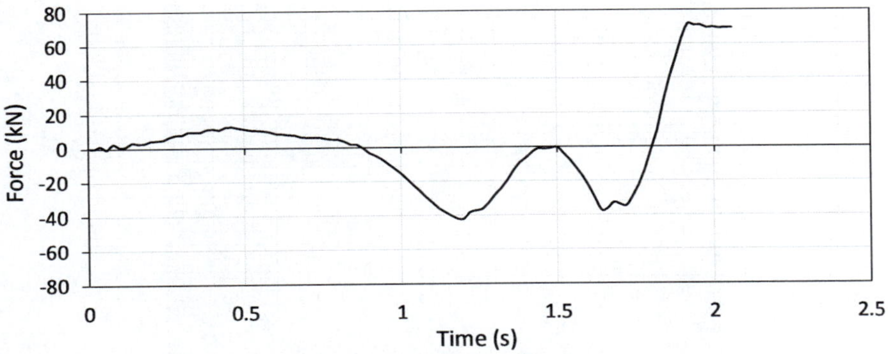


**Fig. 14.** 15 Force vs time graph under El Centro Earthquake scaled to 2 g

**Fig. 15.** Resultant damage to the reinforced concrete footing with outward buckling of the reinforcement

- Increasing the amplitude of the El Centro ground motion record showed that there is a maximum PGA that can be sustained by the footing before failure occurs. For example, the pseudo-dynamic test that was undertaken by amplifying the El Centro ground motion record to a PGA of 2 g only managed to achieve a maximum PGA of 1.21 g before failure occurred.

**Fig. 16.** Resultant damage to the reinforced concrete footing with reinforcement fracturing

**Fig. 17.** Plastic hinge formation at the base of the column at collapse

**Fig. 18.** Plastic hinge formation at the base of the column at collapse

**Table 1.** Results from pseudo-dynamic analysis

| PGA | Cracking | | Yielding | | Maximum | | Damage state |
|---|---|---|---|---|---|---|---|
| [g] | $F_c$ [kN] | $u_c$ [mm] | $F_y$ [kN] | $u_y$ [mm] | $F_m$ [kN] | $u_m$ [mm] | |
| 0.34 | 22 | 1.7 | – | – | 48 | 5.7 | Serviceable |
| 0.68 | 41 | 2.5 | 67 | 6.9 | 70 | 36.3 | Extensive |
| 0.78 | 46 | 2.8 | 75 | 7.2 | 75 | 62.7 | Collapse |
| 1 | 41 | 3.4 | 62 | 7.4 | 74 | 62.7 | Collapse |
| 2 (1.21) | 36 | 2.5 | 72 | 7.8 | 72 | 62.2 | Collapse |

## 4 Conclusions

The pseudo-dynamic experimentation method provides a viable approach to correlate damage to a physical test model of a reinforced concrete footing, which forms part of an overall structure, with increasing earthquake intensity. By using the pseudo-dynamic method, the hysteretic response of the footings could be related to earthquake intensity and observations could be made on the extent of damage incurred by the footing due to the applied earthquake loading. The extent of damage to the reinforced concrete footing is governed predominately by the yield strength of the reinforcement with the rate of damage increasing substantially once the reinforcement had yielded. The research showed that lower PGAs tend to result in the cyclic behaviour of the footing governing the failure of the footing, whereas at large PGAs the fracturing of the reinforcement tends to dominate the failure of the structure.

## References

1. Mosalam, K., White R., Gergely, P.: Seismic evaluation of frames with infill walls using pseudo-dynamic experiments. Technical Report NCEER-97-0020. Cornell University, Ithaca, NY (1997)
2. Takanashi, K., Udagawa, K., Seki, M., Okada, T., Tanaka, H.: Non-linear earthquake response analysis of structures by a computer-actuator on-line system. Earthquake Engineering Research Centre, University of California, Berkeley, CA (1975)
3. Kurt, E.: Investigation of strengthening techniques using pseudo-dynamic testing. Middle East Technical University, Ankara (2010)
4. Thewalt, C., Mahin, S.: Hybrid solution techniques for generalised pseudo dynamic testing. Report No. UCB/EERC-87/09. Earthquake Engineering Research Centre, University of California, Berkeley, CA (1987)

# The Durability of Sustainable Materials and Structures

# A Critical Review on Durability of Sustainable Materials and Structures

Dashnor Hoxha(✉), Naima Belayachi, Xavier Brunetaud, and Sébastien Rémond

University of Orléans, University of Tours, INSA CVL, LaMé EA 7494, France, Polytech Orléans 8 rue Léonard de Vinci, 45072 Orléans Cedex 2, France
dashnor.hoxha@univ-orleans.fr

**Abstract.** This paper presents a critical review of the state-of-the-art on the context of sustainable renovation of buildings and infrastructures. After a short presentation of the method and the context of the study a chronological description is used to evocate the principal fields of research and engineering practices for design of sustainable and durable materials and structures, and for evaluation of the damage state materials and structures. The new trends on the durability of materials and structures developed or reinforced in a context of increasing societal requirements in terms of sustainability are treated along with traditional mechanistic approaches such as mechanics of failure, mechanics of damage, fatigue, environmental impact on the durability including thermal, hydric and chemical interactions. In particular, some technics used for the diagnostic of historical stone buildings are cited as an example of research/engineering practice on the durability of structures and materials. The new trends on sustainable materials design and their industrial use, including the green materials, bio-resourced materials for construction and insulation of buildings, recycled materials from buildings and infrastructures are evocated in conclusion.

**Keywords:** Durability · Life long · Cracking · Damage · Renovation · Environmental interaction · Green materials · Recycling

## 1 Introduction

In the framework of ERASMUS + program KA2 – HIGHER EDUCATION STRATEGIC PARTNERSHIP entitled "Rehabilitation of the Built Environment in the Context of Smart City and Sustainable Development Concepts for Knowledge Transfer and Lifelong Learning" (RE-BUILT), this study aims to give a state of art and an analytical review of existing written resources mapping the knowledge triangle transfer on sustainable rehabilitation of built environment, more specifically in the field of durability of materials and structures. The durability of materials and structures as a research topic is at the same time an old trend and a new dynamic field of technical and scientific research. Nowadays, this old research topic is placed in the framework of sustainable development of society and is seen more and more not as a simple technical issue but also as an economic and environmental one. The main objectives of this work are to:


A. Rotaru (Ed.): CRIT-RE-BUILT 2019, SSGG, pp. 181–199, 2021.
https://doi.org/10.1007/978-3-030-61118-7_16

1. Analyze, describe and establish where necessary the place of durability in the general framework of sustainability
2. Give an updated vision of bases of the durability of materials and structures beyond the mechanical damage and ageing of structures, stability and failure
3. Picturing the new trends on durable and sustainable materials for construction and renovation of infrastructures
4. Presenting new trends and emerging technics on monitoring, evaluation of service state of infrastructures, estimation of the life-time of infrastructures, renovation of infrastructures, preservation and restoration of historical buildings.

The methodology used for this work is a critical analysis of the abundant literature on the durability of construction materials and structures, beginning with general overview articles on each domain. Nowadays, modern approaches place the durability of materials and structures in the framework of a holistic approach to sustainability. Consequently, this review treats the problem in this framework including the sustainability of residential buildings, construction materials, new methods of monitoring and assessment of service state, design of new materials, conservation and restoration of historical buildings and life cycle assessment. The historical review of the durability is used in the first part of this study to explain the dominant material-based works in durability (durability of stones, concretes, steels, woods). While the most of this review is based on the literature analyses, two particular points: bio-sourced new materials and holistic approach in conservation and restoration of stone's historical buildings are treated more in local bases since they constitute some of the principal research themes of the authors of this document. The durability is a research and a technological field with a reach history; to establish a concise document and yet a full critical review (as much as possible), our literature research was concentrated on the scientific literature databases: Scopus, Web of Sciences, ScienceDirect as well as other peer-reviewed publications (conference/congress proceedings, open access referenced journals). In the same time, the normative materials (Eurocodes, Technical documents or guidelines) known and used by construction industry have been consulted in their online version from AFNOR (French Agency for NORmalization https://www.afnor.org/) but the English references are cited. And finally, some worldwide known websites with scientific, technical or social contents were consulted and cited here. Most of the time the English resources are selected and analyzed, even if the original works could be in other languages (few French documents are exceptions of this rule). While many aspects of durability will be discussed in details in this document, the sustainability questions are only present in its first part. A good initial point for sustainability-related literature would be these, always up-to-date, web sites https://sustainabledevelopment.un.org and http://www.fondation-2019.fr that authors suggest to the interested reader.

## 2 Durability Versus Sustainability, Place of Durability in Sustainable Development

While the word "durability" is simple to use, the concept of durability is often more difficult to define [1, 10], and different authors might not mean the same thing when using the same word. Misused sometimes in everyday language as a synonym to the "sustainability" the durability concept is limited to only the longevity of a given product, its ability to remain functional (without excessive maintenance), under normal conditions of use, i.e. the conditions it has been designed for.

When dealing with sustainability two approaches are distinguished: "the weak sustainability" postulates the full substitutability of natural capital whereas the strong conception demonstrates that this substitutability should be severely limited due to the existence of critical elements that natural capital provides for human existence and well-being" [1, 2]. It is clear however that, whatever weak or strong sustainability is adopted, it involves three aspects: economic, social and environmental. It promotes an approach of social development at present such that any of these three dimensions is not compromised for the next generations. Since 1990, Daly [6] defined the criterion for an economic or material flux to be considered as sustainable:

- The consumption rate of renewable resources should not exceed the rate of regeneration of these same resources
- The consumption rate of non-renewable resources should not exceed the rate of developing renewable resources that could substitute these resources
- The rate of pollution issued by the use of any resources should not exceed the capacity of the nature to assimilate this pollution.

This vision of sustainable development implies in particular that the longevity (durability) of materials and structures should match these economic and environmental exigencies. The way this matching could be done, the tools and methods for evaluation of the ability of an infrastructure to satisfy the objectives it is constructed for, the technics to protect or to extend the life of structures, to renovate them are some central topics on durability.

The main topic on durability in Civil Engineering has been, for longtime, design and selection of durable constructive materials (i.e. constructive materials that could assure mechanical stability of civil engineering structures). With time, and because of recent economic challenges, the domain of interest of durability in civil engineering has continuously enlarged and enriched including nowadays various fields such as building physics and hygrothermal behavior of construction materials, durability approach for historical and old buildings, service life prediction methodologies based on big-data analyses. [122, 123]. There is not only a natural extension of the topics covered by the durability but also a shift of the principal domain of research. At least two major societal facts have conditioned this shifting:

Firstly, the increasing pressure from civil society to minimize the anthropic impact on nature. Having long-life materials or structures is not anymore sufficient, the mechanical stability of infrastructure is not anymore the only, neither the principal criterion of limit design approach and life service prediction of structures. Well-being, thermal comfort, air quality, health impact become major criteria for building design in the top of mechanical stability [11, 12]. Consequently, the life of a building is not anymore equivalent to its mechanical life and durability science is now opened to the physics of buildings and complex thermo-hydro-mechanical coupled processes that it is governed by.

Secondly, the massive development of numerical technologies (ongoing numerical revolution) leads to cloud-connected databases and relevant numerical technics for their treatment. In particular, BIM technology and Artificial Intelligence appear today as unavoidable and very powerful tools in analyzing complex situations in diagnostic, monitoring and renovation [13–15]. An important feature on recent developments in the field that has been possible due to the extensive use of numerical tools is the interoperability among technical concepts of durability and other concepts of sustainability: specific tools are developed to give the durability its place in general framework of sustainability including in the assessment of performance of materials and/or structures also the capacity to be recycled and the possibility of resource renovation [5, 14, 16].

## 3 Advanced Durability on the Frame of Sustainability: Brief Historical Review of Fundamentals of Durability and Recent Trends

While the estimation of the duration of the life of structures has always been a question raised from constructors and users of all times from the very beginning of the history of humanity (when the man moves from occasional shelters and caverns towards permanent habitations), The answer has progressively evolved and modernized. During historical times, this answer was given in terms of techniques and construction materials that procured to buildings (castles, religious construction, artworks) and infrastructures (mainly bridges and roads) the longest life possible. Some of the most famous constructions that went through history are here witnesses of human genius to capitalize the experience cumulated by generations even though a scientific response of the whole problem is yet to be found. In the earlier works, the durability of structures (and materials constituting these structures) was posed as a simple question of the lifetime of a damaged structure before its (mechanical) failure. A damaged structure/material in this context was just a structure/material with (mechanical) degraded properties as compared to the properties of a healthy state of this same structure/material [25, 26]. The origins of durability science are closely linked to those of failure mechanics whose bases were founded with studies on crack growth and failure of structures performed in the early 20th century, by Inglis and the theories for crack propagation by Griffith and Irwin [30, 31, 33]. Moving away from classical mechanical design approach considering that the failure of a structure comes as soon as the applied stress is higher than limit stress identified in the laboratory, these works introduced an energetical approach establishing the conditions for which a defect in a material could lead to unstable crack propagation in brittle materials. The stress intensity factor (SIF) and the critical stress intensity ($K_{IC}$, known also as an intrinsic property of a material called toughness or tenacity) become tools for life length estimation and risk failure assessment of structures.

The crack propagation researches on brittle materials were gradually adopted, adapted and extended to other classes of materials (ductile, heterogenous, fibrous, composite materials), in various loading conditions (various modes of crack propagation, modes I, II, and III as well as mixed modes) including, in fine, the effect of the structure (characteristic length of the structure) on the rate of crack growth [31]. The phenomenon of subcritical crack growth, extensively studied from the middle of the last century, with the works of Paris, Atkinson and many other researchers makes it possible to understand and explain the failure of structures under loads lower than critical ones, when these loads are maintained for some periods of times [34, 35]. Fatigue, crack evolution, and failure due to cyclic loadings even under lower stresses as compared to yield stress of materials, were studied almost at the same time than the subcritical crack growth. The well-known theory of fatigue developed by Paris [36] and various other models/theories developed since then have definitely transformed the design approach not only on the mechanical industry but above all, the approach design of civil structures. Remarkably, research works in the field have demonstrated the notable impact of environmental conditions on the rate of crack propagation constituting the bridges with new research domains: thermo poromechanics, multiphysics coupling, design of composite materials, nanomaterials, additives, and so on (see Sect. 4).

The intensive use of man-manufactured materials in constructions such as various metallic materials (steel and aluminium for example just to mention the most widespread among them), concretes, composite materials and so on, brought the question of the durability of materials and structures in the domain of material conception and reuse: how to design the structure and microstructure of material in order to fulfil specific requirements in terms not only of the mechanical stability of structures but also, more and more in terms of comfort, economy, ecological footprint and reuse. In the practice of civil infrastructures and building design the limit design approach includes norms for both Serviceability Limit State (SLS) and Ultimate Limit State (ULS) [37, 42]. It was widely accepted that the long-term behavior of structures is closely related to the time-dependent behavior of materials constituting the structure (they could manifest a brittle or a viscous behavior depending on the micro-mechanisms governing at microstructure level), their interaction with the environment and the conditions of use. For years now, the studies in durability are mostly material-driven: worldwide scientific and technological networks deal with the durability of concrete [39], steel [56], stones [55].

Nowadays, the durability of materials and structures is an everyday-evolution-perimeter field of scientific research and engineering applications with complex interactions to many other scientific and technical disciplines such that it could be, first of all, defined as a multidisciplinary activity. Among modern trends on durability, one could point out the most dynamic ones that shape the nowadays image of scientific and technological research and covers the whole life of a structure from its design to its recycling:

- Design of green, high performant, locally adapted, recycled, bio-sourced, low carbon new materials as a response to the rarefication of raw materials, climate change and social worldwide promotion. From the durability point of view, these materials reveal a double challenge: in one hand, that of technics for the assessment of the durability of these non-conventional materials and, on the other hand, that of long-term compatibility of these materials with traditional materials co-used in construction. [7, 10–12, 19–21, 53, 54]
- Developing of modern technics for continuously, in the field, monitoring of materials and structures state evolution, for durability assessment and maintenance of buildings,

expert systems for structural, thermal, hygrothermal and aeraulic evaluation, numerical technics for restoration and retrofitting including artificial intelligence (AI) [10, 11, 20, 44, 48, 50–52].

- Holistic approaches in design, maintenance, restoration, reuse of materials for buildings and infrastructures. While infrastructures and buildings are seen as essential for mankind, they have an impact and contribute to environmental degradation, including air, water and land pollution, localized health issues and resources depletion [57]. The durability in constructions becomes a part of the global problem that should be resolved as a whole in the framework of sustainable buildings and infrastructures. These approaches express a design philosophy that strives to enhance effective resource efficiency, and reduce negative impacts on human wellbeing and dignity, in a cost-effective manner [57–59].

## 4 Environmental Impact on Durability of Materials and Structures

The concept of loads, used in a normative way in structural design, expresses the (mechanical) interaction of a structure with its environment. Various norms and standards have been established in various parts of the world to harmonize the quantification of this interaction (Eurocodes, ASTM, and so on). Following these approaches, the environmental impact and its evolution in time are established following some consensual guides governing all design processes of structures and eventually their evolution state. For example, wind and/or snow loads are counted in engineering European practice using the guidelines of Eurocodes [37, 42], through so-called characteristic values. However, the impact of the environment on the durability of materials and structures is much more complex and determinant than the mechanical impact given in these standardization codes and guidelines. From a theoretical point of view, poromechanics theory developed since the middle of last century and most of the last decades, gives a solid framework to treat these interactions [26–28, 60, 61]. Without going into details of the theory, some outstanding outcomes and consequences are discussed hereafter. The most distinguished feature of this theory is the mutual impact on the overall response of a material, of its solid matrix, fluid that fills its porous space and the state of each phases constituting the material. The fluids filling the pore space and solid matrix interact with each other mechanically, thermally, chemically, leading to an evolution of the material and structure, eventually to its ageing, and definitely impact its durability.

Within the range of temperatures to which construction materials, buildings and civil engineering infrastructures are exposed (much lower than the fusion temperature of materials), the impact of *thermal conditions* of the environment on the durability is manifested, at least, in three ways:

Firstly, through the strains/stresses generated from the variation of temperature. In the simplest form, the free thermal strains are related linearly to the variation of the temperature through the thermal expansion coefficient [60, 61, 65]. Where the thermal strains are partially or fully obstructed, stresses are generated, whose amplitude depends also on mechanical properties of the material.

- Secondly, in long time when a structure is submitted to fluctuations of temperature due to environmental thermal changes - convection, diffusion, radiation - because of the mechanism of thermal expansion cited before, the local stresses at least up to some depth of the surface, are also susceptible to fluctuate. These fluctuations could lead to the decay and failure of the materials and structures through a mechanism of thermal fatigue [26, 27, 62, 64–66]

- Thirdly, the temperature could significantly impact the creep rate of all materials, especially the construction materials. The most known mechanism through which this impact performs represents the decrease of activation energy of slip lines where the temperature increases (known as Arrhenius law), leading to higher creep rates [41, 66]. In the same way, the temperature increase leads to modification of the rate of critical and subcritical crack growth, impacting directly the life duration of a given structure.

The impact of temperature in the case of porous media (most of the geomaterials used in constructions – stone, concrete, wood, and so on) is a more complex phenomenon, some aspects are treated later on this document.

Hygro - (and more often *hygrothermal*) impact on the durability of materials and structures is known for a very long time. However, the complexity of the problem is such that, nowadays makes the field a domain of intensive research. The impact of nonreactive wetting fluids, saturating the pore space of a material, could be taken into account through the effective stress concept, even if the definition of this concept itself is yet an open question [60, 62–64]. Yet, the mechanisms of hygrothermal impact on the durability of materials and structures are more complexes and multiple:

- When the wetting fluid (most of the time water) fills only part of the pore space, then a capillary meniscus is formed governing the equilibrium of fluid pressure in the pores. Since, in the framework of perfect gases, the capillary pressure (capillary pressure is the difference between the air pressure and fluid pressure) is linked to relative humidity (Kelvin's law), the evolution of the humidity in the environment where the structure is located leads also to the evolution of the "effective stress" (much the same as when the temperature fluctuates). It should be noted that these effective stress fluctuations are produced on the top of all other dead and live loads and could significantly shift the equilibrium point of the structure (equal to the working point for a system that works stably). Then a fatigue mechanism is geared up affecting the durability of the structures [26, 28, 35, 38, 62–64].
- A second way through which hygrothermal state would impact the durability of a material (and the durability of a structure) is the modification of its physical and mechanical properties: a significant decrease of the strength and Young's modulus when the humidity increases is reported by many studies in a great number of porous materials. In some stones, for example, the strength is decreased by 40% in range of in-field humidity variations [45, 62, 64, 65, 69]. The impact of the humidity in the subcritical crack growth rate was also indicated since the very first works [34, 36]
- A third way through which the hygrothermal state of a material and structure would impact its durability is the creation of favorable conditions for chemical and biological activity on/into the structure. In fact, some ranges of humidity are very favorable to the birth and developing of various kinds of dissolution/precipitation mechanisms or

even growth of (micro)organisms and generally biological activity, either in the pores of the material or on its surface [50, 68]

When the question of the impact of environmental conditions on the durability of materials and structures is discussed, the very first factor that gets valuable insight is *the chemical interactions of the structure with its environment*. These factors are activated or not in terms of environmental conditions AND the constituents of the material. Without being exhaustive on this point, some typical chemical interactions that play a crucial role in the durability of materials are shortly commented here:

- Oxidation-reduction reactions: the main concern of these interactions is the oxidation of steel components of structures. The corrosion could dramatically affect the durability of steel, reinforced concrete, pretension cables, mixt structures and so on [26, 56, 71]. The oxidation has at least two consequences on the durability of the structures that suffered because of it: the corrosion products replacing the initial steel have an almost nil strength, so that from some levels of oxidation the stability of the structure is put in question; the products of oxidation have a specific volume higher (from 2 to 3 times) as compared to the volume of the steel itself – such that a spalling of concrete would happen around corroded rod of reinforced concrete structures. The other types of metallic structures (aluminium, stainless steels) are also concerned by this kind of chemical interaction, even if the rate of oxidation is lower than normal steels. While all possible precautions on protection of these structures from corrosion are taken in the design phase of the structures (for example, by coating the elements with non-oxidant material, or by creation of an electric potential preventing oxidation) the technics of assessment of the oxidation state and repairing of these structures are also a topic of perpetual development [46, 56, 71]. In normal conditions, the alkalinity of concrete (pH around 13) prevents the steel from corrosion. However, at lower pH levels, steel attains a high corrosion potential. The development of active corrosion in reinforced concrete results from two mechanisms: chloride diffusion in concrete and carbonation of concrete. The common feature of these two mechanisms is the diffusion of external agents through the pores. Both mechanisms are of great concerns, especially for maritime and off-shore structures.
- The dissolution, mineralogic transformations, and crystallization of new minerals in the pores of materials constituting structures: this physicochemical process is enhanced by the fluid saturating the pores of materials that have a high dissolvable potential (limestones, for example). The anions (positive ions) of dissolved material could combine with cations dissolved in the saturating fluid to eventually create some new minerals. In the case of limestones for example, the Ca++ anions are combined with $SO_4^-$ that could come from the polluted atmosphere enriched with $SO_2$ [55, 61, 70]. Crystallization of new minerals in the one hand modifies the mechanical properties of the structure, and on the other hand, could sometimes be associated with so-called crystallization pressure, thought to be responsible for crack initiation from pores when crystallization takes place.

While thermo-hydro-chemical interactions of the structure with the surrounding environment are cited here separately, in reality, they take place at the same time and in many cases by amplifying the impact of each other. For example, in a saturated porous medium, temperatures below the freezing temperature could lead to cracking of the material, up to its complete failure, by a combination of retraction of material (due to negative variation of the temperature) and crystallization pressure generated by the formation of ice. Likewise, the high temperature could enhance the diffusion in porous media (because of the reduction of the superficial tension of fluids) leading to the acceleration of the oxidation process in concrete as described above. These multiphysics coupled phenomena call for appropriate multiphysics approaches for characterization of their impact [26, 27, 38, 39, 44, 51, 60, 66].

## 5 Stonework Structures Durability: Assessment of State of Historical Buildings, Monitoring, Maintenance and Restoration

Historical buildings (in many cases stonework structures and masonry) are the object of interest not only from cultural heritage and economic point of views but also from the scientific and technical point of views, especially for durability studies [28, 45, 55, 72, 74]. Many of the technics cited hereafter could be used for other types of structures.

The first step of the stonework diagnosis is the state of structure report, which aims at detecting and documenting the state of degradations [68, 70, 75]. Are qualified as degradations all alterations which, from some historical and sociological criteria, are not tolerated and considered to be a damage to the material or its appearance.

So, not all changes in terms of aspect and/or properties imply degradation. The development of a patina, for example, generates a surface hardening, a local decrease in water transport properties, and usually a slight color change. However, these induced changes are rather beneficial to the durability of the masonry and could be considered as roughly uniform for all stone. Hence, patina may not be considered as degradation although it is actually an alteration. Likewise, alterations that are exclusively aesthetic, such as soiling or certain biological colonization, may also not be considered as degradations. The state report shall denominate these alterations and their extent, so as to detect a progression by comparison between two different temporal states. For this comparison to be reliable, it is essential that each report hold the same rigor. Various technics for characterization and quantification of degradations could be used the most frequently used in practice are briefly commented here.

### 5.1 Visual Observations of Degradations

This inspection is currently unavoidable because the criteria for the classification of stone degradations, documented and classified in an illustrated glossary ICOMOS [55], are based exclusively on the appearance of the surface of the stone. This inspection requires a close to surface view, and observations from several angles which could be an operational problem for masonry in elevation, above moats, or masks. Nowadays, the use of drones makes it easier to observe hard-to-access facings at a low cost. Still, several overflights should be undertaken not only to detect degradations through photos but also to have photos from various angles in order to specify the nature and intensity of those degradations.

## 5.2 Photographic Survey

The photographic survey remains today the most common and the simplest solution be applied to document the state of degradation of masonry facings. In order to produce metric mappings, i.e. on which reliable quantitative measurements can be made, the mapping medium must be free from any distortion or perspective effects. Such a mapping medium cannot, therefore, be simple photos and should be treated using various technics such as for 3D models or ortho projections [70, 87].

Ortho projections are produced by projecting a 3D model onto a primitive surface such as a reference plane or cylindrical surface. They, therefore, correspond to an orthographic view (to be opposed to a perspective view), that is to say, that every point of the resulting image comes from a ray perpendicular to the observed surface, much the same as architect plans or elevations, but it reflects the appearance of the surface as a photograph or a laser scanner can appreciate it. It can be realized by terrestrial or aerial photography, or a laser scanning survey, or a combination of both, generally supplemented by a topographic survey and geolocate data and provides a scale if no laser survey is performed. The resulting ortho projection can then serve as cartographic support for informing the health status of the masonry walls. The use of primitive surfaces to store conservation data into a collaborative interface represents the first attempt to create an HBIM (Heritage Building Information Modeling) [85, 86].

Since ortho projection does not allow to observe the 3rd dimension and because some degradations could evolve in-depth at the same time or independently to their extension on the surface, it may be relevant to store and use locally the raw 3D model obtained from laser scanning or the one used to perform orthophotography to spatially compare different temporal states.

## 5.3 Non-destructive Characterization

The only observation of the external aspect of a masonry facing is not enough to ensure the reliability of a health diagnosis, and characterization of structures state by nondestructive technics is nowadays a common practice [77], [78]. These technics complete the purely external visual control by quantifying mechanical and physical properties of the subsurface. They allow in particular to appreciate the extent of desquamation beyond its limits visible to the naked eye.

These technics could be as simple as the “knock test”, consisting to evaluate the state of stone using the different sounds produced when ones knock to its surface, to more sophisticated ones [82].

Among the more inexpensive technics, the application of which requires direct access to the surface tested, are the electrical resistivity profiling and the hardness test. The electrical surface resistivity measurement by tip is generally used in a comparative manner within the same material to differentiate the wet areas from dry ones [83, 84]. The hardness tester estimates the local hardness of stone and makes it possible to evaluate its state relative to other measured points [79]. If an analogue material calibration in the laboratory has been previously performed, this test can become quantitative. Compared to the concrete hardness test, the lower rebound energy allows it to be used on softer materials but also limits its study area to a much lower depth. Although considered as non-destructive, the test may, to some extents, damage the surface appearance. Because stones and mortars generally exhibit a decrease in hardness as moisture increases, variations in electrical resistivity and hardness measurements as a function of moisture changes may exhibit some degree of correlation.

Portable spectrometry analyzes powders from a stone placed on a dedicated strip to identify and quantify certain chemical elements [82]. Nowadays, the test could be performed using a smartphone equipped with the *ad hoc* accessory. Although the collection of the powder to be tested can be considered as destructive, the small quantities of powder required for a test make it possible to put into perspective and often to classify this technique among the non-destructive control.

Karsten's tube measures the water flow time through the surface of a tested stone [81]. This measurement, which requires direct access to the surface tested, is therefore well correlated with the transport properties of the stone.

The microdriller, which consists in measuring the drill resistance of a drill according to the depth, makes it possible to generate a resistance profile of the tested material in situ. Thus, this drilling test is to be placed in the control rather nondestructive than really nondestructive. However, the hole generated by the drill (diameter between 3 and 5 mm and max length 5 cm) is small enough to be plugged by a thin mortar which must allow to recover the initial surface appearance. This test makes it possible to reveal induration profiles [80] or, conversely, damaged areas, provided that the piercing device can be perfectly maintained and produce noise-free measurments.

Infrared thermography (IR) is based on the measurement of infrared radiation emitted by an observed surface. This radiation depends mainly on the temperature of the object, which allows this technique to generate a temperature map of the observed area. It is then possible, using obtained temperature map, to highlight areas of thermal heterogeneity such as thermal bridges, or damaged parts of a structure that for some reasons manifest a thermal difference [76]. For the same environmental conditions, stones of different thermal properties also have distinct surface temperatures so the IR thermography could help to distinguish stones of different nature on a façade.

## 6 Sustainable Materials for Construction and Renovation of Infrastructures and Buildings

The renovation of buildings and infrastructures is a process conditioned by many technic and socio-economic factors. The choice of materials to be used in the restauration of historical buildings is mostly guided by historical, ethical and cultural considerations. However, in the renovation of urban buildings, this choice, certainly constrained by socio-economic considerations, is strongly impacted by technical considerations. As an example, all urban renovations today are subject of strict requests on energetical performance increase as well as on grey energy reduction [87]. In France, a third of the housing stock is concerned by this thermal renovation [88]. One of the objectives of the Energy Transition Act is to renovate 500,000 homes per year by 2050. Consequently, the regulations provide a framework for improving energy efficiency, determine the performance to be respected during insulation work, and encourage the use of materials and equipment that respect the comfort and health of the occupant.

In another aspect, the construction and renovation of public infrastructures is also concerned by the smart choice of materials since this sector, a big consumer of raw materials already, will increase its consumption of materials to 100 billion tonnes by 2030 [90]. Meanwhile, the access to non-renewable raw materials has become increasingly limited, calling for urgent solutions and new alternatives. Among the best sustainable alternatives, recycling of industrial waste and by-products (slag, foundry sand, etc.), materials from deconstruction in the building sector, or agricultural co-products, are already developed worldwide to various extents.

In France, for example, a "Low Carbon Building Renovation" label was introduced in the first half of 2018 [91] aiming to reduce CO2 emissions by half compared to the original building taking into account the carbon storage capacities of constructions [80].

Consequently, there has been growing interest in the development of eco-materials for the energy rehabilitation of old buildings [21]. Numerous works in the literature have been performed on materials with low environmental impact by using different vegetal aggregates and binders [8, 19, 20, 93–95]. Various types of natural fibres are used, depending on the biomass production and climate of regions and countries in the world. An increasing range of plant by-products properties are being studied in order to obtain a wide variety of aggregates for vegetal concrete purposes. Reviews of the availability of vegetal aggregates have been reported in countries such as Spain [96], France [97], UK [98] and more globally in the world [99], as creating local industries with materials made with crude earth raw earth or crude oil and vegetal aggregates is an objective in many different parts of the world [100, 101]. Other advanced research has shown that it is possible to produce high performance natural fiber composite based on clay matrix [102], cement [103, 104] and lime or polymer matrix capable of meeting any engineering demand in terms of strength and energy absorption capability [106, 107]. The choice of the binder type in this application is of great importance and complexity. The binder must comply with an easy mixing, a perfect coating of fibers, a correct molding and a final composite with good mechanical properties. A number of authors agree that hydraulic binder can produce unacceptable results in the case of hemp-based composite [108, 109]. However, other results show that mixtures with higher ratio of cement have higher compressive strength [110]. Several investigations have been performed to study the effect of humidity and temperature on the one hand on thermal conductivity [112] and on the other hand on equilibrium moisture content [113] of coating plaster used for straw bales construction. The results indicate that the thermal conductivity of the coating plaster decreases with increasing fibre content.

In the context of building insulation, other researchers have been interested in evaluating the developed materials with in-situ tests under environmental climatic changes in order to ensure effective implementation in the real case [113–116], or by using accelerated aging conditions in the laboratory [11, 104, 118, 119]. It is also important to study the implementation of rehabilitation that makes possible the preservation of the building identity and its history [7, 120, 121].

Despite the growing interest in this type of materials and the recognition of their thermal, water and comfort performance [21], their use remains hampered by the absence of a normative framework and implementation rules specific to these materials to move towards a control of their use and their durability. For some materials/technics (hemp block for example) several technical recommendations [22, 23], implementation guides and technical data sheets are already in use in the artisanal and industrial practice. The situation is however more disparate for other bio-resourced materials with a level of integration quite different for one material to another or even from one country to another.

## 7 Conclusion

The durability of materials and structures is still nowadays a dynamic field of scientific and technical research and of permanent industrial innovation. The sustainability as a societal approach has however greatly modified and extended the principal fields of application of these researches. In the design of new materials and structures or in renovation of existing ones the criteria of evaluation of durability of materials end structures are not any more exclusively mechanical. While traditional topics such as fracture mechanics, time dependent behavior, interaction of materials and structures with environmental factors such as temperature, humidity, chemical reactive agents, are in continuous development and take advantages from numerical progress, new trends occupied nowadays the center of durability research and technological progress: design of new green sustainable materials, multicriteria assessment of their durability, reducing of raw materials in use and replacing with recycled and renewable ones. The technics for assessment of state of structures evolve too and include along with traditional technics of diagnostics new nondestructive technics that are called to be reinforced and developed in the future.

## References

1. SDKP: Sustainable Development Knowledge Platform (2019). https://sustainabledevelopment.un.org/. Accessed 01 Oct 2019
2. Pelenc, J., Deduerware, T.: Weak Sustainability versus Strong Sustainability (2015). https://doi.org/10.13140/rg.2.1.3265.2009
3. European Commission: Life cycle indicators for resources, products and waste. Report by the *Joint Research Centre of the Europen Commiss*ion. Luxembourg: Publications Office of the European Union (2012)
4. Funtas, G., Polette, M., Rye, T., Tischer, V.: Environmental and economic assessment of traffic-related air pollution using aggregate spatial information: a case study of Balneário Camboriú, Brazil. J. Transp. Health **14**, 100592 (2019)
5. Leonard, A., Limbourg, S., Merchan, A.L., Mostert, M.: Life cycle externalities versus external costs: the case of inland freight transport in Belgium. Transp. Res. Part D: Transp. Environ. **67**, 576–595 (2019)
6. Daly, H.E.: Toward some operational principles of sustainable development. Ecol. Econ. **2**(1), 1–6 (1990). https://doi.org/10.1016/0921-8009(90)90010-R
7. Belayachi, N., Boulnois, J., Hoxha, D.: Réhabilitation thermique d'une maison d'habitation en utilisant le biocomposite béton-paille. 35$^{\text{ème}}$ Rencontres Universitaires de Génie Civil, 22–24 Mai 2017, Nantes (2017)
8. Broard, Y., Belayachi, N., Hoxha, D., Ranganathan, N., Méo, N.: Mechanical and hygrothermal behaviour of clay-sunflower (Helianthus annuus) and rape straw (Brassica napus) plaster bio-composites for building insulation. Constr. Build. Mater. **161**, 196–207 (2018)

9. Siegesmund, S., Snethlage, R. (eds.): Stone in Architecture, 4th edn. Springer, Heidelberg (2014). https://doi.org/10.1007/978-3-642-14475-2_1
10. Pacheco-Torgal, F., Lourenço, … P. Chindaprasir, Eco-efficient Masonry Bricks and Blocks, Design, Properties and Durability. Woodhead Publishing (2015). ISBN 978-1-78242-305-8, 548 p. https://doi.org/10.1016/C2014-0-02158-2
11. Belayachi, N., Hoxha, D., Slaimia, M.: Impact of accelerated climatic aging on the behaviour of gypsum plaster-straw material for building thermal insulation. Constr. Build. Mater. **125**, 912–918 (2016)
12. Pacheco F., Jalali T. S. (2016) Ecoefficient construction and building materials. Spring Edition, ISBN 978-0-85729-892-8, https://doi.org/10.1007/978-0-85729-892-8
13. Joblot, L.: Contribution à la mise en œuvre du BIM en rénovation: proposition d'un Modèle de Maturité BIM spécifique, Ph.D. Ecole Nationale des Arts et Metiers ENSAM (2018). (in french)
14. Volk, R., Stengel, J., Schultmann, F.: Building Information Modeling (BIM) for existing buildings: literature review and future needs. Autom. Constr. **38**, 109–127 (2014)
15. https://www.ademe.fr/sites/default/files/assets/documents/fiche-ravalement-refection-toiture-amenagement-travaux-isolation.pdf
16. Pavanelli, D.D., Voulvoulis, N.: Habitat Equivalency Analysis, a framework for forensic cost evaluation of environmental damage. Ecosyst. Serv. **38**, 100953 (2019). https://doi.org/10.1016/j.ecoser.2019.100953
17. He, G., Li, J., Tam, V.W.Y., Wang, G., Zuo, J.: Stakeholders' willingness to pay for the new construction and demolition waste landfill charge scheme in Shenzhen: a contingent valuation approach. Sustain. Cities Soc. **52**, 101663 (2019). https://doi.org/10.1016/j.scs.2019.101663
18. Bouasker, M., Belayachi, N., Hoxha, D., Al-Mukhtar, M.: Physical characterization of natural straw fibers as aggregates for construction materials applications. Materials **7**(4), 3034–3048 (2014)
19. Belayachi, N., Bouasker, M., Hoxha, D., Al-Mukhtar, M.: Thermo-mechanical behaviour of an innovant straw lime composite for thermal insulation applications. Appl. Mech. Mater. **390**, 542–546 (2013)
20. Ismail, B., Belayachi, N., Hoxha, D.: Optimisation de la performance thermique d'un bio-composite à base de fibres végétales: étude expérimentale et numérique. Conférence Internationale Francophone NoMaD 2018, Liège Université 7–8 Novembre 2018 (2018)
21. Amziane, S., Arnaud, L.: Les bétons de granulats d'origine végétale: application au béton de chanvre. Edition Lavoisier, Paris (2013). ISBN 978-2-7462-3809-1
22. https://www.vicat.fr/content/download/91008/838050/file/CP_BIOSYS.PDF
23. https://www.biofib.com/files/BIOFIB_TRIO-Avis_technique_CSTB_Murs.pdf
24. Lemaitre, J., Chaboche, J.L.: Mécanique des matériaux solides. Dunod (1996)
25. Grasberger, S., Meschke, G.: Thermo-hygro-mechanical degradation of concrete: From coupled 3D material modelling to durability oriented multifield structural analyses. Mater. Struct. Concrete Sci. Eng. **37**, 244–256 (2004)
26. Zhao, C., Hobbs, B.E., Hornby, P., Ord, A., Peng, S., Liu, L.: Theoretical and numerical analyses of chemical-dissolution front instability in fluid-saturated porous rocks. Int. J. Numer. Anal. Meth. Geomech. **32**, 1107–1130 (2008). https://doi.org/10.1002/nag.661
27. Eslami, J., Hoxha, D., Grgic, D.: Estimation of the damage of a porous limestone using continuous wave velocity measurements during uniaxial creep tests. Mech. Mater. **49**, 51–65 (2012)
28. Michel, L., Do, D.P., Coignard, B., Hoxha, D.: On the numerical evaluation of historical stone sculpture artwork restoration. Eur. J. Environ. Civ. Eng. (2014). https://doi.org/10.1080/19648189.2014.891471

29. Bieniawski, Z.T.: Stability concept of brittle fracture propagation in rock. Eng. Geol. **2**(3), 149–162 (1967). https://doi.org/10.1016/0013-7952(67)90014-2
30. Sun, C.T., Jin, Z.H.: Fracture Mechanics, Academic Press (2012). Elsevier, ISBN 978-0-12-385001-0, 336 pg. https://doi.org/10.1016/C2009-0-63512-1
31. Irwin, G.R.: Fracture Dynamics. Fracturing of Metals, American Society for Metals, Cleveland (1948)
32. Griffith, A.A.: Philosophical Transactions. Series A **221**, 163–198 (1920)
33. Atkinson, B.: Subcritical crack growth in geological materials. J. Geophys. Res. **89**(B6), 4077–4114 (1984)
34. Stavros, K.K. (ed.): Fracture and Failure of Natural Building Stones: Applications in the Restoration of Ancient Monuments, Springer Science & Business Media (2007), 592 pg
35. Paris, Paul, C., Erdogan, F.: A critical analysis of crack propagation laws. J. Basic Eng. Trans. Am. Soc. Mech. Eng. 528–534 (1963)
36. NF EN 1990 Eurocode– Basis of structural Design – Annex A1, for Buildings
37. Yao, Y., Wang, L,. Wittmann, F., De Belie, N., Schlangen, E., Gehlen, C., Wang, Z, Eguez, H., Cao, Y., Yunus, B., Li, J.: Recommendation of RILEM TC 246-TDC: test methods to determine durability of concrete under combined environmental actions and mechanical load. Mater. Struct. **50**, 155 (2017). https://doi.org/10.1617/s11527-017-1000-3
38. RILEM Technical Committee: Recommendation of RILEM TC 212-ACD: acoustic emission and related NDE techniques for crack detection and damage evaluation in concrete. Mater. Struct. 43(9), 1187–1189 (2010). 10.1617/s11527-010-9640-6
39. RILEM TC 177-MDT: Test method recommendations of RILEM TC 177-MDT 'Masonry durability and on-site testing' -D.5: In-situ stress - strain behaviour tests based on the flat jack. Mater. Struct. **37**(271), 497–501 (2004)
40. TC 129-MHT:Test methods for mechanical properties of concrete at high temperatures Recommendations: Part 7: Transient Creep for service and accident conditions. Mater. Struct. **31**(209), 290–295 (1998)
41. EN 1991-2:2003 Eurocode 1: Actions on structures -Part 2: Traffic loads on bridges
42. Katpady, D.N., Hazehara, H., Soeda, M. et al.: Int. J. Concr. Struct. Mater. **12**, 30 (2018). https://doi.org/10.1186/s40069-018-0260-9
43. Zaharieva, R., Buyle-Bodin, F., Skoczylas, F., Wirquin, E.: Assessment of the surface permeation properties of recycled aggregate concrete. Cement Concr. Compos. **25**, 223–232 (2003)
44. Molina, E.G., Sebastián, E., Alonso, F.J.: Evaluation of stone durability using a combination of ultrasound, mechanical and accelerated aging tests. J. Geophys. Eng. **10**, 035003 (2013). https://doi.org/10.1088/1742-2132/10/3/035003
45. Cheng, Y.C., Guo, H.B., Wang, X.Q., Jiao, Y.B.: Durability assessment of reinforced concrete bridge based on fuzzy neural networks. Adv. Mater. Res. **838–841**, 1069–1072 (2014)
46. EL-Bashir, S., Althumairi, N., Alzayed, N.: Durability and mechanical performance of PMMA/Stone sludge nanocomposites for acrylic solid surface applications. Polymers (Basel), **9**(11), 604 (2017). https://doi.org/10.3390/polym9110604
47. Purcell, C.E.: Deep Energy Renovation of Traditional Buildings: Addressing Knowledge Gaps and Skills Training in Ireland (2018). https://www.heritagecouncil.ie/
48. DIRECTIVE 2010/31/EU: European Council Directive of the European Parliament and of the Council on the Energy Performance of Buildings (Recast) PE-CONS 15/10, ENER 131 ENV 255, CODEC 382 (2010)
49. EN 16883:2017: Conservation of cultural heritage - Guidelines for improving the energy performance of historic buildings (2017)
50. EN 15026:2007: Hygrothermal Performance of Building Components and Building Elements -Assessment of Moisture Transfer by Numerical Assessment (2007)

51. EN 15978:2011: Sustainability of Construction Works -Assessment of Environmental Performance of Buildings -Calculation Method (2011)
52. Lee, J., Mahendra, S., Alvarez, P.J.: Nanomaterials in the construction industry: a review of their applications and environmental health and safety considerations. ACS Nano. **4**(7), 3580–3590 (2011). https://doi.org/10.1021/nn100866w
53. Hincapiéa, I., Caballero-Guzmana, A., Hiltbrunnerb, D., Nowack, B.: Use of engineered nanomaterials in the construction industry with specific emphasis on paints and their flows in construction and demolition waste in Switzerland. Waste Manag **43**, 398–406 (2015)
54. ICOMOS, International Council for Monuments and Sites https://www.icomos.org/en
55. SCI Steel Construction Institut https://www.steel-sci.com/
56. Janjua, S.Y., Sarker, P.K., Biswas, W.K.: A review of residential buildings' sustainability performance using a life cycle assessment approach. J. Sustain. Res. **1**, e190006 (2019). https://doi.org/10.20900/jsr20190006
57. Interaction Design Foundation: https://www.interaction-design.org/
58. Marjaba, G.E., Chidiac, S.E.: Sustainability and resiliency metrics for buildings—critical review. Build. Environ. **101**, 116–125 (2016)
59. Coussy, O.: Poromechanics. Wiley (2004). ISBN 0-470-84920-7, 312 pp
60. Dormieux, L., Kondo, D., Ulm, F.: Microporomechanics. Wiley Blackwell (2006). ISBN 9780470031889, 344 pg
61. Hoxha, D., Homand, F., Auvray, C.: Deformation on natural gypsum rock: mechanisms and questions. Eng. Geol. **86**(1), 1–17 (2006)
62. Hoxha, D., Homand, F., Giraud, A., Auvray, C.: Saturated and unsaturated behaviour modelling of Meuse-Haute/Marne argillite. Int. J. Plast **23**, 733–766 (2007)
63. Auvray, C., Homand, F., Hoxha, D.: The influence of relative humidity on the rate of convergence in an underground gypsum mine, Int. J. Rock Mech. Min. Sci. **45**(8), 1454–1468 (2008)
64. Belayachi, N., Do, D.P., Hoxha, D.: Thermo-hydro-mechanical behavior of a tuffeau stone masonry. Eur. J. Environ. Civ. Eng. **16**(5), 557–570 (2012)
65. Hoxha, D., Belayachi, N., Do, D.P.: On Thermo-Hydro-Mechanical (THM) fatigue damage of historical stone buildings. Adv. Mater. Res. **891**, 36–41 (2014)
66. Belayachi, N., Hoxha, D.: Damage of historical stone masonry buildings: combined effects of spatial variability of stone properties and environmental condition. J. Civ. Eng. Archit. **10**, 743–754 (2016)
67. Janvier-Badosa, S., Beck, K., Brunetaud, X., Al-Mukhtar, M.: Historical study of chambord castle: basis for establishing the monument health record. Int. J. Archit. Heritage **7**(3), 247–260 (2013). https://doi.org/10.1080/15583058.2011.634959
68. Al-Omari, A., Brunetaud, X., Beck, K., Al-Mukhtar, M., Street, M.: Hygrothermal stress and damage risk in the stones of the Castle of Chambord-France. Int. J. Civ. Struct. Eng. **4**(3), 402–18 (2014)
69. Brunetaud, X., Luca, L.D., Janvier-Badosa, S., Beck, K., Al-Mukhtar, M.: Application of digital techniques in monument preservation. Eur. J. Environ. Civ. Eng. **16**(5), 543–556 (2012)
70. Deacon, D.H.: The durability of steel structures in different environments. In: Durability of materials and structures in Buildings and Civil Engineering, Whittles Publishing (2006). ISBN 1-870325-58-3, 474 pp
71. Janvier-Badosa, S., Beck, K., Brunetaud, X., Al-Mukhtar, M.: The occurrence of gypsum in the scaling of stones at the Castle of Chambord (France). Environ. Earth Sci. **71**(11), 4751–4759 (2014)
72. Delgado, J.M.P.Q. (ed.): New Approaches to Building Pathology and Durability, Building Pathology and Rehabilitation 6, https://doi.org/10.1007/978-981-10-0648-7_2

73. Siegesmund, S., Snethlage, R. (eds.): Stone in Architecture: Properties and Durability, vol. 227, 4th edn. Springer, Heidelberg (2011). https://doi.org/10.1007/978-3-642-14475-2_4
74. Robador, M.D., Arroyo, F., Perez-Rodriguez, J.L.: Study and restoration of the Seville City Hall façade. Constr. Build. Mater. **53**, 370–380 (2014). https://doi.org/10.1016/j.conbuildmat.2013.11.088
75. Tortora, M., Sfarra, S., Chiarini, M., Daniele, V., Taglieri, G., Cerichelli, G.: Non-destructive and micro-invasive testing techniques for characterizing materials, structures and restoration problems in mural paintings. Appl. Surf. Sci. **387**, 971–985 (2016). https://doi.org/10.1016/j.apsusc.2016.07.023
76. Kilic, G.: Using advanced NDT for historic buildings: towards an integrated multidisciplinary health assessment strategy. J. Cult. Heritage **16**(4), 526–535 (2015). https://doi.org/10.1016/j.culher.2014.09.010
77. McCann, D.M., Forde, M.C.: Review of NDT methods in the assessment of concrete and masonry structures. NDT and E Int. **34**(2), 71–84 (2001). https://doi.org/10.1016/S0963-8695(00)00032-3
78. Sýkora, M., Diamantidis, D., Holický, M., Marková, J., Rózsás, Á.: Assessment of compressive strength of historic masonry using non-destructive and destructive techniques. Constr. Build. Mater. **193**, 196–210 (2018). https://doi.org/10.1016/j.conbuildmat.2018.10.180
79. Sena da Fonseca, B., Ferreira Pinto, A.P., Piçarra, S., Montemor, M.F.: Artificial aging route for assessing the potential efficacy of consolidation treatments applied to porous carbonate stones. Mater. Des. **120**, 10–21 (2017). https://doi.org/10.1016/j.matdes.2017.02.001
80. Duarte, R., Flores-Colen, I., de Brito, J., Hawreen, A.: Variability of in-situ testing in wall coating systems - Karsten tube and moisture meter techniques. J. Build. Eng. **27** (2020). https://doi.org/10.1016/j.jobe.2019.100998
81. Manohar, S., Bala, K., Santhanam, M., Menon, A.: Characteristics and deterioration mechanisms in coral stones used in a historical monument in a saline environment. Constr. Build. Mater. **241** (2020). https://doi.org/10.1016/j.conbuildmat.2020.118102
82. Hassine, M.A., Beck, K., Brunetaud, X., Al-Mukhtar, M.: Use of electrical resistance measurement to assess the water saturation profile in porous limestones during capillary imbibition. Constr. Build. Mater. **165**, 206–217 (2018). https://doi.org/10.1016/j.conbuildmat.2017.12.238
83. Alderete, N., Villagrán Zaccardi, Y., Snoeck, D., Van Belleghem, B., Van den Heede, P., Van Tittelboom, K., De Belie, N.: Capillary imbibition in mortars with natural pozzolan, limestone powder and slag evaluated through neutron radiography, electrical conductivity, and gravimetric analysis. Cement Concr. Res. **118**, 57–68 (2019). https://doi.org/10.1016/j.cemconres.2019.02.011
84. Khodeir, L.M., Aly, D., Tarek, S.: Integrating HBIM (Heritage Building Information Modeling) Tools in the Application of Sustainable Retrofitting of Heritage Buildings in Egypt. Procedia Environ. Sci. **34**, 258–270 (2016). https://doi.org/10.1016/j.proenv.2016.04.024
85. Fryskowska, A., Stachelek, J.: A no-reference method of geometric content quality analysis of 3D models generated from laser scanning point clouds for HBIM. J. Cult. Heritage **34**, 95–108 (2018). https://doi.org/10.1016/j.culher.2018.04.003
86. Monego, M., Menin, A., Fabris, M., Achilli, V.: 3D survey of Sarno Baths (Pompeii) by integrated geomatic methodologies. J. Cult. Heritage **40**, 240–246 (2019). https://doi.org/10.1016/j.culher.2019.04.013
87. Islam, S., Bhat, G.: Environmentally-friendly thermal and acoustic insulation materials from recycled textiles. J. Environ. Manage. **251** (2019). https://doi.org/10.1016/j.jenvman.2019.109536
88. https://onpe.org/news/precarite_energetique_combien_de_personnes_peinent_chauffer_leur_logement

89. https://www.ademe.fr/sites/default/files/assets/documents/fiche-ravalement-refection-toiture-amenagement-travaux-isolation.pdf
90. https://ec.europa.eu/environment/integration/research/newsalert/pdf/26si_en.pdf
91. https://www.batimentbascarbone.org/renovation-bas-carbone/
92. Sierra-érez, J., et al.: Environmental implications of cork as thermal insulation in façade retrofits. 10th Conference on Advanced Building Skins, 126 (2015)
93. Millogo, Y., Morel, J.-C., Aubert, J-E., Ghavami, K.: Experimental analysis of pressed adobe blocks reinforced with *Hibiscus cannabinus* fibers. Constr. Build. Mater. **52**, 71–78 (2014)
94. Tonoli, G.H.D., Santos, S.F., Savastano, H., Delvasto, S., Mejiade Gutierrez, R., Lopez de Murphy, M.-del-M.: Effects of natural weathering on microstructure and mineral composition of cementitious roofing tiles reinforced with fique fibre. Cement Concr. Compos. **33**, 225-232 (2011)
95. Alix, S., Philippe, E., Bessadok, A., Lebrun, L., Morvan, C., Marais, S.: Effect of chemical treatments on water sorption and mechanical properties of flax fibres. Bioresour. Technol. **100**, 4742–4749 (2009)
96. Palumbo, M., Avellaneda, J., Lacasta, A.M.: Availability of crop by-products in Spain: new raw materials for natural thermal insulation. Resour. Conserv. Recycl. **99**, 1–6 (2015)
97. Recherche, F., Bewa, H., Angers, A.: Evaluation de la disponibilité et de l ' accessibilité de fibres végétales à usages matériaux en France Assessment of natural fibres availability and accessibility for material uses in France Remerciements. Recherche (2011)
98. Yates, T.: The use of non-food crops in the UK construction industry. J. Sci. Food. Agr. **86**, 1790–6 (2006)
99. Asdrubali, F., D'Alessandro, F., Schiavoni, S.: A review of unconventional sustainable building insulation materials. Sustain. Mater. Technol. **4**, 1–17 (2015)
100. Sharma, V., Marwaha, B.M., Vinayak, H.K.: Enhancing durability of adobe by natural reinforcement for propagating sustainable mud housing. Int. J. Sustain. Built Environ. **5**(1), 141–155 (2016)
101. Dieye, Y., Sambou, V., Faye, M., Thiam, A., Adj, M., Azilinon, D.: Thermo-mechanical characterization of a building material based on Typha Australis. J. Build. Eng. **9**, 142–146 (2017)
102. Calatan, G., Hegyi, A., Dico, C., Mircea, C.: Determining the optimum addition of vegetable materials in adobe bricks. Procedia Technol. **22**, 259–265 (2016)
103. Ali, M., Chouw, N.: Experimental investigations on coconut-fibre rope tensils strength and pullout from coconut fibre reinforced concrete. Constr. Build. Mater. **41**, 681–690 (2013)
104. Ramli, M., Kwan, W.H., Abas, N.-F.: Strength and durability of coconut-fiber-reinforced concrete in aggressive environments. Constr. Build. Mater. **38**, 554–566 (2013)
105. Guozhong, L., Yanzhen, Y., Jianquan, L., Changchun, L., Yingzi, W.: Research on adaptability between crop-stalk fibers and cement. Cem. Concr. Res. **34**, 1081–1085 (2004)
106. Madhoushi, M., Nadalizadeh, H., Ansell, M.P.: Withdrawal strength of fasteners in rice straw fibre-thermoplastic composites under dry and wet conditions. Polym. Test. **28**, 301–306 (2009)
107. Chen, X., Yu, J., Zhang, Z., Lu, C.: Study on structure and thermal stability properties of cellulose fibers from rice straw. Carbohydr. Polym. **85**, 245–250 (2011)
108. Cerezo, V.: Propriétés mécaniques, thermiques et acoustiques d'un matériau à base de particules végétales: approche expérimentale et modélisation théorique. Ph.D. thesis of INSA-Lyon, p. 242 (2005)
109. Bevan, R., Woolley, T.: Hemp lime construction- A guide to building with hemp lime composites. Build. Res. Establ. 110 (2008)
110. Bruijn, P.B., Jeppsson, K.H., Saudin, K., Nilsson, C.: Mechanical properties of lime hemp concrete containing shives and fibres. Biosys. Eng. **103**, 474–479 (2009)

111. Ashour, T., Wieland, H., Georg, H., Bockisch, F.J., Wu, W.: The influence of natural reinforcement fibers on insulation values of earth plaster for straw bale buildings. Mater. Des. **31**, 4676–4685 (2010)
112. Ashour, T., Goerg, H., Wu, W.: An experimental investigation on equilibrium moisture content of earth plaster with natural reinforcement fibers for straw bale buildings. Appl. Therm. Eng. **31**, 293–303 (2011)
113. Van de Lindt, J.W., Carraro, J.A.H., Heyliger, P.R., Choi, C.: Application and feasibility of coal fly ash and scrap tire fiber as wood wall insulation supplements in residential buildings. Resour. Conserv. Recycl. **52**, 1235–1240 (2008)
114. Haapio, A., Viitaniemi, P.: Environmental effect of structural solutions and building materials to a building. Environ. Impact Assesment Rev. **28**, 587–600 (2008)
115. Almalkawi, A.T., Soroushian, P., Shrestha, S.S.: Evaluation of the energy efficiency of an aerated slurry-infiltrated mesh building system with biomass-based insulation. Renew. Energy **133**, 797–806 (2019)
116. Huang, H., Zhou, Y., Huang, R., Wu, H., Sun, Y., Huang, G., Xu, T.: Optimum insulation thicknesses and energy conservation of building thermal insulation materials in Chinese zone of humid subtropical climate 52 (2020). https://doi.org/10.1016/j.scs.2019.101840
117. Marouen, S., Naima, B., Dashnor, H.: In situ performance assessment of a bio-sourced insulation material from an inverse analysis of measurments on a demonstrator building. EBUILT 2016, 16–19 November, Iasi (Romania) (2016)
118. Wei, J., Meyer, C.: Degradation rate of natural fiber in cement composites exposed to various accelerated aging environment conditions. Corros. Sci. **88**, 118–132 (2014)
119. Toledo Filho, R.D., Scrivener, K., England, G.L., Ghavami, K.: Durability of alkali-sensitive sisal and coconut fibers in cement mortar composites. Cement Concr. Composit. **22**, 127–143 (2000)
120. Lideôw, S., Ôrn, T., Luciani, A., Rizzo, A.: Energy-efficiency measures for heritage buildings: a literature review. Sustain. Cities Soc. **45**, 231–242 (2019)
121. Robert, W., Piotr, K.: On rehabilitation of building with historical. Energy Procedia **132**, 927–932 (2017)
122. Rathorea, M., Ahmada, A., Paula, A., Seungmin Rhob, S.: Urban planning and building smart cities based on the Internet of Things using Big Data analytics. Comput. Netw. **101**, 63–80 (2016)
123. Mathew, P.A., Dunn, L.N., Sohn, M.D., Mercado, A., Custudio, C., Walter, T.: Big-data for building energy performance: Lessons from assembling a very

# Mechanical Characteristics of Polymer Concrete with Different Waste Replacements

Marinela Bărbuță, Ancuța Rotaru(✉), and Traian-Dănuț Babor

"Gheorghe Asachi" Technical University of Iaşi, 67 Prof.dr.doc. D. Mangeron Street, 700050 Iaşi, Romania

ancuta.rotaru@tuiasi.ro

**Abstract.** The article analyses the influence of wastes as component of polymer concrete on the mechanical properties. Five types of wastes were used for preparing epoxy resin concrete: fly ash, sun flower, corn, saw dust, polystyrene granules and chopped plastic bottles. Fly ash was used as filler in all mixes. The other wastes were used as aggregate substitution with a percentage of 25% of the weight.

The mechanical properties such as: compressive strength, flexural strength and split tensile strength were experimentally determined and compared with a control mix prepared with epoxy resin and aggregates. Generally the values of compressive strength were smaller than that of the control mix. The flexural strengths presented bigger values than the control mix for all types of wastes. The split strength presented higher values than control mix only for concrete with saw dust. The minimum values of mechanical strengths were obtained for concrete with corn cob as aggregate substitution. The density of polymer concrete with wastes generally corresponded to lightweight concrete.

**Keywords:** Polymer concrete · Fly ash · Sunflower waste · Corn waste · Sawdust · Polystyrene · Chopped plastic bottles

## 1 Introduction

The accumulation of unmanaged wastes from the food industry, particularly in developing countries, is becoming increasingly problematic. New applications for such waste comprise the creation of environmentally friendly materials and composites in the road-building and construction industries, for instance [2]. In the entire world, the construction industry is rapidly developing and a significant objective for the building industry in the forthcoming years is to use sustainable materials which conserve natural resources. This is particularly pertinent given the rising cost and chronic shortages of conventional materials. Wastes, a problem for the environment, are of great interest in the manufacture of green construction materials [5, 7, 8, 23]. Waste materials such as plastics and glass, which present possible environmental hazards, are landfilled and often used in concrete for different applications. Fly ash, silica fume, ground granulated blast furnace, used tire, plastic bottles, wastes of glass, polystyrene, wood, agro-wastes, sludge, etc. are studied


A. Rotaru (Ed.): CRIT-RE-BUILT 2019, SSGG, pp. 200–206, 2021.
https://doi.org/10.1007/978-3-030-61118-7_17

as components of concretes and mortar for ecological reasons but also for improving their properties [1, 4, 6, 9–14, 25]. It has been shown that the properties of these materials are suitable to produce new concrete up to a certain limit.

Using wastes in a high-performance concrete is an accepted practice, yet they are also usable in geopolymers and polymer concrete production. Wastes - such as fly ash, slag, etc. help to replace the cement [24] in geopolymer composition while polymer concrete includes resin and aggregates. Wastes behave as fillers or replace resin or aggregates [3, 15].

The use of those wastes in specific applications may solve two problems, namely, elimination of an environmental pollutant and provision of an alternative material for the construction industry, namely, the lightweight concrete. In 1971, The One Shell Plaza, Texas, a 52-storey building, broke the record as the world's tallest building made of lightweight concrete. Nowadays, this record has been surpassed by Renzo Piano's The Shard, a 300 m high 95-storey supertall tower topped out in London in 2012.

Suspended floors have been benefiting from lightweight concrete for years, as it reduces the weight on the structure and increases fire resistance. Because lightweight concrete is typically thinner is made of different ingredients than traditional concrete, floors can be up to 45% lighter than regular concrete floors.

However, lightweight concrete costs more to produce. Replacing 25% by volume of its mineral aggregates with lightweight waste aggregates could offer benefits to contractors. On this account, this study discusses some properties of the concrete using five types of light wastes as aggregates.

The article is a comparative study that presents the results of experimental studies on polymer concrete prepared with five different types of waste that substitute the aggregates. The addition of such materials can significantly affect the properties of concrete, in some instances altering its strength, density and water resistance.

The analysis of the effects of the analysed kind of wastes on these types of polymer concrete reveals differences in terms of physical and mechanical characteristics of the lightweight concrete.

## 2 Experimental Program

### 2.1 Materials

The study uses polymer concrete, fly ash and five types of wastes.

*Control Polymer Concrete.* The experimental study uses a control polymer concrete composed of epoxy resin and two sorts of aggregates. Polymer concrete is a type of concrete that uses polymers to replace lime-type of cement as a binder. The epoxy resin is type ROPOXID, made in Romania by POLICOLOR Bucharest. The resin represents 12.4% of the mixture weight for all types of polymer concrete. The hardener is type ROMANID 407, also made by POLICOLOR Bucharest. The natural aggregates consist of two sorts: 0–4 mm (sand) and 4–8 mm (gravel), with continuous granulometry according to Romanian standard [18]. In the control mix, both aggregates have a dosage of 43.8% from the mix weight. The control mix is marked C0.

*Fly ash.* Fly ash from Iași-Holboca Power Plant, Romania is used as a cement-like filler in a dosage of 12.8% from the mix weight [21, 22]. The fly ash properties: dark grey, spherical particles of sizes ranging from 0.01 μm to 400 μm, specific area 480–520 $m^2$/kg, the density of 2400–2550 kg/$m^3$, chemical components type Si (18.3%), C (17.15%), Al (13.9%), etc. [21, 22]. Fly ashes are present in all mixes with waste substitution, in the same dosage.

The wastes are as follows:

- *Sunflower wastes* substitute 25% of fine mineral aggregate by volume, sand sort 0–4 mm. Sunflower aggregate preparation consists of cutting the stem of the dried plant in pieces with sizes smaller than 4 mm. The small pieces dried out in a natural ventilated room at 25°C. The density of the sunflower waste experimentally determined is 136 kg/$m^3$. The polymer concrete with sunflower is marked C1. The test specimens were cube moulds with sides of 100 mm for compressive strength testing, 100 × 100 × 500 mm prism moulds for flexural tensile, and cylinders of 100 mm Ø × 200 mm for the split tensile strength test, three samples prepared for each test. Tests come after 28 days of curing, according to Romanian standards [16, 17, 19, 20].
- *Corn wastes* use as a substitution by volume of 25% of fine aggregate, sort 0–4 mm. The preparation of the corn cob aggregate consists of cutting the dried cob in pieces with sizes between 0–4 mm. The experimentally determined density of corn cob waste is 266 kg/$m^3$. The polymer concrete with corn waste is marked C2.
- *Sawdust* use as a substitution of 25% by volume of fine aggregate, sort 0–4 mm. It is a residue from the wood industry. The density of sawdust waste experimentally determined is 168 kg/$m^3$. The polymer concrete is marked C3.
- *Polystyrene* (PS) granules substitute 25% by volume of fine aggregates, sort 0–4 mm. The polystyrene granules are wastes from the construction industry. The density of polystyrene waste experimentally determined is 1.6 kg/$m^3$. The polymer concrete is marked C4.
- *Chopped plastic bottles* (PET) use as substitution of 25% by volume of fine aggregate, sort 0–4 mm. Waste plastics are a threat to the global environment so that recycling may be a solution. Only about 25% of plastic waste is recycled around the world, while 75% frequently find their way into rivers, oceans, coast, beaches, and the land. The percentage of recycled plastic can increase by transforming waste plastic into products suitable for construction materials. The plastic waste may consist of mixed organic (food remains) and inorganic (attached paper level) fractions. Its utilization as an aggregate in concrete for the construction industry could be an efficient solution as long as the contamination of plastic waste has not considerable impact on its properties. Resin-based types of waste plastic like polystyrene (PS) and PET have the highest rate of use for concrete production [9]. PET is a sub-product obtained from plastic bottles recycling process. It should be noted that the processing of the PET supposed crushed plastic to a size between 1.18 mm and 4.75 mm, with angular shape. The density of PET waste experimentally determined is of 433 kg/$m^3$. The polymer concrete is marked C5.

### 2.2 Experimental Procedure

For preparing polymer concrete, the components were mixed, firstly the dried components after that the epoxy resin combined with hardener was added.

The other mixes were noted as: The samples types: cubes of 7 mm sizes and prisms of 70 × 70 × 210 mm were poured. After 24 h the samples were demoulded and kept at a temperature of 20 °C for 14 days. The density of hardened polymer concrete was determined [20]. The cubes were tested in axial compression for determining the compressive strength [21], the prisms were tested in flexure and splitting for determining the flexural strength and split tensile strength according to standards [22, 23]. All tested were done on three samples for each type of concrete.

## 3 Results and Discussions

The experimental results obtained for polymer concrete with wastes are given in Table 1.

**Table 1.**

| Sample | $f_c$ [MPa] | $f_{ti}$ [MPa] | $f_{td}$ [MPa] | Density, [kg/m$^3$] |
|---|---|---|---|---|
| 0 | 69.93 | 12.26 | 6.82 | 2116 |
| C1 | 39.30 | 16.30 | 6.63 | 2027 |
| C2 | 28.96 | 13.61 | 5.15 | 1973 |
| C3 | 56.60 | 16.59 | 6.98 | 1919 |
| C4 | 44.59 | 14.84 | 5.28 | 1861 |
| C5 | 47.21 | 16.94 | 6.14 | 1946 |

The experimental results are compared in the graph presented in Fig. 1.

### 3.1 Compressive Strength

From Fig. 1, the values of compressive strength varied from 56.6 MPa for polymer concrete with corn as aggregate substitution to 28.96 MPa for concrete with saw dust as aggregate substitution. All values of fc were smaller than that of control mix with percentages between 19.1% for C3 and 58.6% for C2. For the same aggregate substitution, the wastes influenced differently the compressive strength. All values of fc indicate that the studied polymer concrete with wastes can be used for structural applications.

### 3.2 Flexural Strength

From experimental results, the values of flexural strengths for all mixes with wastes were bigger than that of the control mix with percentages between 38.2% and 11.1%. The maximum flexural strength was obtained for concrete C5 and the minimum value for C2.

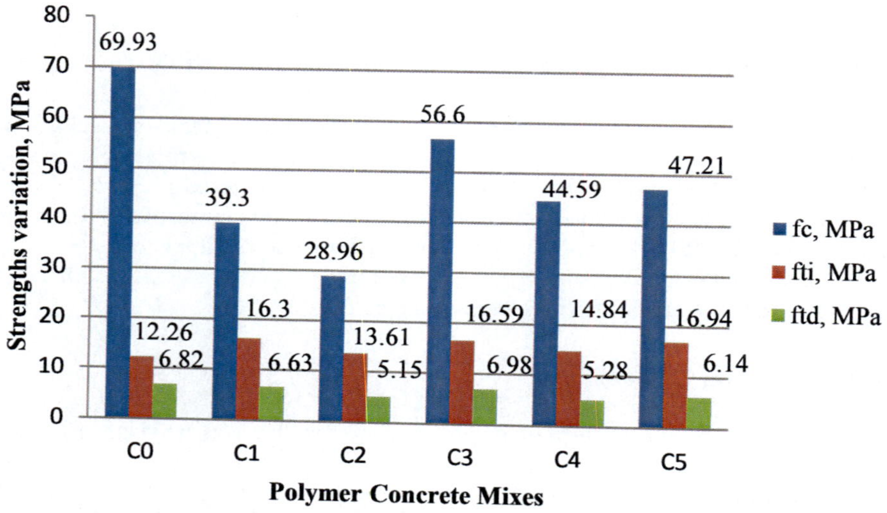


**Fig. 1.** Variation of mechanical properties of fly ash polymer concrete with different wastes

### 3.3 Split Tensile Strength

From Fig. 1, the values of split tensile strength for mixes with wastes were smaller than that of the control mix, except for mix C3, for which a small increase of the strength was obtained. The values of $f_{ti}$ varied between 6.98 MPa for C3 and 5.15 MPa for mix C2, that means an increase of 2.3% for C3 and a maximum decrease of 24.5% for C2.

### 3.4 Density

The values of density for hardened concrete are given in Table 1. The type of waste influenced differently the density. All densities of polymer concrete with waste were smaller than that of control mix. For mixes C2, C3, C4 and C5 the densities were smaller than 2000 kg/m$^3$; this fact indicates that these are lightweight concretes. The minimum value of density was obtained for mix C4, for which mechanical strengths presented relatively higher values. C4 can be considered as a lightweight concrete that can be used for structural applications.

Generally, the polymer concrete with aggregated substitution of agro-wastes presented smaller values of mechanical strengths in comparison with that for the other types of wastes, for the same dosage of substitution. The densities in the case of agro-wastes were bigger than that with other types of wastes.

For each mechanical strength, the maximum value was obtained for different mix. The type of waste influenced differently each mechanical property. Only in the case of mix C3 the maximum value was obtained for $f_c$ and $f_{td}$.

In the case of minimum value of mechanical strengths, these were obtained for the same mix prepared with corn cob.

## 4 Conclusions

Five types of wastes were used for preparing epoxy resin concrete: fly ash, sun flower, corn, saw dust, polystyrene granules and chopped plastic bottles. Fly ash was used as filler in all mixes. The other wastes were used as aggregate substitution with a percentage of 25% of the weight. A control mix prepared with epoxy resin and aggregates was used for control.

For the same aggregate substitution, the wastes influenced differently the mechanical strengths.

The density was smaller than that of control mix for all types of wastes. The concretes with corn waste, saw dust, polystyrene granules and chopped PET are lightweight polymer concrete.

The values of compressive strength study of concrete reported a decrease in the compressive strength than that of control mix for all types of wastes. The higher value of compressive strength was obtained for mix with saw dust aggregate substitution. The flexural strengths were for all mixes bigger than that of control mix. The higher value of flexural strength was obtained for mix with chopped PET aggregate substitution. The split tensile strengths were generally smaller than that of control mix, only the mix with saw dust presented a higher value than that of control mix.

The mechanical strengths of polymer concrete with different types of wastes as aggregate substitution are not very much diminished in comparison with that of polymer concrete without wastes, especially in the case of tensile strengths, case in which the wastes improved the behavior in flexure.

In conclusion, in the case of studied wastes as aggregate substitution the polymer concrete presented some special characteristics:

- mechanical properties are comparable with that of structural concretes;
- higher flexural strength than control polymer concrete;
- the densities were closed to lightweight concrete.

In the mix of polymer concrete different types of wastes can be used contributing to the consuming of wastes that polluted the environment. Studying the properties of polymer concrete with wastes is a possibility of application in construction industry of new eco-materials.

## References

1. Alkaysi, M., El-Tawil, S., Liu, Z., Hansen, W.: Effects of silica powder and cement type on durability of ultra high performance concrete (UHPC). Cem. Concr. Compos. **66**, 47–56 (2016)
2. Bărbuță, M., Marin, E., Câmpeanu, S.M., Paraschiv, G., Lepădatu, D., Bucur, R.D.: Statistical Analysis of the tensile strength of coal Fly ash concrete with fibers using central composite design. Adv. Mater. Sci. Eng. **15**(7), 1–7 (2015)
3. Bărbuță, M., Timu, A., Bejan, L., Bucur, R.D.: Mechanical properties of Fly ash polymer concrete with different fibers. Rev. Rom. Mat. Plast. **55**(3), 405 (2018)

4. Bignozzi, M.C., Saccani, A., Barbieri, L., Lancellotti, I.: Glass waste as supplementary cementing materials: the effects of glass chemical composition. Cement Concr. Compos. **55**, 45–52 (2015)
5. Bolden, J., Abu-Lebdeh, T., Fini, E.: Utilization of recycled and waste materials in various construction applications. Am. J. Environ Sci. **9**(1), 14–24 (2013)
6. Chowdhury, S., Mishra, M., Suganya, O.: The incorporation of wood waste ash as a partial cement replacement material for making structural grade concrete: an overview. Ain Shams Eng. J. **6**, 429–437 (2015)
7. Ding, G.L., Chen, J.: Risk assessment of green construction based on grey system theory. J. Eng. Manag. **24**(2), 182–185 (2009)
8. Franzoni, E.: Materials selection for green buildings: which tools for engineers and architects. Procedia Eng. **21**, 883–890 (2011)
9. Islam, M.J., Meherier, M.S., Rakinul, A.K.M.: Effects of waste PET as coarse aggregate on the fresh and harden properties of concrete. Constr. Build. Mater. **125**, 946–951 (2016)
10. Kaya, A., Kar, F.: Properties of concrete containing waste expanded polystyrene and natural resin. Constr. Build. Mater. **105**, 572–578 (2016)
11. Monzo, J., Paya, J., Borrachero, M.V., Girbes, I.: Reuse of sewage sludge ashes (SSA) in cement mixtures: the effect of SSA on the workability of cement mortars. Waste Manag **23**, 373–381 (2003)
12. Nazari, A., Riahi, S.: The Effects of ZnO2 Nanoparticles on properties of concrete using ground granulated blast furnace slag as binder. Mater. Res. **14**(3), 299–306 (2011)
13. Prusty, J.K., Patro, S.K., Basarkar, S.S.: Concrete using agro-waste as fine aggregate for sustainable built environment – a review. Int. J. Sustain. Built Environ. **5**, 312–333 (2016)
14. Rafieizonooz, M., Mirza, J., Salim, M.R., Hussin, M.W., Khankhaje, E.: Investigation of coal bottom ash and Fly ash in concrete as replacement for sand and cement. Constr. Build. Mater. **116**, 15–24 (2016)
15. Reis, J.M.L.: Effect of textile waste on the mechanical properties of polymer concrete. Mater Res. **12**(1), 63–67 (2009)
16. Romanian Standard Association, SR EN 12390–3:2005, (2011), "Testing hardened concrete," Part 3: "Compressive strength of test specimens"
17. Romanian Standard Association, SR EN 12390–5:2005, (2009), "Testing hardened concrete," Part 5: "Flexural strength of test specimens"
18. Romanian Standard Association, SR EN 12390–5:2005, Tests for geometrical properties of aggregates. Part 1.Determination of particle size distribution-Sieving method
19. Romanian Standard Association, SR EN 12390–6:2010, "Testing hardened concrete," Part 6: "Split tensile strength of test specimen"
20. Romanian Standard Association, SR EN 12390–7:2005, Testing hardened concrete. Part 7: Density of hardened concrete
21. Rotaru, A., Boboc, V.: A material used in substructure and road works: physical characteristics of pozzolana Fly ash from thermal power plant of Iasi, Romania. WSEAS Trans. Environ. Dev. **6**(6), 427–436 (2010)
22. Rotaru A., Boboc V.: Physical properties of pozzolana Fly ash from thermal Power Plant of Iasi, Romania – A Cement-like Material for Substructure Works. In: Recent Advances in Electrical Engineering, WSEAS Press (eds.), 187–193 (2010)
23. Siddique, R.: Waste Materials and By-products in Concrete. Springer, Berlin Heidelberg (2008)
24. Singh, B., Ishwarya, G., Gupta, M., Bhattacharyya, S.K.: Geopolymer concrete: a review of some recent developments. Constr. Build. Mater. **85**, 78–90 (2015)
25. Thomas, B.S., Gupta, R.C.: Properties of high strength concrete containing scrap tire rubber. J. Cleaner Prod. **113**, 86–92 (2016)

# Analysis of Stress and Strain in Flexible Pavement Structures Comprised of Conventional and High Modulus Asphalt Using Viscoelastic Theory

H. T. Tai Nguyen[1,2(✉)], Tran Vu Tu[1], Van-Rin Phan[3], and Ba-Gio Phan[2]

[1] Ho Chi Minh City University of Technology and Education, Ho Chi Minh City, Vietnam
{tainht,tutv}@hcmute.edu.vn

[2] BMT Construction-Investment J.S.C., Ho Chi Minh City, Vietnam

[3] Ho Chi Minh City University of Transport, Ho Chi Minh City, Vietnam

**Abstract.** Permanent deformation of pavements has become a persistent issue to be solved in Vietnam because of hot climatic conditions and high traffic volume. In order to deal with the problem, many attempts have been made by local authorities, researchers and practitioners, e.g. modification of the mixture gradation, improving bitumen rheology and high-quality control of manufacturing and construction of asphalt concrete (AC). SBS-modified bitumen or polymer modified bitumen (PMB) has been applied in Vietnam for many years and very good results in preventing rutting distress have been gained. However, in very high trafficked road and slow-speed area like intersection, PMB did not perform very well. Recently, high modulus asphalt concrete (HMAC) has been considered in the country as an alternative for PMB asphalt concrete. Since the thickness of flexible pavement structures in Vietnam is normally very small, ranging from 12 to 15 cm for asphalt layers while HMAC is stiffer than conventional AC, there are still concerns about the fatigue life of pavement structures comprised of HMAC. The aim of this study is to characterize the linear viscoelastic (LVE) behavior of HMAC and conventional AC and to analyze the LVE behavior of pavement composed of HMAC or (and) AC to evaluate its possible use in Vietnam. The results show that the use of HMAC will decrease the values of stress and strain at the bottom of the second asphalt layer, which increases the resistance to bottom-up fatigue cracking of the pavement.

**Keywords:** High modulus asphalt concrete · Viscoelastic behavior · Stress-strain analysis · Flexible pavement

## 1 Introduction

Permanent deformation (rutting) and cracking distress are two common distresses of flexible pavement in Vietnam. Since rutting distress significantly reduces the traffic safety, it attracts much attentions of local authorities, researchers and practitioners in Vietnam.


A. Rotaru (Ed.): CRIT-RE-BUILT 2019, SSGG, pp. 207–219, 2021.
https://doi.org/10.1007/978-3-030-61118-7_18

Many attempts have been made in the country to predict [1, 2] and prevent rutting distress, e.g. modification of the mixture gradation, bitumen rheology improvement using SBS-polymer and others additives or improving quality control in manufacturing and construction of AC [3–6]. The application of PMB in national highway and expressway in Vietnam has gained very good results in preventing rutting distress and increasing fatigue life of pavement. However, PMB pavement did not perform very well when subjected to severe conditions, e.g. overloading or slow speed traffic.

In recent years, high modulus asphalt concrete (HMAC) has been considered as an alternative for PMB asphalt concrete in the country to prevent rutting distress [7, 8]. HMAC has been longtime recognized by its high stiffness and resistance to fatigue cracking compared to conventional asphalt mixtures. Considering the state-of-the-art relating HMAC pavement and practical issues in Vietnam, the use of HMAC can be a good solution. However, there is still concerns about the fatigue cracking of flexible pavement structures comprised of HMAC in the country because of its high stiffness.

This work is an attempt to demonstrate the possible use of HMAC in the conditions of Vietnam, which are characterized by: thin asphalt layers, ranging from 12–15 cm, and hot climatic conditions [9, 10]. In order to improve the analysis results, the behavior of both HMAC and AC was characterized using the complex dynamic modulus test [11]. The well-known 2S2P1D [12] model was used to predict their linear viscoelastic (LVE) behavior, which allows the LVE analysis of pavement structures. It should be noted that LVE theory is still not recommended in the actual design code for flexible pavement due to the lack of dynamic testing devices for LVE characterization. Therefore, the analysis results obtained with linear elastic (LE) theory was also presented in this study for comparison purpose.

The overall objective of this work is to demonstrate the possible use of HMAC in the conditions of Vietnam. To do that, the stress and strain of asphalt pavement comprised of HMAC was analyzed to prove that the use of HMAC will not increase the risk of fatigue cracking. Two specific objectives are defined:

- LVE characterization of HMAC and conventional AC using complex dynamic modulus test;
- LVE analysis of pavement structures comprised of HMAC and (or) conventional AC using the approximate approach in [13].

## 2 Materials and Testing Methods

### 2.1 Materials

In this work, we will investigate the viscoelastic behavior of a conventional AC and a HMAC which has been recently introduced in Vietnam. Since the dense gradation with nominal maximum particle size of 12.5 mm is widely used in Vietnam, it will be chosen for preparing the mixture samples for this study. The gradation curve is designed such that it is in the limitation specified by [1] as shown in Fig. 1. The 60/70 pen bitumen was used to produce conventional AC samples while a hard bitumen of performance grade PG76 was used for production of HMAC. The aggregate is collected from a quarry in the South of Vietnam. Both filler and binder are provided by BMT Construction Investment

J.S.C. Table 1 gives a variety of fundamental engineering characteristics of the aggregate used in this research while properties of 60/70 bitumen are the same as those reported in [5].

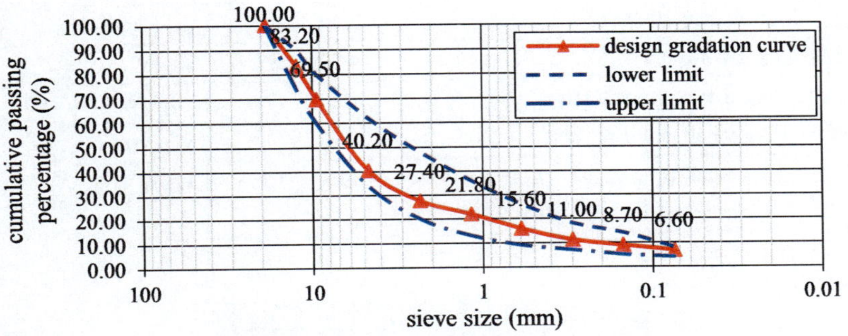


**Fig. 1.** Design gradation curve used for preparing AC and HMAC

**Table 1.** Fundamental properties of conventional AC and HMAC used in this research.

| Properties | Results | Limitations |
|---|---|---|
| Los Angeles abrasion (%) | 12.3 | Max 30 |
| Flat and elongated particle (%) | ≤6 | Max 10 |
| Water absorption (%) | 1.41 | Max 2 |
| Specific gravity of coarse aggregate | 2.77 | – |
| Specific gravity of fine aggregate | 2.78 | – |
| Specific gravity of filler | 2.73 | – |

The design of asphalt mixtures was performed based on the Marshall method to find the optimal binder content (5% of total mixture weight for AC, 5.1% for HMAC). The fundamental properties of these mixtures are summarized in Table 2.

## 2.2 Sample Preparation and Test Set-up for Complex Dynamic Modulus

The asphalt mixtures produced with the chosen gradation curve and the optimal binder content were compacted by a Roller Sector Compactor to obtain AC slabs of dimensions: $320 \times 260 \times 80\ \text{mm}^3$. The AC slabs were then sawed along the width of the slabs to obtain prismatic samples of dimensions: $70 \times 70 \times 150\ \text{mm}^3$ as shown in Fig. 2. In order to assure the precision of dimensions, the sawing machine is automatically controlled by computer.

In this work, the Pavetest hydraulic UTM equipped at Ho Chi Minh City University of Technology and Education (HCMUTE) was used for evaluating the complex dynamic modulus of the samples and the method used is inspired from AASHTO T342-11 [11].

**Table 2.** Fundamental properties of conventional AC and HMAC used in this research.

| Properties \| Asphalt mixtures | AC | HMAC | Limitations |
|---|---|---|---|
| Maximum specific gravity | 2.56 | 2.55 | – |
| Bulk specific gravity | 2.44 | 2.42 | – |
| Air void (%) | 4.92 | 5.5 | 3 ÷ 6 |
| Void in aggregate (%) | 16.82 | 17.5 | Min 14 |
| Marshall stability (kN) | 12.2 | 17.4 | Min 10 |
| Marshall flow (mm) | 3.8 | 3.3 | 3 ÷ 6 |
| Indirect tensile strength at 25 °C (MPa) | 1.1 | 1.7 | – |

**Fig. 2.** Test set-up for dynamic complex modulus testing. (1) Load platen. (2) Displacement transducers. (3) Control temperature sensor.

The test set-up is illustrated in Fig. 2, in which two displacement transducers was installed along the length of the specimen with the aid of metal jigs glued to the specimens. The investigating temperatures were: −10, 0, 10, 20, 30, 40, 50 and 60 °C and the frequency swept from 0.1 Hz to 5 Hz.

### 2.3 Definition of Dynamic Modulus and Phase Angle

When applying a sinusoidal wave shape loading with frequency $f$ to the sample, the measured relative displacement between the two jig points has a sinusoidal wave shape with the same frequency yet with a time lag $\delta$ compared to the load signal. The stress in the sample is thus a sinusoidal wave shape of frequency $f$ with amplitude $\sigma_0$ and phase angle $\varphi_L$, or $\sigma = \sigma_0 \sin(\omega t + \varphi_L)$, and the strain in the sample is a sinusoidal wave shape of frequency $f$ with amplitude $\varepsilon_0$ and phase angle $\varphi_D$, or $\varepsilon = \varepsilon_0 \sin(\omega t + \varphi_D)$ as illustrated in Fig. 3. Since a sinusoidal signal can be represented by a complex number, the stress and strain signal can be rewritten as in Eq. (1)

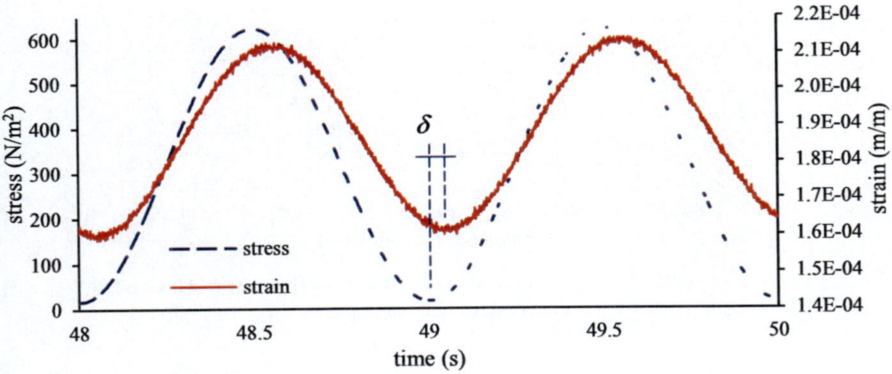


**Fig. 3.** Illustration of complex stress, strain and phase angle definition.

$$\sigma* = \sigma_0 e^{i(\omega t + \varphi_L)};\ \varepsilon* = \varepsilon_0 e^{i(\omega t + \varphi_D)} \tag{1}$$

The complex modulus of the sample is defined by the ratio of complex stress over the complex strain

$$E* = \frac{\sigma*}{\varepsilon*} = \frac{\sigma_0}{\varepsilon_0} e^{i(\omega t + \varphi_L - \varphi_D)} \tag{2}$$

The dynamic modulus is the modulus of the complex number $|E*| = \sigma_0/\varepsilon_0$ and the phase lag angle is thus $\varphi = \varphi_L - \varphi_D = \omega.\delta$.

## 3 Results of LVE Characterization of Materials

### 3.1 Dynamic Modulus

The obtained dynamic modulus of both conventional AC and HMAC are presented in Figs. 4 and 5. The horizontal shifting technique was applied to construct the master curve of dynamic modulus. Based on the time-temperature correspondence principle, the complex modulus of the material at frequency $f$ and temperature $T$ is equivalent to that at frequency $f' = f \times a_T(T')$ and temperature $T'$, or $E^*(f, T) = E^*[f \times a_T(T'), T']$, and $a_T(T')$ is called the shift factor. The shift factors of conventional AC and HMAC are presented in Fig. 6.

It is worth noting that the shift factor is an intrinsic property of a viscoelastic material because the curve of shift factors depends only on the material itself. The well-known William-Landel-Ferry law [14] can be used to model the curve of shift factors in function of temperature as follows:

$$\log(a_T) = \frac{C_1(T - T_{ref})}{C_2 + T - T_{ref}} \tag{3}$$

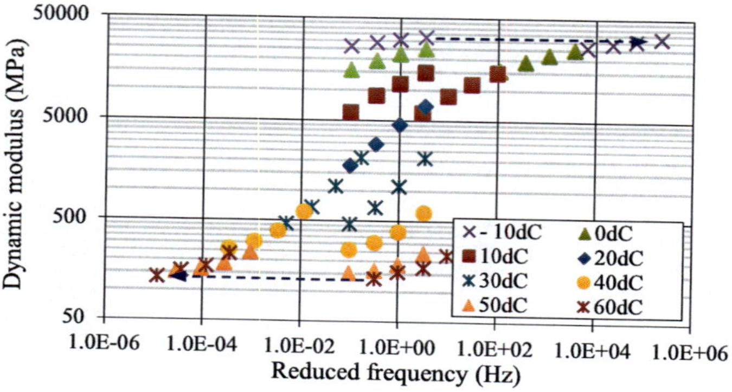


**Fig. 4.** Dynamic modulus of conventional AC at various temperatures (−10 to 60 °C) and its master curve constructed using horizontal shifting technique.

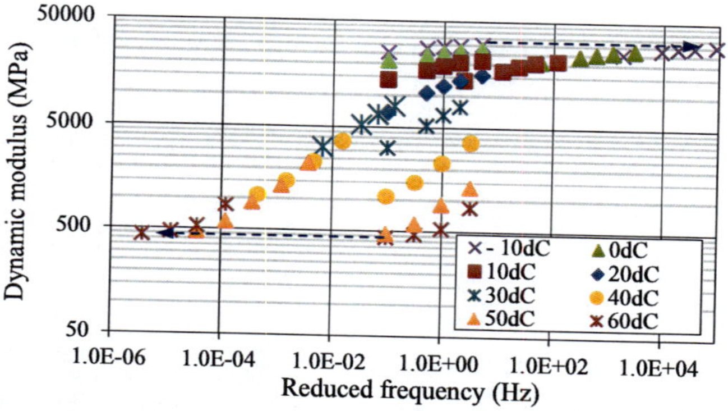


**Fig. 5.** Dynamic modulus of HMAC at various temperatures (−10 to 60 °C) and its master curve constructed using horizontal shifting technique.

where $T_{ref}$ is a chosen temperature of reference and $C_1$ and $C_2$ are two material parameters determined by fitting the model to the tested shift factors. The WLF parameters were determined using nonlinear least-square technique—namely: $C_1 = 27.8144$, $C_2 = 200.9701$ at 20 °C for conventional AC, and $C_1 = 34.5874$, $C_2 = 269.1829$ for HMAC.

## 3.2 Phase Angles

The obtained phase angles of both conventional AC and HMAC at different frequency and temperature were shifted using the same shift factor determined above and presented in Figs. 7 and 8. It can be found that the data of phase angle form a single master curve, which means that both the conventional AC and HMAC obey fully the time-temperature corresponding principle. At a specified temperature, the phase angle tends to zero when the frequency tends to zero or infinite at which the viscous phase of the material losses their stiffness or the viscous phase almost become quasi-elastic, respectively. As a result,

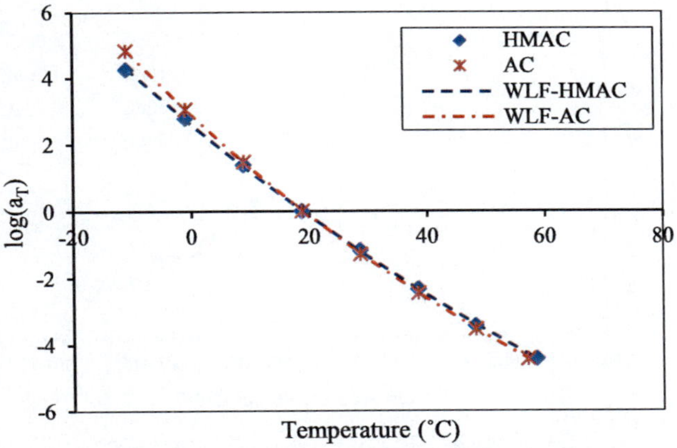


**Fig. 6.** Shift factors of AC and HMAC.

there exists a critical value of the reduced frequency at which the phase angle is maximal. The maximal phase angle of AC is approximately 41° at 0.2 Hz while that of HMAC is 34.5° at 0.005 Hz at the same temperature (20 °C). It can be seen that the phase angle of HMAC is always smaller than that of AC at all reduced frequency, which confirms that the HMAC is more elastic than AC and the resistance to rutting of HMAC is therefore higher than that of conventional AC.

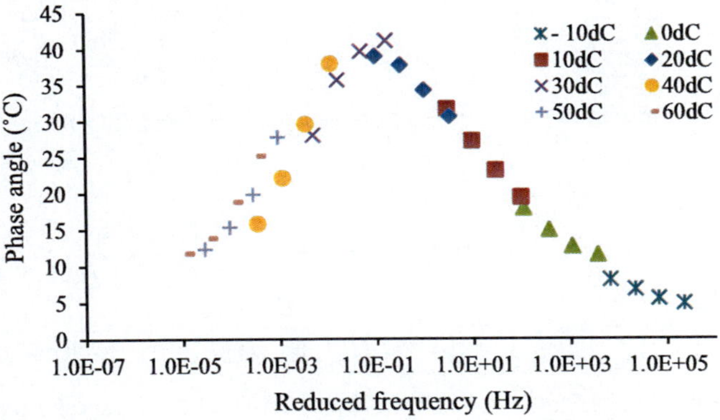


**Fig. 7.** Master curve of phase angles of conventional AC.

### 3.3 Modelling the Viscoelastic Behavior Using 2S2P1D Model

The rheological model of Huet and Sayegh (HS) [15, 16] and its generalization version, 2S2P1D model [12], are well-known for their excellent fitting of viscoelastic behavior of asphalt mixtures. It comprises one spring linked in parallel with a series of one spring,

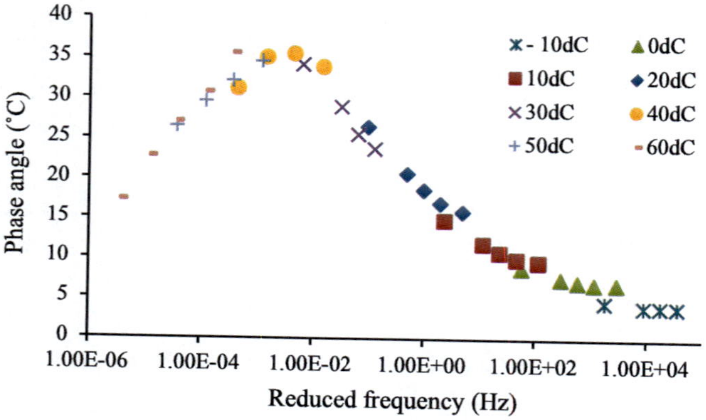


**Fig. 8.** Master curve of phase angles of HMAC.

two parabolic dashpots (and a linear dashpot for 2S2P1D model) as illustrated in Fig. 9. The complex modulus of these two models are presented in Eq. (4)

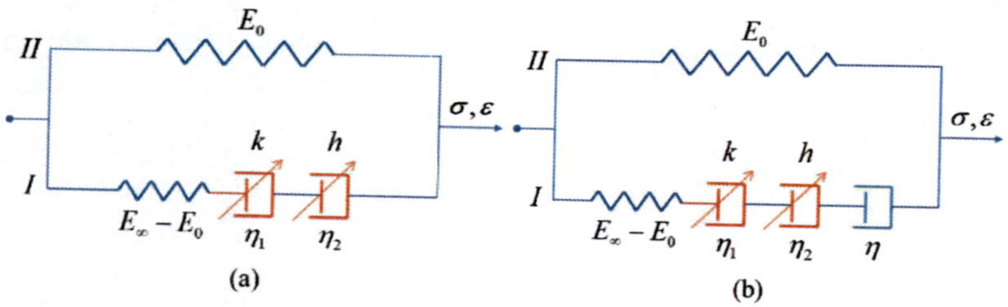


**Fig. 9.** The rheological model of Huet – Sayegh model (a) and 2S2P1D model (b).

$$E^*_{HS}(\omega) = E_0 + \frac{E_\infty - E_0}{1 + \delta[i\omega\tau(T)]^{-k} + [i\omega\tau(T)]^{-h}}$$
$$E^*_{2S2P1D}(\omega) = E_0 + \frac{E_\infty - E_0}{1 + \delta[i\omega\tau(T)]^{-k} + [i\omega\tau(T)]^{-h} + [i\omega\beta\tau(T)]^{-1}} \quad (4)$$

By using curve fitting techniques provided by MATLAB, the model parameters of 2S2P1D models that best fit the material behavior can be found and presented in Table 3. The fitting curves are also presented in Figs. 10 and 11.

**Table 3.** Model parameters for AC and HMAC at the reference temperature 20 °C.

| Mixture | $E_\infty$ (MPa) | $E_0$ (MPa) | $k$ | $h$ | $\delta$ | $\beta$ | $\tau$ ($T_{ref} = 20$ °C) |
|---|---|---|---|---|---|---|---|
| AC | 150.00 | 37338.32 | 0.32 | 0.80 | 5.72 | – | 0.10 |
| HMAC | 424.63 | 34227.28 | 0.21 | 0.59 | 2.07 | – | 0.63 |

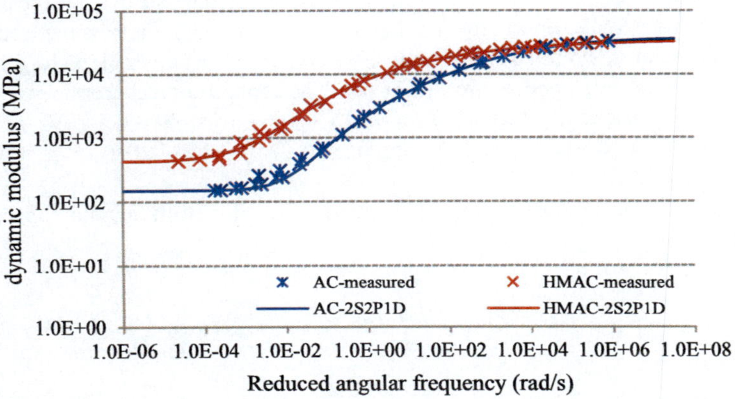


**Fig. 10.** 2S2P1D model prediction. Dynamic modulus vs. reduced angular frequency.

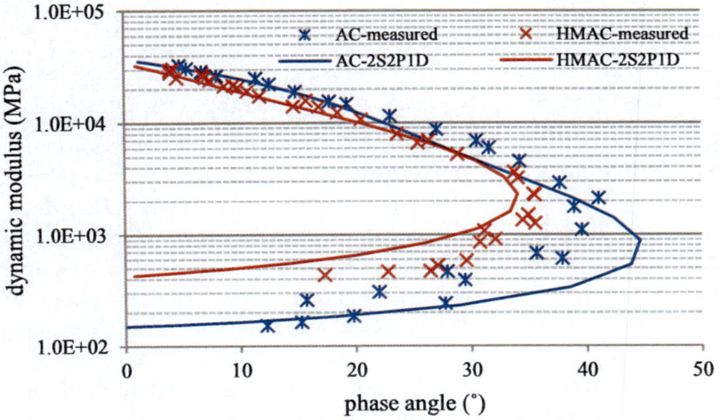


**Fig. 11.** 2S2P1D model prediction. Cole-Cole diagram.

## 4 Viscoelastic Analysis of a Typical Pavement Structure

The asphalt thickness in flexible pavement structure in Vietnam is very thin, normally ranging from 12 to 15 cm and comprising of two asphalt layers, a graded aggregate layer and the road base. In this study, the asphalt layers are 5-cm thick HMAC and 7-cm thick conventional AC layers. The HMAC is in the surface course to maximize the rutting resistance of the whole pavement [17].

The viscoelastic properties of HMAC and AC characterized in Sect. 3 will be used for analyzing the LVE behavior of the pavement. The graded aggregate layer and the road base are assumed to elastic materials of 120 MPa of stiffness modulus. The Poisson ratio is equal to 0.35 for all layers.

The application of one half of a 12-ton axle on the pavement is assumed to be a uniformly distributed force $P$ over a circle of 36-cm diameter. The contact pressure $p(t)$ is assumed to be 0.6 MPa during the loading periods. As a result, the pavement structures can be analyzed in axisymmetric conditions. We consider five pulses of loading, in which the loading time and resting time are equal to 5 s. The applied contact pressure is therefore $p(t) = 0.6MPa$ for time $t$ ranging from 0 ÷ 5 s, 10 ÷ 15 s, 20 ÷ 25 s, 30 ÷ 35 s and 40 ÷ 45 s and $p(t) = 0$ for time $t$ ranging from 5 ÷ 10 s, 15 ÷ 20 s, 25 ÷ 30 s and 35 ÷ 40 s.

The viscoelastic response can be obtained using the Boltzmann's superposition principle [13]:

$$R^{ve}(\mathbf{X}, t) = \int_0^t R_H^{ve}(\mathbf{X}, t - \tau) dI(\tau) \tag{5}$$

where $R^{ve}(\mathbf{X}, t)$ is the LVE response at coordinates $\mathbf{X}$ and time $t$, $R_H^{ve}(\mathbf{X}, t - \tau)$ is the LVE response of the pavement to a step input function at time $t - \tau$, and $dI(\tau)$ is the variation of the input at time $\tau$. The discretized form of Eq. (5) can be obtained:

$$R^{ve}(\mathbf{X}, t_i) = \sum_{j=0}^{i} R_H^{ve}(\mathbf{X}, t_i - t_j) \Delta I(t_j). \tag{6}$$

In Eq. (6), $R_H^{ve}(\mathbf{X}, t_i - t_j)$ is effectively the solution of a quasi-elastic problem using the relaxation modulus at time $t_i - t_j$. Therefore, the viscoelastic solution can be approximated based on a sequence of quasi-elastic solutions, which can be solved by using Burmister multi-layered elastic theory. Interested readers are referred to a detailed introduction of the theory in Ref. [18]. By using this approach, the viscoelastic analysis results of the pavement due to prescribed contact pressure previously defined (values associated with suffix "VE"), e.g. the maximal displacement at the surface of the pavement, the maximal stress and strain at the bottom of the binder course were obtained and presented in Figs. 12, 13 and 14.

In pavement analysis, the dynamic modulus at 10 Hz is widely used for analyzing the behavior of pavement structures using elastic theory. Therefore, linear elastic (LE) solutions (values associated with suffix "LE") obtained with dynamic modulus at 10 Hz are also included in the figures for comparison purpose. It can be found that although the maximal displacements obtained with LE and LVE analysis are comparable at the first loading pulse, the magnitude of the maximal stress and strain obtained with LVE analysis are larger than those obtained with LE analysis. While the magnitude of loading pulse is a constant, the LVE strain increases its magnitude with increasing number of loading pulses. Interestingly, LVE stress changes its sign from positive (tensile stress) to negative (compressive stress) during resting period while LE stress is totally recovered

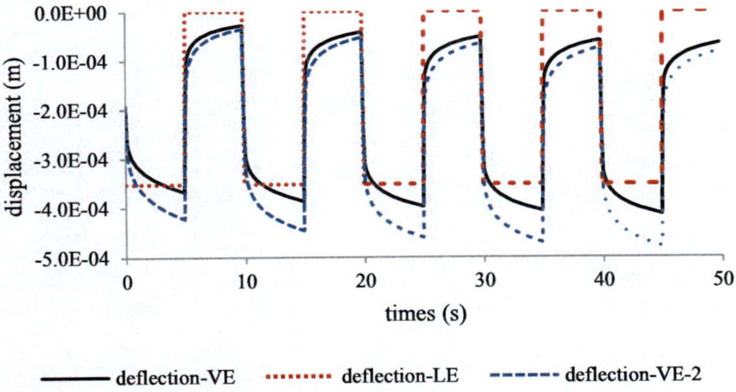


**Fig. 12.** Deflection of the pavement versus time.

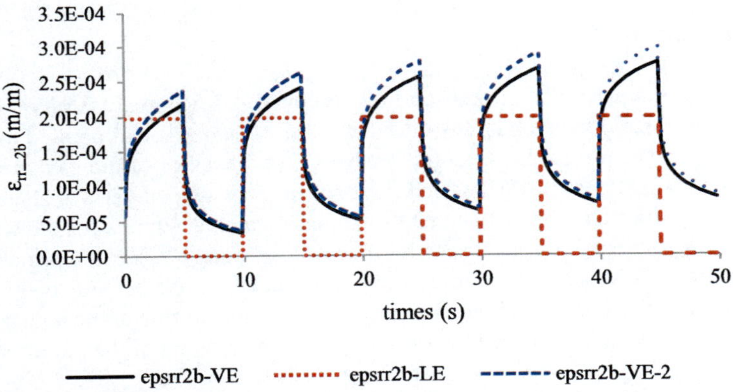


**Fig. 13.** Strain ($\varepsilon_{rr}$) at the bottom of the binder course versus time.

during this period (Fig. 14). This phenomenon can be explained by the fact that the "neutral plane" which separates the compressive and tensile zone changes during LVE analysis because of changes in the relaxation modulus of both HMAC and AC layers.

In order to evaluate the possible use of HMAC, the behavior of the conventional asphalt pavement comprising two layers (5 cm + 7 cm) of conventional AC was also analyzed. The analysis results are also included in the Figures, in which their values are associated with a suffix "VE-2". It can be found that due to the stiffer HMAC at the wearing course, the stress and strain at the bottom of the lower layer (AC layer) decreases. Therefore, the use of HMAC will not increase the risk of bottom-up fatigue cracking.

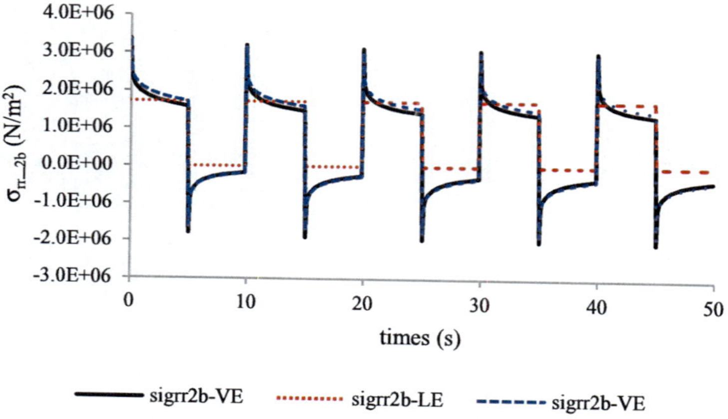


**Fig. 14.** Stress ($\sigma_{rr}$) at the bottom of the binder course versus time.

## 5 Conclusions

The viscoelastic behavior of HMAC and conventional AC was successfully characterized using local equipment and the well-known 2S2P1D model was applied to predict their LVE behavior for use in structural analysis. The behavior of the pavement comprised of two asphalt layers, HMAC and (or) conventional AC, was also analyzed using viscoelastic approach. It was shown that the displacement and strain obtained with LVE theory increase their maximal values with an increase of loading repetition, which cannot be captured with LE analysis. The results also show that the use of HMAC in the surface course gives rise to a reduction in the stress and strain at the bottom of the binder course, which will increase the resistance to bottom-up fatigue cracking of the pavement.

## References

1. Le, A.T., et al.: Development of a permanent deformation model of asphalt mixtures for South Korean pavement design guide. Transp. Res. Rec. **2095**(1), 45–52 (2009)
2. Nguyen, H.T.T.: Modelling the mechanical behaviour of asphalt concrete using the Perzyna viscoplastic theory and Drucker-Prager yield surface. Road Mater. Pavement Des. **18**(S2), 264–280 (2017)
3. Decision 858-MOT: Applying current technical standards to strengthen the quality of management of design and construction of hot mix asphalt pavement of high traffic roadway. MOT, Hanoi (2014)
4. Nguyen, H.T.T., Tran, T.N.: Laboratory study for the use of stone mastic asphalt in Vietnam. Sci. J. Thu Dau Mot Univ. **1**(32), 57–62 (2017). in Vietnamese
5. Nguyen, H.T.T., Tran, T.N.: Effects of crumb rubber content and curing time on the properties of asphalt concrete and stone mastic asphalt using dry process. IJPRT **13**(3), 238–244 (2018)
6. Tran, T.N., Nguyen, H.T.T., Nguyen, K.S., Nguyen, N.T.H.: Semi-flexible material: the sustainable alternative for the use of conventional road materials in heavy-duty pavement. In: Tran-Nguyen, H.-H., Wong, H., Ragueneau, F., Ha-Minh, C. (eds.) CIGOS 2017. LNCE, vol. 8, pp. 14–24. Springer, Singapore (2018). https://doi.org/10.1007/978-981-10-6713-6_54

7. Nguyen, H.T.T., Tran, T.N.: Fatigue characterisation of high modulus asphalt mixture for evaluating its possible use in Vietnam (Technical report). BMT Construction-Investment J.S.C. (2016)
8. Tran, D.H.: Research on high modulus asphalt mixture using for flexible pavements of high-class road in Vietnam. Ph.D. thesis, University of Transport and Communications, Hanoi (2018)
9. Nguyen, H.T.T., Nguyen, N.H.: Using a non-local elastic damage model to predict the fatigue life of asphalt pavement structure. In: Nguyen-Xuan, H., Phung-Van, P., Rabczuk, T. (eds.) ACOME 2017. LNME, pp. 47–63. Springer, Singapore (2018). https://doi.org/10.1007/978-981-10-7149-2_4
10. Nguyen, H.T.T., Tran, V.T.: Analysing the interlayer shear stress of asphalt pavement composed of conventional and high modulus asphalt. J. Sci. Technol. Civil Eng. **13**(5V), 85–92 (2019)
11. AASHTO, T: 342–11: determining dynamic modulus of hot mix asphalt (HMA). AASHTO, Washington (2011)
12. Di Benedetto, H., Olard, F., Sauzéat, C., Delaporte, B.: Linear viscoelastic behaviour of bituminous materials: From binders to mixes. Road Mater. Pavement Des. **5**(S1), 163–202 (2004)
13. Chatti, K. et al.: Enhanced analysis of falling weight deflectometer data for use withmechanistic-empirical flexible pavementdesign and analysis and recommendations for improvements to falling weight deflectometers. Report No. FHWA-HRT-15-063. U.S. Departement of Transportation, Georgetown Pike (2017)
14. Williams, M.L., Landel, R.F., Ferry, J.D.: The temperature dependence of relaxation mechanisms in amorphous polymers and other glass-forming liquids. J. Am. Chem. Soc. **77**(14), 3701–3707 (1955)
15. Huet, C.: Etude par une méthode d'impédance du comportement visco-élastique des matériaux hydrocarbons Ph.D. thesis, Faculté des Sciences de Paris, Paris (1963)
16. Sayegh, G.: Contribution à l'étude des propriétés visco-élastiques des bitumes purs et des bétons bitumineux Ph.D. thesis, Sorbonne University, Paris (1965)
17. Phan, T.N., Nguyen, H.T.T.: Evaluating the possible use of high modulus asphalt mixtures in flexible pavements in Vietnam. In: Reddy, J.N., Wang, C.M., Luong, V.H., Le, A.T. (eds.) ICSCEA 2019. LNCE, vol. 80, pp. 701–708. Springer, Singapore (2020). https://doi.org/10.1007/978-981-15-5144-4_67
18. Charyulu, M.K.: Theoretical stress distribution in an elastic multi-layered medium Ph.D. thesis, Iowa State University of Science and Technology, Ames (1964)

# Improvements of Strength and Dynamic Elastic Characteristics of Mortars by Using Carbon Nano-tubes

Daniel Covatariu, Mihai-Sergiu Alexa-Stratulat, and Ionuț-Ovidiu Toma(✉)

"Gheorghe Asachi" Technical University of Iasi, Iasi 700050, Romania
ionut.ovidiu.toma@tuiasi.ro

**Abstract.** The paper presents the preliminary results in terms of strength and dynamic elastic properties obtained on mortar prisms. The bending tensile and the compressive strength of regular and carbon nano-tubes (CNT) mortar were determined by means of standardized tests at the age of 28 days. Prior to assessing the strength value, non-destructive testing was used to determine the dynamic elastic characteristics, namely dynamic longitudinal modulus of elasticity and material damping ratio. The use of CNT in the mortar mix led to a reduction in the self-weight. However, the surfactant used to ensure the uniform dispersion of CNTs had a detrimental influence on the mechanical properties of mortar.

**Keywords:** CNT mortar · Strength · Dynamic modulus of elasticity

## 1 Introduction

Cement is the key ingredient of the modern concrete that binds together the sand and the aggregates, in the presence of water, into a hard material known as concrete. Concrete is considered to be the most used construction material in the world due to its versatility and relatively low production cost. Consequently, concrete production industry, and by extension the construction industry, are among the largest consumers of natural resources with a significant impact on the environment [1].

From the very beginning it was clear that even though concrete can sustain high compressive loads it is very weak when it comes to tensile stresses. Solutions were found in the form of steel reinforcing bars and later on in the form of disperse fibers made of different materials [2, 3].

Perhaps one of the most important characteristics of concrete resides in its ability to incorporate various other materials. Extensive research was conducted during the past decades to make concrete a sustainable material [4, 5]. Supplementary cementitious materials, some of them by-products of other industries, have proved to enhance the durability [6] and strength properties of concrete [7, 8] while attempting so solve the issue of landfills and reduce the burden on the environment [9, 10].

Ever since the early 1960s when the production of nano-materials was possible, they sparkled the interest of many researchers from all fields of science. In construction


A. Rotaru (Ed.): CRIT-RE-BUILT 2019, SSGG, pp. 220–228, 2021.
https://doi.org/10.1007/978-3-030-61118-7_19

industry, with concrete being such a widely used building material, this opportunity paved the way for better cement-based materials with significant improvements at the level of the cementitious matrix.

The nano particles used in concrete have the size less than 500 nm. Such particles are used to improve the packing model structure of concrete with direct effects on the strength of conventional concrete. This is due to the fact that nano particles act as filler and decrease the micro-porosity in the mineral matrix [11]. As with supplementary cementitious materials, nano materials can be synthesized from a variety of sources, even industrial wastes [12]. Nano silica is one of the newest technologies in nano process which has been used as an alternative to silica fume [13] together with carbon nano tubes [14, 15].

Carbon nanotubes (CNTs) extraordinary strength and stiffness characteristics prompted many researchers to mix them with cement paste and mortars in order to enhance their mechanical properties. A wide CNT to-cement weight fraction range could be found in the scientific literature with percentages from 0.02 to 1 wt%. The general conclusion that could be drawn from the reported results was that cement mixes containing higher CNTs percentages showed a reduction in strength compared to those blended with a relatively lower CNTs content. This was partly attributed to the lower dispersion quality associated with incorporating high CNTs' weight fractions into the cement/ mortar mix [16].

The paper presents some preliminary results in terms of strength and dynamic elastic properties obtained on CNT-mortar prisms. The bending tensile and the compressive strength of regular and carbon nano-tubes (CNT) mortar were determined by means of standardized tests at the age of 28 days. Prior to assessing the strength value, non-destructive testing was used to determine the dynamic elastic characteristics, namely dynamic longitudinal modulus of elasticity and material damping ratio.

## 2 Materials and Experimental Program

### 2.1 Cement and Sand

A CEMI-42.5 N type of cement was considered at this stage of the research. It is a general purpose cement being suitable for all uses in works requiring high strength values and it complies with the specifications of SR EN 197-1:2002 standard [17].

The maximum grain size of the sand was 2.0 mm. The sand was used in its dry state for each mortar mix considered at this stage of the research.

### 2.2 Carbon Nano Tubes

The carbon nano tubes had the dimensions of 100–200 μm in length and 30–40 nm in diameter. The CNTs were used in a 0.05% by weight of the cement. Prior to mixing with cement and sand, the CNTs were set in water with an sodium-dodecyl sulphate (SDS) surfactant and sonicated for 1 h. This would ensure their uniform distribution into the mixing water and, consequently, into the mortar. This uniform distribution is a key factor for improving the mechanical properties of cement-based mineral matrices [18, 19].

## 2.3 Experimental Program

The mortar mixes considered at this stage of the research are shown in Table 1, where CM stands for *cement mortar* and CNTM stands for *carbon nano-tubes mortar*. Three samples were cast for each mix proportion.

**Table 1.** Mix proportions.

| Mix | Cement | Water | W/C | Sand | CNT |
|---|---|---|---|---|---|
| | [kg/m$^3$] | [L] | [%] | [kg/m$^3$] | [g/m$^3$] |
| CM | 450 | 225 | 0.5 | 1350 | – |
| CNTM | 450 | 225 | 0.5 | 1350 | 225 |

The mortar paste was cast in 160 × 40 × 40 mm (Length × Thickness × Width) prisms. The molds were covered by wet cloth and kept for 24 h at constant room temperature of 23 °C. After demolding the specimens were placed in an environmental chamber, at 23 °C and 80% relative humidity, for curing until the age of 28 days in order to prevent shrinkage cracking.

During the 28 days of curing, a thermal gauge was placed in each prism for each mortar mix shown in Table 1. This was done in order to monitor the heat generated during the first 24 h of the hydration process of cement and see whether the use of CNTs would have any influence on the heat of hydration.

Before loading, the prisms were measured and weighed. This was done both to check for changes in dimension and to determine the density of the hardened mortar.

The dynamic longitudinal modulus of elasticity and the material damping ratio were obtained based on the First Resonant Frequency (FRF) as described in ASTM C215 standard [20] and given in Eq. (1):

$$E_d = D \times M \times f_n^2 \quad (1)$$

where M is the mass of the prism (kg), $f_n$ is the fundamental longitudinal frequency of vibration (Hz) and D is a coefficient that depends on the measured dimensions of the prism, Eq. (2):

$$D = 4 \times \frac{L}{b \times t} \quad (2)$$

with L being the length of the prism (m), b and t being the cross-sectional dimensions (m).

The material damping coefficient was calculated based on the Full Width at Half Maxima (FWHM) Method and it was based on the longitudinal fundamental frequency of vibration, $f_n$ and the frequencies $f_1$ and $f_2$ of the points located on the response spectrum graph, corresponding to an amplitude equal to half the amplitude at the resonance, Eq. (3).

The Forced Vibrations Method, Fig. 1, was used to determine the longitudinal resonant frequency of the prisms. The vibrations were introduced via a vibrations generator

**Fig. 1.** Experimental set-up for the Forced Vibrations testing method

and the response of the prism specimen was recorded by means of an accelerometer placed longitudinally at the opposite end of the specimen, as shown in Fig. 1.

$$\zeta_d = \frac{1}{\sqrt{3}} \times \frac{f_2 - f_1}{2f_n} \tag{3}$$

In the case of forced vibrations, a constant and small amplitude vibration of 0.01 g was introduced, the sweeping range being between 30 Hz–18000 Hz, at a rate of 10000 Hz/min. The ASTM C215-14 standard [20] does not give any specifications on the sweeping rate but earlier research works showed that this parameter plays little influence on the obtained fundamental frequency of vibration [21]. This however, depends on the homogeneity of the material and the sampling speed and accuracy of the measuring equipment.

The prisms were subjected to a three-point bending test, as shown in Fig. 2, with a loading rate of 50 N/s according to the standard specifications [22] in order to determine the flexural strength of mortar prisms. The resulting half prisms were examined for signs of visible cracks and then subjected to uniaxial compression test, Fig. 3, with a loading rate of 2400 N/s [22].

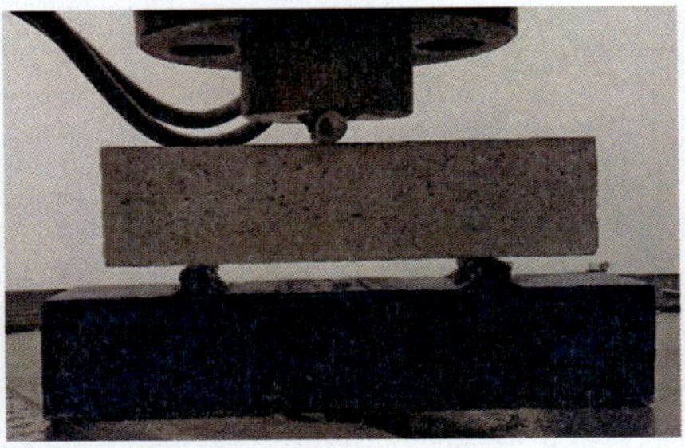

**Fig. 2.** Three-point bending test

**Fig. 3.** Uniaxial compression test

## 3 Results and Discussions

### 3.1 Generated Heat

When Portland cement is mixed with water heat is generated as a result of the chemical reactions between the constituents of cement and water. This heat of hydration is mostly influenced by the cement composition but the water-to-cement ratio, fineness and curing temperature were also proven to be significant factors. The temperature profile in each mortar prism during the first 24 h is shown in Fig. 4. It can be observed that the mortar mixes with CNTs exhibited a lower temperature generated during the hydration process of cement. This might be due to the fact that CNTs are good heat conductors [23] and this helped in dissipating some of the generated heat.

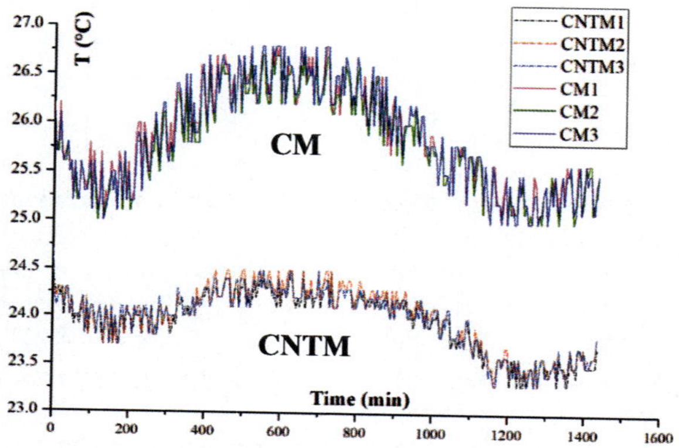


**Fig. 4.** Temperature variation at early age

Moreover, the variation from the maximum to the minimum recording temperatures is significantly smaller for CNT mortar than for cement mortar meaning smaller thermal stresses.

### 3.2 Dynamic Modulus of Elasticity and Damping Ratio

Figure 5 presents the response spectra, average of 3 measurements, for the cement and CNT mortars, respectively.

It can be observed that CNT mortar exhibits a higher fundamental frequency of vibration and, according to Eq. (1), should have a larger value for the dynamic modulus of elasticity compare to cement-based mortar. Additionally, the amplitude of the peak corresponding to the fundamental longitudinal period of vibration for CNT mortar is 57% larger than for the cement mortar with possible indication of a stiffer internal structure due to the presence of CNTs. The obtained results are summarized in Table 2.

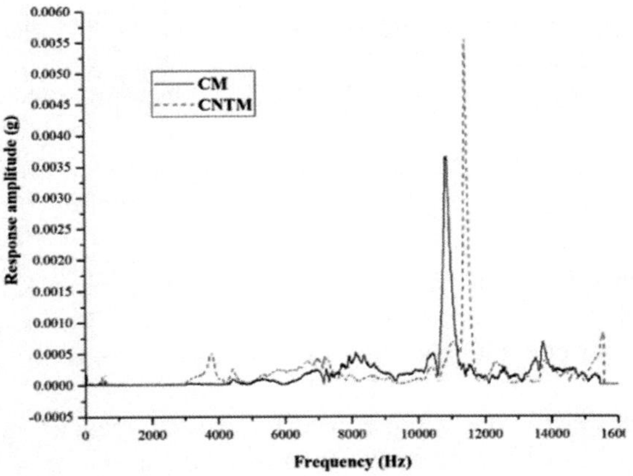


**Fig. 5.** Response spectra

**Table 2.** Dynamic properties of mortars.

| Sample | Mass [kg] | Dimensions | | | Resonant frequency [Hz] | $E_d$ [GPa] | $\zeta_d$ [%] |
|---|---|---|---|---|---|---|---|
| | | L [m] | b [m] | t [m] | | | |
| CM | 0.5 | 0.1597 | 0.041 | 0.0401 | 10871.3 | 23.099 | 0.643 |
| CNTM | 0.325 | 0.1608 | 0.042 | 0.0402 | 11428.7 | 16.284 | 0.335 |

Even though CNT mortar showed a higher fundamental frequency of vibration the average mass was 35% smaller than that of cement mortar. Considering the shape of Eq. (1) it follows that the dynamic modulus of elasticity is 29.66% smaller for the CNT mortar than for cement mortar. The advantage of CNT mortar seems to be its light-weightiness, in this case. The damping ratio is larger for cement mortar due to the high rigidity of carbon nano-tubes in CNT mortar which allows for the vibrations to pass with virtually no damping at all. The cement and sand from the CNT mortar manage to dampen some of the vibrations. This observation is in line with the higher amplitude on the response spectrum corresponding to the resonant frequency recorded for the CNT mortar.

### 3.3 Flexural and Compressive Strengths

The flexural and compressive strengths were determined as described in Sect. 2.3. The results are summarized in Table 3. The value of the flexural tensile strength was the average of 3 determinations whereas the value of the compressive strength was the average of 6 determinations.

**Table 3.** Strength properties of mortars at the age of 28 days.

| Sample | Flexural tensile strength, $f_t$ [MPa] | Compressive strength $f_c$' [MPa] |
|---|---|---|
| CM | 3.48 | 53.55 |
| CNTM | 0.58 | 4.38 |

Contrary to what has been reported in the scientific literature [24], the CNT mortar seems to be very weak compared to the cement mortar. The only explanation for this unusual behavior could be obtained if the microstructure of the CNT mortar was investigated.

Figure 6 shows the microstructure of the CNT mortar where it can be observed that there are smooth surfaces between the mineral matrix and sand aggregates. Additionally, a lot of voids could be observed. Figure 7 represents a detailed view of the microstructure. It can be seen that CNTs coalesced during the mixing process and, therefore, were not uniformly distributed in the mortar prism. The only possible cause that could help explain both the presence of voids and the coalescence of CNTs was the surfactant that was used to actually ensure a uniform distribution of CNTs in the mixing water.

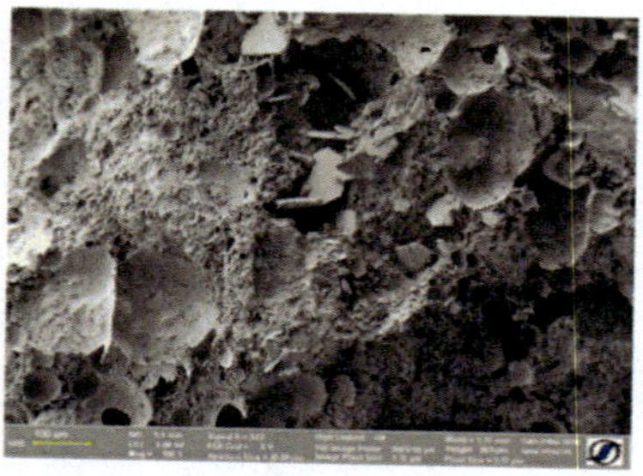

**Fig. 6.** Microstructure of CNT mortar

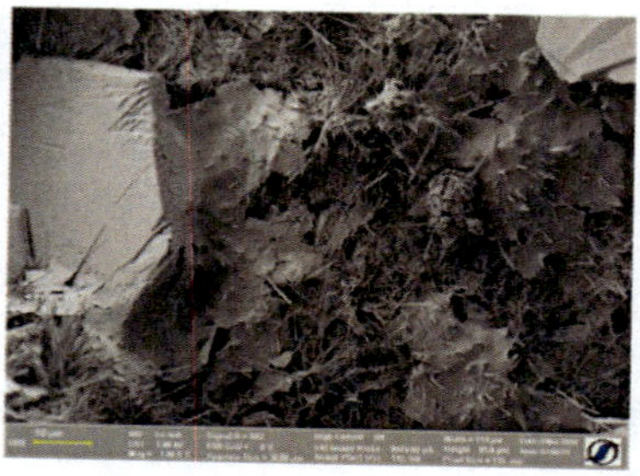

**Fig. 7.** CNT tubes in the mortar mix

## 4 Conclusions

Based on the obtained results, the following conclusions could be drawn:

The use of CNTs in mortar can help reduce the temperature during the hydration process of cement. Another benefit of using CNTs in mortar is the reduced mass, by as much as 35%, compared to cement mortar.

The type of surfactant seems to play an important role in the process of obtaining CNT mortars because an incorrectly chose surfactant may lead to the occurrence of large volume of voids in the hardened mortar with direct consequences on the strength properties.

The dynamic modulus of elasticity is 29.66% smaller for the CNT mortar than for cement mortar. The damping ratio is larger for cement mortar, possibly due to the high rigidity of carbon nano-tubes in CNT mortar which allows for the vibrations to pass with virtually no damping at all.

## References


1. Maddalena, R., Roberts, J.J., Hamilton, A.: Can Portland cement be replaced by low-carbon alternative materials? a study on the thermal properties and carbon emissions of innovative cements. J. Cleaner Prod. **186**, 933–942 (2018)
2. Toledo Filho, R.D., Koenders, E.A.B., Formagini, S., Fairbairn, E.M.R.: Performance assessment of ultra high performance fiber reinforced cementitious composites in view of sustainability. Mater. Des. **36**, 880–888 (2012)
3. Nili, M., Afroughsabet, V.: Property assessment of steel-fibre reinforced concrete made with silica fume. Constr. Build. Mater. **28**(1), 664–669 (2012)
4. Bassioni, G.: A study towards "greener" construction. Appl. Energy **93**, 132–137 (2012)
5. Gartner, E., Sui, T.: Alternative cement clinkers. Cement Concr. Res. **114**, 27–39 (2018)
6. O'Connell, M., McNally, C., Richardson, M.G.: Performance of concrete incorporating GGBS in aggressive wastewater environments. Constr. Build. Mater. **27**(1), 368–374 (2012)
7. Scrivener, K.L., John, V.M., Gartner, E.M.: Eco-efficient cements: potential economically viable solutions for a low-CO2 cement-based materials industry. Cement Concr. Res. **114**, 2–26 (2018)
8. Juenger, M.C.G., Snellings, R., Bernal, S.A.: Supplementary cementitious materials: new sources, characterization, and performance insights. Cement Concr. Res. **122**, 257–273 (2019)
9. Miller, S.A., John, V.M., Pacca, S.A., Horvath, A.: Carbon dioxide reduction potential in the global cement industry by 2050. Cement Concr. Res. **114**, 115–124 (2018)
10. Naqi, A., Jang, J.: Recent progress in green cement technology utilizing low-carbon emission fuels and raw materials: a review. Sustain. **11**(2), 537 (2019)
11. Norhasri, M.S.M., Hamidah, M.S., Fadzil, A.M.: Applications of using nano material in concrete: a review. Constr. Build. Mater. **133**, 91–97 (2017)
12. Samaddar, P., Ok, Y.S., Kim, K.-H., Kwon, E.E., Tsang, D.C.W.: Synthesis of nanomaterials from various wastes and their new age applications. J. Cleaner Prod. **197**, 1190–1209 (2018)
13. Yu, R., Spiesz, P., Brouwers, H.J.H.: Effect of nano-silica on the hydration and microstructure development of ultra-high performance concrete (UHPC) with a low binder amount. Constr. Build. Mater. **65**, 140–150 (2014)
14. Morsy, M.S., Alsayed, S.H., Aqel, M.: Hybrid effect of carbon nanotube and nano-clay on physico-mechanical properties of cement mortar. Constr. Build. Mater. **25**(1), 145–149 (2011)
15. Sharma, S., Kothiyal, N.C.: Facile growth of carbon nanotubes coated with carbon nanoparticles: a potential low-cost hybrid nanoadditive for improved mechanical, electrical, microstructural and crystalline properties of cement mortar matrix. Constr. Build. Mater. **123**, 829–846 (2016)
16. Mohsen, M.O., Taha, R., Abu Taqa, A., Shaat, A.: Optimum carbon nanotubes' content for improving flexural and compressive strength of cement paste. Constr. Build. Mater. **150**, 395–403 (2017)
17. ASRO (Romanian Standards Association): SR EN 197-1: Cement. Part I: Composition, Specifications and Conformity Criteria for Normal Use Cements (2011)
18. Jarolim, T., Labaj, M., Hela, R., Michnova, K.: Carbon nanotubes in cementitious composites: dispersion, implementation, and influence on mechanical characteristics. Constr. Build. Mater. **2016**, 1–6 (2016)
19. Gillani, S.S.-H., Khitab, A., Ahmad, S., Khushnood, R.A., Ferro, G.A., Saleem Kazmi, S.M., Qureshi, L.A., Restuccia, L.: Improving the mechanical performance of cement composites by carbon nanotubes addition. Procedia Struct. Integr. **3**, 11–17 (2017)
20. ASTM International: ASTM C215-14 - Standard Test Method for Fundamental Transverse, Longitudinal, and Torsional Resonant Frequencies of Concrete Specimens (2014)

21. Alexa-Stratulat, M.-S., Pişta, R.C., Petcu, O.A., Toma, A.M., Toma, I.-O.: Assessment of plain concrete material damping by means of half-power bandwidth method and full-width at half maximum method. In: C65 International Conference - Tradition and Innovation - 65 Years of Constructions in Transilvania, Cluj-Napoca, Romania (2018)
22. ASRO (Romanian Standards Association): SR EN 196-1:2016, Methods of Testing Cement - Part 1: Determination of Strength (2016)
23. Kumanek, B., Janas, D.: Thermal conductivity of carbon nanotube networks: a review. J. Mater. Sci. **54**(10), 7397–7427 (2019)
24. Wang, B., Han, Y., Liu, S.: Effect of highly dispersed carbon nanotubes on the flexural toughness of cement-based composites. Constr.Build. Mater. **46**, 8–12 (2013)

# Mechanical Bevahiour of an Unsaturated Soil Associated with Seepage

Luis José Andrade Pais[1], Paulo Eduardo Maia de Carvalho[1], Luis Manuel Ferreira Gomes[1], and Ancuța Rotaru[2](✉)

[1] University of Beira Interior, Covilhã, Portugal
[2] "Gheorghe Asachi" Technical University of Iași, Iași, Romania
ancuta.rotaru@tuiasi.ro

**Abstract.** Frequently, residual soils form under unsaturated conditions because of volatile fluids with different absorption coefficients at the chemical interface between the skeleton and fluid. That leads to the large requirement of controlling the stress increment derived from suction potentials by performing constant shear tests to water, leachate and gasoline infiltrations. Thus, the saturation with fluids having different specific weights and viscosities than water may occur in the unsaturated gaseous zone, adsorbed according to their physical and chemical properties (volatility, solubility, adsorption coefficient) as well as the time that has gone after contamination. Analysing these phenomena is of particular interest to evaluate the mechanical parameters and changes in shear strength.

**Keywords:** Shear strength · Unsaturated soil · Residual soil · Consolidated Undrained triaxial test · Degree of saturation

## 1 Introduction

Even though classical Soil Mechanics developed predominantly from studies in saturated soils, the soil above the water table keeps unsaturated in some temperate zones. Water trapped above the water table is in a reduced pressure state (below atmospheric pressure) called groundwater suction or matric suction. Suction means the force wherewith the soil holds water. It can vary up to thousands of kPa. The larger the force is, the drier is the soil. When it refers to air pressure in the voids, the matric suction sometimes contributes to slope stability in residual soils [11, 15].

New concepts for unsaturated soils developed based on the approach of practical results acquired on modified shear tests of residual soils based on independent variables of the stress state [2, 3, 7, 8]. The definition of the shear stress versus matric suction envelope explains the mechanical behaviour of the soil generating the suction effect [6, 9, 14, 18]. Several authors propose an extension to the Mohr-Coulomb envelope theory through three-dimensional evaluation (shear stress: confinement stress: matric suction - $\tau$: $\sigma'_{30}$: s), defining the boundary between possible and impossible states by a flat or curved surface.


A. Rotaru (Ed.): CRIT-RE-BUILT 2019, SSGG, pp. 229–240, 2021.
https://doi.org/10.1007/978-3-030-61118-7_20

Under this approach and for all values of the imposed suction, the friction angle is constant and equal to its value under complete saturation. The relation between the matric suction ($s = u_a - u_w$) and the shear stress ($\tau$) assumed linearity, but it was confirmed later nonlinear [9] assuming the variation of angle $\phi^b$. Angle $\phi^b$ indicates the ratio of the increment of $\tau$ in translational terms relative to the suction $(u_a - u_w)f$.

When the partially saturated soil gets wet to an ideal saturation condition, a small transformation concerning its strength comes up in the shear stress equation. The relation changes in respect to translation and rotation of the effective cohesion, and it relates to the cohesion as follows [14]:

$$c = c' + (u_a - u_w)f \text{ tg } \Phi^b \tag{1}$$

where:

$c$ – the intersection of the Mohr-Coulomb failure envelope in the plane defined by the matric suction and the axis of the shear stress;

$c'$ – the effective cohesion read off for an extended Mohr-Coulomb model, achieved at the cross point of the envelope and $\tau$ axis, where not only the normal stress but also the suction stress at failure is equal to zero;

$(u_a - u_w)f$ - matric suction in the failure plane;

$\phi^b$ - angle indicating the ratio of the increment of $\tau$ in translational terms relative to the suction $(u_a - u_w)f$.

Different combinations of $\sigma$, $u_a$ and $u_w$ are in usage, however, $(\sigma - u_a)$ and $(u_a - u_w)$ are now generally recognized. Under normal conditions in situ, $u_a = p_a$, where $(\sigma - u_a)$ is the net normal stress and $(u_a - u_w)$ is the matric suction where $u_a$ and $u_w$ are the air pressure and the pore water pressure, respectively meaning that the water pressure is negative $(-u_w)$. Generally, it is more convenient to work in terms of the mean stress (p) and shear stress (q) invariants in the triaxial space [4].

Provided an invariant coordinate system sets out the shear strength parameters (assuming the sum of the principal stresses), whether the shear stress vs. the average stress or the deviator stress (q) vs. the octahedral stress (p′), the following equation expresses the critical state values [4]:

$$q = M(S) + \mu(S) \tag{2}$$

The infiltration of water in the soil causes a decrease in suction forces of the soil matric up to instability, assuming the importance of this variable in the evaluation of slope stability [16, 17]. Generally, both normal stress and shear stress keep constant in the process of slope infiltration, yet initially, the failure goes with a decrease in suction stress by increasing soil saturation.

The experimental evaluation of this subject is carried out by conducting conventional triaxial tests, modified to allow the translation of the shear plane or stress plane. The tests which the paper analyses use this technique, that is, with the imposed and controlled suction ($s = 40$ kPa), performing the shear tests with the constant minimum principal stress ($\sigma_3$) and carried out in remoulded specimens. The sections were developed in two ways: i) keeping the water content constant throughout the test and ii) drying the specimen followed by infiltration by a few different fluids and for imposed deformation.

Environmental geotechnics also studies the infiltration and contamination of soil masses by pipe and sediment failure, to which are associated with the effects of volatility and suction tensions that develop in the pores. In this work, a light approach or comparative interpretation is only made in cases of soil infiltration by gasoline and leachate in a given deformation state.

## 2 Experimental Program

### 2.1 Materials

Tests samples run away with the residual granite soil of Covilhã. During sampling, the in situ degrees of saturation ranged between 50 and 65% and in the natural state it tended to form aggregates frequently larger than 20 mm in diameter. Corresponding to this degree of saturation, the interstitial or matrix suction, which is a function of the water-soil characteristic curve, is approximately 40 kPa [5]. The residual granitic soil is non-plastic silty sand in its natural state, and it is classified in the SW-SM in the same group with the rubble [1, 13].

The soils were unstructured and reshaped in terms of compactness, void ratio and degree of saturation. The group A-nc (series I) takes shape in the dry branch of the compaction curve for the compactive effort (energy $E_2$) = 112 kJ/m$^3$. The samples are separated according to their compactness and initial degree of saturation: series I (A-nc: n = 0.550 to 0.630; Si = 52% to 68%). The groups of tests, thus defined, are presented in Fig. 1a and Fig. 1b.

The water content represents the amount of water in the soil, independent of the void ratio, yet the residual saturation degree arises as a fundamental parameter that depends on the void ratio.

In unsaturated soils, the soil fabric is of greater importance than in saturated soils. It is considered that the compacted soil on the dry branch of the compaction curve or at optimum water content can generate aggregates. On the wet side of the compaction curve, this aggregation virtually does not exist. Porosity inside the aggregates may not vary, but the void ratio of the sample as a whole may vary considerably, depending on the degree of aggregation and the size of the inter-aggregate voids. This becomes important for the interpretation of the mechanical behaviour of the soil [12].

### 2.2 Experimental Procedure

The series of Consolidated Undrained triaxial tests (CU tests) were performed on saturated and unsaturated soil samples to establish a reference state to interpert seepage tests. Conducting shear tests with constant water content in undrained conditions is such as to study the pressure of water generated in the pores during the collapse of the structure, which represents a critical scenario [10].

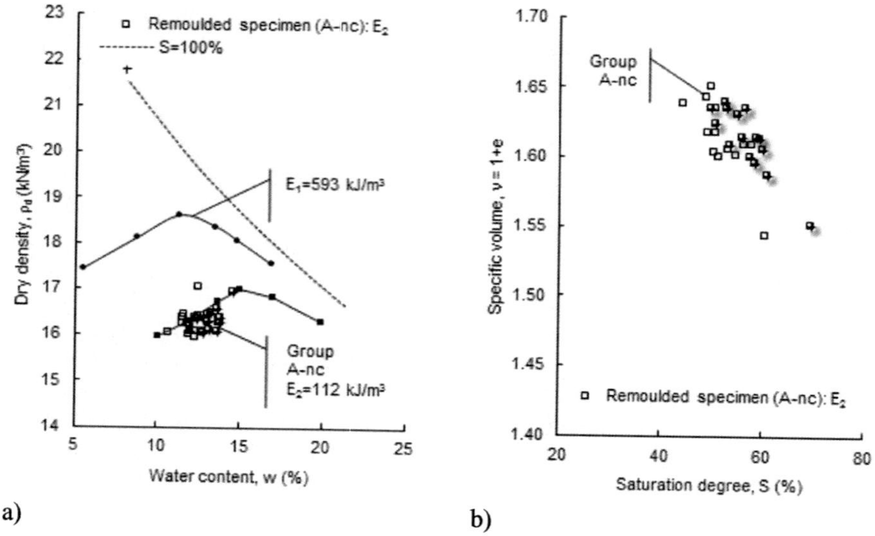


**Fig. 1.** Residual granite soil of Covilhã: a) relative position of remoulded samples (*A-nc*) in function of compaction curves defined by the compactive efforts $E_1 = 593$ kJ/m$^3$ and $E_2 = 112$ kJ/m$^3$; b) spatial distribution *e:S* for a group of samples

Table 1 presents the physical indexes at different stages of the tested series of samples to establish the stress plane at residual suction s = 40 kPa.

**Table 1.** Physical characteristics of Consolidated Undrained series of samples with constant water content tested in triaxial at different stages, group A-nc (SERIES I)

| Series (*I*) *A-nc* | Initial condition | | | | Consolidation phase | | Residual (Ultimate) Shear stress | | |
|---|---|---|---|---|---|---|---|---|---|
| | $\sigma'_{30}$ (kPa) | $\gamma_d$ kN/m$^3$ | $S_r$ (%) | $e_0$ | $e_{fc}$ | $s_{fc}$ (kPa) | $e_f$ | $S_f$ (%) | $s_f$ (kPa) |
| CU | 50 | 16.4 | 57.1 | 0.599 | 0.528 | 39.6 | 0.517 | 76.7 | 33.6 |
| CU | 50 | 16.3 | 66.7 | 0.608 | 0.554 | 39.9 | 0.544 | 68.6 | 32.7 |
| CU | 100 | 16.5 | 67.0 | 0.553 | 0.510 | 39.7 | 0.480 | 68.9 | 15.3 |
| CU | 200 | 16.3 | 62.5 | 0.614 | 0.479 | 39.7 | 0.430 | 84.3 | −3.1 |

The samples consolidate at $\sigma'_{30} = 50$, 100, 200 and 400 kPa and shear test fulfils in an undrained manner keeping $\sigma_{30}$ constant. Series I samples have residual suction from $s_r = 15$ to 40 kPa. The tests were carried out for s = 40 kPa, a typical *in situ* value in these types of soils as $S_r = 55$ to 60%. Initially, manoeuvers equalized the pressure decrease of the sample at 40 kPa with water entering or leaving the soil. As for the shear tests conducted with constant water content and imposed suction, the sample reaches the suction of 40 kPa, that is, water inlet or outlet (O´X´), consolidated until the pressure equalization (X′Y′) and sheared off in an undrained manner with constant $\sigma_{30}$ (Y′Z′). The final suction ($s_f$) is different because it dependends on the volumetric variation of the sample, see Fig. 2.

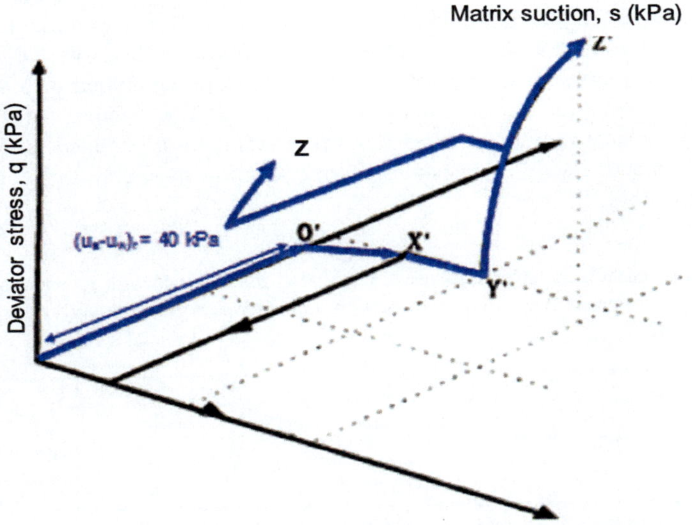


**Fig. 2.** Prediction of the stress paths for triaxial tests on samples of residual granitic soil [8]

Tests conducted in partial saturation conditions establishes three stages: 1) Induction of suction, which consists of percolating water through the interior of the sample, opening the water channel at the base of the sample and, with a decrease of pressure at its top, getting a low hydraulic load (10 kPa); 2) "Consolidation" or compression and equalization of pressures inside the specimen, that is, application of the average stress $\{p' = [(\sigma_1 + 2\sigma_3)/3 - u_a]\}$, with $\sigma_1 = \sigma_{30}$ in different samples waiting for equalization of pressures at the top and bottom $[s_{top} = s_{base} = (u_a - u_w) = 40\ kPa]$; and 3) Shear off the constant moisture content in the s = 40 kPa plane with a speed of 0.04 mm/min reaching the 25% axial deformation of the specimen (Y′Z′).

The seepage effect during the shear test shows the soil strength when it breaks due to the infiltration of water, leachate and gasoline, at relatively high stresses. The programme included soils after drying, and the stress paths followed during the equalization process are the same as in the shear tests with initial suction of 40 kPa, Fig. 2.

The specimen follows the established path and action mode at point X', i.e. the imposition of suction (O′), followed by the required compression stage ($\sigma'_3 = 100$ kPa) and pressure equalization (Y′). Upon reaching Y′, the shearing process starts with constant suction to certain axial stress, when the shear test breaks for the time required to infiltrate the fluid. As soon as the specimen is saturated, the compression stage continues in a drained way to Z´´. Figure 2 shows the various stress paths in the {q: p´: s} space for triaxial tests carried out on residual granitic soil with small to moderate compactness [8].

Percolating the sample one and a half times its volume ensured 70 to 75% saturation and generated a suction of 10 kPa. At this time, the effective compressive stress changes by increasing the water pressure in the pores. The suction decreases and the shearing process continues in a drained manner and the horizontal force must decrease. The water input ratio must be low and depends on the permeability ($k = 10^{-5}$ m/s, for this type of samples). The specimens used for the seepage test were remoulded with the same physical characteristics as the specimens used in series I. The summary of the stress typology and physical indexes of stage A, corresponding to the drained sample with constant suction and stage B corresponding to the shearing stage after the seepage are systematized in Table 2.

**Table 2** Physical indexes in different phases of modified triaxial tests with seepage, in stage I corresponding to drained test with constant suction and stage II corresponding to the shearing stage after seepage

| Stage | Reference | Initial condition | Residual stage | $S_r$ (%) | $e_0$ | $e_f$ | $S_f$ (%) | $s_f$ (kPa) |
|---|---|---|---|---|---|---|---|---|
| | | $\sigma'_{30}$ (kPa) | $\gamma_d$ kN/m$^3$ | | | | | |
| *A* | Cid100-s40i-nc-A2 Infiltration – Water | 100 | 16.0 | 43.4 | 0.642 | 0.495 | 56.20 | 18.9 |
| *B* | | 60 | 18.3 | 56.2 | 0.495 | 0.499 | 83.85 | −4.1 |
| *A* | Cid100-s40i-nc-LA1 Infiltration - Leachate | 100 | 16.0 | 62.8 | 0.642 | 0.538 | 74.90 | 38.1 |
| *B* | | 60 | 17.1 | 74.9 | 0.538 | 0.496 | 78.50 | −2.5 |
| *A* | Cid100-s40i-nc-GA1 Infiltration - Gasoline | 100 | 16.0 | 62.8 | 0.642 | 0.641 | 71.10 | 38.5 |
| *B* | | 60 | 16.8 | 71.1 | 0.561 | 0.564 | 73.65 | 6.4 |
| *CID* | Cid100-s40-A1 ($s = 40$ kPa) | 100 | 16.1 | 52.5 | 0.631 | 0.534 | 69.35 | 39.0 |

The wetting test with the water changed its density from stage A to B and presented the highest degree of saturation at the end. Infiltrations with leachate and gasoline show more subtle changes in density and saturation increase. Thus, a significant reduction in suction reflected in the shear strength occurs when infiltrating water.

## 3 Results and Discussions

### 3.1 Stress Paths in Critical State

The true critical point (critical state) will not be reached in unsaturated tests as all samples continue to expand even over large extensions ($\varepsilon a > 25\%$). However, the critical state paths will plotted assuming the closest value to the critical state.

Figure 3 shows the paths defined in the space of deviatoric stresses vs. mean stress (*q:p´*) for the failure criterion $[(\sigma'_1/\sigma'_3)_{ult}]$ of the series I of samples, unsaturated $[s = (u_a - u_w = 40\ kPa]$, and the corresponding samples tested in the same way, but saturated $[s = (u_a - u_w) = 0\ kPa]$. The gradient is essentially similar $[M(s) = M]$. This real parallelism becomes more evident when reported to the failure criterion $(\sigma'_1 - \sigma'_3)_{ult}$, with the respective results of tests on saturated samples having the same initial physical characteristics. This may suggest that the effective friction angle ($\phi$´) in the *q:p´* plane is constant and independent with the s = 0 kPa or s = 40 kPa plane, which means the independence of $\phi$´ with the suction applied.

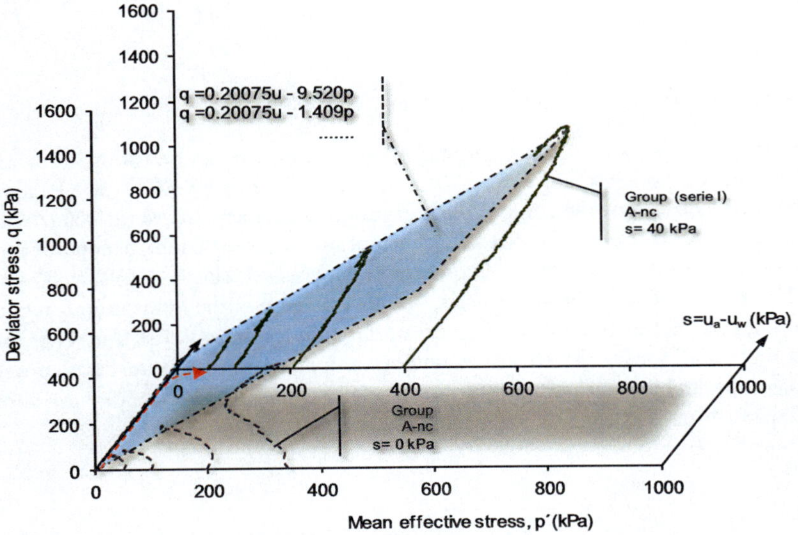


**Fig. 3.** Critical state lines of residual granitic soil: representation in the {q: p′: s} space of the stress paths followed in the planes s = 0 kPa and s = 40 kPa and the respective failure plans for the series I.

The plans assume the degree of saturation of 100%, that is $\phi_b = \phi'$ for the first interpolation, and for the second interpolation, $\phi_b < \phi'$ is assumed and associated with the residual degree of saturation for a suction of 40 kPa. The geometry is flat although it is an approximation, as it would require shearing tests in different imposed suction planes.

Table 3 presents the equations describing the failure planes for the failure criteria $(\sigma'_1 - \sigma'_3)ult$ in the {q: p: s} space.

The values of M(s) decrease and their variation is smaller between planes, which supposes a greater independence of $\phi$ from the imposed suction (*series I:* $\phi s = 40\ kP = 36$ *to* $39^o$ *e* $\phi s = 0\ kPa = 35$ *to* $39^o$), in contrast to cohesion (c) (*series I:* $cs = 40\ kPa = 2.3$ *to* $2.6$ *and* $cs = 0\ kPa = 0$). The value of $\phi_b$ is between 3.4° and 3.8°. This interval is small because it is close to the intrinsic behaviour and increasingly dependent only on the friction effect and the suction in the sample.

**Table 3.** Equations that describe the failure planes and parameters for the failure criterion [$(\sigma'1-\sigma'3)$ *ult*] in the {q: p: s} space

| Serie | Suction (kPa) | Failure criterion $[(\sigma'_1-\sigma'_3)_{ult}]$ | | | |
|---|---|---|---|---|---|
| | | Mean Plan Failure envelope | Shear parameters | | |
| | | | c′ (kPa) | ϕ′ (°) | $\phi^b$ (°) |
| *I* | 0 | q = 0.20075u–5.4645p | 0 | 35 | 3.8 |
| | 40 | | 2.6 | 36 | |
| | 40 | | 2.3 | 39 | |

## 3.2 Shear Behaviour with Seepage Effect

The maximum shear stress at the beginning of the seepage can be estimated from the results of the total shear stress with σ′30 and constant suction (σ′30 = 100 kPa and s = 40 kPa) and by observing the curve in the logEtang: logεa space. In tests affected by seepage, the shearing stage is carried out with constant suction (s = 40 kPa) except for the fact that this occurs until 3 to 4% of axial compression increase is reached, followed by seepage. The increase of the axial stress used for the beginning of the infiltration is related to the moment when the abrupt discontinuity starts in the evolution of stiffness (E), marking the first yield, for the dry shear test without seepage, throughout its entire progress. Figure 4 shows the discontinuity of the tangent deformability modulus for 3 to 4% of the axial deformation.

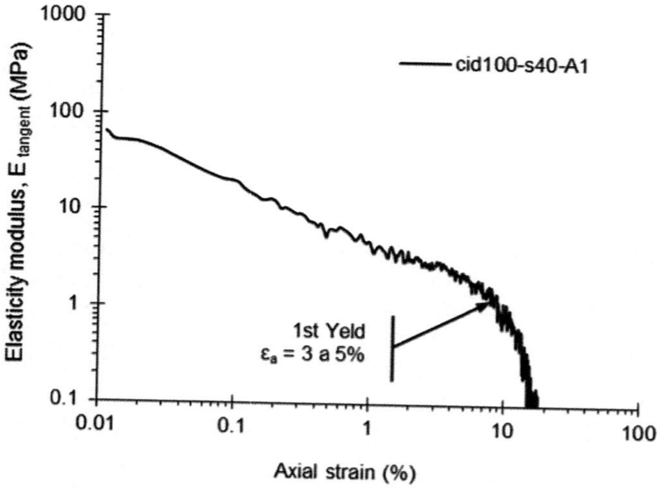


**Fig. 4.** Variation of the stiffness module for drained tests with constant suction and sheared off to $\sigma'_3 = 100$ kPa

When the test stops, the creep effect must be separated to proceed with the seepage. That is, the change in the volumetric stress before the infiltration begins, verified by the deformation of the soil specimen under constant shear and suction stresses. The creep effect is negligible in this type of soil concerning maintaining or correcting the vertical tension when seepage acts. The absence of the creep at this stage ensures that the material does not yield during seepage. The displacement during the seepage is caused by the water pressure variations in pores.

In the first stage of tests considering infiltration, the stress level fixes at the first yield ($\varepsilon_a = 3\ to\ 4\%$), and the shear stress reduces by reducing the suction during the infiltration process. The strength reduction is gradual, starting at the base and moving towards the top while water flowed into the soil. The yield strength gradually decreases and a greater shear surface is expected in tests without seepage. This is in line with the need to increase the axial stress up to failure.

Samples which have undergone water infiltration and leachate have shrinkage behaviour, and it appears that the volumetric growth relates to the failure of the soil skeleton. However, the magnitude of the volumetric growth involved in the failure is less than that induced by tests with constant moisture content. This relates to the wetting process (or suction reduction) carried out at high-stress ratio or the level of acquired axial stress increase, suppressing the negative dilation.

In the case of the sample infiltrated with gasoline, it has a virtually dilating behaviour, which is largely due to the gradual corrosion of the membrane. Although it is difficult to keep separate the membrane effect from the physical effect inherent in the evolution of the air-fluid-solid skeleton system (coagulation of fines, particle bonds, gasoline volatility and lubrication), it is certain that this infiltration provides the highest shear stress for small axial stresses.

In the representation of the space defined by the planes *{q:p:s}* (Fig. 5), the stress paths of infiltration tests follow the pathway defined by the imposition of suction, compression and pressure equalization of the specimen, drained shear test with $\sigma'_{30} = 100$ kPa, seepage, and final shear under the same conditions but with the average shear stress altered by the decreased suction. The tests proceed as the drained shear tests with $\sigma'_{30} = 50$ to 55 kPa. The maximum shear stress show loss of shear strength in all samples tested after the seepage by the various fluids when compared with the compression shear tests $\sigma'_{30} = 50$ kPa. The stress paths follow the break envelope defined in the stress plane for $s = 0$ kPa without touching it.

Table 4 reports the values of shear stress vs. mean stress at failure for different criteria, for samples with infiltration of different fluids and for samples with constant suction without infiltration.

Following different criteria, the soil strength after seepage is lower than those without infiltration or wetting. The average difference of shear stress of samples infiltrated with water and leachate is approximately 70 kPa. This difference is smaller when compared to the infiltration of gasoline. The axial stress required at failure is generally lower when the soil is infiltrated.

During the shearing stage and after infiltration, the suction still exists, leading to the conclusion that the failure occurs under conditions of negative pore water pressure

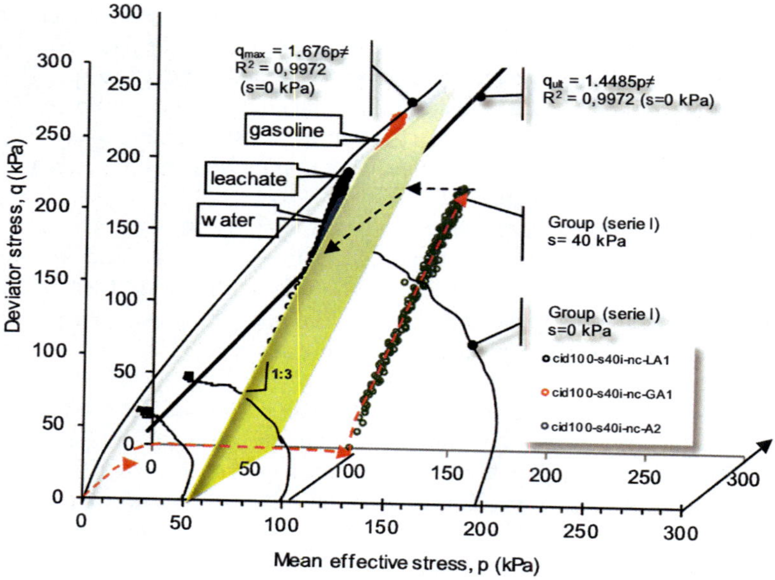


**Fig. 5.** Stress paths for shear tests with initial $\sigma'_{30} = 100$ kPa and infiltration with leachate water and gasoline

**Table 4.** Shear test for $\sigma'_{30} = 100$ kPa, with dry samples followed by infiltration (water or leachate or gasoline) in soil type A-nc: stress values for different failure criteria

| Fluid | | Infiltration | | Without infiltration | |
|---|---|---|---|---|---|
| | | Water | Leachate | Gasoline | |
| $\sigma_3$ (kPa) | | 60 | 60 | 60 | 60 |
| $s_{initial}$ (kPa) | | 40 | 40 | 40 | 40 |
| $u_{last}$ (kPa) | | 1.5 | 2.5 | −6.4 | 40 |
| p′(kPa) | $(\sigma'_1/\sigma'_3)_{max}$ | 128.4 | 126.0 | 148.6 | 200.1 |
| q (kPa | | 206.1 | 213.56 | 245.7 | 279.3 |
| $\varepsilon_a$ (%) | | 11.8 | 14.0 | 13.5 | 15.1 |
| p′(kPa) | $(\sigma'_1-\sigma'_3)_{max}$ | 127.9 | 126.1 | 150.9 | 202.9 |
| q (kPa) | | 206.5 | 213.8 | 249.76 | 278.4 |
| $\varepsilon_a$ (%) | | 13.3 | 13.6 | 13.0 | 14.4 |
| p′(kPa) | $(\sigma'_1-\sigma'_3)_{ult}$ | 124.6 | 123.2 | 148.4 | 194.8 |
| q (kPa) | | 191.1 | 198.6 | 239.6 | 254.2 |
| $\varepsilon_a$ (%) | | 25.4 | 24.9 | 25.2 | 25.1 |

and the soil strength decreases exactly how much the suction decreases. In tests with infiltration, the failure area is larger than in tests with constant suction or saturated.

## 4 Conclusions

The convergence of yield lines in case of high stresses mean that M (from the expression $q = Mp' + d$) tends to be constant and $\phi$ depends only of frictional characteristics of the soil, admitting $\phi'(s = 40\ kPa) = 37^{\circ}$ and $\phi'(s = 0\ kPa) = 36^{\circ}$ and the cohesion grows due to the small variation of $\phi_b$. The value of $\phi_b$ allows to conclude that it is practically constant ($\phi_b \approx 3{,}5^{\circ}$). This will not be so clear as it depends on Sr and fabric, fact that makes it impossible to build a single plan for the limit state in the {q:p´:s} space.

The soil shear strength for $p'_0 = 100$ kPa, after infiltration or wetting with water or leachate, decreases with approximately 70 kPa, this difference being smaller in case of infiltration with gasoline. In the first stage of the shear test at constant moisture content, the particles and aggregates act as individual particles, but with the reduction of suction by seepage the structure of aggregates becomes unstable and the collapse of the structure guarantees the decrease of strength and volume.

In tests with infiltration, the soil collapse at a lower saturation level than 100%, and the suction decreases with the increase of the effective mean stress. This fact is also evident for tests at constant water content by increasing the volumetric compression. The magnitude of the volumetric stress in the collapse during wetting is less than that induced by shear stress at constant water content, and this is due to the infiltration process taking place at high stresses already installed which suppressed some contraction.

The displacement during the seepage stage is essentially caused by the change in pore water pressure as the creep is negligible. Even so, this moment can be used as a failure criterion for slopes subjected to infiltration. The failure can be considered when the displacement ratio increases rapidly. The suction measured at this point still exists and should not be overlooked. In other words, this means that the slope may break in an unsaturated condition, that is, there is negative water pressure in the pores.

In the first shearing stage, the imposed suction is high, but the aggregates act as individual particles. The soil typical behaves as granular justifying the actual grain-size distribution of the soil. When the suction reduces at the seepage, the aggregate structure becomes unstable and produces a decrease in strength and volume due to the collapse of the created structure. In other words, it appears that steep slopes may be less vulnerable than softer slopes.

## References

1. ASTM: Classification of soils for engineering purposes. In: 1989 Standards Annual Book D2487-85, vol. 04, pp. 288–297. ASTM, Philadelphia (1989)
2. Bishop, A.W., Blight, G.E.: Some aspects of effective stress in saturated and partly saturated soils. Geotechnique **13**, 177–197 (1963)
3. Buzzi, O., Fityus, S., Sheng, D. (eds) : Proceedings of the 4th Asia Pacific Conference on Unsaturated Soils, Newcastle, Australia, 23–25 November 2009, CRC Press, Taylor & Francis Group (2009)

4. Charles, W.W., Abraham, C.F.: Laboratory study of loose saturated and unsaturated decomposed granitic soil. J. Geotech. Geoenvironmental Eng. ASCE **129**(6), 550–559 (2003)
5. Escario, V., Sáez, J.: The shear strength of partly saturated soil. Géotechnique **36**(3), 456–463 (1986)
6. Fredlund, D.G., Rahardjo, H.: Soil Mechanics, for Unsaturated Soils. John Wiley & Sons, New York (1993)
7. Fredlund, D.G., Rahardjo, H., Fredlund, M.D.: Unsaturated Soil Mechanics in Engineering Practice. John Wiley & Sons, Hoboken (2012)
8. Fredlund, D.G., Morgenstern, N.R., Widger, R.A.: The shear strength of unsaturated soils. Can. Geotech. J. **15**, 313–322 (1978)
9. Gan, J.K.M., Fredlund, D.G.: Shear strength characteristics of two saprolitic soils. Can. Geotech. J. **33**, 595–609 (1996)
10. Karube, D., Kawai, K.: The role of pore water in the mechanical behavior of unsaturated soils. Geotech. Geol. Eng. **19**(3), 211–241 (2001)
11. Milatz, M., Törzs, T., Grabe J: Settlements in unsaturated granular soils induced by changes in saturation and suction. In: E3S Web of Conferences, vol. 9, pp. 1–7, E-UNSAT (2016)
12. Mofiz, S.A., Sarker, D.C., Taha M.R.: Experimental and modeling of unsaturated soil behavior under stress path tests. In: Romero & Cui (eds.) Advanced Experimental Unsaturated Soil Mechanics, Tarantino, Taylor & Francis Group, London (2005)
13. Pham, K., Kim, D., Lee, I.-M., Choi, H.: Hydraulic-mechanical properties of unsaturated granite-weathered residual Soil in Korea. Vadose Zone J. **18**(1), 1–13 (2019)
14. Rahardjo, H., Lim, T.T., Chang, M.F., Fredlund, D.G.: Shear - strength characteristics of a residual soil. Can. Geotech. J. **32**, 60–77 (1995). Canada
15. Rahardjo, H., Kim, T., Satyanaga, A.: Role of unsaturated soil mechanics in geotechnical engineering. Int. J. Geo-Engineering **10**(1), 1–23 (2019). https://doi.org/10.1186/s40703-019-0104-8
16. Rotaru, A., Babor, T.D., Răileanu, P., Rotaru, P.: The time influence of stress particularities on the soil strength and deformations.In: Scientific Session CIB 2004, Braşov, 18 – 19 November 2004, vol. 2, pp. 161-166. Transilvania University Publishing House (2004)
17. Rotaru, A., Răileanu, P., Rotaru, P.: Expressions of stress-strain-time functions that characterize the soil creep. In: International Technical-Scientific Conference "Current Problems of Urbanism and Spatial Planning", vol.3, Roads, Materials and Construction Mechanization Section, 30 September – 1 October 2004, Kishinev (2004)
18. Toll, D.G.: A framework for unsaturated soil behaviour. Géotechnique **40**(1), 31–44 (1990)

# Identification of Thermal Properties of Construction Materials Using Far-Field Contactless Temperature Measurements

Nabil Younes[2], Ikram Mouchrik[2], Marouen Slamia[1,2], Naima Belayachi[1,2], and Dashnor Hoxha[1,2(✉)]

[1] Univ Orléans, Univ Tours, INSA CVL, LaMé, 45072 Orléans, France
dashnor.hoxha@univ-orleans.fr
[2] Univ Orléans, Polytech Orléans, 45072 Orléans, France

**Abstract.** The IR thermography of buildings is used extensively in the diagnostic of buildings for thermal renovation, but is limited more often in detection of hot/cold regions and thermal bridges. In this paper we propose an alternative use of IR temperature measurements to retrieve effective local thermal diffusivity of building walls. The approach is inspired from known technics for thermal measurement such as hot-disc method, but the heat source is supplied by a laser beam and field temperatures are measured by a camera. So, the setup for contactless measure of thermal diffusivity combined a laser spot of diameter of 1.2 cm as a source of adjustable heat power and a camera IR able to record infrared images with a chosen frequency (up to 200 Hz). An adapted numerical method is proposed to retrieve the effective radial thermal diffusion value from the temperature evolution recorded on some points on the wall surface. Since the setup is designed for in field measurement a robust algorithm for parameter identification problem such as a genetic algorithm (GA) has been used. Although reputed for its robustness, GA does a massive call to the direct problem leading to calculation time some time prohibitive. In this paper GA use a fast estimator of direct problem that use a metamodel issued from Proper Orthogonal Decomposition of a beforehand built-up snapshot matrix combined with a Radial Basis Function interpolation. Some results obtained using the proposed setup and inversion procedure are presented as un example.

**Keywords:** Infra red camera · Inverse problem · Global minimum · Genetic algorithm · Proper orthogonal decomposition · Radial basis functions

## 1 Introduction

Thermal renovation of buildings has become nowadays more and more synonym of building renovation and in many countries program-msters are setup in order to match the exigencies on renovation ([4], [8]). In France, for example, a third of the housing stock is concerned by this thermal renovation and one of the objectives of the Energy


A. Rotaru (Ed.): CRIT-RE-BUILT 2019, SSGG, pp. 241–248, 2021.
https://doi.org/10.1007/978-3-030-61118-7_21

Transition Act is to renovate 500,000 homes per year by 2050 [8]. Along with developing of new materials and proceedings for building insulations the thermal diagnostic technics have seen a considerable increase taking advantage also to numerical revolution and increase of capacity and rate operation of recent computers ([1], [9]). In most of cases the thermal diagnostic is performed by checking in situ the materials used in constructions/insulation and then use some material databases to retrieve the thermal parameters of components before making some reglementary analyses or in a better case some dynamic simulation to quantify the thermal performance of a given building ([4, 7]). In some other cases Infra Red (IR) thermography is used in order to identify thermal bridges or to evaluate the efficiencies of thermal insulation ([13]). In this last case the operator of IR thermography obtains a temperature field that could be further used to perform various analyses. In the best knowledge of these authors the problem of in field thermal properties identification from these measured fields of temperatures has never been addressed. In some rare cases, where, in field observations have been used to identify thermal properties of wall buildings, these analyses are performed in new constructions, pre-equipped with adapted sensors of temperature, humidity and/or thermal flux installed beforehand ([6, 9]). Otherwise the technics of measuring thermal properties (specific heat capacity, thermal conductivity, thermal diffusion, thermal effusion) are well developed in laboratory using either steady state or transient situation of diffusion problem ([1–3, 10, 11]). Nowadays these techniques of thermal measurements have large applications in industry and could be used for classification and control of industrial proceedings ([11, 12]). Yet, these technics are rarely adapted for characterization of thermal properties of in field heterogenous materials.

The purpose of this paper is to present a new, relatively simple, contactless method that could be used for measuring of in field thermal properties of heterogenous materials, namely thermal properties of wall buildings. This method, inspired from "hot disk" technique of thermal measures, consists in applying a thermal load of known intensity usually of a regular shape (circular here) and monitoring the full field of temperatures. Then these measured data are used in an inverse problem to deduce thermal parameters. As compared with classical hot-disc method of measurements the thermal load is applied on the surface of a material and in difference to flash method the thermal load is maintained during several seconds. This is important in particular when the material to be characterized is heterogenous where a significant volume of material, superior to Representative Elementary Volume (REV), should be characterized.

This paper represents firstly a laboratory setup that could be considered as an protype yet, not sufficiently versatile to be used in practice for in field measurements since some optimization in weight and dimensions are yet necessary. Nevertheless, we describe in details the principles of measurements and we perform measurements on 50 cm × 50 cm × 10 cm samples to validate the setup and the specific numerical procedure developed for this device.

# 2 Materials and Methods

## 2.1 Laboratory Setup

The laboratory setup used for these measurements is constituted by a $CO_2$ laser system with adjustable power (0–25 W) producing an invisible laser beam of diameter 1.2 mm, a beam reshaping lent assuring a zoom of X10. An infrared camera FLIR A655sc (resolution 480 × 640) is used to record temperature field able to take photos with a frequency up to 200 Hz. The camera and the laser generator are switch on simultaneously in order to assure the synchronization of images and the thermal heating history. The sequences of infrared videos are firstly saved in an external storage device in order to be treated using a specific numerical method of filtering and inversion (Fig. 1).

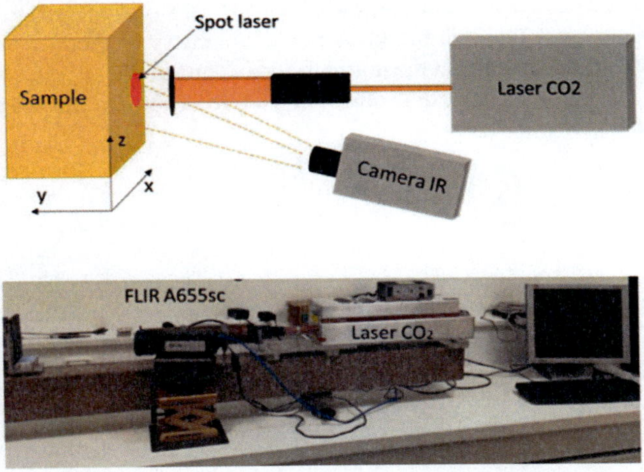


**Fig. 1.** Schematic presentation and view of experimental setup used for measurements

The use of a continuous heating source and a opening window of laser beam up to 18 mm are two singular features of this device. The choice of these characteristics are motivated by the necessity to heat up samples of materials with dimensions up to some decimeters without excessively increasing local temperature of samples in order to avoid any burning of bio-resourced insulating materials. In particular in order to be used for in field measurements, the temperature field of in front face are measured and used. These choices induced a special treatment of measured fields of temperature that differs from known similar methods such as hot-disc or flash method [11, 13]. The theoretical bases of this treatment are shortly given here.

## 2.2 Direct Problem of Heat Diffusion Due to a Circular Laser Spot

The authors recommended to interested reader the references [2, 3] for a detailed description of fundaments of heat transfer in solids. The solution of the problem of heat diffusion in a solid described in Fig. 2, is given by ([1, 2]):

$$T\,(q, r, a, \mu) = \int_{\theta=\pi}^{\theta=2\pi} \int_{\rho=0}^{\rho=R} \frac{q[\rho, 0]}{4\pi ar} ercf\left[\frac{(x-\xi)^2 + (y-\eta)^2 + (z-\xi)^2}{\sqrt{4\pi ar}}\right] d\rho\, d\theta \tag{1}$$

In this equation $T$ stands for temperature rise, $x$, $y$ and $z$ are the cartesian coordinates of point where is measured the temperature, x, h, z are coordinates of a point inside the circular laser spot of rayon R, a is the thermal diffusivity of the solid $a = \frac{\lambda}{\mu}$ with λ beign the thermal conductivity, and μ the volumetric heat storage (equal to specific heat capacity multiply by volumetric mass). A radial symmetry of thermal flux density is assumed within the spot such that the normalized heat power $q[\rho, \theta]$ is a radial function only. In turn, since at this stage the solid is considered isotropic this leads to a radial symmetry of field temperature.

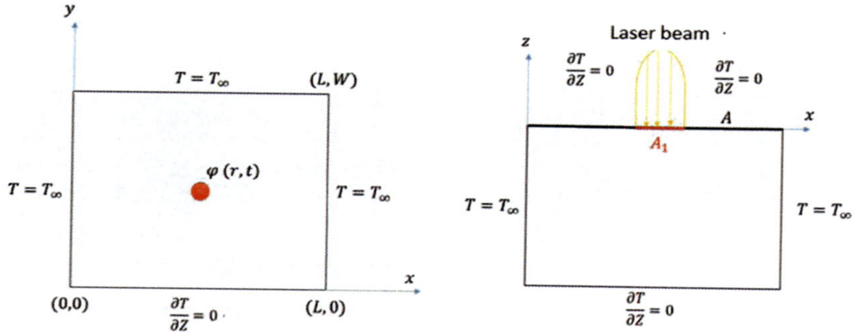


**Fig. 2.** Definition of direct problem of diffusion in a solid of heat produced by a laser spot on its surface and hypothesis of boundary conditions

If the temperature rising in two different instant of time $t_1$ and $t_2$ is $T_1$ and $T_2$ the ratio of temperatures rising do not depend any more from the source intensity ([1]):

$$R_T(r, t_1, t_2, a) = \frac{T(q, r, a, t_1)}{T(q, r, a, t_2)} \tag{2}$$

For the method proposed here one construct the field of temperature ratios from measured temperatures in some selected instants. Then inverse problem of parameter estimation is solved in order to retrieve the thermal diffusion will all other quantities known.

### 2.3 Thermal Diffusivity Retrieval from Inverse Problem Using POD-RBF Approach

Retrieval of thermal diffusivity from temperature fields measured in different instants is solved here as an inverse problem of parameter estimation that could stated as below: given a theoretical model of a measurable quantity $A(\underline{p})$ depending on a set of parameters $\underline{p}$ find the vector $\underline{p^*}$ that minimize the functional:

$$\chi(p) = \frac{1}{2}\sum_{i=1}^{nmes} w_i(A_i(p) - \gamma_i^{mes})^2 \tag{3}$$

In this expression $w_i$ are some weighting function and $\gamma_i$ indicates a measured value from the overall number of measures $n_{mess}$.

Various strategies could be used to solve the problem, with advantages and drawbacks to be evaluated for each problem. The gradient-based methods are the most used because of their efficiency and their quadratic convergence on the vicinity of solution point. However their "blindness" to distinguish local minima calls for prudence and posterior verification of obtained solution if any. To avoid as much as possible the pitfall of local minima, more sophisticated, but more time-consuming methods such as Neural Networks (NN) or Genetic Algorithm (GA) could be used ([1]).

### 2.4 POD-RBF Based Metamodel for Diffusion Problem

In this work a GA algorithm from standard Matlab toolboxes is used combined with POD-RBF (Proper Orthogonal Decomposition – Radial Bases Functions) approach. Basically, the POB-RBF approach is used to a fast evaluation of temperature to a point of sample. Without going in mathematical details, one could say this is possible because POB-RBF approach could be proposed as a meta model to any theoretical model. This fast evaluation is based on interpolation performed on the space of singular vectors of a snapshot matrix, this last one represents some theoretical results obtained beforehand. Let's suppose that one collect in a matrix $\underline{\underline{U}}$ (snapshot matrix), M vectors of N elements, each representing results of a run of theoretical model (N points when the theoretical model is evaluated). Denoting by $\underline{\varphi_1}, \underline{\varphi_2}, \ldots, \underline{\varphi_K}$ a base of orthonormal vectors, it is obvious that each vector of matrix $\underline{\underline{U}}$ (so each results) could be expressed as a linear combination of this base, i.e.:

$$\underline{u_i} \approx \overline{\underline{\underline{\Phi}}}.\overline{a_i} \tag{4}$$

with $\overline{\underline{\underline{\Phi}}}$ being the matrix of all the orthonormal basis of the new space $\underline{\varphi_j}$, and $\overline{\underline{\underline{a_i}}}$ are the amplitudes corresponding to the $i^{th}$ vector of the new basis. Both the matrix $\underline{\underline{\Phi}}$ and amplitudes $\overline{a_i}$ are truncated and only a few vectors are used to obtain the approximation (4). The vectors $\underline{\varphi_i}$ of proper orthogonal decomposition of matrix $\underline{\underline{U}}$ are calculated using the eigenvalues and eigenvectors $\{\lambda_i\}_{1\le i\le M}$ and $\{\underline{v_i}\}_{1\le i\le M}$ of the covariance matrix $\underline{\underline{C}}$ issued by the snapshot matrix $\left(\underline{\underline{C}} = U\,U^T\right)$:

$$\underline{\varphi_i} = \underline{\underline{U}}v_i\lambda_i^{-\frac{1}{2}} \tag{5}$$

Once the proper vectors for POD are calculated one could reproduce back the snapshot matrix if necessary by a simple matrix multiplication:

$$\underline{\underline{U}} = \underline{\underline{\Phi}}\,\underline{\underline{A}} \tag{6}$$

with $\underline{\underline{A}}$ being the amplitude matrix. This allows to find the amplitude matrix by inversion of Eq. (6), i.e.:

$$\underline{\underline{A}} = \underline{\underline{\Phi}}^T\,\underline{\underline{U}} \tag{7}$$

In the last equation the orthogonality of the matrix $\underline{\underline{\Phi}}$ is used so that $\underline{\underline{\Phi}}^T = \underline{\underline{\Phi}}^{-1}$.

The Eqs. (4) to (7) makes it possible to calculate the amplitude coefficients for given snapshot result, or, inversely, for some fixed amplitude coefficients to calculate the response in the matrix $\underline{\underline{U}}$ by just a simple matrix operation. At this point one recalls that each vector (column) $\underline{u_i}$ of the matrix $\underline{\underline{U}}$ is obtained using theoretical model and a realization of a set parameters (vector) $\underline{p_i}$. Now, if one would like to know what the response of the system would be if a vector $\underline{p}$*, is used in theoretical model, a fast approximation could be obtained using Eq. (7) with amplitude coefficient obtained by an interpolation of the amplitude coefficient over the support point of parameters. While this interpolation could be whatever, in practice Radial Base Function interpolation has been proved to be an excellent robust technic. A radial basis interpolation of amplitude coefficients for the amplitude vector will be a linear combination of some functions of Euclidian distances between the vector $\underline{p}$* the other parameters vectors used in snapshot matrix

$$\underline{A}^*(P^*) = \sum_{j=1}^{M} \lambda_j \phi\left(\left\|\underline{p}* - \underline{p_j}\right\|\right) \tag{8}$$

The coefficient $\lambda_j$ of RBF interpolation are calculated from the interpolation condition applied on parameter vectors and amplitude coefficients used or issued from snapshot matrix. If $\underline{\underline{F}}$ is the matrix of interpolation, i.e.

$$\underline{\underline{F}} = \begin{bmatrix} \phi(\|p_1 - p_1\|) & \dots & \phi(\|p_M - p_1\|) \\ \vdots & \ddots & \vdots \\ \phi(\|p_1 - p_M\|) & \dots & \phi(\|p_M - p_M\|) \end{bmatrix} \text{ then } \underline{\underline{\lambda}} = \underline{\underline{F^{-1}}}\,\underline{\underline{A}} \tag{9}$$

## 3 Results and Discussion

Field temperature measurements are performed on a porous stone used in historical building using the setup described in the Sect. 2.1 The IR images are recorded with a frequency of 4 Hz during 125 s (see Fig. 3).

Once recorded the field temperatures are treated in order to create the vector of observations and the temperatures are extracted in number of observation points (Fig. 4).

With measured data the inverse problem of diffusivity identification is resolved using two algorithms: classical Levenberg Marquardt and GA. In both cases the POD-RBF approximation is used in order to accelerate the identification procedure. Results of these analyses are given in Table 1. The comparison of time execution of used algorithms justifies the use of GA in combination with POD-RBF for thermal diffusivity retrieval from IR images for in field observation as an alternative to classical gradient methods.

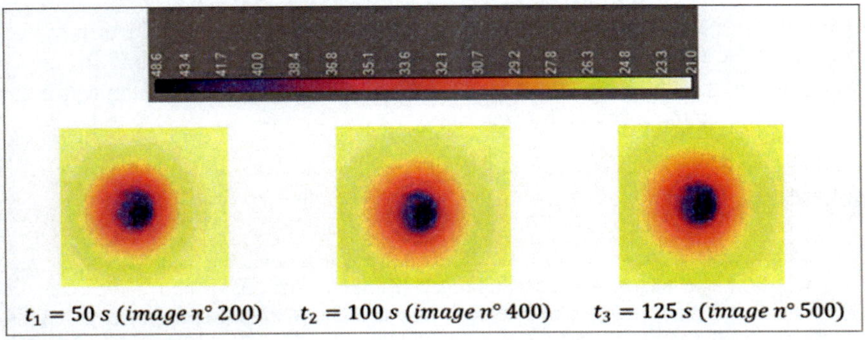


**Fig. 3.** IR images recorded during heat test on Richmond stone

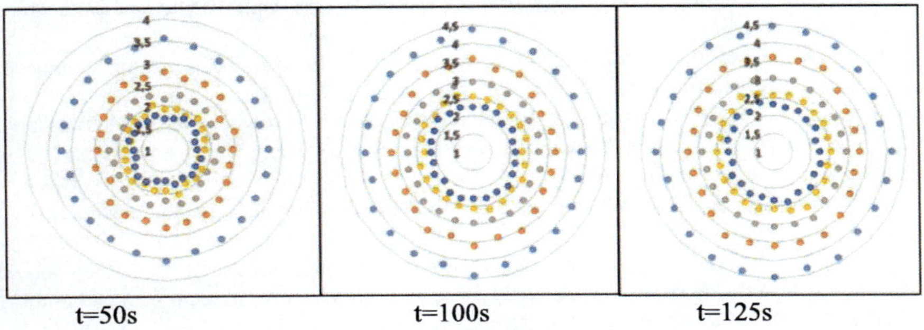


**Fig. 4.** Evolution of temperature ratio in observation points as a function of time

**Table 1.** Time of for diffusivity identification using various strategies for inverse problem

| Algorithm | Time in seconds of procedure using POD-RBF | Time in seconds of procedure using POD-RBF |
|---|---|---|
| Levenberg – Marquardt | 0.0017 | 0.0159 |
| Genetic Algorithm | 0.0714 | 3.6234 |

## 4 Conclusions

A method for retrieval of thermal conductivity using contactless measures has been presented. As compared to other laboratory technics used for this kind of measurements the method presented here is based on the measures of temperature field on the in front face of samples, and so make it possible to perform in field measurements since it do not needs to access the back face sample as it is necessary for known methods. However two technical problems should be resolved: the first problem is the possibility to perform an inverse analysis in an acceptable time, all by avoiding the pitfall of local minima and the second problem is the optimization of the design of the setup in order to transform it in a mobile device. Only the first problem is treated here and a GA inverse method is proposed. The use of POD-RBF interpolation speed-up the overall time for inverse problem and reduce it in acceptable time for in field operation. As for the second problem, it is a subject of ongoing research by authors.

## References

1. Adamczyk, W.P., Bialecki, R.A., Kruczek, T.: Retrieving thermal conductivities of isotropic and orthotropic materials, Appl. Math. Modell. **10**(28) 56–62 (2015)
2. Carslaw, B.H.S., Jaeger, J.C.: Conduction of Heat in Solids, Oxford Clarendon Press, Oxford, 2nd Ed. (1959)
3. Cole, K.D., Beck, J.V., Haji-Sheikh, A., Litkouhi, B.: Heat Conduction Using Green's Functions, 2nd Edition, CRC Press (2010)
4. Sustainable Development Knowledge Platform (2019). 01 October 2019. https://sustainabledevelopment.un.org/
5. Hoxha, D., Belayachi, N., Do, D.P.: On thermo-hydro-mechanical (THM) fatigue damage of historical stone buildings. Adv. Mater. Res. **891**, 36–41 (2014)
6. Islam, S., Bhat, G.: Environmentally-friendly thermal and acoustic insulation materials from recycled textiles. J. Environ. Manage. **251** (2019). https://doi.org/10.1016/j.jenvman.2019.109536
7. https://onpe.org/news/precarite_energetique_combien_de_personnes_peinent_chauffer_leur_logement
8. Slaimia, M., Belayachi, N., Hoxha, D.: In situ: performance assessment of a bio-sourced insulation material from an inverse analysis of measurements on a demonstrator building. EBUILT2016, 16–19 November, Iasi (Romania) (2016)
9. Adamczyk, W.P., Białecki, R.A., Kruczek, T.: Retrieving thermal conductivities of isotropic and orthotropic materials. Appl. Math. Model. **40**, 3410–3421 (2016). https://doi.org/10.1016/j.apm.2015.10.028
10. He, Y.: Rapid thermal conductivity measurement with a hot disk sensor. Thermochim. Acta **436**, 122–129 (2005). https://doi.org/10.1016/j.tca.2005.06.026
11. Hemminger, W., Jugel, R.: A guarded hot-plate apparatus for thermal conductivity measurements over the temperature range − 75 to 200 C. Int. J. Thermophys. **6**, 483–498 (1985). https://doi.org/10.1007/BF00508892
12. Degiovanni, A.: Diffusivité et méthode flash. Rev. Générale Therm. **185**, 420–442 (1977)
13. Slamia, M.: Caractérisation thermo-hydrique et évaluation in situ des matériaux de construction hétérogènes à base de fibres végétales. Ph.D. in Civil Engineering, University of Orleans (in french) (2018)

# Tensile Testing of a Biocomposite Material – "Liquid Wood"

Tudor-Cristian Petrescu(✉), Petru Mihai, and Ionuț-Ovidiu Toma

"Gheorghe Asachi" Technical University of Iași, Iași, Romania
tudor.petrescu@tuiasi.ro

**Abstract.** Modern society has faced significant economic development in the previous decades. In this, the construction industry played (and continues to play) a central role, contributing an important amount to the GDP of both developed and developing countries. However, development comes with negative consequences as well – pollution, waste and excessive $CO_2$ emissions often being mentioned. For the construction industry, this means that new solutions should be found that can minimize, as much as possible, the impact upon the environment. One such solution may reside in the development and employment of biocomposites. In short, biocomposites are nothing else than "classic" composite materials when it comes to their structure (fibers held together by a matrix), but the fibers are naturally – sourced and so are the matrix resins. In theory, they are up to 100% biodegradable and completely eco – friendly. "Liquid wood" is a relatively new biocomposite, which is compliant with the above – mentioned principles, but it is thermo – injectable (just like plastics). This has the potential to open up some interesting ways of employing it. The following article is organized as follow: in the introduction, "liquid wood" is presented in more detail. After that, some wood – specific degradations that can be solved by using "liquid wood" are discussed. Then, the "liquid wood" samples are presented. Following this, tensile testing in laboratory is explained. Finally, some conclusions may be drawn related to the obtained results, as well as following up with more research.

**Keywords:** Biocomposite · Liquid wood · Testing

## 1 Introduction

The necessity of finding new, sustainable materials, with their use ranging from everyday items to construction materials and other advanced applications (ex.: automotive and aero industries), represents a significant direction of research for modern societies. The massive pollution and subsequent damage to the environment caused by the production of "classic" building materials, along with the ubiquitous use of plastics, are driving factors for the discovery of new materials.

In the construction industry, the first "modern" material was considered to be what actually turned out to be a new class of materials – composite materials. A composite material is made of matrix and reinforcement. The matrix is usually a type of resin (epoxy,


A. Rotaru (Ed.): CRIT-RE-BUILT 2019, SSGG, pp. 249–253, 2021.
https://doi.org/10.1007/978-3-030-61118-7_22

vinyl ester etc.), which binds together a type of fiber (Kevlar, aramid etc.). Composites were adopted in the construction industry at a later time, as the aero industry was the premier user of such materials, followed by the automotive industry.

In time, composites have proven to be suitable in rehabilitation work, being quite easy to apply and proving resilient to some forms of degradations. Nevertheless, composite materials share a common problem with "classic" building materials: sustainability. Usually, they require significant amounts of energy to be produced and some of their base "ingredients" may include petroleum – derived products.

Logically, the answer to the aforementioned problem is the development of biocomposites. Biocomposites is a term describing biodegradable composite materials. The term implies that the matrix is biodegradable and the reinforcement is made of natural fibers (Pilla 2011). Furthermore, the matrix can be made from polymers sourced from recyclable, renewable resources. This is important in protecting the fibers from degradation, as well as in binding them together, the end result being an increase in mechanical and structural properties (Nägele et al. 2013). The fibers have biological origins and can be made out of various plant species (hemp, cotton, flax), recycled wood or regenerated cellulose.

## 2 A Short Description of "Liquid Wood"

"Liquid wood" is a relatively new polymeric biocomposite material, initially developed by a German company, Tecnaro GmbH. The "composite" aspect means that it has a matrix which holds together the reinforcement. In the case of "liquid wood", the matrix is made of lignin (a naturally – occurring substance that lends strength to plants) and the reinforcement is made of various natural fibers (hemp, flax, straw – depending on the recipe). The "polymeric" aspect of this product is not a reference to plastics, but rather to the material's behavior, which is akin to thermoplastics – it is possible to inject it and mold it into various objects. Finally, the "bio" aspect of the product is a result of the constituent materials present in "liquid wood". Depending on the presentation form (or recipe), it can be up to 100% biodegradable. Sustainability is ensured thanks to the fact that its constituents are naturally – sourced and can be obtained from renewable resources and the manufacturing procedure has a low environment impact (according to the manufacturer).

As such, "liquid wood" appears to be a good candidate for a modern, sustainable material with a good range of applications.

## 3 "Liquid Wood" Injection – Material and Technology

Everything written in the previous chapter holds true on a still mostly theoretical level. To make sure that "liquid wood" is a valid material for the construction industry, there is only one way to do that – laboratory testing. Because "liquid wood" has, as stated by the producer, the same processing characteristics as thermoplastics (Tecnaro GmbH 2019), Arbofill and Arboblend material was injected into dumbbell shapes (according with technological sheets provided by the manufacturer) and then it was tested to tension.

Characteristics of dumbbell shapes: L = 150 mm; T = 4 mm; W = 8 mm; M = 8–10 g.

Following the installation of tensiometric marks (red color in Fig. 1), the samples were installed and tested up to failure, one by one, in the tension press (Figs. 2 and 3).

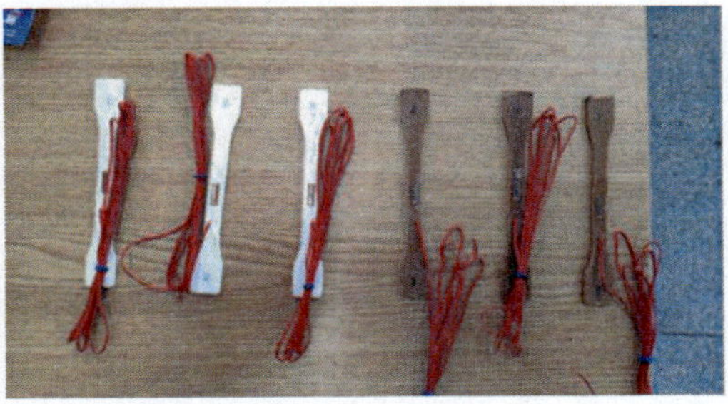

**Fig. 1.** Arbofill and Arboblend dumbbell samples

**Fig. 2.** Dumbbell sample fixed in the press

**Fig. 3.** Dumbbell sample after failure

# 4 Experimental Results

It has been observed that the samples failed in a rather brittle way, although they did suffer a slight elongation. This would be in accordance with the manufacturer's stated 3% elongation ratio. The results shown in Fig. 4 have been obtained on Arbofill (brown) and Arboblend (white) samples.

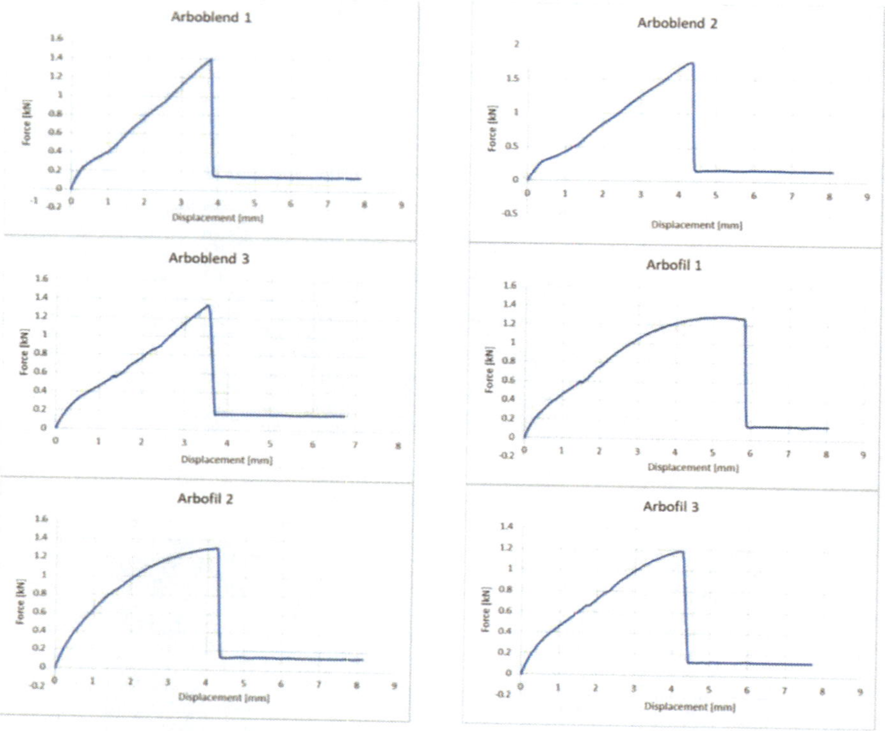


**Fig. 4.** Charts describing Arbofill and Arboblend tensile performance

## 5 Conclusions

From the obtained diagrams, it can be observed that both Arbofill and Arboblend present a good behavior in tension (40.6 MPa), in comparison with traditional wood (average value under 30 MPa). The results indicate that "liquid wood", from a structural point of view, could be used to rehabilitate existing degraded wooden structures.

As an average of values, breakage occurred at a force of approximately 1.3 kN (40.6 MPa), at a maximum displacement of approx. 4–4.5 mm. Correlated with the fact that the mid – cross section has a dimension of 8 mm × 4 mm, it is suggested that the material has enough strength to be used in conjunction with existing wooden structures.

Finally, the brittle fracturing of the material depends a lot on the intended use of "liquid wood", so it can be viewed either as an advantage or a disadvantage, depending on its purpose. This effect might also matter less, as it is more likely that the wooden element will be the first to fail. In studying the two recipes, Arbofill has been observed to present a bit more elasticity than Arboblend, but Arboblend has a slightly lower melting point. Overall, both materials are fairly balanced in terms of performance.

## References

GmbH, T.: Applications - tecnaro.de. Tecnaro.de (2019). https://www.tecnaro.de/en/applications.html. Accessed 20 Feb 2019

Nägele, H., Pfitzer, J., Ziegler, L., Inone-Kauffmann, E.R., Eckl, W., Eisenreich, N.: Lignin matrix composites from natural resources – ARBOFORM®. In: Kabasci, S. (ed.) Bio-Based Plastics (2013)

Pilla, S.: Engineering Applications of Bioplastics and Biocomposites – An overview, pp. 1–14. Wiley, New Jersey (2011)

# Sustainable Rehabilitation in Architecture and Urban Development

# Sustainable Rehabilitation in Architecture and Urban Development

Kaja Pogačar(✉), Vesna Žegarac Leskovar, Miroslav Premrov, and Peter Šenk

University of Maribor, Maribor, Slovenia
kaja.pogacar@um.si

**Abstract.** The issues of sustainability in architecture and urban development (usually designated as urban regeneration) imply a strategic approach and thinking as well as careful planning, development and management. The presented approaches to the rehabilitation of buildings refer to structural improvements, material reinforcements and program adaptations, generally divided into four directions: preservation, restoration, renovation and remodeling or adaptive reuse.

On the other hand, the topic of sustainable rehabilitation in urban development relates to urban regeneration in the broadest sense, a multidisciplinary field of research, scholarship, public policy and practice, which includes the elements of city planning, housing, transport, infrastructure, political economy, urban design, urban tourism, community development, sustainability and cultural industries studies with the aim to improve the quality of life. Some of the exposed and presented approaches include creative urban regeneration, the flagship projects approach, the temporary use approach, participatory practices in urban regeneration, the smart city approach and urban regeneration through gentrification.

To achieve the sustainability goals of rehabilitation in architecture and the urban environment, the approaches and tools range from the classic or traditional top-down planning or design to the more recent contemporary, informal, creative and adaptive bottom-up approaches involving direct physical interventions.

The scope of the presented topics concerns not only architects and urban planners, but all disciplines that deal with issues of our built environment; therefore, interdisciplinary engagement and cross-sectorial connections are necessary.

**Keywords:** Sustainable rehabilitation · Architecture · Urban development

## 1 Introduction

Sustainable rehabilitation in architecture and urban development is considered one of the most important topics within the discipline, as the extent of already built buildings is notably higher compared to the newly built ones [1] (i.e. in Germany approximately 75% of all construction activities are executed upon already built buildings, including their demolition, and they are expected to increase. Only 25% of all construction activities concentrate on the entirely newly built objects. In addition, almost three quarters of housing stock is between 15 and 60 years old).


A. Rotaru (Ed.): CRIT-RE-BUILT 2019, SSGG, pp. 257–272, 2021.
https://doi.org/10.1007/978-3-030-61118-7_23

Rehabilitation addresses architecture with historic and cultural legacy, the 19th and 20th century industrial buildings as well as the huge share of housing estates and other typologies of the built environment. According to Oliveira and Bragança [1] "to rehabilitate the urban environment – public and built space – implies inhabiting existing buildings, so that they can acquire life and thus assigning movement to the public spaces. To inhabit these buildings, it is necessary to recover them, providing good living conditions, space functionality and environmental effectiveness, that is, rehabilitate existing buildings, improving their efficiency".

Addressing the issues of sustainability in architecture and urban development (usually designated as urban regeneration) implies a strategic approach and thinking as well as careful planning, development and management, as the viewpoints are many and diverse. The topic has to be taken seriously, since more and more people live in cities (today more than half of the world population lives in cities, by 2050 more than 68% of the world population is expected to live in urban areas) [2]. As the trend of the rapid urbanization process will continue, according to the U.N. General Secretary Ban-Ki Moon "our struggle for sustainability will be won or lost in cities" [3].

Sustainable rehabilitation in architecture and urban development will be addressed separately below, since approaches, methodologies, tools and practical examples may vary.

## 2 Sustainable Rehabilitation in Architecture

Architecture, defined both as the process and the product of planning, designing, and constructing buildings or any other structures [4], has its beginnings in the Neolithic when men began to design and construct the built environment to protect themselves. It varies according to the functionality and purpose, the location where it stands, the materials and constructions used, different styles resulting from historic periods and influencing cultures, habits, social and economic differences. Architecture as such is also defined as the art and technique of designing and building [4]. When speaking of rehabilitation in architecture as we know it today, it is considered a rather young discipline originating from the late 19th century. However, we have to be aware that humans have sought to improve and renew buildings throughout history ever since the first settlements were built (with archeological evidence as early as 8000 BC, i.e. Jericho in Levant or Çatalhöyük in Anatolia). Traditional approaches to the rehabilitation of buildings refer to structural improvements, material reinforcements but also program adaptations. In general, they are divided into four different directions [5]:

- *preservation* fixes the building in its found state, making no attempt to repair or improve it but ensuring that, so far as it is possible, it is immune to further decay;

- *restoration* returns the building to its as-built state using period materials and techniques to create the illusion that it has been untouched by time;

- *renovation:* renews and updates the building to make it suitable for contemporary life;

- *remodeling* (referred to as Adaptive Reuse in America) locates an entirely new function within an existing building, which may be substantially modified to accept that interjection [5].

*Preservation* and *restoration* will not be elaborated further, as these two approaches demand specific knowledge of art historians and experts of the traditional use of materials and constructions. In architecture, the other two approaches play an important role, i.e. *renovation* and *remodeling*, which involve a certain degree of artistic freedom and the implementation of new programs and represent the fundamental mission of architectural endeavors. Using the material and cultural capital of existing buildings to prolong their lifespan and livability, they can be considered those with the most sustainable principles today. It is mainly because reuse, which encompasses renovation and remodeling, saves energy wasted in new construction, and generates less construction waste. In these approaches all three main pillars of sustainability - the economic, environmental and social are taken into account, while the environmental is especially emphasized. Surprisingly, in the first set of guidelines, the engineers drafting the LEED criteria ignored existing buildings [6].

## 2.1 Renovation and Remodeling

Adaptive reuse is a noticeable recent phenomenon in which architects use the cultural and material capital intrinsic to the building shell to connect the old and the new. There are many examples, but the most impressive ones include former industrial buildings transformed into contemporary public programs such as galleries, museums, libraries, etc. In practice, the LocHal library in Tilburg, an iconic former locomotive shed of the Dutch National Railways redesigned by Mecanoo Architecten is a good example [7]. The building was rehabilitated by combining characteristic historical elements with new additions. Another famous example is the Tate Modern gallery in the former Bankside Power Station in London, whose renovations were completed in 2000 by Swiss architects Herzog and de Meuron. The building, which was originally built between 1947 and 1963, was at risk of being demolished by developers after it stopped operating (Fig. 1). At the time, also the application to list the building was refused, which shows the lack of understanding to the approach of preserving and remodeling buildings at the time. Moreover, this was not a solitary example, as a substantial part of industrial heritage was demolished at the end of the 20$^{th}$ century (the phenomenon can be compared with the destructions of medieval structures in the times of modernization at the end of the 19$^{th}$ and at the beginning of the 20$^{th}$ century). As mentioned, the paradigm has changed and nowadays we can notice many sustainable examples of the rehabilitation of existing buildings, supporting Peter Blake's statement of regarding the sustainability impacts of reuse, saying that "all over the world, buildings that have been recycled from an earlier function to a new one seem to serve their users better today than they ever did before…" [5]. In addition to the functional aspect, the appearance of sustainability has become extremely important in contemporary culture of co-modification [8]. The aesthetic aspect has to be taken into account, as it may reflect and unmask the ideology at work in sustainable rehabilitation.

The practice of interior architecture as a discipline is also heavily involved with the remodeling and repurposing of existing buildings, therefore bearing an important role in the sustainable reuse of the built environment [5]. Furthermore, energy-efficient renovations of building stock present a large share in overall rehabilitation activities. A promising trend in the dense urban environment is also renovation with vertical building extension.

**Fig. 1.** Herzog & de Meuron, exterior and interior of Tate Modern, London, UK (left photo: © Peter Šenk / right photo: © Jan Kranendonk I dreamstime.com)

**Energy-Efficient Renovation**
Energy efficiency is one of the leading perspectives within the environmental sustainability aspect of building renovation (Fig. 2). Forcefully imposed by the Energy Performance of Buildings Directive [9], the intensive trend of energy efficient renovation has been indicated in Europe in the last two decades. Since the majority (80–90%) of negative environmental impacts occur because the building consumes energy during its use stage, the basic aim of energy-efficient renovation is to reduce energy consumption for building operation. Other important benefits of building renovation, such as better indoor environmental quality and higher economic property value, are parallel goals in addition to the energy efficiency aspect.

In scientific literature, a considerable effort has lately been made to provide early-stage decision support tools and strategies to guide the energy-efficient renovation process [10–12]. Regarding the scale of consideration, the existing scientific literature addresses building renovation at different levels, the urban scale category [13, 14] comprising the building stock, the building scale category [15, 16] including different building typologies, and the component scale category [17, 18] in which several renovation measures are discussed.

**Renovation with Vertical Building Extension**
A part from the renovation of existing buildings, the increasing trend of renovation with vertical building extension (Fig. 3) is seen in large European cities, which is mostly encouraged by the growing trend of population and simultaneously the growing demand for new usable surfaces. The vertical building extension leads to the provision of new usable surfaces, and in some cases also increases the energy efficiency of existing buildings. Due to the limited load-bearing capacity of existing buildings, the vertical extension is mainly carried out with light-weight timber construction. One of the first studies exploring the possibility of integrating additional floors within energy-efficient building renovation was conducted by Konstantinou and Knaack [19, 20]. The first complex

**Fig. 2.** Studio Stratum, Bled Kindergarten - before and after energy efficient renovation, Bled, Slovenia (photo: © Virginia Vrecl)

research treating the building extension and predominately focusing on the development of timber-glass upgrade modules was conducted by Špegelj et al. [21]. The optimal lightweight timber-glass modules from the viewpoint of energy efficiency regarding possible geometries of existing buildings were developed. In the following research [22], four different upgrade modules covering different shares of the roof of the two case study buildings in the Slovenian town of Velenje were applied in order to verify their influence on the energy need for heating and cooling of the entire renovated building. The results of the research indicate a positive impact of different upgrade modules on the total energy consumption of the renovated building. The development of the presented researches can be found in a study [23] analyzing energy efficiency effects of building extension with additional types of timber-glass modules in terms of the number of floors, the size of net floor areas, aspect ratios (a/b), thermal transmission values (U) and glazing sizes (AGAW) applied to three case study buildings of different geometric typology. The study points out several conclusions related to the optimal timber-glass module design, which depends on the existing building size, geometry and orientation.

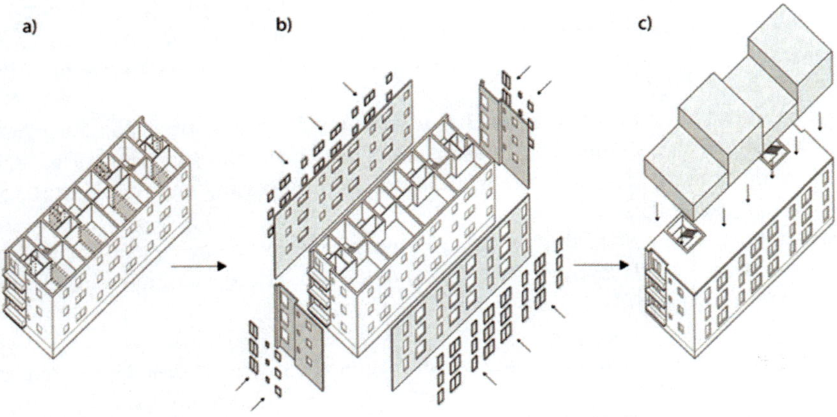


**Fig. 3.** Existing building stock (a), energy efficient renovation (b) and renovation with vertical building extension (c) [24]

The study also shows that a comprehensive building renovation can provide overall energy savings, whereby the efficiency decreases with the increasing number of existing building floors. A very complex overview of building renovation including literature overview, the classification of renovation approaches and latest scientific findings on energy efficient building renovation measures can be found in [24].

## 3 Sustainable Rehabilitation in Urban Development

Urban development deals with complex systems of the built urban environment. It refers to urbanization with its different dimensions and perceptions: physical (land use change such as urban sprawl and an increase in artificial surfaces), geographical (population and employment concentration), economic (markets, agglomeration economies and knowledge spillovers) and societal (social and cultural change).

Sustainable rehabilitation in urban development refers to urban regeneration in the broadest sense [25–28]. In the introduction to The Routledge Companion to Urban Regeneration, the editors Michael E. Leary and John McCarthy [29] map the phenomenon: "although it may be called different things in different countries: revitalization, renewal or urban renaissance…, urban regeneration remains a high profile and important field of government intervention in a host of countries around the globe."

Urban regeneration is a multidisciplinary field of research, scholarship, public policy and practice, including the elements of city planning, housing, transport/infrastructure, political economy, urban design, urban tourism, community development, sustainability and cultural industries studies [29]. Therefore, any improvement in quality life by transforming urban areas can be generally described as urban regeneration.

Urban regeneration aims to improve decaying urban areas in comprehensive and integral way by integrating new urban development with the existing one. It comprehensively addresses the economic, social, spatial and environmental transformation of urban areas. Because of its complexity, urban regeneration is often managed by public institutions with participation of different private and civil stakeholders. Urban regeneration is an area-based intervention, which is initiated, funded, supported or inspired by the public sector, aimed at producing significant sustainable improvements in the living conditions of local people, communities and places suffering from deprivation, often multiple in nature [29].

Spatial focus of urban regeneration lies in former industrial districts and other functionally deprived areas, suburban housing areas and historical city centers as most characteristic examples in need of rehabilitation, each addressing various and different questions and challenges.

Broadly speaking, according to key commentators, the main thematic narratives (although not necessarily mutually exclusive) to area-based regeneration and renewal, include the following [30]:

- Property-led physical approach, where, for example, a major retail-led or mixed-use scheme is expected to have multiplier effects on the local economy [31, 32].
- Business-driven approach, which highlights the importance of 'underserved markets' particularly in inner-city areas as important foci for regeneration through business investment [33].

- Urban form and design perspective, which highlights the importance of the relationship between sustainable development (SD) and urban form [34].
- Health and well-being perspective, which highlights the role that well-designed spaces can have on the neighborhood health and livability [35].

The previously mentioned approaches to urban regeneration are considered rather traditional, although they are also related to more contemporary examples. A thorough overview of recent acknowledgments and research is presented amongst others in the series of books from the international conferences on urban regeneration and sustainability entitled The Sustainable City [3]. Summing up some of the approaches and characteristics of the new millennium:

- *creative urban regeneration* (involving for example creative industries, event-formats), identified as the cultural industries approach [30, 36–39];
- *the flagship projects approach* (operating as the iconic project regeneration approach and urban activator), similar to urban form and design perspective) [40–44];
- *the temporary use approach* (reviving urban spaces by temporary activities) [45–49];
- *participatory practices in urban regeneration*, identified as the community-based, social economy approach, which highlights the importance of involving local communities in decision making and developing social capital networks [30, 50–52];
- *the smart city approach* (applying the smart city model to rehabilitation projects);
- *urban regeneration through gentrification* (involves different models, from economic implying new retail, to social leading and segregation…), similar to the property-led physical and business-driven approach [53, 54].

### 3.1 Creative Urban Regeneration

So-called creative urban regeneration is a contemporary model which integrates creative industries as a tool to promote sustainable urban regeneration [36]. According to Florida, "the need for urban regeneration occurs as an outcome of broader socio-economic processes, like transition into the post-industrial age and suburbanization, which both result in empty city centers and deserted industrial sites. These abandoned neighborhoods are often the scene of bottom-up regeneration stimulated by the creative class" [37]. Incorporating culture as a change maker can also be traced in the works of Charles Landry's *Creative city* [38]. The work is a call for imaginative action in the development and running of urban life, and a clear and detailed toolkit of methods by which our cities can be revived and revitalized. Culture as the driving force of urban change in many European cities can be best practiced in the cases of European Capitals of Culture. These are a unique opportunity for cities to generate considerable cultural, social and economic benefits, but most importantly to foster their urban regeneration as can be noticed in the case of Liverpool in 2008 [39] or Graz in 2003 and other cities (Fig. 4). In recent decades, major sports and cultural events such as the Olympic Games have emerged as significant elements of public policy, particularly in efforts to achieve urban regeneration.

**Fig. 4.** European Capital of Culture – Graz 2003 with Kunsthaus art museum (photo: © Erix2005 | dreamstime.com)

### 3.2 Flagship Projects Approach

Flagship development is defined as a significant and a high-profile development, which plays an important (influential or catalytic) role in urban regeneration, and is justified by attracting other investments [40]. It comprises characteristic elements: a development in its own right, which may or may not be self-sustainable, a marshalling point for further investment and a marketing tool for an area or a city [41].

The flagship projects may also have an ambiguous role in sustainable urban regeneration. While transforming residential, commercial, entertainment or tourist areas, they usually foster gentrification with economic and social stratification in the urban areas.

Its main approach is through the *urban form and design perspective* and subsequent marketing strategies related to new economic models for the urban regeneration area as well as social relations change. There are many iconic architectural projects which follow the example of the Guggenheim Museum flagship project in Bilbao as a pioneering example. Resulting in the "Bilbao effect," these urban regenerations are usually based on cultural investment with spectacular architecture planned with a clear focus on their economic benefits for the city and even the region.

Well-known examples of this type of urban regeneration also include many waterfront regeneration projects from pioneering Baltimore's Inner Harbor in 2007 to Seoul's Cheonggyecheon (Fig. 5), a new leisure and recreation area along a historic stream [42], and from London's Canary Wharf skyscraper city [43] to Hamburg's HafenCity with an emphasis on its Elbphilharomonie concert hall (Fig. 6).

These projects also follow the *health and well-being perspective.* Although waterfront projects with Mediterranization and gentrification may be oriented towards urban mix, culture-led or conceptualized as brownfield regeneration, the flagship approach emphasized by specific projects remains one of the key elements and can result in a comprehensive sustainable development. It can be designated through different principles [44]:

- secure the quality of water and the environment,
- be a part of the existing urban fabric, subject to historic identity which gives it character,
- prioritize mixed use,
- enable public access,
- make the waterfront a community priority,
- plan in public private partnerships, which speeds the process,
- secure strategic public investment and attract private resources,
- enable public participation as an element of sustainability
- take into account that waterfronts may be long-term projects with regeneration as an ongoing process,
- require multidisciplinary work.

**Fig. 5.** Cheonggyecheon Stream in downtown Seoul, South Korea (photo: © Siavramova | dreamstime.com)

**Fig. 6.** Hamburg Hafen City with iconic Elbphilharmonie designed by Herzog & de Meuron, Hamburg, Germany (photo: © Saiko3P | dreamstime.com)

### 3.3 Temporary Use Approach

Encouraging the revitalization and redevelopment of urban areas, many city authorities in Europe and North America are facing the lack of resources, power and control to implement formal masterplans [45]. Instead, a rather experimental approach with 'looser planning visions', engulfing temporary initiatives and activities, often making the use of vacant properties has been noticed in the past decade; to name a few significant authors and publications: Franck and Stevens – Loose space [46], Chase and Crawford and Kaliski - Everyday urbanism [47], Hou – Insurgent public space [48]. This approach has been noticed by many initiatives in the UK, Germany and also in the USA where it is known under the term 'tactical urbanism' [49]. Bishop and Williams [45] claim that it could be assumed that anything temporary is not particularly sustainable because the term can imply a short-lived attitude. On the other hand, it could also suggest a desire to leave minimal traces on the Earth. After analyzing concrete examples, many of the temporary activities embody at least some of the principles of sustainable development (Fig. 7). Temporary structures are usually built by re-using or recycling materials at hand, while temporary activities represent a more efficient use of land and buildings. What has been an everyday practice in third world countries seem to have become an alternative approach in western urban metropolises.

**Fig. 7.** Demonstration project - bottom-up proposal for pedestrian crossings and temporary installations in Koroška Street in Maribor, Slovenia, 2015 (photo: © Kaja Pogačar)

### 3.4 Participatory Practices

The participatory approach to urban regeneration projects can be observed since the 1960s by several isolated examples. However, in recent decades, stronger engagement is noticeable in that field. Agendas on sustainability – i.e. National Environmental Policy Act from 1969, Agenda 21 from 1992, Aalborg Charter from 1994, and Aarhus Convention, 1998 – were among the forerunners that embraced public participation in legislation and first exposed the need for more public involvement in protecting our environment but also in shaping our cities [55]. Co-creating urban life has been introduced as a social urban regeneration tool [51]. According to Lopez, the need to rethink any action in the city from the point of view of citizen involvement and participation is clear.

It is understood as the need to analyze, develop, work and build hand in hand with social and citizen groups towards the improvement of habitability conditions in their cities and neighborhoods at a smaller scale [52]. Urban regeneration processes often imply building rehabilitation. There are numerous examples and networks (such as Actors of urban change or CityToolBox network) (Fig. 8).

**Fig. 8.** Vivacidade - participatory rehabilitation of the urban void in Aveiro [56]

### 3.5 Smart City Approach (Involving for Example Sensors and Technological Improvements)

The smart city approach is a primarily technological improvement-based urban regeneration. It also encompasses BIM technologies in planning. By definition, a smart city is a city seeking to address public issues via ICT-based (information and communications technology) solutions through a multi-stakeholder municipality-based partnership [57]. According to its definition, a smart city addresses one or more of the following characteristics: smart governance, smart people, smart living, smart mobility, smart economy and smart environment. ICT links and strengthens networks of people, businesses,

**Fig. 9.** The concept of smart city and internet of Things (photo: © Elnur, dreamstime.com)

infrastructures, resources, energy and spaces, and provides intelligent organization and governance tools [57]. While there may be economic and social benefits to the approach, there are also negative consequences. The main arguments against a smart city are in its non-compatibility with an informal character of the city, and the facts that it subjects a city to corporate power and reproduces social and urban inequalities [58] (Fig. 9).

### 3.6 Urban Regeneration Through Gentrification

Nowadays, the urban transformation process is part of economic strategies, especially when there is market-driven regeneration at work. The *property-led physical approach and business-driven approach* inevitably create new economic and social conditions. The result of urban regeneration is a transformed physical structure of the cityscape which impacts the social structure of the residents. Private investors and wealthier newcomers enjoy the benefits of urban regeneration, while the old not that wealthy residents are excluded. The described phenomenon – gentrification refers to social, economic and cultural changes which are a result of urban transformation [53, 54]. Old residents and culture are replaced by wealthy newcomers. Initially, it used to be a sporadic consequence of urban redevelopment but has become generalized world-wide as a global urban strategy.

## 4 Tools for Urban Rehabilitation

Many tools may be used to achieve the sustainability goals of rehabilitation in architecture and the urban environment. Approaches and tools range from the traditional top-down planning or designing to the more recent contemporary, informal, creative and adaptive bottom-up approaches involving direct physical interventions (in the urban environment) and participation of citizens [59, 60]. The traditional tools, with the top-down, formal or traditional point of view, are [60]:

- a strategic document which outlines the vision and goals of future development;
- urban planning, a masterplan which defines future spatial development;
- an urban-design competition and architectural competition, which are public/private calls for the best design solution.

On the other hand, the alternative approach with 'tactical urbanism' [49], bottom-up, informal or emerging perspective, encompasses several tools which may be effective in certain rehabilitation processes [60]:

- an urban planning workshop, urban design workshop and charrette explore development possibilities in groups and seek reflections upon them;
- exhibitions are presentations of the planning issues to the wider public;
- street art is artistic intervention in public space (murals, graffiti, street painting, etc.);
- urban hacking is an illegal individual or group action in public space;
- creative licensing supports the use of public space in a more flexible way, e.g.'park(ing) day' license;

- a demonstration project tests and measures the effects of program changes in real-world situations (one day to one week);
- pilot projects are preliminary small-scale experiments conducted to evaluate feasibility, time, cost, etc. and attempting to improve full-scale implementation prior to implementation (one month to one year);
- interim design projects focus on the improvement of public spaces in the short term, when funds are limited (one to five years);
- temporary events mean a temporary change in the use of public space by festivals, marathons, fairs, etc.;
- critical mass gatherings, protests, referendum are gatherings of larger groups of people in an action or decision-making format;
- a discussion group/meeting addresses groups of different stakeholders to meet (including conferences, lectures, presentations with round tables, etc.) and discuss ideas and critical issues;
- web-based community planning involves communities in planning by using information- and web-based technologies;
- urban hackathon is a compact format involving group work on IT, programming and data mining to find creative solutions for urban problems.

In addition to the mentioned tools, there are also supporting tools (research and development projects, metrics, evaluation, opinion polls, etc.) and structured in-depth data-based analyses related to a specific topic (e.g. traffic study, spatial study, program study, public opinion, etc.). These tools serve as arguments for political and professional decisions, and their results are indispensable for their legitimations and public presentations [60].

Contemporary tools are often characterized by the immediate physical intervention in space. It demands the involvement of a wider range of actors and can be effective in the architectural and urban environment scales. It involves the participation of citizens, experts, public authorities, etc., and enables identification and identity formation. Participation is considered the key element of contemporary approaches [61], since the social fabric can be built along with physical interventions through the involvement, cooperation and decision-making of communities., However, the decisive factors for implementation usually remain a clear vision and a determination by the city government to facilitate change [62]. In the alternative approaches all three main pillars of sustainability – the economic, environmental (spatial) and social – are taken into account in a more balanced way, since, contrary to classic approaches, the social aspect is especially emphasized.

## 5 Conclusions

The scope of the presented topics affects not only architects and urban planners, but all disciplines that deal with issues of our built environment. Therefore, interdisciplinary engagement and cross-sectorial connections are necessary. There are still several research gaps and unresolved questions or challenges in the field to name only few very exposed ones:

- challenges connected to the ICT development, particularly smart city;
- challenges related to technological solutions in densification of the city: especially the renovation and remodeling of existing buildings, better energy efficiency and renovation with vertical building extension;
- challenges related to different modes of participation in architecture and urban regeneration;
- challenges related to the transformation of design and planning approaches – merging traditional and alternative approaches;
- challenges related to the aesthetics of sustainable rehabilitation in architecture;
- challenges related to the aesthetics of sustainability in urban regeneration.

Thinking about the sustainable rehabilitation of architecture and the urban environment means thinking about tomorrow and the mechanisms, processes and tools that will increase our sensitivity towards the most important issues on the planet. Above all, we need to develop a healthy relationship between our communities and the built and natural environment, which will hopefully sustain us in the long term.

## References

1. Oliveira, P., Luís, B.: Urban rehabilitation towards sustainability. In: Seminário Reabilitação Energética de Edifícios, pp. 129–136. Universidade do Minho, Minho (2012)
2. United Nations: Department of Economic and Social Affairs, Population Division: World Urbanization Prospects, The 2018 Revision. United Nations, New York (2019)
3. Marchettini, N., Brebbia, C.A., Pulselli, R., Bastianoni, S. (eds.): The Sustainable City IX: Urban Regeneration and Sustainability. WITPress, Southampton, Boston (2014)
4. Architecture definition. https://www.britannica.com/topic/architecture. Accessed 19 Jan 2020
5. Coles, J., House, N.: The Fundamentals of Interior Architecture. AVA Academia, Lausanne (2007)
6. Hewitt, M.A.: Why reusing buildings should - and must - be the next big thing. https://www.archdaily.com/909863/why-reusing-buildings-should-and-must-be-the-next-big-thing. Accessed 19 Jan 2020
7. LocHal Library, https://www.mecanoo.nl/Projects/project/221/LocHal-Library?t=1. Accessed 19 Jan 2020
8. Lee, S. (ed) Aesthetics of sustainable architecture. 010 Publishers, Rotterdam (2011)
9. Directive 2010/31/EU - Energy performance of buildings directive — European Environment Agency. https://www.eea.europa.eu/policy-documents/energy-performance-of-buildings-directive. Accessed 19 Jan 2020
10. Thuvander, L., Femenías, P., Mjörnell, K., Meiling, P.: Unveiling the process of sustainable renovation. Sustainability **4**, 1188–1213 (2012). https://doi.org/10.3390/su4061188
11. Nielsen, A.N., Jensen, R.L., Larsen, T.S., Nissen, S.B.: Early stage decision support for sustainable building renovation - a review. Build. Environ. **103**, 165–181 (2016). https://doi.org/10.1016/j.buildenv.2016.04.009
12. Lee, S.H., Hong, T., Piette, M.A., Taylor-Lange, S.C.: Energy retrofit analysis toolkits for commercial buildings: a review. Energy **89**, 1087–1100 (2015). https://doi.org/10.1016/j.energy.2015.06.112
13. Wu, Z., Wang, B., Xia, X.: Large-scale building energy efficiency retrofit: concept, model and control. Energy **109**, 456–465 (2016). https://doi.org/10.1016/j.energy.2016.04.124

14. Brøgger, M., Wittchen, K.B.: Estimating the energy-saving potential in national building stocks – a methodology review. Renew. Sustain. Energy Rev. **82**, 1489–1496 (2018). https://doi.org/10.1016/j.rser.2017.05.239
15. Economidou, M., Atanasiu, B., Despret, C., Economidou, M., Maio, J., Nolte, I., Rapf, O.: Europe's Buildings Under The Microscope: A Country-by-country Review of the Energy Performance of Buildings, Bussels (2011)
16. Ruparathna, R., Hewage, K., Sadiq, R.: Improving the energy efficiency of the existing building stock: a critical review of commercial and institutional buildings. Renew. Sustain. Energy Rev. **53**, 1032–1045 (2016). https://doi.org/10.1016/j.rser.2015.09.084
17. Ferrari, S., Beccali, M.: Energy-environmental and cost assessment of a set of strategies for retrofitting a public building toward nearly zero-energy building target. Sustain. Cities Soc. **32**, 226–234 (2017). https://doi.org/10.1016/j.scs.2017.03.010
18. Hilliaho, K., Nordquist, B., Wallentèn, P., Hamid, A.A., Lahdensivu, J.: Energy saving and indoor climate effects of an added glazed facade to a brick wall building. Case study (2016). https://doi.org/10.1016/j.jobe.2016.07.004
19. Konstantinou, T., Knaack, U.: Refurbishment of residential buildings: a design approach to energy-efficiency upgrades. In: Procedia Engineering, pp. 666–675. Elsevier Ltd (2011). https://doi.org/10.1016/j.proeng.2011.11.2063
20. Konstantinou, T., Knaack, U.: An approach to integrate energy efficiency upgrade into refurbishment design process, applied in two case-study buildings in Northern European climate. Energy Build. **59**, 301–309 (2013). https://doi.org/10.1016/j.enbuild.2012.12.023
21. Špegelj, T., Žegarac Leskovar, V., Premrov, M.: Application of the timber-glass upgrade module for energy refurbishment of the existing energy-inefficient multi-family buildings. Energy Build. **116**, 362–375 (2016). https://doi.org/10.1016/j.enbuild.2016.01.013
22. Špegelj, T., Premrov, M., Leskovar, V.Ž.: Development of the timber-glass upgrade module for the purpose of its installation on energy-inefficient buildings in the refurbishment process. Energ. Effi. **10**(4), 973–988 (2016). https://doi.org/10.1007/s12053-016-9494-z
23. Lešnik, M., Premrov, M., Žegarac Leskovar, V.: Design parameters of the timber-glass upgrade module and the existing building: Impact on the energy-efficient refurbishment process. Energy **162**, 1125–1138 (2018). https://doi.org/10.1016/j.energy.2018.08.108
24. Žegarac Leskovar, V., Premrov, M. (eds.): Integrative Approach to Comprehensive Building Renovations. GET. Springer, Cham (2019). https://doi.org/10.1007/978-3-030-11476-3
25. John, M.: Partnership, Collaborative Planning and Urban Regeneration. Ashgate, Basingstoke (2007)
26. Jones, N.R., Jones, A., van Sluijs, E.M.F., Panter, J., Harrison, F., Griffin, S.J.: School environments and physical activity: the development and testing of an audit tool. Heal. Place. **16**, 776–783 (2010). https://doi.org/10.1016/j.healthplace.2010.04.002
27. Jones, P., Evans, J.: Urban Regeneration in the UK: Theory and Practice. SAGE, London (2008)
28. Tallon, A. (ed.): Urban Regeneration and Renewal. Routledge, London (2010)
29. Leary, E., Michael, M.J. (eds.): The Routledge Companion to Urban Regeneration. Routledge, London, New York (2013)
30. Colantonio, A., Dixon, T.J.: Urban regeneration and social sustainability: best practice from European cities. Wiley-Blackwell, Oxford (2011)
31. Dixon, T.J.: The role of retailing in urban regeneration. J. Local Econ. Policy Unit. **20**, 168–182 (2003). https://doi.org/10.1080/13575270500053266
32. Claxton, R., Siora, G., DTZ Consulting and Research: Retail-led regeneration : why it matters to our communities. DTZ, London (2008)
33. Porter, M.E.: the competitive advantage of the inner city. Harv. Bus. Rev. **73**, 55–71 (1995)
34. Jenks, M. (Michael), Burton, E., Williams, K.: The Compact City : A Sustainable Urban Form? Routlege, London (1996)

35. Barton, H., Grant, M., Guise, R.: Shaping Neighbourhoods, A Guide for Health. Spon Press, London (2003)
36. IPOP: Potentials of creative urban regeneration, Ljubljana (2011)
37. Florida, R.L.: The Rise of the Creative Class, Revisited. Basic Books, New York (2012)
38. Landry, C.: The Creative City: A Toolkit for Urban Innovators. Comedia, New York (2008)
39. Liu, Y.D.: Cultural event and urban regeneration: lessons from liverpool as the 2008 european capital of culture. Eur. Rev. **24**, 159–176 (2016). https://doi.org/10.1017/S1062798715000265
40. Bianchini, F., Dawson, J., Evans, R.: Flagship projects in urban regeneration. In: Healey, P., Davsudi, S., Toole, M.O., Tavsanoglu, S., Usher, D. (eds.) Rebuilding the City: property-led urban regeneration, pp. 245–255. E. & F. N. Spon, London (1992)
41. Smyth, H.: Marketing the City : The Role of Flagship Developments in Urban Regeneration. Taylor & Francis (2005)
42. Křižnik, B.: Selling global Seoul: competitive urban policy and symbolic reconstruction of cities. Sociol. Rev. **41**, 291–313 (2011). https://doi.org/10.5613/rzs.41.3.2
43. Sudjic, D.: The Language of Cities. Penguin Random House UK (2017)
44. Pekin, T.U.: Urban waterfront regenerations. In: Advances in Landscape Architecture. InTech (2013). https://doi.org/10.5772/55759
45. Bishop, P., Williams, L.: The Temporary City. Routledge, London (2012)
46. Franck, K.A., Stevens, Q.: Loose Space : Possibility and Diversity in Urban Life. Routledge (2007)
47. Chase, J., Crawford, M., Kaliski, J.: Everyday urbanism. Monacelli Press (2008)
48. Hou, J.: Insurgent Public Space: Guerrilla Urbanism and The Remaking of Contemporary Cities. Routledge, London (2010)
49. Lydon, M., Garcia, A.: Tactical Urbanism: Short-term Action for Long-term Change. Island Press, Washington (2015)
50. Duncan, P., Thomas, S.: Neighbourhood Regeneration: Resourcing Community Involvement. Policy Press, Bristol (2002)
51. Yasminah, B. (ed.) The Participatory City. JOVIS Publishers (2016)
52. López, de A.A.: Participatory Approach to Urban Regeneration Processes (2016)
53. Palen, J.J., London, B.: Gentrification, Displacement, and Neighborhood Revitalization. State University of New York Press, Albany (1984)
54. Lees, L., Slater, T., Wyly, E.K.: Gentrification. Routledge, London, New York (2008)
55. Pogacar, K.: "Urban Activator" - participatory and bottom-up tool of urban change. Sch. J. Archit. Urban Plan. **22**, 189–199 (2014)
56. Platform for Social Inovation: CIty Tool Box. https://www.citytoolbox.net/. Accessed 05 June 2020
57. Manville, C., Europe, R., Millard, J., Technological Institute. In: Liebe, D., Massink, A.R.: Mapping Smart Cities in the EU (2014)
58. Krivý, M.: Towards a critique of cybernetic urbanism: the smart city and the society of control. Plan. Theory. **17**, 8–30 (2018). https://doi.org/10.1177/1473095216645631
59. Lisa, H. (ed.): New Approaches to Urban Planning - Insights from Participatory Communities. Aalto University Publication Series, Helsiniki (2013)
60. Pogačar, K., Šenk, P.: Alternative approaches and tools for the transformation of streetscapes direct physical interventions and different modes of participation. Prostor **26**, 170–183 (2018)
61. Mueller, J., Lu, H., Chirkin, A., Klein, B., Schmitt, G.: Citizen design science: a strategy for crowd-creative urban design. Cities **72**, 181–188 (2018). https://doi.org/10.1016/j.cities.2017.08.018
62. Šenk, P., Pogačar, K.: Tracking contemporary streetscape transformation processes – two case studies from slovenia sledovanie aktuálneho procesu transformácie ulíc – dve prípadové štúdie zo Slovinska. Archit. A Urban **1–2**, 60–73 (2019)

# Sustainable Transformation of City Streets – Towards a Holistic Approach

Kaja Pogačar(✉) and Peter Šenk

University of Maribor, 2000 Maribor, Slovenia
kaja.pogacar@um.si

**Abstract.** Streets are complex urban spaces that occupy a fairly large piece of land in cities, i.e. up to 50% of the built-up area. In the past century, city streets have been subjected to the gradual increase in motorisation, while their many important aspects and potentials have been neglected. In line with the sustainability agenda, there is a need for a gradual transformation of streetscapes into more humane and ecological environments. The paper addresses different parameters that need to be taken into account when talking about the sustainable transformation processes of streetscapes, including environmental, social and economic aspects. The holistic approach focuses not only on the issues of mobility, which reveals more frequent use of alternative means of mobility and an adaptive use of street space, but also on factors such as the street ecosystem which is facing new requirements due to climate change. The social aspect is promoted by participatory approaches, as streets are considered the most important public spaces. In the paper, two examples of streetscape transformation from Slovenia and Austria are presented. The paper concludes with the idea of integrating different disciplines in order to achieve the holistic sustainable transformation of city streets in the future.

**Keywords:** Sustainability · Transformation processes · Streets

## 1 Introduction

### 1.1 City Streets – From Failures to Potentials

Streets are important and complex urban spaces that occupy a fairly large piece of land in our cities, from one-third to even a half according to Mehta [1]. According to Furchtlehner and Lička "today's focus on this fundamental type of urban space is based on three main considerations: it creates social spaces, provides structural resources, and has important environmental functions" [2]. However, in many cases city streets still do not meet the necessary requirements and often fail to provide their surrounding communities with basic needs, such as a space where people can safely walk, cycle, ride, transit, and socialize [3]. As the number of people living in cities will continue to increase, the pressure on "primary infrastructure for movement, access and connectivity" [1] will also increase. It is therefore necessary to seriously address the failures of the past century, especially in regard to the extensive increase in individual motorization


A. Rotaru (Ed.): CRIT-RE-BUILT 2019, SSGG, pp. 273–282, 2021.
https://doi.org/10.1007/978-3-030-61118-7_24

that has taken up such a valuable space. On the other hand, it is necessary to look to the future, and identify relevant aspects and potentials for a more sustainable development of this important public space.

Following the sustainability agenda, the need to gradually transform streetscapes into more humane, livable and even circular environments is necessary. The paper addresses different parameters that need to be taken into account when talking about sustainable transformation processes of streetscapes, including environmental, social and economic aspects. The holistic approach focuses not only on the issues of mobility, which reveals more frequent use of alternative means of mobility and the adaptive use of street space, but also on factors such as the street ecosystem facing new requirements due to climate change. The social aspect is promoted by participatory approaches, which have become increasingly important in recent decades. The research question deals with the components of the holistic streetscape transformation process, either as an evaluation method or as planning support.

### 1.2 Sustainability Parameters

The paper is based on the assumption that problems and failures of the past century in connection with the development of city streets are also related to the fragmented planning and the isolated disciplines that looked into and dealt with the phenomenon of the street. If we look at recent literature, the topic of the street has been linked to, and addressed by, many different research fields, including infrastructural and mobility issues (traffic flows and regulation), the architectural and spatial context (the structuring of the streetscape's surface and buildings along them), environmental issues related to the ecosystem of city streets, arguing their huge potential for a better climate in cities (also addressing water drainage, light pollution, etc.), economic evaluation, cultural capital as well as historic values of streets. Furthermore, the social values of streets are exposed, different behaviors and social patterns related to the environment of the city street and its users as well as the issues of their interaction with smart technologies. Similar to the way the street has been analyzed on the basis of various aspects, the street is also treated and planned in general practice on a sectoral basis. Such an approach certainly neglects the potentials and needs that these spaces have. It should also be mentioned that in the past century streets have been treated more as technical premises controlled by numerous regulations, which, while giving priority to motorized traffic, have destroyed other spatial relationships important for the quality of life. Sustainability cannot be achieved by partial solutions because things are interrelated. One of the most important definitions of sustainable development is that development meets the needs of the present without compromising the ability of future generations to meet their own needs [4]. The desired result is a state of society in which living conditions and resources are used to continue to meet human needs without undermining the integrity and stability of the natural system [4]. How can all aspects be integrated to achieve sustainability criteria and to move towards a holistic approach?

Below, we will briefly describe the selected sustainability parameters with regard to mobility, urban design, environment, economy, culture, social aspects and smart technologies.

### 1.2.1 Mobility (Means and Infrastructure)

Mobility issues are an important topic when it comes to sustainability, as transport accounts for almost 27% of all $CO_2$ emissions in the EU [5]. There are many agendas aimed at promoting more sustainable modes of transport, especially within cities, while travel by bicycle, public transport and walking is the most sustainable. The aim of reducing dependence on private motorized vehicles goes hand in hand with improving public transport infrastructure and redesigning streetscapes as well [6].

### 1.2.2 Urban Design

The contribution of urban design to sustainability measures can be seen in many successful examples of urban rehabilitation and transformation of public spaces in European cities [7]. Urban design defines urban connectivity and is related to social values expressed in the choice of materials, the scope of interventions in terms of human scale, friendliness towards the elderly, children and the disabled, the inclusion of green elements, etc. Urban design contributes most directly to the character of the streetscape, emphasizes its symbolic value, and also enables, directs or suggests special uses (e.g. if benches are available, people have the opportunity to sit down, etc.) [3].

### 1.2.3 Environmental Aspect

The environmental aspect is becoming an important issue in the context of streetscapes, as streets are more and more understood as parts of the ecosystem in contrast to the past when they were regarded as mere technical infrastructure spaces [8]. Accordingly, much research is being done on water drainage systems, especially regarding stormwaters, but also for water collection [9], light pollution and emissions, urban gardening, biocomposting, beekeeping, etc. are also becoming tangible topics. This new forward thinking goes hand in hand with more greenery along streets to establish a network of green spaces in the city to allow biodiversity, provide shade, reduce the risk of overheating, collect dirt, embrace the living habitat, etc., in order to mitigate the negative effects of climate change [8].

### 1.2.4 Economy

Although the transformation of streets into more pedestrian-friendly zones generally triggers rejection among the shopkeepers, studies show that high streets economically benefit from walking. When streets are regenerated to encourage walking, this in turn affects turnover, property values and rental yields [10]. Complete streets improve the vitality of the street's economy, which also has a positive effect on tourism (some cities unfortunately already have to deal with overtourism). An important aspect of the street economy are also micro-economies in the form of small kiosks that sell food, drinks, newspapers, etc. [11].

### 1.2.5 Culture

Culture is understood in a very broad sense – applied to the streets, it can have different meanings – from cultural heritage in the form of architecture and urban complexes to art in public spaces (sculptures, video projections…), performing arts, such as street theater, musical performances in the street, etc. All these substructures contribute significantly to the basic culture of a space and the behavior of the people. Indeed, the European Spatial Development Perspective defines the preservation of cultural diversity as one of the potentially most important development factors for the EU [12].

#### 1.2.6 Social Aspect

Since Jane Jacobs' observations of street life in the 1960s, experts in urban planning and social sciences have been putting emphasis on streets as public spaces to support social behavior [13]. The quality of life can be measured by the number of social interactions, such as greetings, conversations in the street, even passive observation has meaningful social effect [1]. In addition, it is also necessary to reveal the participatory commitment of citizens to the changes taking place on the street, in terms of information, but also participation in the shaping of programs or future needs. Such democratic planning is already being implemented in many cities.

#### 1.2.7 Smart Technologies

This aspect deals with the smart city technologies that could improve the quality of services on city streets, from smart traffic lights, streetlights control and monitoring, air pollution monitoring, traffic management to public safety. The sustainability aspect lies in the optimization processes and thus in reduced energy consumption. The topic also has a speculative problematic connotation in terms of privacy and data protection.

## 2 Methodology

To address the transformation of city streets in a sustainable way, we need to understand what characterizes sustainability and what does not. The sustainability parameters were compiled from different sources and are presented in Table 1. We created a list of selected representative criteria. In addition, two streets were studied in their transformation by presenting their state before the transformation and the impacts and improvements afterwards. We observed the process of change during the course of transformation to find out whether the results contributed to sustainability or not. We collected data from previous studies and fieldwork in Slovenska cesta in Ljubljana and Mariahilfer Strasse in Vienna. Both are rather important streets in the capital cities of Slovenia and Austria, respectively. At the same time, their transformation was quite intensive and long-term in both cases. Although they are not comparable in terms of street profile and length, they represent the new wave of streetscape transformations – these are transformations that usually take place outside the historical city centers and embody Jan Gehl's philosophy of 'cities for people'.

## 3 Research

Based on the selected parameters, we evaluated two cases of streetscape transformations that had been carried out in the capitals of Slovenia and Austria in the past decade.

### 3.1 Transformation of Slovenska Cesta, Ljubljana

Since ancient times, Slovenska cesta has been the main transport road in Ljubljana [14]. With its transformation throughout history, it has become the city's main artery and a transit corridor in the north-south direction [15], with an average of 17,000 vehicles per day in 2012 [16]. Increasing motorization and the decline in pedestrian numbers also caused the loss of many public programs along the street. The observed segment of the central part of Slovenska cesta is about 400 m long. The middle section is about 22 m wide, while it is 8 to 10 m wider at both edges. Its redesign was tracked from 2007 until its completion in 2015 and its use by 2017. The long and complex process involved many stages, actors, agents and target groups. From the interim design project, the modification of traffic measures through many temporary events and demonstration projects to a workshop/competition that resulted in a comprehensive urban design and its implementation. The vision of the *Ljubljana 2000* [17] plan from 1986 to abolish motorized traffic aside from public transport and create urban space for pedestrians and cyclists was realized and made possible in 2015. The central part of Slovenska cesta with a new traffic regime, the shared space, a new row of trees, street furniture, lighting, bus stops, the original paving with regard to the blind and the disabled, and other implemented measures and elements became one of the main attractions in the center of Ljubljana. The change in the traffic regime resulted in a 58% reduction in the concentration of black carbon [21]. By promoting sustainable urban mobility in the city center, Ljubljana implemented one of the most important steps in its transport policy and also one of the key elements of the European Green Capital 2016 project [18] (Figs. 1 and 2).

### 3.2 Transformation of Mariahilfer Strasse, Vienna

Mariahilfer Strasse is located in the 6$^{th}$ and 7$^{th}$ districts of Vienna and extends from the outer Ring towards the west (in the direction of the Schönbrunn palace). The 3.7 km long street is divided into two parts, the so-called outer and inner parts; the latter is further investigated and presented (ger.: Innere Mariahilfer Strasse). The street has developed into an important shopping street, but also a living and working space that attracts many people every day. It is a street with a rich history spanning from the Iron Age, the Roman period until today. As an important street leading to the city center, it was plastered in 1825, and in 1869 the first tram ran on it. After 1950, cars began to dominate, and in 1993 it was thoroughly renovated with a new metro beneath it, new design and a tree alley, but motorized traffic and the tram line continued to occupy the street. However, in 2010 it was decided that the street should be transformed again to create better conditions for all its users. The 1.8 km long section of the street – the inner part of Mariahilfer Strasse – was partly transformed into a pedestrian zone and partly into a shared space (ger. Begegnugszone) [19]. The project "Mariahilfer Strasse Neu" comprised many stages from the architectural competition, interim design phase, a referendum and the implementation phase. The whole process having a strong focus on the participatory involvement of citizens from the beginning to the completion [20]. The project triggered an intense discussion and political debate at the city level (Figs. 3 and 4).

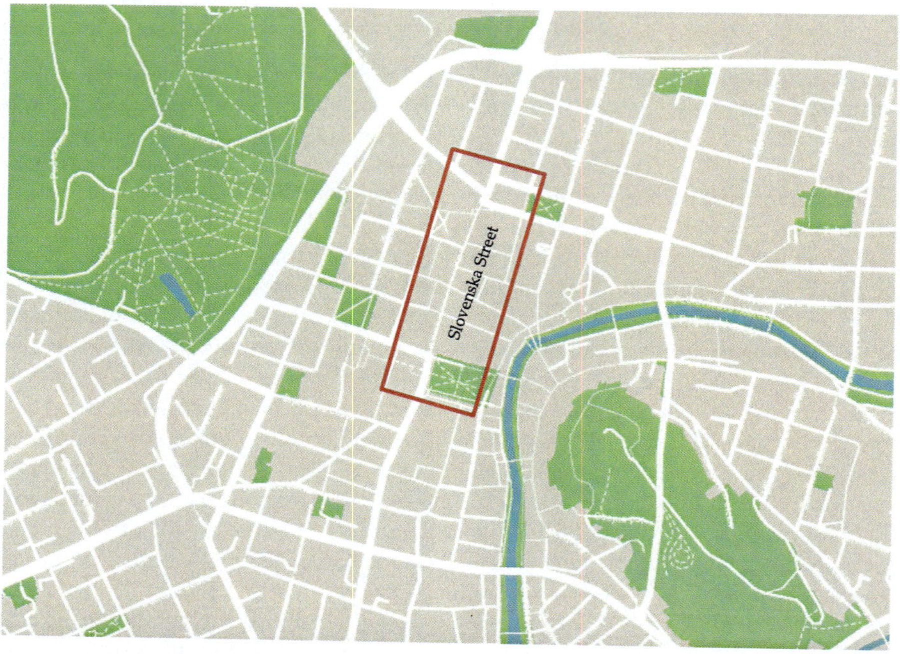


**Fig. 1.** Location of the central part of Slovenska cesta in Ljubljana

**Fig. 2.** Slovenska cesta in Ljubljana – before and after transformation (Source: left: prostorož/right: http://dekleva-gregoric.si/slovenska-cesta/)

Before the transformation, the street was organized as a two-way street for motorized traffic with a tram line and sidewalks; there were 63% of pedestrians (25–70.000 per day), 29% of cars (6.000–16.000 per day), 4% of bicycles and 4% of motorcycles every day. In addition to the pedestrian zone, the new design brought a new set of urban furniture (82 new sitting possibilities), more greenery (24 additional tree gardens), better

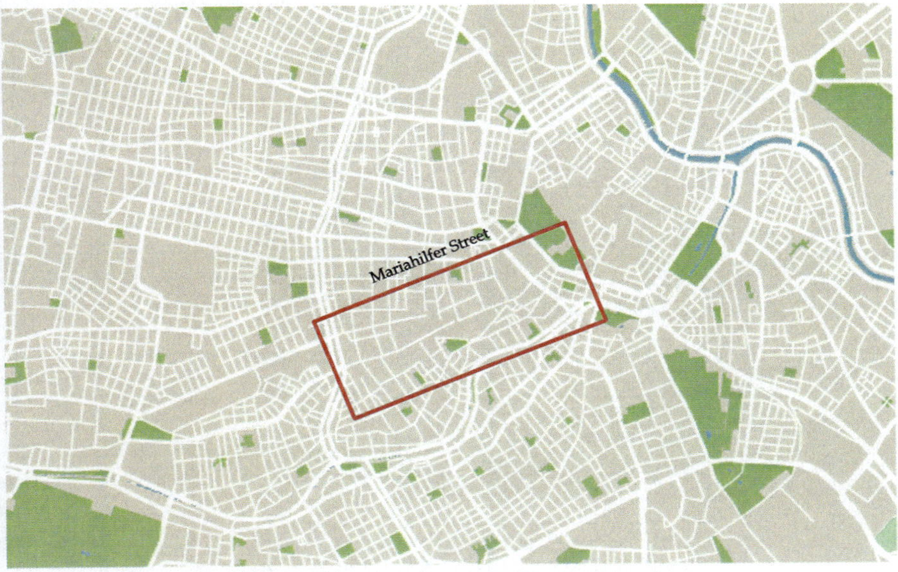


**Fig. 3.** Location of the inner Mariahilfer Strasse in Vienna

**Fig. 4.** Mariahilfer Strasse in Vienna before and after transformation (Source: left: https://kurier.at/chronik/wien/mariahilfer-strasse-vassilakou-laedt-zu-gipfeltreffen/54.665.196/right: https://www.vienna.at/mariahilfer-strasse-fertiggestellt-grosses-eroeffnungsfest-am-samstag/4404035)

conditions for trees (better drainage), energy efficient public lighting (111 new lamps), 4 water tables, a barrier free area with a 3 km long line for the blind, 12 WLAN stations, public toilets, space for dogs (ger. Hundezone), etc. [19]. All this contributed to the transformation into the biggest and most vibrant pedestrian shopping street in Austria.

### 3.3 Sustainability Parameters Overview

LEGEND:

+ new

**Table 1.** Sustainability parameters overview for two case-studies

| | Characteristic parameters | SLOVENSKA CESTA | MARIAHILFER STRASSE |
|---|---|---|---|
| MOBILITY MEANS | cars, trucks | −(*) | ⊖ / − (*) |
| | public transport | ○ | ⊕ / metro |
| | cyclists | ⊕ | ⊕ |
| | pedestrians | ⊕ | ⊕ |
| MOBILITY INFRASTRUCTURE | bicycle stands | ⊕ | ⊕ |
| | bicycle/scooters rental system | ○ | + |
| | parking along the street | ○ | − |
| | shared space | + | + |
| | covered bus stops / metro stations | ⊕ | ⊕ |
| URBAN DESIGN | benches | ⊕ | ⊕ |
| | other sitting opportunities | + | + |
| | city info boards | ⊕ | + |
| | technical street design (asphalt) | − | x |
| | ambiental street design (paving) | + | ⊕ |
| ENVIRONMENT | trees (shade) | ⊕ | ○ |
| | small trees (no shade) | + | + |
| | green surfaces | ○ | x |
| | water drainage | ⊕ | ⊕ |
| | public lightning (LED) | + | + |
| | gardening opportunity | ○ | + |
| ECONOMY | shops along the street | ⊕ | ⊕ |
| | cafés and restaurants along the street | ⊕ | ⊕ |
| | micro economies (kiosks…) | x | ⊕ |
| | city light boxes | ⊕ | ⊕ |
| CULTURE | concert / events held on the street | ⊕ | ⊕ |
| | street art | ○ | ⊕ |
| SOCIAL ASPECTS | people talk on the street | ⊕ | ⊕ |
| | people sit and eat/drink on the street | + | ⊕ |
| | citizens organize activities on the street (under permission) | + | + |
| TECHNOLOGY | smart monitoring | x | + |
| | telephone boxes / free Wi-Fi | ○ | ○ / + |

⊕increase
○equal
⊖decrease
− elimination
* delivery and maintenance purpose
× not available.

## 4 Discussion

In this research, we traced the sustainability-oriented parameters of two streets before and after transformation. We found that according to the analyzed parameters the sustainability of the two streetscapes had improved. In all groups considered – mobility, urban design, environment, economy, culture, social aspects and smart technologies – most sustainability parameters increased in the desired direction or even new elements appeared, such as new seating that can positively affect social behavior or new LED lighting that reduces energy consumption. Traditional parameters, such as noise levels and fine dust particles, were deliberately excluded from the research because other selected parameters are related to them. Furthermore, since it is not always possible to obtain all the information for the respective before-and-after state of affairs in each case, there are some elements which deserve to be considered. However, the sustainability parameters must be critically evaluated, since the apparent improvement in sustainability from of one aspect may also have negative consequences on the sustainability of another aspect or even on the overall sustainability of a particular streetscape. We can give some examples, such as the surveillance mechanisms technology which is closely linked to the technologically supported concept of the "smart city", and may be contradictory to spontaneous behavior and the social aspect; the use of LED light is economically viable, while due to their light specter, they are not favorable by environmentalists; another topic is the appropriation and commodification of the streetscape for economical means which leads to the private use of public space and in some cases also contributes to gentrification; there is the question of the risk resulting from the "democratization of movement" on the street – i.e. free flow enabled by the shared space concept can also bring risks of unpredicted (extreme or violent) behavior which can be dangerous for users of the street, etc. Nonetheless, the streetscape transformation should be seen as an attempt to democratize space and improve the quality of life in cities.

The presented topic addresses a trend of streetscape transformations, which is not only found in Europe but all around the world. More and more cities have been preparing sustainability agendas, pilot projects or have even created areas where motorized traffic is reduced and other sustainability measures have been implemented. In Hamburg, for example, a large pilot project to close a certain residential area to cars is being reported, similar to those in Barcelona, Paris, Brussels, etc. In both researched cases, we also noted the implementation of the strategy to reduce inner city traffic to reclaim public space for people with the aim to increase its attractivity and vitality. It is no surprise that Vienna has been announced for many years as the city with the highest quality of life in the world (according to the Mercer ranking), and Ljubljana was declared European Green Capital 2016. Both cities are trying to implement sustainability measures in their urban design.

Finally, the research has shown that less motorized traffic in cities must be accompanied by a number of urban, technical, environmental and social improvements to achieve sustainable impacts. Moreover, sustainability should be seen as a highly complex undertaking that needs to be approached holistically to achieve the desired results. The presented transformations may be seen as a step towards more sustainable streets, although in the future, we can expect this development to intensify not only in scope but also in the selected parameters.

## References

1. Mehta, V.: The Street: A Quintessential Social Public Space. Routledge, New York (2013)

2. Furchtlehner, J., Lička, L.: Back on the Street: Vienna, Copenhagen, Munich, and Rotterdam in FOCUS. J. Landsc. Archit. **14**, 72–83 (2019). https://doi.org/10.1080/18626033.2019.1623551
3. NACTO: Urban Street Design Guide. Island Press, New York (2013)
4. IISD. https://www.iisd.org/topic/sustainable-development. Accessed 15 Sep 2019
5. European Environmental Agency. https://www.eea.europa.eu/data-and-maps/indicators/transport-emissions-of-greenhouse-gases/transport-emissions-of-greenhouse-gases-11. Accessed 12 Sep 2019
6. Zavetoski, S., Agyeman, J.: Complete Streets. What's missing? In: Zavetoski, S., Agyeman, J. (eds.) Incomplete Streets: Processes, Practices, and Possibilities, pp. 1–13. Routledge, New York (2015)
7. Antoni, R. (ed): The Art of Urban Design in Urban Regeneration. Interdisciplinarity, Policies, Governance, Public Space. Edicions i Publicacions de la Universitat de Barcelona, Barcelona (2016)
8. Mehta, V.: The Street as Ecology. Incomplete Streets: Processes, Practices, and Possibilities, pp. 94–116. Routledge, New York (2015)
9. NACTO: Urban Street Stormwater Guide. Island Press, New York (2017)
10. Lawlor, E.: Pedestrian pound. https://www.livingstreets.org.uk/media/3890/pedestrian-pound-2018.pdf. Accessed 15 Sep 2019
11. Vallianatos, M.: Comple(eat)ing the streets: legalising sidewalk food wending in Los Angeles. In: Zavetoski, S., Agyeman, J. (eds.) Incomplete Streets Processes, practices, and possibilities, pp. 205–224. Routledge, New York (2015)
12. Commission, E.: European Spatial Development Perspective: Towards Balanced and Sustainable of the European Union. European Commission, Potsdam (1999)
13. Gehl, J.: Cities for People. Island Press, Washington DC (2010)
14. Korošec, B.: Ljubljana skozi stoletja: mesto na nacrtih, projektih in v stvarnosti. Založba Mladinska knjiga, Ljubljana (1991)
15. Dešman, M.: Introduction: If we are to find a way forward, we have to know where we come from, pp. 197–198. AB – Architect's Bulletin, Ljubljana (2013)
16. Koblar, S.: Prometna obremenjenost cest v Ljubljani. Urbani izziv. Special edn., pp. 140–144 (2016)
17. Pojasnila oddelkov mestne uprave glede prometne politike Mestne obcine Ljubljana. Mestni svet Mestne obcine Ljubljana, 032-2/2015-21 (9.1.2015)
18. Matić, B.: Slovenska street (2017). http://www.studiokristof.com/projects/slovenska/data.html
19. MAHU - Magazin zu Umgestaltung und Neuorganisation der Wiener Mariahilfer Strasse. Magistrat der Stadt Wien (2015)
20. Wien, S.: Beteiligungsprocess Mariahilfer Straße., Wien (2014)
21. Civitas Ljubljana. Na Slovenski cesti se laže diha: koncentracije crnega ogljika nižje za 58%. http://www.civitasljubljana.si/novice/na-slovenski-cesti-se-laze-diha-koncentracije-crnega-ogljika-nizje-za-58. Accessed 24 Sep 2017

# Predicting Service Life of External Wall Painting for the School Buildings in Indonesia

Peter F. Kaming[1](✉), Zofar S. C. Sihotang[1], Ferianto Rahajo[1], and Necta P. Pramesti[2]

[1] Universitas Atma Jaya Yogyakarta, Yogyakarta 55281, Indonesia
peterkaming82@gmail.com
[2] Universitas Gajah Mada, Yogyakarta 55281, Indonesia

**Abstract.** Service life in accordance with ISO 15686:2008 as a period of time after construction in which the building and its materials are the same or exceed the minimum performance requirements. Service life is needed for the building operation and maintenance. If routine maintenance and repairs is carried out properly, it would provide building durability and minimize building maintenance cost in the long run. This research was conducted to determine the service life of paint works from various brands on the market applied in the school buildings in Indonesia. The data used in this study were obtained directly from the building maintenance officers. Authors did some direct observation accompanied by the officers, take photos of damaged walls and measure the area of the external walls. Data obtained are grouped according to degradation level by using regression curves. The results of the data analysis obtained the higher square of determination in the four orientations of the building. Degradation of services in external wall are proportional non linear to the age of the paint work. In general, the appearance of a paint in an external building would decrease along with its age. The level of degradation due to type of damage such as dirt, cracks, bloating and peeling of the paint on wall. The study suggests that service life for external wall paint for the school buildings on maximum 5 years.

**Keywords:** Age of paint · Degradation · Maintenance · Service life · Indonesia

## 1 Introduction

There is a growing awareness in building maintenance from various construction sectors to increase building durability. Various studies have been carried out aimed at evaluating the decline in function related to paint work and durability other material of buildings in Indonesia [1–3]. At present there are various methods and approaches has existed for predicting service life or live service for the buildings [4]. Service life is defined in accordance with ISO 15686 (ISO 2008) [5] as a period of time after construction in which the building and its materials are the same or exceed the minimum performance requirements. Its definition involves previous definitions of acceptance criteria that vary in time and place, according to the measuring point of view in addition to all economic, political, aesthetic, environmental, or normative contexts that frame the assessment of


A. Rotaru (Ed.): CRIT-RE-BUILT 2019, SSGG, pp. 283–297, 2021.
https://doi.org/10.1007/978-3-030-61118-7_25

construction [4]. For the sake of simplicity, it is usually considered the end of the building of service life to occur through functional obsolescence, lack of economic viability, or physical wear and tear of the key material [7]. One of the purposes of this study is the development of service life predictions for painted exterior wall surfaces based on the definition of degradation regression curves that are intended to model the performance of materials and components over time [8]. The degradation curve applied to model of the external wall paint performance basically depends on the nature of the phenomenon being modeled and must be adjusted to the general trend of the degradation scatter plots. Based on the explanation that has been stated in the background, the formulation of the problem in this research are: what is the reference services life of paint from various brands on the market in Indonesia, and what are the factors that influence the service life of the external wall of the building?

The study has limited to as follows: 1) the location of the study was conducted at 30 school buildings in Yogyakarta; and 2) respondents were from school maintenance officers in Yogyakarta, Indonesia.

Studies on the service life of external wall paint in Indonesia has not been carried out. The objectives of this study are: 1) to identify the age of the paint reference service from various brands on the market; and 2) examine the relationship of the service life of the exterior wall paint and its degradation of service. The results of this study are expected to provide learning about predicting the age of paint services in buildings and subsequently the it can be further developed by the next study in Indonesia.

## 2 Literature Review

### 2.1 Paint Terminology

Paint is a liquid that is used to coat the surface of a material with the aim of beautifying, reinforcing or protecting the material. After being applied to the surface and drying, the paint will form a thin layer that is firmly and densely attached to the surface. Gluing the paint to the surface can be done in many ways: rubbed (wiping), smeared, brushed, sprayed, or dipped. Paint is a general term used for a family of products that is used to protect and give color to an object or surface by coating it with a pigmented layer. Paint can be used on almost all types of objects, such as to produce works of art (by painters to make paintings), industrial coatings (industrial coatings), driver assistance (road markings), or preservatives (to prevent corrosion or damage caused by water) [9].

### 2.2 The Phenomenon of Declining Paint Surface and Its Levels

Paint is an important element to cover the outside of a building. In general, it is used in most buildings. Environment and weather are the main cause of damages that subsequently to decrease the appearance on the surface of the external wall of the building. The decrease in appearance can be influenced by various factors that contribute to the reduction in service life of the paint surface. These factors can normally be predicted and are known to be able to find the best way to handle the treatment process of the external wall of paint. The level of reduction for each damage can be identified through the physical and visual scale of the external wall paint surface of the building. It also can identify the level and type of damages. According to BS EN ISO 4628-1 2005 [10], there were four types of damage that affect the surface of the exterior wall of the building, namely: 1. Dirt and discoloration generally affect the visual quality of building walls and usually appear in the early years of service life on the exterior walls of buildings. 2. Calcification or whitish spots usually occur after the loss of a smooth surface on the wall paint and cause wear, peeling of the paint material. 3. The level of reduction and cracks in visual form and frequent cracks in the paint. When compared with other materials cracks that occur on the surface of the outer walls of buildings are usually less than 1 mm in width. Sometimes it is very difficult to measure the width of these cracks. 4. Loss of attachment generally appears to be influenced by various factors. In general there are two factors that influence the occurrence of the loss of attachment in paint such as bloating and flaking.

## 2.3 Models for Estimating the Service Life of Paint Surfaces

Gaspar and de Brito [11], Gaspar et al. [12], and Chai et al. [13] devised a number index to reveal the process of decreasing the surface of the paint wall. This reduction process generally occurs in the outer covering of the building is the process of comparing the ratio between the damaged area (balance as a function of the status being repaired) and the area that is an example of the same thing with all the outer areas of the building with the maximum area that is experiencing decrease.

This method was initially developed for cement wall walls and later developed into a working model that was adapted to the types of the external wall coverings such as ceramics (Bordola et al. [14] and natural stone (Silva et al. [15]. To determine the service life of paints for the outer building layers using the formula:

$Sw, p = \sum ⟦(An \times Kn \times Ka, n)⟧) / (A \times K)$....................................Eq. (1)

Where:

$Sw, p =$ degradation of service life,

$Kn =$ multiplication factor of deviation or anomaly n,

$Ka, n =$ weighting factor related to the relative weight of the deviation ($m^2$),

$K =$ multiplication factor relationship to decrease level {0, 1, 2, 3, 4},

$A =$ total area of the outer wall taken, and

$An =$ area affected by paint damage.

The service life prediction for paints is also closely related to the ability and durability of paints against the weather and climate of the surrounding environment. The condition of the location of a building to the orientation of the wall and the level of humidity and high rainfall intensity mostly affect the adhesion of the paint to the wall surfaces.

## 2.4 Type of Severity in Paint Damage

Changes in color or damage that occurs on the surface of the painting is not always the same, especially when compared to the loss of viscosity of paint on the external wall. Based on a research study conducted by Gaspar [3], differences in the state of the wall could be analyzed by various weight and light tests between damaged parts. The results of the analysis of the study served to reveal the physical state. The study did not include all the parts tested, only one part that represents all the best parts, namely reducing dirt, stains or colors that change dramatically and increasing the value of attachment loss. The next values for ka, n where the numbers used to describe the above conditions include: 1) 0.25 for stains or discoloration; 2) 1.00 for the crack; 3) 1.00 for the Chalking or whitening process; and 4) 1.5 for flaking and blistering.

The level of degradation depends of the visibility of the damage quantity in five level: level 0 = no visible of paint damage, level 1 = good condition, level 2 = moderately declined, level 3, severely declined, and level 5 very severely declined.

## 2.5 Inspection of Paint Quality Degradation on the External Wall

According to Gaspar and de Brito [8] with scattered data revealing the results obtained on the buildings being observed, this makes it possible to determine the degradation lines of elements that have decreased over time using statistical techniques as presented with linear and polynomial curves. This degradation line represents the average reduction. Quadratic analysis using the Pearson correlation coefficient ($R^2$) method to evaluate with the proportion of the y-axis value (severity) relative to the x-axis value (paint age) reveals that the two regression curves (linear and second derivatives of polynomials). The polynomial curve is one of the best ways to match the shape of the sample decline and the curve explains the tendency for very slow development of the deviation or abnormal. The effect of this is that there is a decline in function that will slowly change rapidly in a short time.

The end of the service period of the material depends on the criteria whether it is acceptable or not. The minimal level of appearance is a relatively different concept according to the context of the surrounding environment. Analysis of all sample points to determine the level of service life of the paint is a method used to predict the level of service. After a certain period, the external wall surface needs repair (repaint) to improve the good appearance level.

## 2.6 Effect of Different Characteristics of Paint Wall Surfaces

In the case study it was found that the painted surface has various characteristics that affect the durability of the paint. According to Gaspar and de Brito 2008, these factors can be divided into three groups, as follows: 1) quality of the material used; 2) quality of execution; and 3) conditions for environmental exposure

By selecting a sample according to these variables, an independent regression curve is obtained which results in a degradation model that is related to each endurance factor.

Furthermore Gasper and de Brito [11], the first factor analyzed, related to material quality, was the type of paint product. For this coat of paint are grouped into three groups, as follows: 1) traditional and non-traditional fine paint; 2) textured paint; and 3) elastic membrane.

The quality of treatment in execution of the surface to be painted would affect the flaking of the paint. Two categories are considered, as follows: 1) repaint the previous coat; and 2) painting directly on the rendering of the new wall. This result is expected because when paint is applied to the previous layer there is a greater likelihood of product incompatibility and worse adhesion conditions of the external wall.

Moisture conditions are unfavorable to the paint showing that a faster decline than those experiencing abnormal conditions. The location of the building also affects the fast or slow peeling of wall paint, such as buildings close to the beach because the level of salt contained in the air accelerates the process of peeling paint on buildings and mountains, because the location of the air has a large humidity. In addition, damage to wall paint can also occur because the building is located close to plants, this is because the dirt on plants can damage the wall paint.

# 3 Methodology

## 3.1 Research Location and Time Frame

This research was conducted in several school buildings spanning in Yogyakarta Special Region, Centre of Java. There were typically school buildings that appear to have a decline in the service life of paint of their external wall. Most of the research locations are highland, not far from Merapi mountain. The schools typically landed house or two stories brick wall buildings with reinforced concrete structures. The preparation of this research began in March 2019. The questionnaire began to be distributed to respondents in May 2019. Data analysis, discussion and drawing conclusions were carried out from June to July 2019.

## 3.2 Research Methods

The method used to predict the service life of external wall paint in buildings.

**Observation Method** The method of observation is the technique of data collection, the researcher makes direct observations regarding matters related to research. Data collected in the form of photos of walls that have decreased service life of paint and measure the area of the outer walls of buildings that have decreased.

**Survey Method.** Survey method is a data collection using a questionnaire that contains a list of questions that have been prepared by researchers in which each question relates to the research problem to be filled in by respondents, and each question is an answer that has meaning to support the research results.

The questionnaire in this study was filled with questions about the age of service efforts to paint the outer walls of school buildings. Data also obtained from the degradation of the exterior paint walls of the building is to calculate the degradation of the outer paint walls of the building. How to calculate degradation with the following formula.

$$Sw.p = \frac{\sum(\text{An} \times \text{kn} \times \text{ka.n})}{A \times k} \quad (1)$$

The degradation severity Sw, p, is applied to the painted surface. Where kn is the anomaly multiplication factor n, as a function of level reduction, with a range k = {1, 2, 3, 4}; ka, n is a weighting factor related to the relative weight of the deviation of n ($m^2$)

ka, n which is used with a range of: (1) 0.25 for stains or discoloration; (2) 1,00 (for cracks); (3) 1.00 (for chalking or bleaching processes and (4) 1.5 (for peeling and blister) A = area of the outer wall area, and k is the multiplication factor of the relationship to the high level of reduction in the wall or paint, as K as above in area A. From all the data obtained, the next step is to conduct data analysis. The model developed in this study is either linear or non linear regression curve analysis. Best results are determined by their coefficient of determinations [11].

### 3.3 Method of Collecting Data

Data collection is the stages that must be done before and conducting research in order to avoid mistakes in collecting data in the field. This research was conducted by taking data from 30 school buildings in Yogyakarta. Research activities that will be carried out include: 1. **Field survey**. The things that need to be considered, namely: *Tools preparation.* The equipment used is questionnaire sheets, stationery, calculators (laptops and calculators) and documentation aids in the form of cameras. *Location Selection.* The locations taken in this study are school buildings that appear to have experienced a decline in the service life of the paint, viewed from the side of the buildings in the North, West, East, and South. *Determination of implementation time.* The time of the research is carried out in March 2019 until August 2019. The research was conducted during working hours. **2. Retrieval of data on site.** Data retrieval was conducted directly by giving a questionnaire to the officers for building maintenance of the school and taking pictures at the location in the area. As the sample in the research can be seen from the level of wall damage as in Table 1.

**Table 1.** Decreased level of crack damage

| Level of Degradation | Level 0 | Level 1 Good Condition | Level 2 Slightly decrease | Level 3 Moderate decrease | Level 4 Severe de-crease |
|---|---|---|---|---|---|
| | | 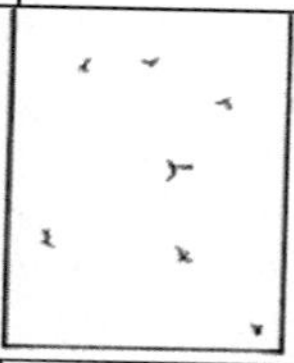 |  |  | 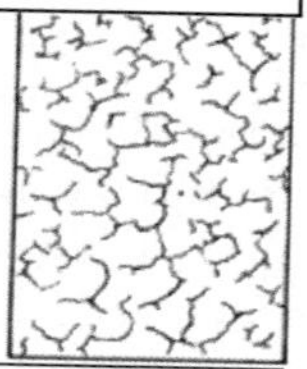 |
| Quantity | No exist | Smallest among of cracks | Smaller of cracks | Moderate cracks | Highly cracks |

### 3.4 Data Analysis

In this study, data analysis was conducted using curved regression analysis. The curve can be generated both in linear and non-linear equation of the relationship the dependent and independent variables. The result of the analysis subsequently also displaced the coefficient of determination for both linear and non-linear analyzes primary and secondary data that will be processed to determine the service life of the external wall paint for the buildings. In this case what will be analyzed was as follows, namely analyzing the service life of the paint calculated by the service life prediction model, predicting the service life of the external wall paints using linear and non-linear curves. To find out the degradation of the external wall paint, the data were then grouped according to the next level of degradation by using a degradation curve in linear and non-linear, the percentage of service life of the paint would then be obtained.

## 4 Result and Discussion

### 4.1 General Data of Respondents

In this study, researchers distributed as many as 30 questionnaire data to respondents directly so as to facilitate respondents in filling out the questionnaire given by asking researchers directly if there is something unclear or poorly understood. In this study, researchers gave questionnaires to school maintenance officers who responded to various types of kindergarten, elementary, junior high and high school/vocational school buildings. Data on building domicile classification can be seen in Fig. 1 below.

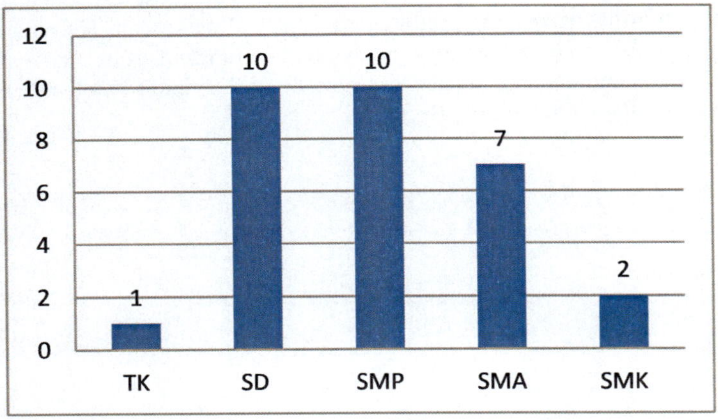


**Fig. 1.** Types of school building notes: TK = Kinder garden, SD = Elementary, SMP = Middle, SMA = High School, SMK = Vocational

Based on Fig. 1 above it can be seen that from a total of 30 samples, there are 1 sample from kindergarten buildings, 10 samples from elementary school buildings, 10 samples from junior high school buildings, 7 samples from high school buildings and also 2 samples from vocational school buildings.

In this study are the people responsible for the school maintenance and the people involved in the building renovation and painting process. Respondents were asked to fill out questionnaires that had been prepared to obtain data to predict the age of paint services in Yogyakarta.

## 4.2 School Building Assessment

In this study, respondents directly assessed schools in Yogyakarta using a questionnaire. In this study there is the age of the building, the last painting, the type of buildings selected were kindergarten, elementary school, junior high school and vocational school located in Yogyakarta. The school building being studied were applied paint of external wall buildings.

## 4.3 Geographical Condition of the Building

Yogyakarta's geographical condition is above an altitude of around under 800 m above sea level. Like most tropical regions, Yogyakarta has two seasons namely the rainy season and the dry season. The weather in the dry season during the day is in the range of 32 to 33 °C while the temperature at night is in the range of 23 to 25 °C. The climate that is in the city of Yogyakarta is a tropical climate like in general in the country.

Based on the rating scale measuring the service life, respondents provide an assessment of the age of the building, knowing when the last re-painting was conducted and the last renovation of the school building. In addition, an assessment by knowing the technique of painting of the building, follows by the stages such as whether the painting is done on certain parts or on all wall of the building, or doing cleaning before re-painting and evaluating some of the techniques of painting. Since the most respondents were not the builders, information regarding how many layers of paint were not obtained. The assessment was also done by looking at the types and brands of paint used on the walls of the school building, then an assessment is made of the condition of the cracks and damage to the walls of the building (Fig. 2).

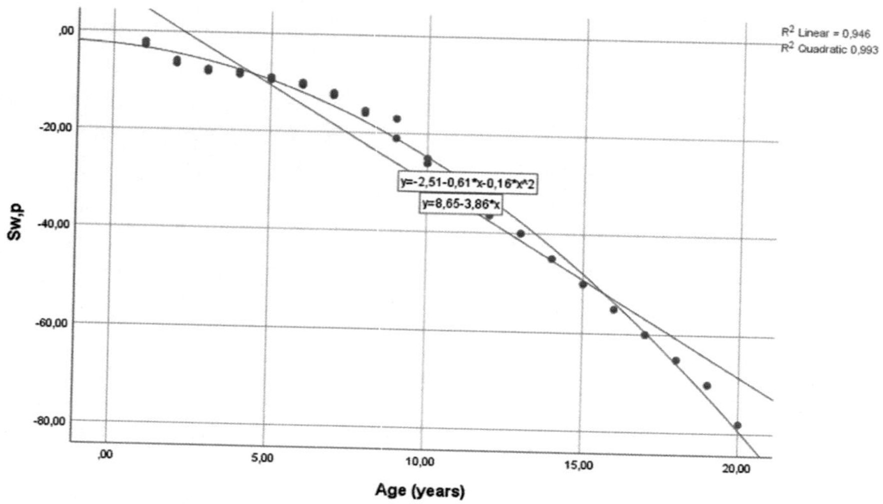


**Fig. 2.** Relationship of age-Sw,p from North of building.

After the data is collected then it is managed using the "service life prediction model" method or the service life prediction model according to Gaspar and de Brito [11]. To determine the service life of some data such as the total number of multiplication factors deviation or abnormality (kn), the weighting factor is related to the relative weight of the deviation (Ka.n), the area of damage (An) divided by the total area of the building part reviewed (A) times multiplication factor relationship to the level of decline (k = 0, 1, 2, 3, 4) One example in calculating the percentage of service life on the north, east, west, south side of SMP N 2 Banguntapan (named of a Middle High of Stated School) as follows:

$$Sw, p = \frac{\sum(\text{An} \times \text{kn} \times \text{ka.n})}{A \times k} = \frac{\sum(0,27 \times 0,6 \times 0,25)}{1 \times 2} = 0,020 = 2\% = \text{North}$$

$$Sw, p = \frac{\sum(\text{An} \times \text{kn} \times \text{ka.n})}{A \times k} = \frac{\sum(0,1 \times 0,4 \times 0,25)}{1 \times 1} = 0,01 = 1\% \text{ West}$$

$$Sw, p = \frac{\sum(\text{An} \times \text{kn} \times \text{ka.n})}{A \times k} = \frac{\sum(0,15 \times 0,4 \times 0,25)}{1 \times 1} = 0,015 = 1,5\% \text{ East}$$

$$Sw, p = \frac{\sum(\text{An} \times \text{kn} \times \text{ka.n})}{A \times k} = \frac{\sum(0,25 \times 0,2 \times 1)}{1 \times 2} = 0,025 = 2,5\% \text{ South}$$

Furthermore, from the calculation exhibits in Table 2. It shows that the service life of paint on the walls of the school section, while the number of schools reviewed is as many as 30 school buildings scattered in various places in the Yogyakarta area. The school building is viewed from four sides namely north, south, east and west. The results of the data analysis found that of the four sides the highest percentage of service age of paint was found in Islamic High School 3 Pakem, namely in the northern part of –78, the western part of –74, the eastern part of –72 and the southern part of –77, while the percentage of the age of paint service the lowest is in the 2nd Banguntapan Middle School, in the northern part –2, the western part –1, the eastern part –1.5 and the southern part –2.5. The minus number in the Sw.p column shows the direction of degradation which has been declining over the years or the percentage decline in the quality of service life of the building has increased. Furthermore, it can be seen in the graph below Sw, paint the walls of buildings outside the building (Fig. 3).

Interpretation of the four graphs above, the highest coefficient of determination $R^2$ for quadratic results of equation were obtained in North orientation of the building by 0,993, West by 0,988, and South by 0,988, and the smallest in East by 0,986. The percentage reduction in the service life of paint in the external wall paint of the buildings was proportional to the age of the paint. In general, the appearance of paint on external wall of the buildings would decrease along with the age of the wall paint (Fig. 4).

Percentage level of degradation was assessed from some damage such as stains, cracks, bloating and peeling paint on the wall. Judging from the results of the data analysis, it obtained that service life has decreased the appearance of paint on the external walls of the school building at intervals of 1 to 5 years, it can be concluded that the repair of external surfaces can be conducted in the interval of maximum of 5th year (Fig. 5).

**Table 2.** Percentage decreased wall paint service life

| No | School Id | Sw,P | | | | Service life (years) |
|---|---|---|---|---|---|---|
| | | North | West | East | South | |
| 1 | SMP N 2 banguntapan | −2 | −1 | −1,5 | −2,5 | 1 |
| 2 | SD N minomaratni 1 | −5,8 | −5,5 | −5 | −6 | 2 |
| 3 | SD perum concat | −7,5 | −6 | −6,5 | −8 | 3 |
| 4 | SMP N 6 | −8 | −7 | −8,5 | −9 | 4 |
| 5 | SMP Muh Pakem | −9 | −8,5 | −8 | −9,5 | 5 |
| 6 | SMP N 4 Pakem | −10 | −9 | −9,5 | −10,5 | 6 |
| 7 | SMP bopkri 3 | −12 | −11 | −10 | −13 | 7 |
| 8 | SD kansius | −15,5 | −15 | −14 | −16 | 8 |
| 9 | SMK bopkri 1 | −17 | −16 | −17 | −18 | 9 |
| 10 | SD N Minomartani 6 | −25 | −26 | −27 | −28 | 10 |
| 11 | SMP karitas ngaglik | −2,7 | −1,5 | −2 | −3 | 1 |
| 12 | SMK N 2 | −6,5 | −6 | −5,5 | −7 | 2 |
| 13 | SMA kanisius 1 | −8 | −6,5 | −7 | −8,5 | 3 |
| 14 | SMP N 4 | −8,5 | −7,5 | −8 | −9,5 | 4 |
| 15 | SD N Babarsari | −9,5 | −9 | −9,5 | −10 | 5 |
| 16 | SMA N 1 | −10,5 | −9,5 | −10 | −11 | 6 |
| 17 | SD karitas | −12,5 | −11,5 | −11 | −12 | 7 |
| 18 | TK karitas | −16 | −15,5 | −15 | −17 | 8 |
| 19 | SMA N 1 Cangkringan | −21 | −19 | −20 | −22 | 9 |
| 20 | SMP N 1 Banguntapan | −26 | −27 | −28 | −29 | 10 |
| 21 | SMA N 9 | −30 | −32 | −33 | −34 | 11 |
| 22 | SMP N 4 Yogyakarta | −36,5 | −35,5 | −36 | −37 | 12 |
| 23 | SD Muh YOGYAKARTA | −40 | −42 | −41,5 | −43 | 13 |
| 24 | SMA N 1 teladan | −45 | −46 | −47 | −48 | 14 |
| 25 | SD N ngupasan | −50 | −52 | −51,5 | −52,5 | 15 |
| 26 | SMA N 6 yogyakarta | −55 | −56,5 | −55,5 | −56 | 16 |
| 27 | SMP N 8 | −60 | −62 | −63 | −64 | 17 |
| 28 | SD N golo | −65 | −66 | −66,5 | −65,5 | 18 |
| 29 | SD kanisius sengkan | −70 | −72 | −71 | −73 | 19 |
| 30 | SMA Islam 3 Pakem | −78 | −75 | −74 | −77 | 20 |

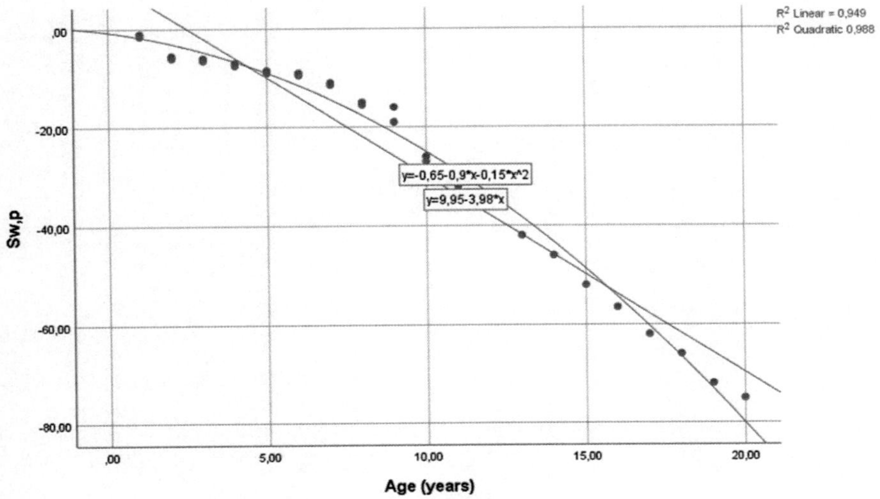


**Fig. 3.** Relationship of age-Sw,p from West of building.

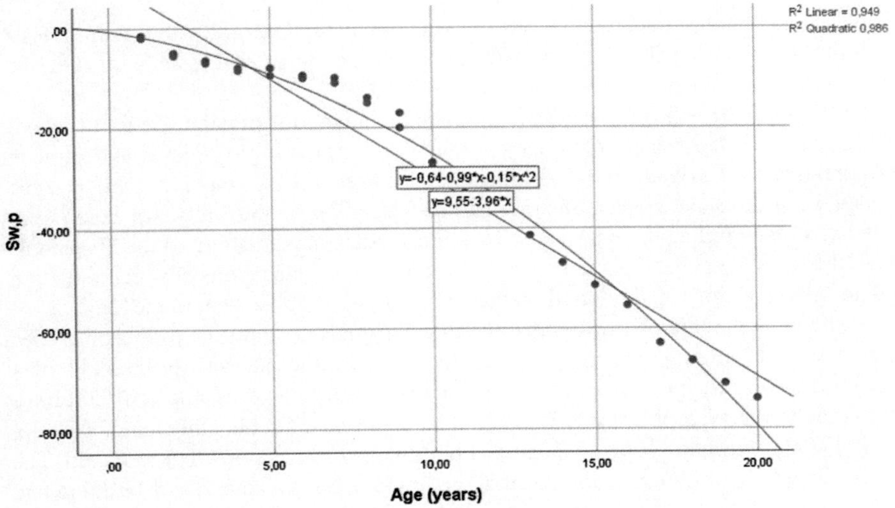


**Fig. 4.** Relationship of age-Sw,p from East of building.

## 4.4 Service Life Based on the Brand of Paint

The service life depends on the brand of paint and the type of paint used on the walls of the building outside. From the results of the study note that the type of paint that is often used in school buildings in the Yogyakarta area is paint with the brand Decolith as can be seen in Table 3.

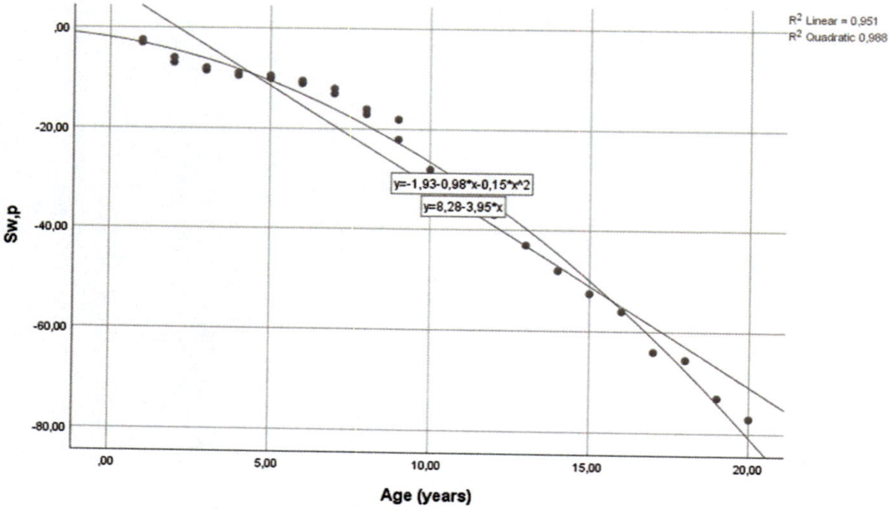


**Fig. 5.** Relationship of age-Sw,p from South of building.

From the Table 3, there are 13 school buildings using Decolith paint, there are 12 school buildings using the Catylac, 3 school using the Dulux and 2 schools using the Toa.

Percentage The level of degradation or reduction in the age of paint service in school buildings is very dependent on the age of the building's painting. The level of decline in the appearance of the walls of the building outside the school building is judged by some damage such as stains, cracks, bloating and peeling paint on the walls. An example of the level of paint reduction can be seen in Table 4. Analysis of data from the 30 schools found there are 4 brands of paint that are often used, so that the results of a recapitulation to determine the age of paint based on the brand of paint can be seen in the Table 5.

Based on the recapitulation results of Table 5. Popular brand such as Decolith used in by 4 school buildings experiencing service life with the intervals in intervals of 1 to 5 years, as is the case with Catylac paint from data known as 4 school buildings have decreased service life at intervals of 1 year to 5 years. On the other hand, a school building that uses Dulux paint as much as 1 building is in an age of 1 to 5 years old. The number of school buildings painted using Toa brand paint is as much as 1 building and is at intervals of 1 year to 5 years.

From the above data, it is found that service life has decreased the appearance of paint on the walls of buildings outside the school building at intervals of 1 to 5 years. Then the change of new paint is done in the 5th year. If the change of paint is done after the age of 5 years, the appearance of paint has decreased significantly.

**Table 3.** Service life of paint based on the brand

| Paint brand | School Ide | Sum | Average percentage (%) |
|---|---|---|---|
| Decolith | SD N minomaratni 1 | 13 | 28,48 |
| | SD perum concat | | |
| | SMP bopkri 3 | | |
| | SD kansius | | |
| | SMK N 2 | | |
| | SD N Babarsari | | |
| | TK karitas | | |
| | SMP N 1 Banguntapan | | |
| | SMP N 4 Yogyakarta | | |
| | SD Muh 3 YOGYAKARTA | | |
| | SMA N 6 yogyakarta | | |
| | SMP N 8 | | |
| | SMA Islam 3 Pakem | | |
| Catylac | SMP N 2 banguntapan | 12 | 25,78 |
| | SMP Muh Pakem | | |
| | SMP N 4 Pakem | | |
| | SD N Minomartani 6 | | |
| | SMA kanisius 1 | | |
| | SMP N 4 | | |
| | SMA N 1 | | |
| | SMA N 1 Cangkringan | | |
| | SMA N 9 | | |
| | SMA N 1 teladan | | |
| | SD N golo | | |
| | SD kanisius sengkan | | |
| Dulux | SMK Bopkri 1 | 3 | 10,35 |
| | SMP Karitas Ngaglik | | |
| | SD Karitas | | |
| Toa | SMP N 6 | 2 | 29,81 |
| | SD N Ngupasan | | |

**Table 4.** Level of degradation from the surface paint condition.

| Level 0 | Level 1 | Level 2 | Level 3 | Level 4 |
|---|---|---|---|---|
| | 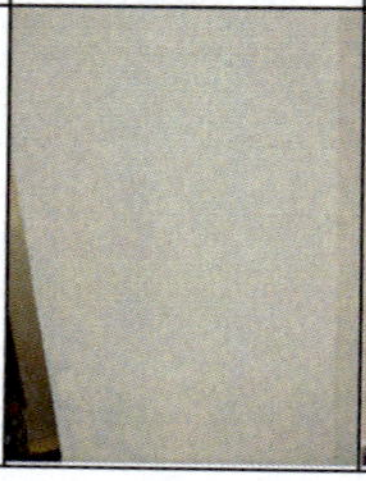 |  |  |  |

**Table 5.** Summary of the interval of the service life based on brand

| Interval of service life | Decolith | Catylac | Dulux | Toa |
|---|---|---|---|---|
| | No of data | No of data | No of data | No of data |
| 1–5 | 4 | 4 | 1 | 1 |
| 6–10 | 4 | 4 | 2 | 0 |
| 11–15 | 2 | 2 | 0 | 1 |
| 16–20 | 3 | 2 | 0 | 0 |

## 5 Conclusions and Suggestions

The study concludes as follows: 1) brand of paint used for painting school buildings from 30 samples of most schools using Decolith; 2) the service life of the external wall paints identified from various brands due to dirt and discoloration, cracking, bloating, peeling. Service life based on orientation of external wall paints seem to be similar has its own relationship in quadratic equations; and 3) in some extend, specifically for stated funded schools, external wall paint work is conducted due to social factor such as yearly budget, ceremonial event, and clean appearance for front wall. there is a correlation between the last year of painting and the service life of paint.

Some suggestions from the authors based on this study are as follows: 1) service life of external wall paint could be obtained longer if the building managers considers regarding painting techniques applied on works, and better quality of paint that would be applied; 2) in the operation and maintenance of buildings, it is better to repaint the external wall in every five year; and 3) further study can be conducted in other building components in deferent environment, such as pollution, sea sites building and other variables that can be influenced service life in building in Indonesia.

Since the studies regarding service life in building component in tropically country such as Indonesia is limited number, the authors suggest acceleration of this topics in the future. The service life obtained from this study can be applied in supporting life cycle cost for long term operation and maintenance of the building in general, particularly for re-painting works.

**Acknowledgement.** Authors are very grateful for the schools' management in Yogyakarta that allows to access to administration offices and collected data for this study. This paper is part of the outcome of the research funded by the grant no 124B-PEN/LPPM/V/2019 from The Ministry of Higher Education and Technology Research (Kementerian Ristekdikti) of Indonesia. The authors are grateful to Universitas Atma Jaya Yogyakarta for its support in the study and publication. The authors are also grateful to the reviewers for their constructive critics for the paper.

## References

1. Hartono, S.: Perawatan cat dinding pada gedung perpustakaan daerah: learning (2018). https://slideplayer.info/slide/13690516/. Accessed Sept 2018. (in Bahasa)
2. Homint, H.: Usia cat dinding, cat besi, cat kayu dan melamin: learning (2017). https://hakimhomint.wordpress.com/2017/03/23/usia-cat/. Accessed 23 Mar 2017. (in Bahasa)
3. Ramadhan, F.A.: Pengertian cat: learning (2011). http://hunter-science.blogspot.com/2011/06/pengertian-cat.html. Accessed 16 June 2011. (in Bahasa)
4. Silva, A., de Brito, J., Gaspar, D.L.: Methodologies for Service Life Prediction of Buildings: With a Focus on Façade Claddings, Green Energy and Technology. Springer, Cham (2016)
5. ISO. Building and constructed assets–Service life planning, part 1: General principles. ISO 15686-1, Geneva (2008)
6. Gaspar, P.: Service life of constructions: development of a method to estimate the durability of construction elements. Application to renderings of current buildings. Ph.D. thesis, Instituto Superior Técnico, Technical Univ. of Lisbon, Lisbon, Portugal. (2009). (in Portuguese)
7. Ang, G.K.I., Wyatt, D.P.: Performance concept in procurement of durability and serviceability of buildings. In: Lacasse, M.A., Vanier, D.J. (eds.) Proceedings of the International Conference on Durability of Buildings Materials and Components. Institute for Research in Construction, Ottawa, ON, pp. 1821–1832 (1999)
8. Shohet, I.M., Paciuk, M.: Service life prediction of exterior cladding components under standard conditions. Constr. Manage. Econ. **22**(10), 1081–1090 (2004). https://doi.org/10.1080/0144619042000213274
9. Sari, M.: Kelompok 10 – cat: learning. Miranda sari (2016). https://www.academia.edu/36182197/Kelompok10-Cat.docx. Accessed 12 Jan 2016. (in Bahasa)
10. BS-EN-ISO4628-1. Paints and varnishes. Evaluation of degradation of coatings. Designation of quantity and size of defects, and of intensity of uniform changes in appearance. Part 1. General introduction and designation system (2005)
11. Gaspar, P.L., de Brito, J.: Assessment of the overall degradation level of an element, based on field data. In: 10DBMC International Conférence on Durability of Building Materials and Components, Lyon, France, April 17–20, 2005
12. Gaspar, P.L., de Brito, J., Silva, A.: Statistical modelling of the service life prediction of painted surfaces. Int. J. Strateg. Prop. Manag. (2015). https://doi.org/10.3846/1648715x.2015.103185
13. Bordalo, R., de Brito, J., Gaspar, P., Silva, A.: Service life prediction modelling of adhesive ceramic tiling systems. Build. Res. Inf. **39**(1), 66–78 (2011)
14. Silva, A., Dias, J.L., Gaspar, P.L., De Brito, J.: Service life prediction models for exterior stone cladding. Build. Res. Inf. Routledge **39**(6), 637–653 (2011)
15. Sekaran, U., Bougie, R.: Research Methods Business, A Skill Building Approach, 7th edn. Wiley, Chichester (2016)

# 3D Printing of Buildings. Limits, Design, Advantages and Disadvantages. Could This Technique Contribute to Sustainability of Future Buildings?

Vitalie Florea(✉), Fideliu Păuleţ-Crăiniceanu, Septimiu-George Luca, and Cristian Pastia

"Gheorghe Asachi" Technical University of Iasi, 700050 Iaşi, Romania
vitalie.florea@tuiasi.ro

**Abstract.** Recently, a phenomenon of "printing" houses and construction 3D printers has taken on or greatly increased in many parts of the world. This phenomenon is intensively shown in various media and in science news and science articles. 3D printers and house printing are praised to be very advantageous from economical, energetic and sustainable point of view. However, there are no norms or rules to cover this type of construction works. In addition, a lack of methods for structural calculation and for certification of materials regarding "printing" is observed. Effects on future design and execution of this type of constructions on the market labor and on education in construction domain are unknown. In addition to some practical applications and their characteristics, this paper tries to present the requirements of 3D-Printing applications in the field of constructions, limitation of current technologies, the advantages and disadvantages of 3D printed houses and the effects on the listed above sub-domains. Future of 3D printing for constructions is analyzed in a sustainability perspective.

**Keywords:** 3D printing · Buildings · Sustainability

## 1 Introduction

During the last years, in media as in professional literature, news and scientific articles about 3D printers and 3D printed houses often appear. 3D printing technology emerged in 1980s [1], when the Japanese Dr. Hideo Kodama had requested a patent on a device for rapid prototyping (RP). This patent had never been issued, however.

In 1986, a first invention patent is received by Charles (Chuck) Hull, the co-founder of 3D Systems Corporation, one of the largest and most prosperous of today's 3D printing organization. In 1987, Hull created a first 3D printer named SLA-1 and sold it in 1988. For a view on the topic beginnings, Table 1 [1] shows a short history of 3D printing. *Additive fabrication, additive processes, additive techniques, additive layer manufacturing, layer manufacturing*, and *freeform fabrication* are some synonyms for AM. This process is


A. Rotaru (Ed.): CRIT-RE-BUILT 2019, SSGG, pp. 298–308, 2021.
https://doi.org/10.1007/978-3-030-61118-7_26

based on virtual model data created in CAD software in order to obtain objects or parts. Organizations are trying to watch the field in order to regulate it. Conforming to the ISO/ASTM 52921:2013 [2] international standard, "Printing 3D" or "Additive Manufacturing – AM" is defined as the process of joining materials to make objects from 3D model data, usually layer upon layer, as opposed to subtractive manufacturing methods.

**Table 1.** A history of 3D printing beginnings [1].

| Year | Achievement |
|---|---|
| 1980 | First patent application for RP technology, filled by Dr. Kodama, in Japan |
| 1986 | The first patent - to Charles Hull for stereolithography apparatus (SLA) |
| 1987 | SLA-1, introduced |
| 1989 | SLS (selective laser sintering) patent issued to Carl Deckard |
| 1990 | EOS company sold its first "Stereos" system |
| 1992 | FDM (Fused Deposition Modelling) patent issued to Stratasys |
| 1996 | Sanders Prototype (later Solidscape) and ZCorporation were set up |
| 1997 | Arcam was established |
| 1998 | Object Geometries launched |
| 2000 | MCP Technologies introduced the SLM technology |
| 2002 | EnvisionTec was founded |
| 2004 | Dr. Bowyer - RepRap concept of open source, self-replicating 3D printer |
| 2005 | ExOne - spin-off from the Extrude Hone Corporation |
| 2007 | The first system under $10.000 from 3D Systems |
| 2008 | Desktop Factory was acquired by 3D Systems |
| 2009 | The first commercially available 3D printer – in a kit form, based on the RepRap concept |
| 2012 | Alternative 3D printing processes introduced at the entry level on the market |
| 2013 | Stratasys acquires Makerboot |

Additionally, according to ISO/ASTM standards [3], AM can be divided in seven process categories according to the techniques used to create those layers (Fig. 1).

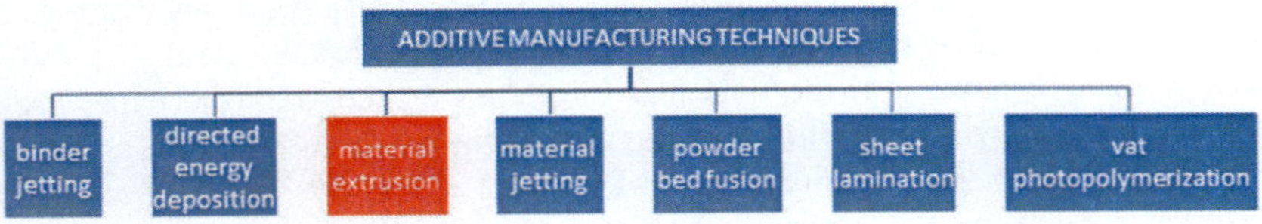


**Fig. 1.** Additive manufacturing techniques [3].

For constructions, from the seven process categories pertaining to 3D printing shown in Fig. 1, the 3D printing extruding material technology is used (for concrete, steel, clay etc.). This method is often named Fused Deposition Modelling (FDM) and sometimes, Fused Filament Fabrication (FFF). Fused Deposition Modelling, or FDM, technology is the most used on the global level. Conforming to statistics from July 2018 [4], this technology is used in 69% of cases (Fig. 2).

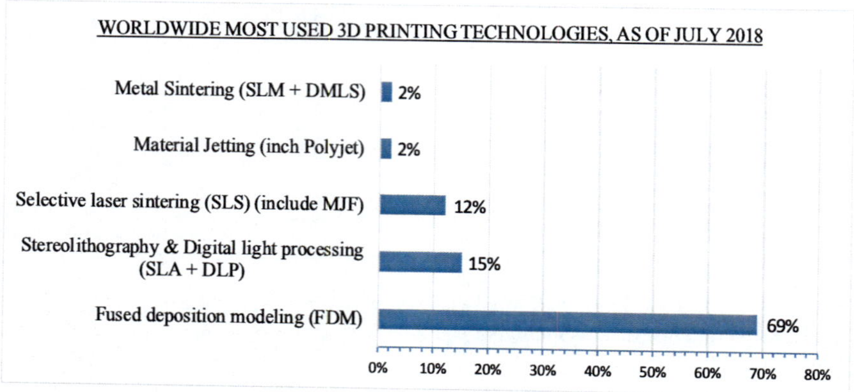


**Fig. 2.** Worldwide most used 3D printing technologies, as of July 2018 [4]. SLM = Selective Laser Melting; DMLS = Direct Metal Selective Sintering; SLS = Selective Laser Sintering; MJF = Multi-Jet Fusion; SLA = Stereo-Lithography Apparatus; DLP = Digital Light Processing; FDM = Fused Deposition Modeling

The FDM 3D printing technology for constructions had been invented in 1989 by Scott Crump, the American co-founder of the Californian Stratasys company, a relevant actor in the field. Finally, in 1992, the brevet for FDM was issued to Stratasys marking the start of an intensive development in 3D technologies. It is worthy to remember that Medicine is one of the first domain to adopt the FDM technology, in early 1990s.

## 2 Goal and Methodology in 3D Printing Analysis

The goal of this research is to establish the role (as a component of sustainability) and the extent of the phenomenon, the need for actual use, the possibility of future use, directions of development and any other professional observations that could be withdrawn. These elements should be wrapped up in series of advantages and disadvantages that must be in the views professionals involved or willing to be involved into this field.

Due to relative novelty of the field and for achieving the goal, the paper uses a mostly intuitive, analytic and critic way to conduct the research based on a series of representative examples of applications worldwide. The analysis is trying to observe the difference between the common, classical way, of building and the new means, devices and realizations. In addition, it should be underlined what could be the benefits for the actual requirements of our society in the context of smart cities, sustainability playing a key place. Therefore, the results of this stage in research of 3D printing for Civil Engineering domain is mainly general and qualitative. Remarks and observations are opening the way to quantitative and normative studies when the field approaches maturity.

## 3 3D Printed Constructions. Examples

In order to get a critical view on the 3D technology in constructions field, it is relevant to show some important practical applications and the year they emerged, as follows. Some data about each application, considered significant, is presented. In addition, a large diversity of construction types, country of applications, methods, materials, designers is implying the scale of the interest for 3D printing in construction field. Of course, the number of real 3D buildings and bridges is already very large.

The examples that follow might be named already as "classical" and became very fast well known for professionals. They are useful for this paper because they offer the frame for a critical view on 3D printing in construction domain.

**2013 - "Grotto I" - The 9th Architecture Exhibition, Archilab, France**

One of the first use of a 3D printed constructions, dated back in 2013, might be considered the architects/programmers' Michael Hansmeyer and Benjamin Dillenburger creation exposed to the French Archilab Exhibition [5, 6]. It is named *Grotto Prototype*, or *Digital Grotesque I*, and represents an immersive space at human scale, totally built from 3D printed sandstone (and a second *Grotto* was made in 2017). A resolution of 0.3 mm on each layer, or 300 dpi, a height of 3.2 m, a surface of 16 $m^2$ and a weight of 5.8 tones are some very impressive characteristics of this piece of art. In order to facilitate the transportation and assembling of *Grotto I*, a modular blocks system had been developed. Each block is about 80 by 120 cm and it is weighting around 100 kg. The virtual model had 260 millions individual surfaces, 42 billion vortexes and needed 78 GB to be digitally stored. The digital design lasted for one year, printing – one months, and assembling – one day (Fig. 3). Michael Hansmeyer had created this *Grotto* under the concept of *Computational Architecture*.

**Fig. 3.** Grotto I [5, 6]

**2014 - "Project EGG" - The Netherlands**
"Project EGG" [6, 7] is conceived by the Dutch architect Michiel van der Kley. The project is composed from 4760 of unique "stones" and was presented at Dutch Design Week in Eindhoven in 2014. Some users of 3D printers from different parts of the world helped printing the 4760 pieces and sent them to be assembled in The Netherlands. The contributors had obtained the STL file and used it. STL is a file format native for the stereolithography CAD software and it was created by 3D Systems in 1987. The "Project EGG" is 4.88 by 3.96 by 3.05 m in size (Fig. 4a).

**Fig. 4.** a - "Project EGG" - The Netherland [6, 7]; b - "First 3D Printed Castle" - USA [6, 8]

**2014 - "First 3D Printed Castle" - USA**
"First 3D Printed Castle" [6, 8] was printed by Andrey Rudenko. This construction is made from printed concrete layers with 10 mm thickness by 30 mm width. Construction's total dimensions are 3 by 5 by 4–6 m (Fig. 4b).

**2014 - "Apartment Building in China" - China**
"Apartment Building in China" [6, 9] is a construction realized in Chinese province of Jiangsu by Winsun company. The walls and other structural elements had been 3D printed outside the construction site.

For walls' printing, a so called "ink", made from constructions' waste and a bonding agent, was used. Finally, a high thermal and structural resistance had been achieved. The 3D printed building is supposed to resist well to earthquakes.

According to Winsun, compared to traditional construction method, some of the new method achievements are: the material costs is reduced by 60%, the work force costs is reduced by 80% and the building time is reduced by 70% (Fig. 5a).

**2015 - "MX3D Bridge" - The Netherlands**
"MX3D Bridge" [6, 10] is a metallic bridge designed by the Dutch Joris Laarman in 2015 and finally printed in 2018. Robot arms attached to the structure were used for

(a) (b)

**Fig. 5.** a - "Apartment Building in China" - China [6, 9]; b - "MX3D Bridge" - The Netherlands [6, 10]

printing. A special characteristic of this system is that it can be used in relatively narrow spaces (Fig. 5b).

**2015 - "Philippines Hotel" - Philippines**

The "Philippines Hotel" [6, 8] – is a 130 m$^2$ of Lewis Grand Hotel in Philippines. It includes 3D printed bedrooms, living room and Jacuzzi bathroom. The owner, Lewis Yakich together with the Russian born Andrey Rudenko were the main designers. For the printing, sand and volcanic ashes were used. After 100 h printing, strong walls had been realized (Fig. 6a).

(a) (b)

**Fig. 6.** a. "Philippines Hotel" - Philippines [6, 8]; b. "Office Building" - U A E [6, 9]

**2015 - "Office Building" - United Arab Emirates**

The Office Building [6, 9], was designed by Thornton Tomasetti and Syska Hennessy from Gensler architecture company for the National Committee of United Arab Emirates. The building is the headquarters of Dubai Futures Foundation and is 240 m$^2$. 3D printed pieces of the building had been made by Winsun company from China, in Shanghai. The printing lasted for 17 days in China and, assembled in two days in Dubai (Fig. 6b).

**2016 - "Apis Cor" - Russia**

"Apis Cor" [6, 11] - is a 38 m$^2$ house printed *in situ.* It is using a mixture of solid elements an liquid polyurethane. The main observation in this case is that the *in situ* printing had reduced transportation and assembling costs (Fig. 7a).

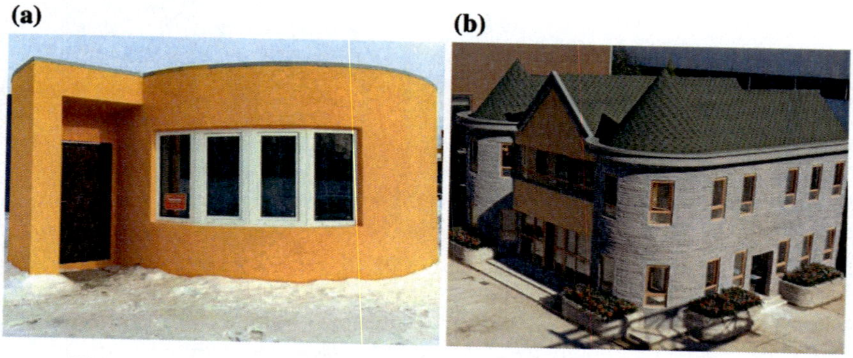

**Fig. 7.** a - "Apis Cor" - Russia [6, 11]; b - "Villa in China" - China [6, 12].

**2016 - "Villa in China" – China**

"Villa in China" [6, 12]. This is a 400 m$^2$ house conceived by the Chinese company named HuaShang Tengda. The walls are 25 cm thick made from C30 concrete. It is estimated that this 3D printed construction would resist to an 8 Richter Local Magnitude major earthquake (Fig. 7b).

**2017 - "Pedestrian Bridge in Madrid" - Spain**

"Pedestrian Bridge in Madrid" [6, 13] - is a bridge designed by Catalunya Advanced Architecture Institute and printed by ACCIONA, a company specialized on infrastructure and renewable energy. The length of the bridge is 12 m while the width is 1.75 m. It is using concrete reinforced by microfibers (Fig. 8a).

**2017 - "Bicycle Bridge" - The Netherlands**

"Bicycle Bridge", [6, 14], is designed mainly by the Technical University of Eindhoven in collaboration with BAM company from United Kingdom. It is 3D printed and it is destined for bicycle users. The bridge is made from pre-tensioned reinforced concrete. This bridge was initially 3D printed in 8 individual parts, each 3.5 m width and then assembled. The concrete used in this case is a non-shrinkable one (Fig. 8b).

**2018 - "Gaia, a 3D Printed House with Earth" - Italy**

**"Gaia"** [6, 15] - is low-cost house that is only 12 m$^2$ surface, conceived by the Italian engineer Alberto Chiusoli, and 3D printed using Crane WASP technology.

(a) (b)

**Fig. 8.** a - "Pedestrian Bridge in Madrid" - Spain [6, 13]; b - "Bicycle Bridge" - Technical University in Eindhoven & BAM [6, 14]

(a) (b)

**Fig. 9.** a - "Gaia" - Italy [6, 15]; b - "DFAB House" - Switzerland [6, 16].

A natural material had been used, a mixture from brut soil and natural wastes of rice crops. This way, the house is becoming ecological and sustainable (see Fig. 9a).

**2019 - "DFAB House" - Switzerland**

"DFAB House" [6, 16] - is realized by ETH Zurich's researches and professors in collaboration with industrial partners using 3D printers and robots. It has three floors and is based on NEST (Next Evolution in Sustainable Building Technologies) system. It has a surface of 200 $m^2$. It is considered the first inhabited digitally planned and build construction (see Fig. 9b).

## 4 Advantages and Disadvantages of 3D Printing in Constructions

Analyzing the methods, real examples, materials, concepts, opinions of professionals, some advantages and disadvantages of 3D printing for constructions had been observed, especially when compared to the classical ones.

On one hand, advantages that emerged after the above analysis are:

- rapidity of construction building compared to classical methods
- reduction of material use due to printed advantageous shapes (a goal of sustainability)
- reduction of wood/steel used for forms and scaffolding (in many cases they are no longer needed) – also a goal of sustainability
- reduced costs of constructions, compared to classical constructions
- 3D printing in construction no longer needs design drawing for forms or reinforcement
- possibility of recycled materials use (goal of sustainability)
- execution details with high precision, improving architecture and strength
- possibility of building constructions parts or objects even in narrow spaces (goal of sustainability through more rational use of space)
- 3D printing of complex constructions shapes within a reduced cost range
- substantial reduction of manpower cost in building
- reduced waste quantity resulted from constructions' building (sustainability goal)
- fuel consumption reduction and assembling time reduction in the case of *in situ* 3D printing. It implies also a reduction in toxic gases coming from constructions machineries and, this way, it is combating the global warming (goal of sustainability)
- dramatic decrease in work and traffic accidents involving construction domain
- repeatability of design, projects and multiple use of the same devices (3D printers) implying reduced costs for social houses and for after-disaster shelters. Conforming [17], "Not only could this (3D printing) revolutionize the construction industry, but the less expensive process could also affect housing affordability"
- integration with BIM related design and technologies [18] (goal of sustainability).

On the other hand, disadvantages that emerged after the above analysis are:

- relatively high price of 3D printers-devices (starting from $250,000 and above)
- the cvasi-impossibility of simultaneous use of more than one material
- 3D printing software and programming may be very expensive
- structural computation for 3D printed constructions is cumbersome, complex and need to be developed
- materials (like concrete) cannot be vibrated, in order to eliminate voids
- irregular shapes of walls resulted from 3D printing, due to layers
- limited height of 3D printed constructions (at least for the moment)
- loosing of work places not only in constructions domain but also in adjacent connected fields of activities
- unknown, yet, consequences of damaged or faulted 3D printers usage
- insufficient information on sustainability of (some) 3D printed constructions
- insufficient information on strength of (some) 3D printed constructions
- work temperature of 3D printing processes must be positive, above freezing point
- possibility of disastrous cyberattacks on 3D printers software or hardware
- insufficient information on long term behavior of 3D printed walls, in case of performing holes throughout them for installing pipes and cables
- a possible monopolist market if 3D printing of constructions is concentrated in a few companies

- impossibility of use or reduced use of low skilled labor force, that could (apparently) lead to higher costs or lack of high shilled manpower
- lack of design, building and verification codes for 3D printed constructions
- lack of formation and training for skilled workers in 3D printing technology field
- high costs due to patented inventions and projects
- lack of scientific studies and researches, technical references that are enough advanced that could reduce and limit the above presented disadvantages
- strong increase of energy consumption (electricity) for 3D printers functioning, even 100 times more than the classical construction methods, conforming [19].

## 5 Conclusions. Future of 3D Printing for Constructions

This paper shows a brief history and examples of 3D printing in construction field. A list of advantages and disadvantages is demonstrating the stage of development in this topic. Acting in amplifying advantages and reducing the disadvantages is leading to a great progress.

Use of 3D printing in constructions domain generates a strong emulation in the field. If, for the moment, mass use of 3D printing in civil and industrial constructions field is not yet seen, certainly it has a very favorable near future in prefabricated construction industry. Appearance of 3D printers simultaneously using more materials, robotization of (prefabricated) parts assembling and other technologies will unavoidable lead to cvasi-complete constructions (embedding concrete, reinforcement, finishing, electrical cables, pipes and so on) with very advantageous production costs not only for investors but for beneficiaries, too. A consequence will be a drastic reduction in the lack of shelters for pour population in many parts of the world leading to a better resilience of our society. Another consequence will be a more sustainable construction domain, [20], that will be environmental friendly and combating global warming.

## References


1. DPI Homepage. https://3dprintingindustry.com/3d-printing-basics-free-beginners-guide
2. ISO/ASTM 52921:2013 Standard terminology for additive manufacturing - Coordinate systems and test methodologies
3. ISO/ASTM 52900:2015 [ASTM F2792] Additive manufacturing - General principles – Terminology
4. Statista Homepage. https://www.statista.com/statistics/756690/worldwide-most-used-3d-printing-technologies/
5. DG Homepage. http://www.digital-grotesque.com
6. All3DP Homepage. https://all3dp.com/1/3d-printed-house-homes-buildings-3d-printing-construction/
7. EGG Homepage. http://projectegg.org
8. DCHP Homepage. http://www.totalkustom.com/home.html
9. Winsun Homepage. http://www.winsun3d.com/En/Product/pro_inner_5/id/102
10. MX3D Homepage. https://mx3d.com
11. Apis Cor Homepage. https://www.apis-cor.com
12. HST Homepage. http://www.hstdgm.com

13. IACC Homepage. https://iaac.net/institute-advanced-architecture-catalonia-designs-first-3d-printed-bridge-world/
14. TUE Homepage. https://www.tue.nl/en/research/research-groups/structural-engineering-and-design/concrete-research-areas/3d-concrete-printing/3dcp-projects/
15. WASP Homepage. https://www.3dwasp.com/stampante-3d-per-case-crane-wasp/
16. DFB Homepage. https://dfabhouse.ch/dfab-house/
17. Johnson, A., et al.: Why transform the building industry? A case for technological change in New Zealand construction. In: Proceedings of the CIB World Building Congress 2019 on Constructing Smart Cities, pp. 203–217, Hong Kong SAR, China (2019)
18. Ding, L., Wei, R., Che, H.: Development of a BIM-based automated construction system. Proc. Eng. **85**, 123–131 (2014)
19. Manju, R., Deepika, R., Gokulakrishnan, T., Srinithi, K., Mohamed, M.I.: A research on 3D printing concrete. Int. J. Recent Technol. Eng. **8**(2S8), 1691–1693 (2019)
20. Sakin, M., Kiroglu, Y.C.: 3D printing of buildings: construction of the sustainable houses of the future by BIM. Energy Proc. **134**, 702–711 (2017)

# Vulnerability, Seismic Survey and HBIM-Based Structural Analysis

# Vulnerability of Buildings: From Heritage Building Information Modelling (HBIM) to Seismic Analysis

Maura Imbimbo(✉), Giuseppe Modoni, Assunta Pelliccio, and Marco Saccucci

University of Cassino and Southern Lazio, Via G. Di Biasio, 43, 03043 Cassino, Italy
mimbimbo@unicas.it

**Abstract.** Many historical urban centres in Europe are showing their fragility, particularly those located in seismic zones, because structures built in ancient times were not specifically conceived to resist earthquakes. Their vulnerability assessment implies a comprehensive knowledge of their construction procedure and history of transformations. There is a minimum level of knowledge intrinsically connected with the analysis to be performed. The object of the UNICAS module is related to the vulnerability assessment of historical and post-industrial buildings in Europe against different seismic hazards (shaking, liquefaction). Specific reference is made to Italian case studies, moving from the information modelling of heritage building (HBIM) to seismic assessment through the subsoil hazard.

**Keywords:** Vulnerability · Hazard · HBIM · Seismic analysis

## 1 Introduction

The European landscape is populated of thousands of small to medium towns, each witnessing a peculiar historical and cultural heritage, altogether preserving like in a polychromatic mosaic the memory of civilization (Fig. 1). Being all scenarios exposed to the injury of time, challenges continuously come for the European scientific community to create new tools and undertake smart strategies to preserve this treasure.

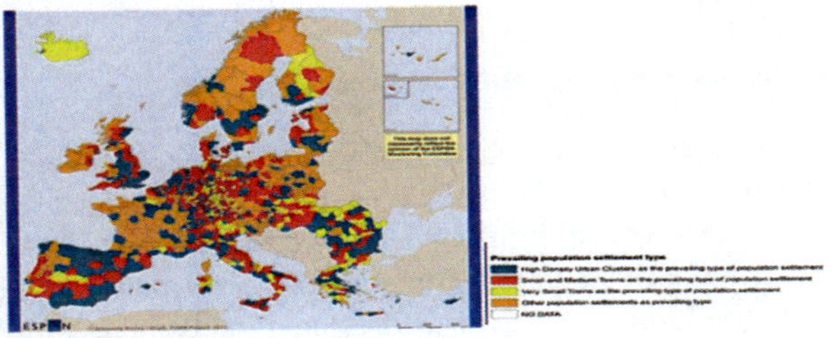

**Fig. 1.** Distribution of towns by size over the European territory (http://www.espon.eu/)


A. Rotaru (Ed.): CRIT-RE-BUILT 2019, SSGG, pp. 311–329, 2021.
https://doi.org/10.1007/978-3-030-61118-7_27

Among the different natural and artificial hazards, earthquakes are probably the most aggressive one, being able to destroy buildings and infrastructures in a few seconds, sweep centuries of historic and cultural stratification, undermine the existence of the whole communities. The loss in the population of sense of belonging to the place and tortuous discouraging recovery processes to get the original life conditions often persuade people to abandon places and move elsewhere (Fig. 2).

In a strategy to improve resilience of communities, the fragility of old urban systems to earthquakes poses continuous challenges for the society, claiming for the contribution of the scientific community and academic world in terms of innovation, technology transfer, education, best practice promotion and standardization, etc. The object of this module is related to the vulnerability assessment of historical and postindustrial buildings in Europe. Considered as typical examples of the European scenario, a path is traced to exploit the contribution of heritage building information modelling (HBIM) to seismic risk assessment.

**Fig. 2.** Images of Central Italy before and after the Aug-Oct 2016 seismic sequence

## 2 The Italian Heritage

### 2.1 Historic Towns, an Environmental Monument

In the last decades, the preservation of building heritage has increasingly faced with the concept of environmental sustainability, reshaping the process of social, architectural and urban regeneration with the goal of limiting new construction and recover old ones. The Italian civil construction heritage gathers three macro categories, historical buildings, modern construction (building and infrastructures) and industrial establishments.

The historical building heritage, in addition to the architectural singularities, is concentrated for its most part in towns, from large to small ones (Pelliccio and Cigola 2015). In particular, the latter are more than 20,000, characterized by a high quality of aggregative life and by a significant aesthetic identity of the urban structure, settled with a strong empathy with the territory. Very often, morphology, spatial distribution, use of materials and construction technology conform to the surrounding natural environment. As a remarkable example, the structure of many buildings is made of masonry composed of local stone (carbonatic rock, sandstone, tuff). These materials combine a cost-effective solution with an extraordinary control of inner temperature and humidity. In conjunction with the narrow conformation of the urban system and the optimal orientation with sun, walls protect houses from very hot or cold climate conditions and exploit brilliantly natural heating.

Together with the orientation along the prevailing wind directions, the colour of the external facades and cover are often designed to reduce solar heat accumulation and improve light reflection. If on one hand historical towns express in their quintessence the concept of environmental sustainability, their deep relationship with the landscape represents a main vulnerability factor (Arpino et al. 2018). Established mostly in the Middle or Renaissance Age, most cities are located along the Apennine chain, classified in terms of seismicity as hazard zone level 1 or 2, i.e. subjected to relatively high macro-seismic intensity. The morphological conformation (Fig. 3) often follows the orography of the territory, sometimes generating real canyons with important effects of wind and shadow degradation (Pelliccio et al. 2017). In addition, the continuous transformations of the architectural organisms with "superfetations" (accretions of building without artistic, architectural and historical value), determines a further increase of seismic vulnerability. Small towns owe the prestige of being considered real environmental monuments to the originality of their urban composition, that deserves to be protected and preserved as an expression of the local cultural identity.

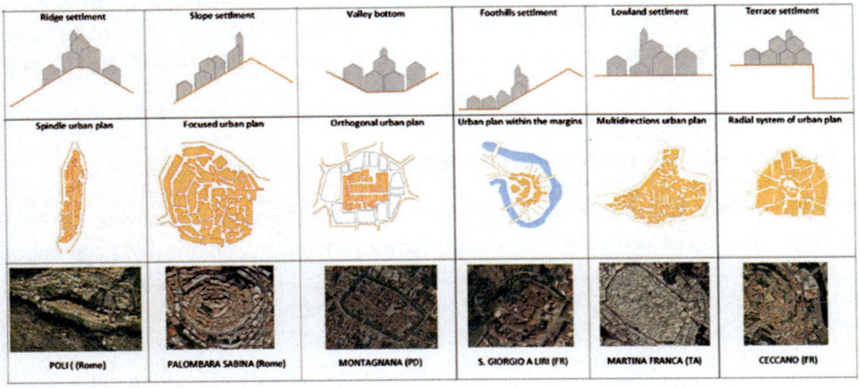


**Fig. 3.** Typical urban shapes of Italian historical towns for different orographic conditions

## 2.2 Post-industrial Settlements

The Italian industrial heritage grew up during the massive industrialization process that involved Europe since the end of the XVIII century. Factories were initially established close to urban centers, sometimes fully incorporated with the residential districts, to facilitate the economic, political and administrative relationships. Sometimes the industrial plants are located near riverbanks or sea shoreline to exploit natural resources or to take advantage of maritime transportation infrastructures. With time, the industrial sites have noticeably expanded to meet new production requirements or to introduce new machineries, often orienting, polarizing or obstructing the urban development. In the later more mature industrialization phase, the tendency took place, favored by the improvement of transportation facilities, to delocalize the industrial districts in the suburbs, covering vast surfaces that in some cases became larger than the same urban centers. From the architectural point of view, the evolution of the production processes and the increase of production required the expansion of buildings with new structural, typological and technological elements. Factories often present a first nucleus of buildings made of masonry and vaulted covers, a subsequent addition of buildings made of steel and masonry, a third more recent portion formed of in concrete and steel (see Fig. 4). After the 70's of the last centuries, following a slow decline in productivity, the factories were dismantled leaving ample abandoned areas full of waste and pollution sources (brownfields). In some regions, the size of the phenomenon reaches impressive level: a survey commissioned by APAT (2006) counts about 12,000 abandoned industrial sites, with 5.000 seriously affected by pollution, all over Italy. In Lazio, one of the most industrialized region of central Italy, 329 abandoned sites are counted, with 220 affected by serious pollution. The southern part of the Region, object of the present study, includes 42 of these cases. Nowadays, the rehabilitation and reuse of brownfields has become an important game for the political management of the territorial asset. The Environmental Sustainability Indices (ESI) classifies the different countries in terms of economic, social and landscape factors. As a consequence, the interest for the regeneration of brownfields has significantly grown pushed by architects and urban planners who want to renew cities and make them more attractive and environmentally sustainable (Pelliccio et al. 2018).

## 2.3 Building Information Modelling (BIM): Dimensions and Loads

The above situation requires more complete analyses for the reduction of risk, based on the management of richer databases grouping heterogeneous information (e.g. geometry, materials, technological, structural, environmental etc.) at the level of single unit, aggregate or whole urban nucleus. To this aim, the modelling and visualization of building seismic vulnerability can be fruitfully manipulated to analyse and consciously reduce risks. In recent years, the introduction of BIM, an acronym for Building Information Modelling, has greatly facilitated the integration between design, representation, production and management of the entire lifecycle of new buildings. These systems combine the multidimensional display of parametric databases, linking design, construction and management of buildings and facilitating communication between designers, executors and other interested parties. BIM projects merge different disciplines, from architectural

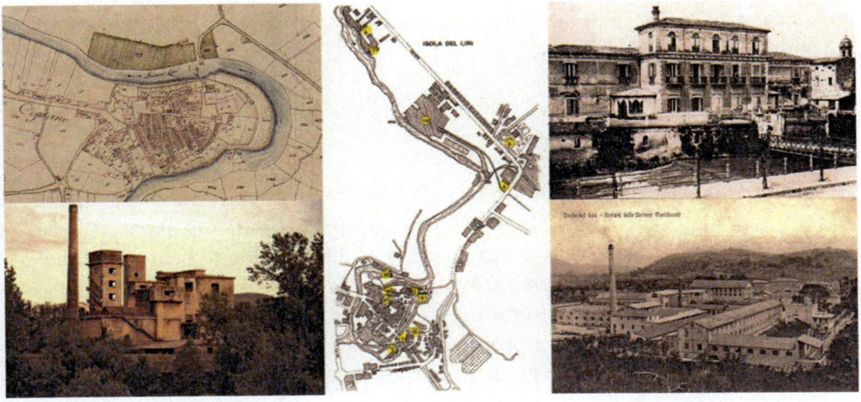

**Fig. 4.** Examples of industrial site in Southern Latium

design to structural calculation to MEP (plant design), with the goal of combining safety, energy performance and cost-effectiveness, in a single word sustainability. Often BIM implies to manage complex digital models obtained from the survey through laser scanning, photogrammetry and GIS, taking full advantage of the most modern technologies and drawing inspiration from the most innovative analysis methods. The graphic model is linked to information on materials, structure, etc., integrating technical documents in a parametric database constantly updated with the progress of building design and construction. In this way the risk of errors is minimised. The core for the interoperability of heterogeneous data is the IFC files (Interchange Industry Foundation Classes). Interoperability supports construction and management throughout the entire life cycle, superintending all building phases, from the feasibility study to construction and maintenance. Exportation of the parametric model in other software is also possible to perform different analyses at different representation scales (Zonta et al. 2008). Thanks to the IFC, BIM systems enable analyses in various sectors, particularly structural and environmental ones. Through the model, it is possible to perform, quite simply, structural analyses of independent skeletal buildings made with different technologies. The analysis for masonry buildings, particularly for cultural heritage identification and restoration, requires a more complex information model, called Historic Building Information Modeling (HBIM). The HBIM is able to manage survey and recovery processes into a unique framework ruling restoration works (Meyer et al. 2007). Moreover, thanks to the georeferenced BIM model, it is also possible to perform a detailed energetic assessment, i.e. energy performance of the building during its lifecycle quantifying sun and wind exposure (light and shadow), climate and solar radiation. Despite BIMs are widely used everywhere, the executive regulations of these systems are slowly issued in great part of Europe, including Italy. In particular, the UNI 11337/2017 Standards concern both the structure of the models and the figures involved and the characteristics of the system. In particular, the UNI 11337-9 standard introduces the concept of Level of Details (LOD) which is particularly important in the context of the analysis of existing buildings (Azhar 2011). In fact, in this case the standard introduces 5 LODS from 100 to 500 and the most significant for the purpose of analysing the built heritage are the lod 400 (construction stage) and the LOD 500 (as built).

## 3 Seismic Vulnerability

"Vulnerability" refers to the inability of a system or unit to withstand the effects of a hostile environment. Referring with mathematical terms to a building exposed to seismic events, vulnerability is defined as the probability of the building to be damaged under a certain seismic input. In a more general application of risk assessment to building assets, different damage levels are defined for different buildings typologies, and vulnerability is expressed with curves giving the probability of that building typology to undergo that damage level for variable seismic input. Therefore, the purpose of a seismic vulnerability assessment for a single building is to estimate the damage level experienced by a given building due to a given seismic action, that is to analyse the physical Damage-to-Ground Motion Intensity Correlation.

The seismic assessment of existing buildings, especially those with historical values, is a complex process being most of them designed without any seismic resistant code provisions, Additionally, the knowledge of as-built information is rarely achievable and this situation prevents to define a single standard modelling strategy.

A common approach, also reported in many recommendations and guidelines (ICOMOS 2005; ISO13822, 2012; CIB 335, 2010; PERPETUATE; etc.), considers essential to adopt a stepwise strategy. A preliminary assessment is compulsory, based on a qualitative analysis of historical information, structural details, response to past events (recorded damages), comparisons with similar structures, etc. Then the analysis can move to a quantitative approach, as defined in codes such as the Italian Code (known as NTC 2018). Complexity becomes higher in case of historical constructions. Many research projects have analysed the issue of preservation of historical buildings. Among these: RISKEU (https://eu-risk.eucentre.it/) developed a procedure for the assessment at urban level; EU-CHIC provided an "identity card" for historical constructions able to tracking environmental changes and human interventions; ONSITEFORMASONRY (https://cordis.europa.eu/project/rcn/58620/en) provided recommendations on the application of test methodologies for the preliminary diagnosis of the state of the structure and the materials; PROHITEC and NIKER analysed different strengthening interventions; PERPETUATE (https://cordis.europa.eu/project/id/244229/reporting) provided a robust framework for the assessment of historical constructions in terms of quantitative information on safety levels and based on a qualitative and historical knowledge within the PBA approach in accordance to the international codes for buildings (CEN, 2005; ASCE-SEI 41-13, 2014).

### 3.1 Vulnerability Assessment Against Ground Shaking

Many methods are available for the vulnerability assessment. A clear scheme of the existing approaches is provided by Meslem and Lang (2017) in their state of art review about physical vulnerability in the earth risk assessment and reported in Fig. 5.

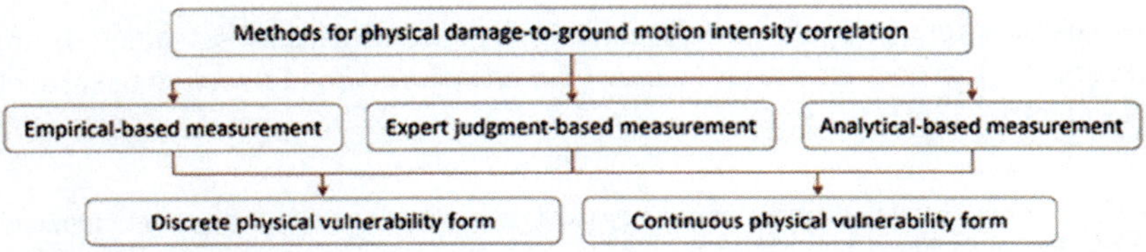


**Fig. 5.** Existing approaches for seismic physical vulnerability (after Meslem and Lang 2017)

The empirical methods are based on field observations and statistical analysis of building damage data (i.e., the seismic performance of the building) from past earthquake events. Some examples are those developed on the base of the use of macroseismic intensities for characterizing the earthquake shaking (among these the European Macroseismic Scale—EMS-98, Grünthal 1998), or those using physical parameters such as the peak ground acceleration, the spectral acceleration (among these the DPM introduced first by Whitman et al. 1973) and later implemented in many regions of the world such as in Italy after the 1980 Irpinia earthquake (Braga et al. 1982) or, finally, those use continuous physical damage-to-ground motion intensity curves (fragility or vulnerability curves) that are based on damage to buildings from past earthquake events.

The analytical-based assessment is focused to theoretical approaches able to predict the structural damage produced by a seismic input. Different methods are available in literature and they vary in terms of complexity and accuracy. A clear scheme of the most commonly used is provided in the state of art paper by (Meslem and Lang 2017) and reported here in Table 1.

The nonlinear static procedures (NSP) procedure is the standard method adopted in the assessment analysis. It works within a performance-based seismic design (PBSD) approach and, thus, it assumes that structural damage is essentially determined by a lateral displacement or drifts. It consists of: i) evaluating a capacity curve of the building (relationship between the base shear force and the displacement of a control point) through a push-over (non-linear quasi static analysis; ii) evaluating the response spectrum converted in the spectral acceleration-spectral displacement domain; iii) identifying the performance point representing the displacement reached under the respective seismic demand and, hence, the mean damage induced in the individual building of this typology.

A simplified version of the NSP procedure is provided by the Nonlinear Static Simplified Mechanism-based Procedures. They are procedure easily applied to a large number of buildings. Among these it is worthy to mention the FaMIVE (Failure Mechanism Identification and Vulnerability Evaluation) procedure proposed in D'Ayala and Speranza (2003) which provided collapse multipliers for in-plane and out-of-plane mechanisms.

The Nonlinear Dynamic Procedures (NDP) which are limited due to the high computational efforts, consists of applying a time dependent loads to the structure and analyze its response in the real time considering damping and inertial effects. Both the NSP and NDP belong to the incremental-iterative analysis approaches.

In addition, to these approaches there are also the limit analysis-based solutions (Heyman 1966; Como 2013) which, on the basis of the static and kinematic theorem, provide the collapse multiplier and the collapse mechanism but with no information on the postpeak response, which is essential in displacement-based seismic assessment procedures for masonry structures note that the first paragraph of a section or subsection is not indented.

**Table 1.** Analytical methods most commonly used for measuring seismic physical vulnerability (after Meslem and Lang 2017)

| Method | | Implementation building typology | Reference |
|---|---|---|---|
| Nonlinear static procedures | | | |
| | Capacity spectrum method (CSM) | Steel, reinforced concrete, timber | ATC-40 (ATC 1996); FEMA-279 (FEMA 1997) |
| | Modified capacity spectrum method (MADRS) | Steel, reinforced concrete, timber | FEMA-440 (FEMA 2005) |
| | Displacement coefficient method (CDM) | Steel, reinforced concrete, timber | FEMA-356 (FEMA 2000) |
| | Improved displacement coefficient method (I-DCM) | Steel, reinforced concrete, timber | FEMA-440 (FEMA 2005) |
| | N2 Method | Steel, reinforced concrete | Fajfar (2000); Solsek and Fajar (2004); Eurocode 8 (CEN 2004) |
| Nonolinear static simplified mechanism based procedures | | | |
| Collapse-based methods (CBM) | | | |
| | Failure mechanism identification and vulnerability evaluation (FaMIVE) | Earthen (historic buildings) | D'Ayala and Speranza (2003) |
| | VULNUS | Unreinforced masonry | Bernardini et al. (1990); Cosenza et al. (2005) |
| Displacement-based methods (DBM) | | | |
| | Displacement-based earthquake loss assessment (DBELA) | Reinforced concrete | Miranda (1999); Crowley et al. (2004) |
| | Mechanical based procedure for the seismic risk estimation (MeBaSe) | Unreinforced masonry | Restrepo-Vèlez and Magenes (2004); Restrepo-Vèlez (2005); Modena et al. (2005) |
| Nonlinear dynamic procedures (NDP) | | | |
| | Incremental dynamic analysis (IDA) | For all building typologies | Shome and Cornell (1999); Vamvatsiokos and Cornell (2002) |

*Modelling Scale Approach.* Many approaches are available for the modelling of buildings in the seismic assessment procedure. The seismic assessment of a masonry building has to be performed by considering its Global Response when it behaves as a whole, and its Local Response related to the possible occurrence of local mechanisms. Furthermore, it is also very important to consider that most buildings of our urban and historical centres are embedded in aggregates which affect the global and local behaviour of each of them and, also, the aggregates constitute urban entities to be analysed in their complexity and completeness Fig. 6.

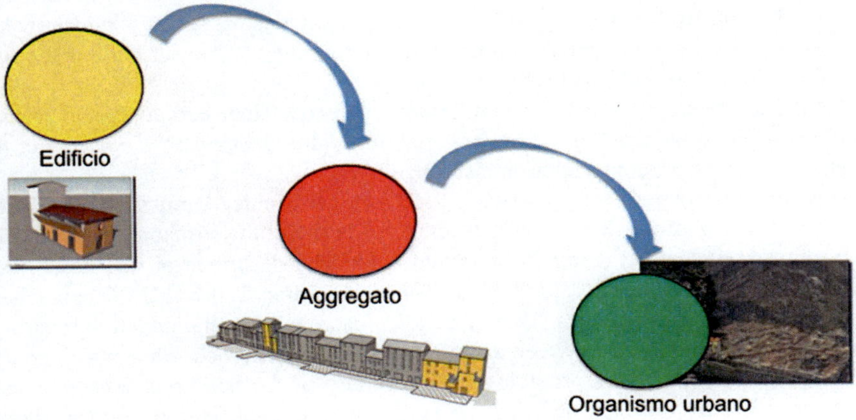


**Fig. 6.** System: building/aggregate/urban centre.

Thus, the quantitative assessment evaluation of a building can be performed by adopting the following types of modelling approaches:

- Local/Macro-element Scale Approach where the building has not anymore a box-type behaviour because of different causes such as poor structural connections, flexible horizontal slabs, sparse internal walls, etc. In this case the building behaves as a set of macro-element (i.e. wall with piers and sprandels, etc.) behaving independently which are individually analysed, generally, by means of 2D models and, then, combined in a proper way to provide the response of the entire building.
- Global Scale Approach where the building, characterized by a box-type behaviour, behaves ad a whole and it is analysed by means of a 3D model. It should be underlined that 3D models should be generally analysed, if possible, in order to account for the structural details and the geometric irregularities characterizing buildings.
- Aggregate Scale approach where the building is part of a system interacting with others in terms of local and global behaviour.
- Urban Scale approach used for territorial analysis without considering the specific response of each buildings.

*Modeling Strategies.* Although the impossibility to provide a whole and complete classification of the modelling strategies, a well comprehensive scheme is provided by (Ghiassi 2019) who suggests four classes for describing the numerical strategies most commonly used, that are: block-based models (BBMs), continuum models (CMs), macroelement models (MMs), and geometry-based models (GBMs).

Each of these strategies can be appropriate for a specific case study on the basis of its complexity, available data, required outputs and, also, level of expertise of the practitioners. BBMs are the most accurate approaches since it analyses masonry at the scale of its texture made of blocks assembled with mortar joints. It requires large computational efforts and an in-depth knowledge on specific characteristics of masonry, which make them appropriate for very special applications. Among the main contributions are for example D'Altri et al. 2018 and Lourenco 2002.

CMs treats masonry as a continuum deformable body. Their computational efforts are less than those required by BBMs although it is quite demanding the necessity of formulating adequate homogeneous constitutive laws.

MMs are the most utilized approaches thanks to their limited computational efforts and easy definition of mechanical properties. They consider the structure as an assemblage of macroelements (generally piers and sprandels) described on the basis of a mechanical or phenomenological behaviour. The most commonly used MMS approaches are the equivalent frame models. They have to be used with special attention to structural details and interaction between out-of-plane and in-plane behaviour which could be neglected in the analyses providing false response of the buildings. And they are very appropriate for ordinary buildings. Among the main contributions are for example (Dolce 1991; Moon et al. 2006; Lagomarsino et al. 2013, 2018; Parisi and Augenti 2013; Parisi et al. 2013, etc.) and with reference to the equivalent beam-based models (Magenes and Fontana 1998; Grande et al. 2011; Addessi and Liberatore 2015, etc.)

Geometry-based models (GBMs) analyse the structure on the basis of equilibrium and collapse by applying static and kinematic theorem limit analysis-based solutions, although the kinematic ones are most extensively used in the recent decades (Giuffre´ 1991; NTC 2018; Circolare NTC 2018). They require the knowledge only of the geometry of the structure. They are very useful in a rapid evaluation of the main vulnerable parts of a masonry building but not in a displacement-based seismic assessment method, since they do not provide any displacement capacity of the structures.

### 3.2 Vulnerability Assessment Against Subsoil Hazard

Subsoil is a primary structural component of buildings being deputed to withstand loads guaranteeing safety and serviceability, exactly as the upper elements of the superstructure (Fang 1991). Novel investigation and analytical tools nowadays enable to understand on a scientific basis the complex soil-structure interaction (Das 1999), perform reliable and accurate predictions and develop smart, cost-effective and sustainable solutions. On the other hand, design and construction of foundations in heritage buildings are based on practical rules, mostly derived from local tradition. Rehabilitation and adjustment to new standards and regulation requires a thorough assessment of the structure, considering the factors that produce hazard and the mechanical conditions (geometrical and mechanical characteristics of the structural components, conservation state of the materials) that determine fragility.

There are several attempts in the literature to classify underground hazard for buildings. A classic example (Sowers and Sowers 1970) distinguishes factors determining settlements of buildings into:

- structural
- environmental.

The former is dictated by the transfer of loads from the superstructure to the subsoil and to the concurrent deformation of both elements. Upon a variation of loads, soil reacts with distortion (induced by shear stress) and volume change (induced by compression stress and distortional volumetric coupling). The relative proportion among those factors is dictated by the characteristics of the material, with sand undergoing predominant distortion and clays developing more volume contraction. The amount of deformation depends on the inherent and state characteristics of the material (relative density of sand, overconsolidation of clays). A paramount issue is the time progress of deformative phenomena, dictated by the groundwater condition, drainage facilities and by the tendency of material to experience creep (e.g. organic soils). In heritage buildings most of the above deformation is arguably expired, unless overload is created for instance with additional construction. Signs of excessive past deformation are frequently noticed on old buildings, especially masonry, in the form of discontinuities and cracks. The mechanical assessment of such phenomena requires the accurate modelling of the soil response (with non-linear, stress history dependent constitutive models), and of the interaction with the superstructure responsible for the redistribution of loads onto the foundation.

The second category (environmental) includes external factors that determine a modification of the boundary conditions. There are plenty examples falling in this category and the following list is just a non-exhaustive summary of the most frequent occurrences.

- Large mass movement
- Landslides (slides, topples, falls, spreads, flows)
- Earthquake (liquefaction, post-earthquake consolidation, lateral spreading)
- Deformation induced by surrounding cavities

  - naturally formed (e.g. karst phenomena, sinkholes)
  - artificially excavated (e.g. tunnels, mining)
  - accidental (sewers damage)

- Modification of the pore pressure regime
- Subsidence due to large scale changes of the groundwater (climatic change, water-course deviation or interruption by dams, reservoir)
- Subsidence induced by groundwater exploitation
- Extraction of gas from deeper strata
- Saturation (or thawing) that induces bulging (or shrinkage) in plastic soil

- Chemical and biochemical alteration/degradation of the groundwater and the soil matrix (salt dissolution, chalk expansion).

The quantification of vulnerability, for buildings as for other structures, implies to establish a relation between demand, expressed by a quantitative parameter and level of damage, referred as loss of functionality. he level of damage is normally expressed in terms of limit states, such as those defined in the Eurocodes (Operativity – Damage – Vulnerability – Collapse) and quantification is achieved associating to each level the direct and indirect costs (e.g. FEMA/NIBS 1998). The engineering demand parameters must be defined from case to case identifying the physical variable best related with damage. Choice is not straightforward as natural phenomena like those listed above can induce multiple negative effects. Deformation of structures is traditionally assumed as one of the most hampering effects for the functionality of buildings and criteria based on limit values of variables like relative deflection or relative rotation (see Fig. 7) are proposed in the literature and codes (e.g. EN 1997).

The damage on buildings, from the formation of small cracks up the structure collapses, is traditionally related to the development of angular distortions (Skempton and Mc Donald 1956) or tensile strains (Burland et al. 1974). The limit values depend on the type of structure and the deformation mode (sagging or hogging) (Fig. 8). Many authors have studied the problem of observed damage on structure in deformed configuration, both through empirical methods or by using theoretical models of continuum mechanics. Skempton and MacDonald (1956), through the examination of a large number of real cases, provide some design guidelines on maximum permissible deflections that can cause damage to the architectural elements and structural problems. The curvature of the profile of the foundation settlements is related to the damage, and the relative rotation parameter $\beta$ is used as an indicator of damage on the building. Another classification of the damage consists in distinguishing the aesthetic, functional and structural damage (Burland et al. 1977). These main classes can be further divided into categories, creating a scale of severity of the damage. This classification of the damage is based on the evaluation of the limiting tensile strain $\varepsilon_{lim}$, which is related to each category of damage with reference to a particular building material.

The $\varepsilon_{lim}$ values for each category, reported in Table 2, have been obtained from the examination of real cases and model of masonry buildings (Burland 1995).

As an indicative examples of vulnerability analysis against environmental factors, Fig. 9 reports the map of subsidence induced in the city of Bologna (Italy) by groundwater extraction (Fig. 9a) together with the prediction of damage on the buildings of a zone of the city exposed to major hazard (Fig. 9b).

| Category of damage | Normal degree of severity | $\varepsilon_{lim}$ [%] |
|---|---|---|
| 0 | Negligible | 0–0.05 |
| 1 | Very slight | 0.05–0.075 |
| 2 | Slight | 0.075–0.15 |
| 3 | Moderate | 0.15–0.3 |
| 4 to 5 | Severe to very severe | >0.3 |

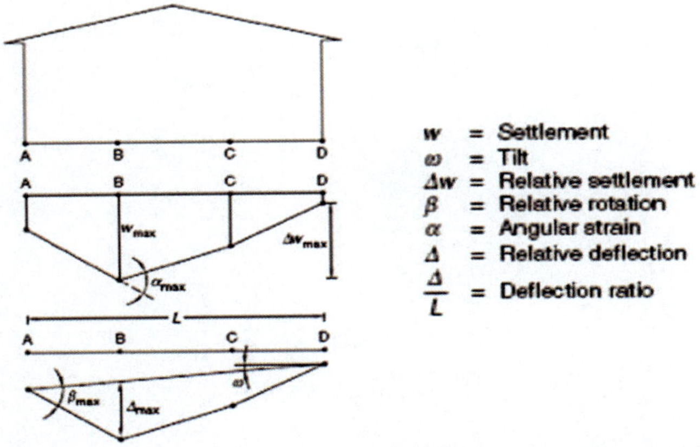


**Fig. 7.** Definition of relative deflection Δ and deflection ratio Δ/L – sagging e hogging (Skempton and Mc Donald 1956).

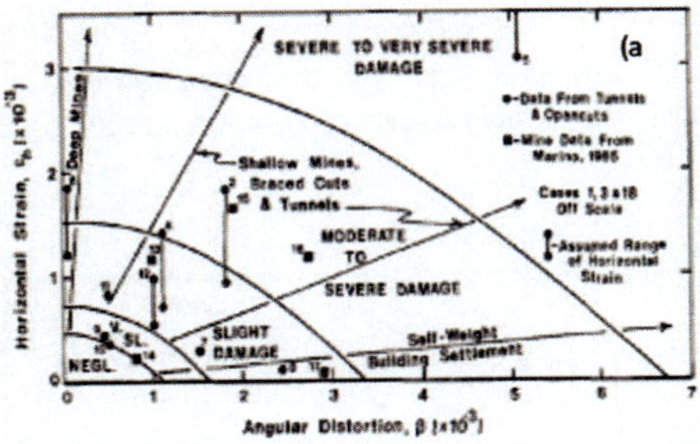


**Fig. 8.** Relationship between category of damage and limiting tensile strains $\varepsilon_{lim}$ (Burland 1995) and Fragility function for buildings affected by settlements provoked by excavation (Boscardin and Cording 1989).

The expression of fragility relation with probabilistic functions requires to quantify the uncertainty connected with grouping of different buildings in homogeneous categories. Figure 10 reports and a recent probabilistic function for fragility of frame buildings subjected to liquefaction-induced differential displacements (Fotopoulou et al. 2018).

Calculation of differential settlements induced by liquefaction often requires a complex analysis of the phenomena taking place upon shearing of sand, only partly detectable at laboratory scale (Modoni et al. 2018; Salvatore et al. 2016; Salvatore et al. 2018). An example of vulnerability assessment carried out using the functions of Fig. 11, for the reinforced concrete buildings of Christchurch (New Zealand) is reported in Fig. 11 (Paolella et al. 2020). The plot reports the Mean Damage Ratio computed for each building with the following formula:

$$MDR = \Sigma_i P_{ds_i} * (rc/RC)_i \tag{1}$$

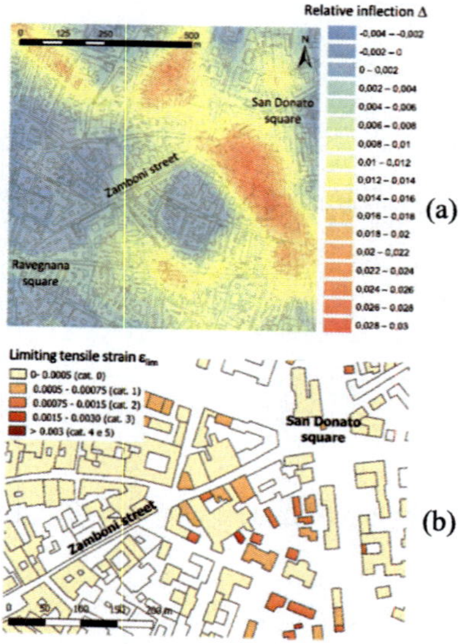


**Fig. 9.** Map of maximum relative inflection Δ in time span 1943–2011 (a) and Map prediction of damage of buildings near San Donato square (b) (Modoni et al. 2013; Spacagna et al. 2017; Spacagna and Modoni 2018)

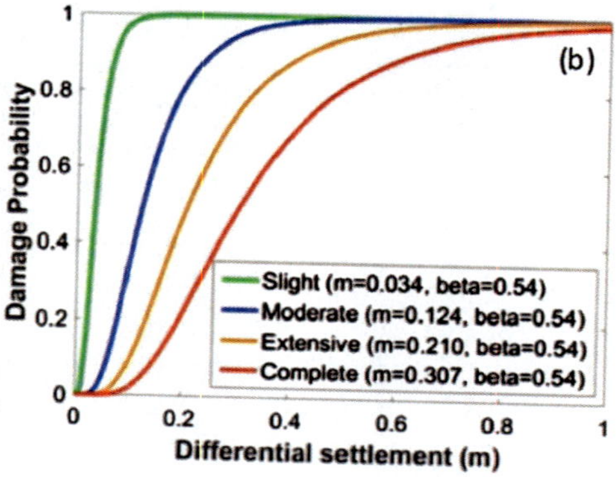


**Fig. 10.** Probabilistic fragility function for frame buildings subjected to liquefaction-induced differential displacements (Fotopoulou et al. 2018).

Where $P_{dsi}$ is the damage probability corresponding to each limit state of Fig. 11, $(rc/Rc)_i$ are weigthing factors assigned as in Table 1.

**Table 2.** Example of loss factor for residential buildings "RES1" (Hazus, FEMA 1998).

| Damage limit state (*ds*) | Loss factor $(rc/RC)_i$ |
|---|---|
| Slight | 0.02 |
| Moderate | 0.10 |
| Extensive | 0.45 |
| Complete | 1.00 |

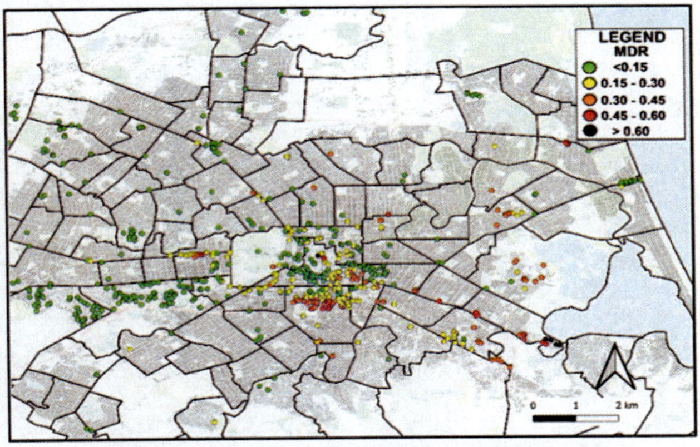


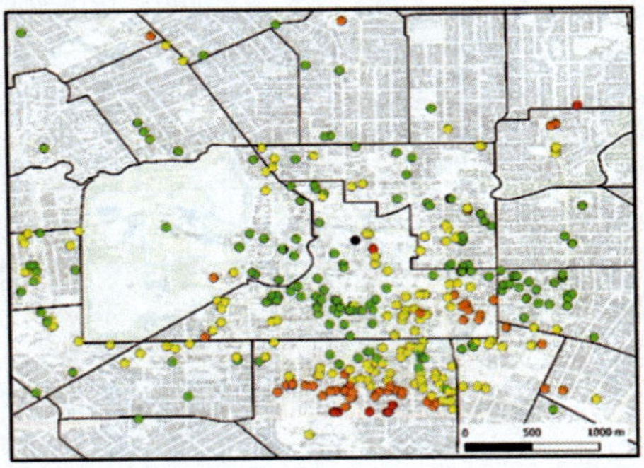


**Fig. 11.** Mean Damage Ratio computed on reinforced concrete buildings for the $M_W$6.2 22nd February 2011 earthquake: (a) general map of the city; and (b). enlargement in the central business district.

## 4 Case Study

BIM is a dynamic 3D model that contains information on geometry, materials, supporting structure, thermal characteristics and energy performance, installations, costs, safety, maintenance, life cycle, demolition, disposal etc. It is a parametric system, thanks to which it is possible to visualize some local collapse mechanisms of the facades of masonry buildings. To this end, a BIM model is tested on a case study, Borgo San Rocco a Sora (Saccucci et al. 2019).

The Borgo consists of two aggregates of masonry buildings with a height ranging from two to four floors that reach up to 12 m. In particular, the wall of the device is irregular, composed of mortar and limestone and covered with a simple plaster. The homogeneity of the external composition contrasts with the rather irregular interior, with types of horizontal elements transformed into wooden, brick and concrete floors, mixed etc. The two blocks are identified as two structural aggregates composed of a total of 37 units, interconnected by the arch which does not change the general structural behavior. The average height of the buildings is 10.15 m; the number of floors varies from 1 to 5. The total length of the road (connecting infrastructure) is 273.7 m (Fig. 12).

**Fig. 12.** 3D HBIM model of Borgo San Rocco in Sora

The geometric BIM model was made on the digital photogrammetric detection by drone and the LOD 400 was subjected to a direct detection. Thanks to the IFC property of the BIM system, the model was imported into SAP software to analyze the kinematics of the facades (Fig. 13). The results make it clear that most buildings, due to their structural characteristics, are vulnerable to seismic activities.

## 5 Conclusion

The vulnerability assessment of historical and post-industrial buildings in Europe, against different seismic hazards, is a complex process involving different fields of knowledge. In this paper it has been analysed through different perspective combining the contribution of heritage information building (HBIM) to seismic assessment and subsoil hazard within an inter-disciplinary vision. A specific reference to the Italian case-studied has been made.

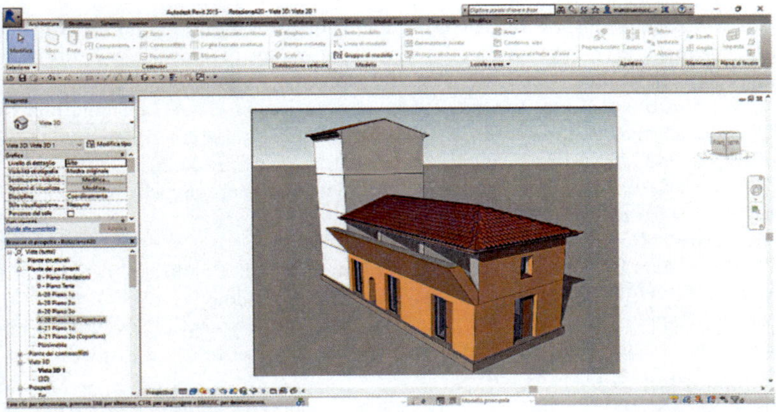

**Fig. 13.** San Rocco village. BIM model of US-A21 and the visualization of the most possible mechanism of collapse

## References


Addessi, D., Liberatore, D., Masiani, R.: Force-based beam finite element (fe) for the pushover analysis of masonry buildings. Int. J. Archit. Herit. **9**(3), 231–243 (2015)

Arpino, F., Pelliccio, A., Cortellessa, G., Frattolillo, A.: Mechatronics in the process of cultural heritage and civil infrastructure management. In: Mechatronics for Cultural Heritage and Civil Engineering, International Series on Intelligent System Control and Automation: Science and Engineering, vol. 92, pp. 1–31 (2018). https://doi.org/10.1007/978-3-319-68646-2. ISBN: 978-3-319-68645-5, ISSN: 2213-8986

APAT, Agenzia per la protezione dell'ambiente e per i servizi tecnici. Proposta di linee guida per il recupero ambientale e la valorizzazione economica dei brownfields. I.G.E.R. s.r.l. Editore, Roma (2006)

Azhar, S.: Building information modeling (BIM): trends, benefits, risks, and challenges for the AEC industry. Leadersh. Manag. Eng. **11**(3), 241–252 (2011)

Braga, F., Dolce, M., Liberatore, D.: A statistical study on damaged buildings and an ensuing review of the MSK-76 scale. In: Proceedings of the Seventh European Conference on Earthquake Engineering, pp. 431–450. Technical Chamber of Greece, Athens (1982)

Boscardin, M.D., Cording, E.J.: Building response to excavation-induced settlement. J. Geotech. Eng. **115**(1), 1–21 (1989)

Burland, J.B., Wroth, C.P.: Settlement of buildings and associated damage. In: Proceedings of Conference on Settlement of Structures, Cambridge, UK, pp. 611–654 (1974)

Burland, J.B., Broms, B.B., de Mello, V.F.B.: Behaviour of foundations and structures – SOA report, session 2. In: Proceedings of 9th International Conference SMFE, Tokyo, vol. 2, pp. 495–546 (1977)

Burland, J.B.: Assessment of risk of damage to buildings due to tunnelling and excavation, Invited special lecture. In: Proceedings 1st International Conference on Earthquake Geotechnology and Engineering - IS-Tokyo, vol. 95 (1995)

Circolare NTC 2018: Circolare n. 7 del 21 Gennaio 2019, "Istruzioni per l'applicazione dell' « Aggiornamento delle "Norme tecniche per le costruzioni" » di cui al DM 17 gennaio 2018". C.S.LL.PP (2019)

Como, M.: Statics of Historic Masonry Constructions. Springer, Heidelberg (2013)

Das, B.: Shallow Foundations, Bearing Capacity and Settlements. CRC Press, Boca Raton (1999)
D'Ayala, D., Speranza, E.: Definition of collapse mechanisms and seismic vulnerability of historic masonry buildings. Earthq. Spectra **19**, 479–509 (2003)
D'Altri, A.M., Messali, F., Rots, J., Castellazzi, G., de Miranda, S.: A damaging block-based model for the analysis of the cyclic behaviour of full-scale masonry structures. Eng. Fract. Mech. **209**, 423448 (2019b). http://dx.doi.org/10.1016/j.engfracmech.2018.11.046
Dolce, M.: Schematizzazione e modellazione degli edifici in muratura soggetti ad azioni sismiche. L'Industria delle Costruzioni **25**(242), 44–57 (1991)
EN1997 Eurocode 7: Geotechnical design
Fang, H.Y.: Foundation Engineering Handbook. Chapman & Hall, London, p. 923 (1991)
FEMA/NIBS, 1998: HAZUS - earthquake loss estimation methodology, vol. 1 (1998)
Fotopoulou, S., Karafagka, S., Pitilakis, K.: Vulnerability assessment of low-code reinforced concrete frame buildings subjected to liquefaction induced differential displacements. Soil Dyn. Earthq. Eng. **110**, 173–184 (2018)
And, G.B., Milani, G. (eds.): Theory to Application, Woodhead Publishing Series in Civil and Structural Engineering. Elsevier, Amsterdam (2019). ISBN 978-0-08-102439-3
Giuffre´, A.: Letture sulla meccanica delle murature storiche. Kappa (1991)
Grande, E., Imbimbo, M., Sacco, E.: A beam finite element for nonlinear analysis of masonry elements with or without fiber-reinforced plastic (FRP) reinforcements. Int. J. Archit. Herit. **5**(6), 693–716 (2011)
Grünthal, G. (ed.): European macroseismic scale 1998 (EMS–98), Cahiers du Centre Européen de Géodynamique et de Séismologie, vol. 15. European Seismological Commission, Luxembourg (1998)
Heyman, J.: The stone skeleton. Int. J. Solids Struct. **2**(2), 249279 (1966)
Lagomarsino, S., Penna, A., Galasco, A., Cattari, S.: Tremuri program: an equivalent frame model for the nonlinear seismic analysis of masonry buildings. Eng. Struct. **56**, 1787–1799 (2013)
Lagomarsino, S., Camilletti, D., Cattari, S., Marino, S.: In plane seismic response of irregular URM walls through equivalent frame and finite element models. In: Recent Advances in Earthquake Engineering in Europe: 16th European Conference on Earthquake Engineering - Thessaloniki 2018, pp. 123–151. Springer (2018)
Lourenco, P.B.: Computations on historic masonry structures. Prog. Struct. Eng. Mater. **4**(3), 301–319 (2002)
Magenes, G., Fontana, A.: Simplified non-linear seismic analysis of masonry buildings. In: Proc. Br. Masonry Soc, no. 8, pp. 190–195 (1998)
Meslem, A., Lang, D.H.: Physical vulnerability in earthquake risk assessment. Oxford Research Encyclopedia of Natural Hazard Science (2017)
Meyer, E., Grussenmeyer, P., Perrin, J.P., Durand, A., Drap, P.: A web information system for the management and the dissemination of cultural heritage data. J. Cult. Herit. **8**(4), 396–411 (2007)
Modoni, G., Darini, G., Spacagna, R.L., Saroli, M., Russo, G., Croce, P.: Spatial-temporal analysis of the subsidence in the city of Bologna. In: Proceedings of the 2nd International Symposium on Geotechnical Engineering for the Preservation of Monuments and Historic Sites, Napoli, Italy, 30 May 2013, pp. 565–572 (2013)
Modoni, G., Albano, M., Salvatore, E., Koseki, J.: Effects of compaction on the seismic performance of embankments built with gravel. Soil Dyn. Earthq. Eng. **106**, 231–242 (2018)
Moon, F.L., Yi, T., Leon, R.T., Kahn, L.F.: Recommendations for seismic evaluation and retrofit of low-rise URM structures. J. Struct. Eng. **132**(5), 663–672 (2006)
NTC 2018: Aggiornamento delle "Norme Tecniche per le Costruzioni" D. M. 17 gennaio 2018 (2018)
Paolella, L., Spacagna, R.L., Chiaro, G., Modoni, G.: Simplified vulnerability model for the extensive liquefaction risk assessment of buildings. Bull. Earthq. Eng., submitted (2020)

Parisi, F., Augenti, N.: Seismic capacity of irregular unreinforced masonry walls with openings. Earthq. Eng. Struct. Dyn. **42**(1), 101–121 (2013)

Parisi, F., Lignola, G.P., Augenti, N., Prota, A., Manfredi, G.: Rocking response assessment of in-plane laterally-loaded masonry walls with openings. Eng. Struct. **56**, 1234–1248 (2013)

Pelliccio, A., Cigola, M.: Geographic information systems (G.I.S.) for the analysis of historical small towns. In: Khosrow-Pour, M. (ed.) Encyclopedia of Information Science and Technology, Third Edition, pp. 3128–3135. IGI Global (2015). ISBN 978-1-4666-5888-2 (hardcover) – ISBN 978-1-4666-5889-9 (ebook) – ISBN 978-1-4666-5891-2 (print & perpetual access)

Pelliccio, A., Saccucci, M., Grande, E.: HT_BIM: parametric modelling for the assessment of risk in historic centers. DISEGNARECON **10**(18), 1–5 (2017). ISNN 18285961

Pelliccio, A., Żmudzińska Nowak, M., Saccucci, M.: Una mappa per la consapevolezza: il GIS per il recupero dei siti industriali dismessi map for awareness: GIS for the revitalization of brownfields. In La Città Altra/The Other City - In Storia e Iconografia dell'architettura, delle città e dei siti europei, pp. 1233–1243 (2018). ISBN:978-88-99930-03-5

Saccucci, M., Cima, V., Grande, E., Imbimbo, M., Pelliccio, A.: Valutazione della vulnerabilità sismica degli aggregati in muratura: il caso di Borgo San Rocco a Sora (Italia). In: Accettato al XVIII Convegno "L'Ingegneria Sismica in Italia", ANIDIS 2019, Ascoli Piceno, Settembre 2019 (2019)

Salvatore, E., Andò, E., Modoni, G., Viggiani, G.: Micromechanical study of cyclically loaded sands with x-ray microtomography and digital image correlation. In: 6th Italian Conference of Researchers in Geotechnical Engineering, CNRIG 2016, Bologna, Italy, 22 September 2016, vol. 158, pp. 92–97 (2016)

Salvatore, E., Andò, E., Proia, R., Modoni, G., Viiggiani, G.: Effect of strain localization on the response of granular materials subjected to monotonic and cyclic triaxial tests. Rivista Italiana di Geotecnica **52**(2), 30–43 (2018)

Skempton, A.W., Mac Donald, D.H.: The allowable settlements of buildings. ICE Proc.: Eng. Div. **5**(6), 727–768 (1956)

Sowers, G.B., Sowers, C.F.: Introductory Soil Mechanics and Foundations. Collier Mac Millan, London (1970)

Spacagna, R.L., Rasulo, A., Modoni, G.: Geostatistical analysis of settlements induced by groundwater extraction. In: 17th International Conference on Computational Science and its Applications, ICCSA 2017, Lecture Notes in Computer Science, LNCS, Trieste, Italy, 3 July 2017, vol. 10407, pp. 350–364 (2017)

Spacagna, R.L., Modoni, G.: GIS-based study of land subsidence in the city of Bologna. Intell. Syst. Control Autom.: Sci. Eng. **92**, 235–256 (2018)

Whitman, R.V., Reed, J.W., Hong, S.T.: Earthquake damage probability matrices. In: Proceedings of the 5th World Conference on Earthquake Engineering, Italy, vol. 2, pp. 2531–2540 (1973)

Zonta, D., Pozzi, M., Zanon, P.: Managing the historical heritage using distributed technologies. Int. J. Archit. Herit. **2**(3), 200–225 (2008)

# The Knowledge Process in the Seismic Assessment of Masonry Building Aggregates – An Italian Case Study

Marco Saccucci[1], Valentina Cima[1], Ernesto Grande[2], Maura Imbimbo[1](✉), and Assunta Pelliccio[1]

[1] University of Cassino and Southern Lazio, Via G. Di Biasio 43, 03043 Cassino, Italy
m.imbimbo@unicas.it

[2] University G. Marconi, Via Plinio 44, 00193 Rome, Italy

**Abstract.** The recent seismic events occurred in Central Italy have rekindled the attention toward the safeguard of historical cities by leading to a renovated debate within the academic community regarding specific aspects of the seismic vulnerability and their protection. The study of masonry building aggregates is certainly one of these aspects, since they represent a typical urban configuration of the Italian but also European historical towns.

Because of their constructive and evolutionary history, masonry aggregates are a complex assemblage composed of units with different periods and techniques of construction, materials and structural details, which, all, contribute to the seismic response of the whole. Therefore, the assessment of the seismic vulnerability of aggregates has to start necessarily from a process of knowledge able to provide a set of reliable input data useful for the subsequent development of numerical models employed for structural analysis.

The main goal of the paper is to present a procedure for the seismic assessment of masonry building aggregates toward the most common out-of-plane failure mechanisms generally observed during seismic events. The main feature of the proposed procedure is a multi–level input-analysis-output approach, combining the level of detail of the input data, deduced from the knowledge get by a survey process, at the level of detail of the output thus obtaining a refinement of the numerical models. The procedure is applied to the case study of Borgo San Rocco in Sora (FR), a historical center located in Central Italy.

**Keywords:** Historical cities · Masonry · Seismic vulnerability

## 1 Introduction

Masonry buildings of Italian historical cities are fascinated witnesses of the ancient past of the country. Nevertheless, despite their historical and cultural value, the majority of masonry buildings are generally characterized by a relevant seismic vulnerability mainly due to the state of decay of materials, the construction process carried out without


A. Rotaru (Ed.): CRIT-RE-BUILT 2019, SSGG, pp. 330–347, 2021.
https://doi.org/10.1007/978-3-030-61118-7_28

considering the importance of structural details toward seismic actions and, in addition, the level of seismic hazard (the majority of Italian historical centers are indeed located along the Apennines, i.e. near to the fault line [1]; Fig. 1).

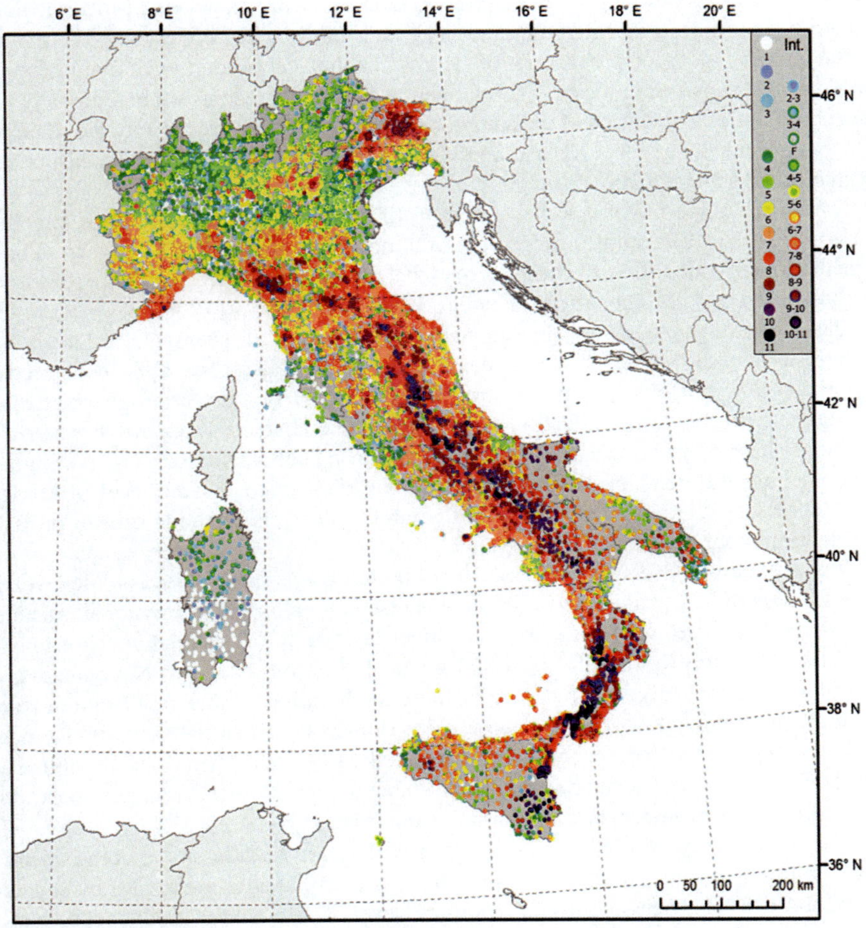


**Fig. 1.** Distribution of the maximum intensities observed for the 15213 Italian locations (source DBMI15).

As a result, the masonry buildings in Italian historical centers are in the majority of cases characterized by a high level of seismic vulnerability, which can be reduced necessarily by starting from their knowledge [2–5]. Indeed, due to the heterogeneity and complexity of the scenario of existing masonry buildings, the evaluation of their vulnerability has to be connected to a responsible process of knowledge. This process becomes particularly useful in case of aggregate configurations where every aggregate is composed of structural units [6] generally built in different periods with different techniques of construction, materials, etc. In addition, the buildings composing an aggregate have undergone superfetations (intended as some elements of the buildings without artistic or historical value) and structural transformations that have changed their original structure: an aspect which makes difficult to identify the structural units of a building aggregate. The above considerations clearly emphasize the role of the knowledge process, which generally is reflected in collecting and analyzing information concerning the history of the building, geometry, construction details, mechanical properties of materials, etc. These aspects have been recently formalized in the Italian technical regulations [7–9]. Indeed, the rules at the basis of the study of existing buildings, are directly based on the concept of knowledge level KL and confidence factor CF which allow to 'measure' the degree of the achieved knowledge.

There are numerous researches which testify the scientific community's attention toward this issue. The studies carried out until now strongly highlight the need to have a methodology that allows to combine the level of detail of survey techniques with the methods used to carry out structural analyses [10–12].

The knowledge process of a building aggregate consists of several phases: territorial classification of the aggregate within the urban fabric, reconstruction of the constructive and evolutionary history, geometric survey, structural survey, detection of the state of damage and degradation, evaluation of the mechanical characteristics of the materials, etc. Each phase of the knowledge process can be carried out quickly or more thoroughly, thus reaching different levels of accuracy of the knowledge. As the level of knowledge increases, it is possible to apply vulnerability analysis methods able to provide information more detailed and more reliable [13].

The evaluation of seismic vulnerability of building aggregates based on the level of knowledge is of great applicative relevance in research fields and professional practice.

For existing masonry buildings, the Italian legislation [7–9] requires to evaluate the seismic safety toward both local and global failure mechanisms. Nevertheless, it is well known that in case of historical masonry buildings, local failure modes are generally predominant. Indeed, the poor quality of masonry (materials and arrangement of blocks) together with deficiencies in terms of structural details (in particular the ones concerning the connection among walls and slabs) do not allow the structure to exhibit a global box-like behavior toward seismic actions [14, 15].

The aim of the paper is then to present a procedure for the assessment of the seismic vulnerability of masonry buildings toward local out-of-plane failure modes by considering a multi-level process which combines the level of detail of the input data deduced from the knowledge process to the level of refinement of the structural model employed for the numerical analyses.

The procedure is applied to the case study of Borgo San Rocco in Sora, a case also accounted in the context of the research project "Technologies for improving safety and building historic centers in the seismic area (SISMI)" of the Technological District for new technologies applied to cultural assets and activities of the Lazio Region (DTC).

## 2 Evaluation Procedure of Seismic Vulnerability

The main feature of the procedure here proposed is to link the level of knowledge, get through a sight or detailed survey process, of existing buildings (and then the set of deducible data) to the methods of structural analysis generally employed for the evaluation of the seismic vulnerability. For this purpose, as explained in the follows, the procedure identifies two approaches, defined qualitative [Qla] and quantitative [Qna] level, applicable to both isolated and enclosed buildings in an aggregate system [6].

The two approaches differ in terms of degree of knowledge. In the qualitative one, the procedure requires as the input data the information relating to geometry, material and technological/structural system of building. In the quantitative approach, the input data are based on an increased level of detail knowledge deduced from the application of the selected methods for the evaluation of the seismic vulnerability (i.e. the output data of the procedure). In detail, the first level [Qla], is based on a low degree of knowledge, or rather on a limited number of input data, and it allows to obtain a qualitative evaluation of the seismic vulnerability. The second level, [Qna], is based on more detailed information and provides a quantitative evaluation of the seismic vulnerability (Fig. 2). Both the levels are described in detail in the following sections.

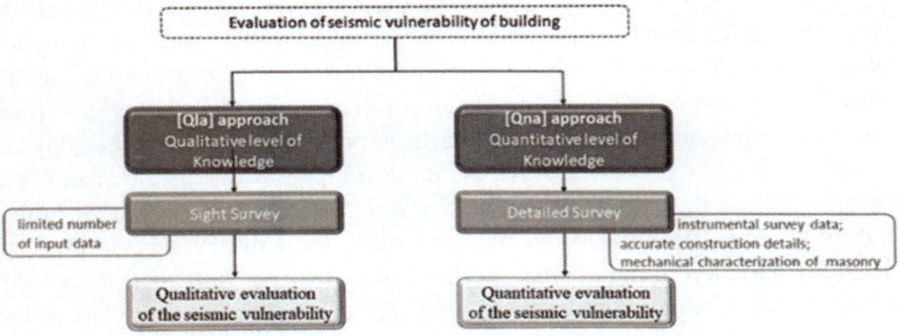


**Fig. 2.** Representative scheme of the procedure.

## 2.1 Qualitative Level [Qla]

The qualitative level consists in a rapid and simplified approach for the evaluation of the seismic vulnerability, aimed at identifying the most probable local collapse mechanism of the building (output data) on the basis of information deductible through expeditious survey techniques (input data).

At this level of analysis, the survey process is mainly finalized to obtain the following information of the building/structural unit:

- the architectural typology together with the geometrical properties such as the number of floors, interfloor height, distribution of openings, misalignment of floors, slenderness ratio of walls subjected to out-of-plane mechanisms;
- the structural properties such as:

  - vertical load-bearing structures, such as masonry system, mixed system, etc.;
  - horizontal structural elements including roofing as vaults, wooden structures, etc.;

- slab orientation, actions of thrusting elements (roof, vaults, arches, etc.);
- connections among orthogonal walls and floors and walls (chains, tie bars, curbs);
- openings in the walls (chimney flues, etc.);
- masonry quality and presence of damages;
- previous interventions (additional stories, extensions, heavy and strong curbs on poor masonry, heavy roof etc.);

– the presence of misaligned facades, misaligned floors along the walls in common between two consecutive units, irregularities of masonry due to the assemblage process, different heights of consecutive structural units, structural modifications with or without structural connections to the existent structure, or with partially connections (structural joints, etc.);
– the constructive and historic transformations.

These data can be obtained through a quick survey on sight supported by historical-archival researches (real estate registry cartographies, project documents related to the original construction of the building and to any subsequent interventions), which allow to perform a historical-critical analysis of the building. Useful information can also be found using interactive online maps or databases that report the typological-structural characteristics of the national building heritage. Among the databases nowadays available, a valid support is provided by CARTIS developed within the research project DPC/ReLUIS 2019–2021 WP2 - Inventory of structural types and existing buildings, promoted by the ReLUIS consortium and funded by the Civil Protection Department [16]. The sight detection of the building surfaces also provides information on the main pathologies of the building's natural and anthropic degradation, helping to define the state of conservation of the materials. The survey is based on the acquisition of ortho-rectified photos on which, it is possible to digitize and quantify deteriorated areas in the post-processing phase. The procedure is carried out in agreement with NorMAL recommendations [17], which attribute an international code and graphic symbology to visualize the different pathologies. Again, according to NorMal standards, it is also possible to provide a graphic cracking image (in plan and in elevation) capable of suggesting any structural failures such as widespread injuries, at an angle, away from lead, disconnection, ejection of the material, etc. The collected information can be used as input parameters for the evaluation of the seismic vulnerability of existing masonry buildings toward out-of-plane local collapse mechanisms. Indeed, according to the approach proposed in [18], the criticisms in terms of construction weaknesses emerged from the knowledge process can be related to the occurrence of specific out-of-plane local mechanisms of the perimeter facades composing the building. This allows to identify the most probable out-of-plane collapse mechanisms which can be activated during a seismic event. For example, the simple overturning mechanism results a potential local mechanism in case of perimeter walls without efficient connections to orthogonal walls and slabs. In case of poor connection among walls and slab at the top of the building and presence of horizontal thrusts induced by the roof or by arches, vaults, slabs, the horizontal bending results a further potential failure mechanism.

In the procedure here proposed, the collected data and the result of the qualitative analysis are summarized in a specific survey module as reported in Fig. 3.

## 2.2 Quantitative Level [Qna]

The quantitative level is defined on the base of detailed and advanced instrumental surveys aimed at deepening the knowledge of the buildings and perform a quantitative evaluation of the vulnerability of the building. At this level, in fact, the objective of the procedure is to evaluate, by using more refined structural models, the spectral acceleration a0 activating the potential out-of-plane local mechanisms detected in the previous qualitative level analysis.

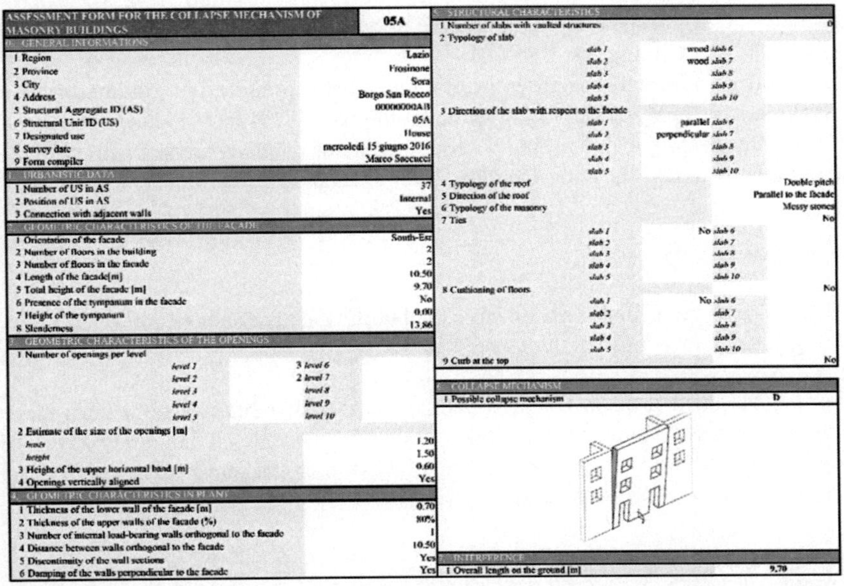

| | | | |
|---|---|---|---|
| ASSESSMENT FORM FOR THE COLLAPSE MECHANISM OF MASONRY BUILDINGS | | | 05A |
| 0. GENERAL INFORMATIONS | | | |
| 1 Region | | | Lazio |
| 2 Province | | | Frosinone |
| 3 City | | | Sora |
| 4 Address | | | Borgo San Rocco |
| 5 Structural Aggregate ID (AS) | | | 0000000AB |
| 6 Structural Unit ID (US) | | | 05A |
| 7 Designated use | | | House |
| 8 Survey date | | | mercoledì 15 giugno 2016 |
| 9 Form compiler | | | Marco Saccucci |
| 1. URBANISTIC DATA | | | |
| 1 Number of US in AS | | | 37 |
| 2 Position of US in AS | | | Internal |
| 3 Connection with adjacent walls | | | Yes |
| 2. GEOMETRIC CHARACTERISTICS OF THE FACADE | | | |
| 1 Orientation of the facade | | | South-Est |
| 2 Number of floors in the building | | | 2 |
| 3 Number of floors in the facade | | | 2 |
| 4 Length of the facade[m] | | | 10.50 |
| 5 Total height of the facade [m] | | | 9.70 |
| 6 Presence of the tympanum in the facade | | | No |
| 7 Height of the tympanum | | | 0.00 |
| 8 Slenderness | | | 13,86 |
| 3. GEOMETRIC CHARACTERISTICS OF THE OPENINGS | | | |
| 1 Number of openings per level | | | |
| level 1 | 3 | level 6 | |
| level 2 | 2 | level 7 | |
| level 3 | | level 8 | |
| level 4 | | level 9 | |
| level 5 | | level 10 | |
| 2 Estimate of the size of the openings [m] | | | |
| width | | | 1.20 |
| height | | | 1.50 |
| 3 Height of the upper horizontal band [m] | | | 0.60 |
| 4 Openings vertically aligned | | | Yes |
| 4. GEOMETRIC CHARACTERISTICS IN PLANT | | | |
| 1 Thickness of the lower wall of the facade [m] | | | 0.70 |
| 2 Thickness of the upper walls of the facade (%) | | | 80% |
| 3 Number of internal load-bearing walls orthogonal to the facade | | | 1 |
| 4 Distance between walls orthogonal to the facade | | | 10.50 |
| 5 Discontinuity of the wall sections | | | Yes |
| 6 Damping of the walls perpendicular to the facade | | | Yes |
| 5. STRUCTURAL CHARACTERISTICS | | | |
| 1 Number of slabs with vaulted structures | | | 0 |
| 2 Typology of slab | | | |
| slab 1 | wood | slab 6 | |
| slab 2 | wood | slab 7 | |
| slab 3 | | slab 8 | |
| slab 4 | | slab 9 | |
| slab 5 | | slab 10 | |
| 3 Direction of the slab with respect to the facade | | | |
| slab 1 | parallel | slab 6 | |
| slab 2 | perpendicular | slab 7 | |
| slab 3 | | slab 8 | |
| slab 4 | | slab 9 | |
| slab 5 | | slab 10 | |
| 4 Typology of the roof | | | Double pitch |
| 5 Direction of the roof | | | Parallel to the facade |
| 6 Typology of the masonry | | | Messy stones |
| 7 Ties | | | No |
| slab 1 | No | slab 6 | |
| slab 2 | | slab 7 | |
| slab 3 | | slab 8 | |
| slab 4 | | slab 9 | |
| slab 5 | | slab 10 | |
| 8 Cushioning of floors | | | No |
| slab 1 | No | slab 6 | |
| slab 2 | | slab 7 | |
| slab 3 | | slab 8 | |
| slab 4 | | slab 9 | |
| slab 5 | | slab 10 | |
| 9 Curb at the top | | | No |
| 6. COLLAPSE MECHANISM | | | |
| 1 Possible collapse mechanism | | | D |
| 7. INTERFERENCE | | | |
| 1 Overall length on the ground [m] | | | 9,70 |

**Fig. 3.** Survey Form (M. Saccucci)

To apply the quantitative analysis, therefore, it is necessary:

- to have information on the geometry of the individual elements;
- to know accurately the construction details;
- to characterize mechanically the masonry.

To obtain a consistent set of geometric information it is necessary to use tools such as laser scanners or digital, aerial and terrestrial photogrammetry able to gather structural details. The most economical and fastest tool is undoubtedly the digital photogrammetry that allows to obtain 3D models, returned as point clouds, from which to get mesh to be used in the structural analysis software.

The digital photogrammetry, in fact, allows create 3D models based on reality starting from high resolution digital photographic images and applying particular image correlation algorithms [19] thanks to which it is possible to acquire accurate alphanumeric information and to analyze any states of deformation with the reading of planes out of plumb compared to the initial geometry. Moreover, on the 3D model it is also possible to integrate data coming from other different survey methods thanks to which be able to know more accurate construction details (e.g. thermal imager, reflectivity of laser scanner etc.).

On the other hand, the mechanical characterization of the masonry can be carried out throughout non-destructive on-site tests and/or laboratory tests or to refer to the characteristics collected by guidelines (table C8.5.I of the Circular no. 7 of 2019 [9]). Alternatively, the method called Masonry Quality Index (MQI) [20] can be used. This method involves evaluating the MQI in order to correlate it with the average values of compressive strength, shear strength and the normal modulus of elasticity provided by the table C8.5.I of Circular no. 7 of 2019.

Data collected from these phases are employed in the proposed procedure to perform a linear kinematic analysis according to the Italian code [8, 9] for evaluating the spectral acceleration activating the identified potential local out-of-plane mechanisms and, then, to quantitatively assess the corresponding level of seismic vulnerability.

## 3 Case of Study

The case of study here analyzed concerns the buildings aggregates of Borgo San Rocco in Sora, a town in central Italy located in Lazio region the province of Frosinone (Fig. 4 a).

Sora is the third town in the province by number of inhabitants and it is located near the Central Apennines, at the meeting point of three valleys: the Valley of Liri, the Roveto Valley and the Comino Valley. The area in which the municipality of Sora falls is characterized by a high seismic hazard (Fig. 4 b).

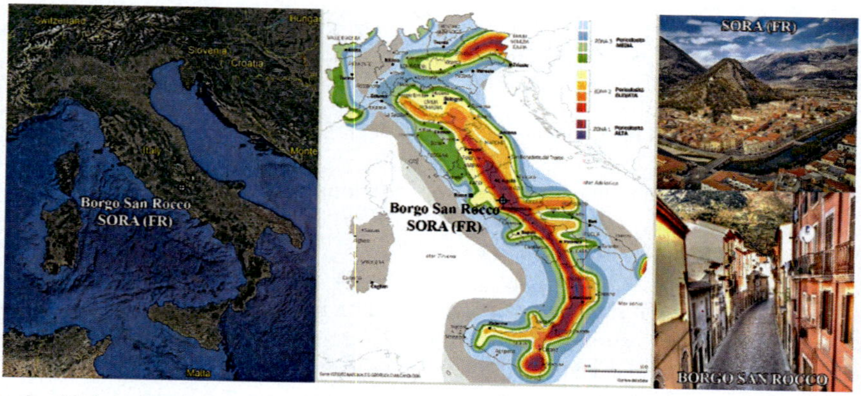


**Fig. 4.** a) Map of Italy with location of the case of study; b) seismic hazard map of the Italian territory; c) up: top view of Sora (ph. E. Roscilli), down: photo of Borgo S. Rocco's street.

Due to its geographical position, indeed, the territory of Sora was stricken by numerous and severe seismic events [2]. Among the most disastrous seismic events, both in terms of damaged buildings and human victims, we recall the earthquake of Lazio-Molise in 1349 [21], the earthquake of Sorano in 1654 and the earthquake of Marsica in 1915 [22]. The seismic event of 1915 was one of the most catastrophic in Italian history (Fig. 5): it caused about 30,000 victims, of which about 500 in Sora [23], where it had significant consequences on the urban, social and economic structure.

While today the inhabited center of Sora occupies almost the entire Liri plain, in the past it was between the slopes of Monte San Casto (546 m) and the banks of the Liri river [24]. Indeed, exactly in this position, within the historic center, is located the case study of Borgo San Rocco, composed of two structural masonry aggregates which develop like a "Street Canyon" and are connected by an arch known as San Rocco's Arch (Fig. 6 a) b)).

**Fig. 5.** Historical photo of Sora after the earthquake of 1915

**Fig. 6.** a) Identification of Borgo San Rocco on ortho-photo; b) photo of Borgo San Rocco with drone (Ph. M. Saccucci).

### 3.1 Qualitative Level [Qla]

Starting from the application of the qualitative level, quick on sight surveys and geometric surveys have been carried out. Further information has been deduced by using interactive online maps and consulting the available archive documents, such as the historical maps and the documents relating to the Project of rehabilitation and hygienic-functional adaptation of the buildings damaged by the earthquakes of May, 7th and 11th, 1984 (Fig. 7). Through the information collected it has been possible to perform an historical-critical analysis, from which it is emerged that Borgo San Rocco dates back to the sixth century B.C. but it assumed its present conformation in the first half of the nineteenth century. Over the years, it has undergone several building renovations, one of the most recent and significant dates back to the post-earthquake of 1984 affecting central-southern Italy. On the occasion of this event, the horizontal elements which had suffered considerable damage were replaced and, in some cases, metal ties were inserted in order to improve the box-like behavior of the buildings.

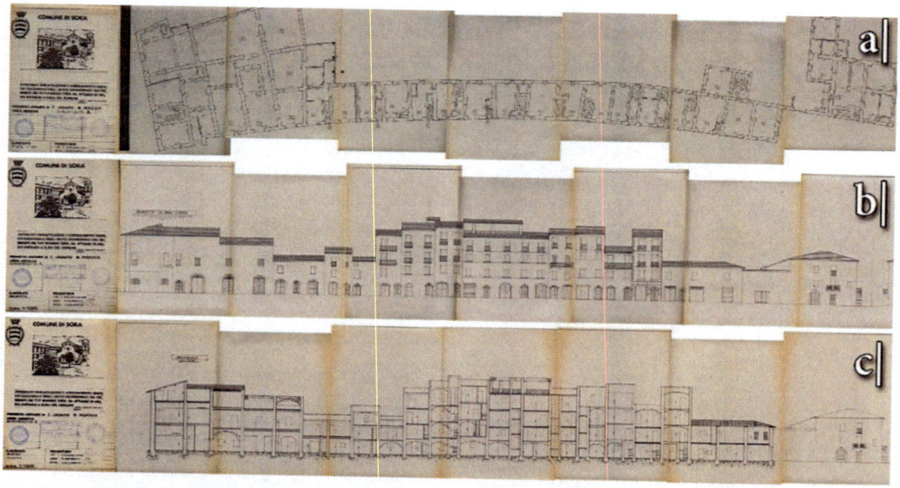


**Fig. 7.** Archive documentation: "Renovation and hygienic-functional adaptation of buildings damaged by the earthquakes of 7th and 11th May 1984 to be implemented in a unitary way by the municipality": a) ground floor plan; b) prospectus; c) section (Municipality of Sora).

Downstream of the survey process has been obtained that the 2 building aggregates of Borgo San Rocco are composed of 39 Structural Units, of which 30 facing the road that separates the two compartments. The two aggregates have been called, for analysis purposes, River Aggregate and Hill Aggregate, and each building has been assigned an identification code (Fig. 8).

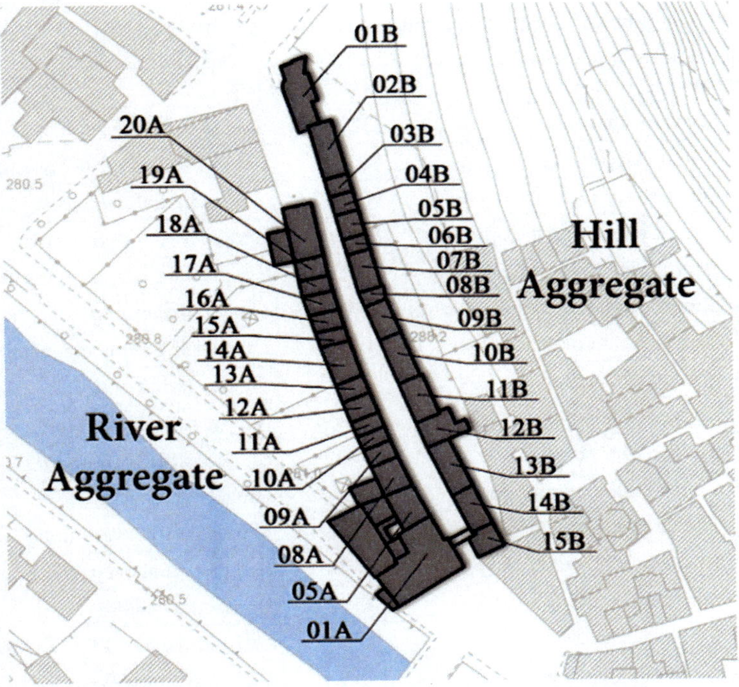


**Fig. 8.** Identification of the structural units and attribution of the identification codes ID (M. Saccucci).

The visual survey has underlined for all the structural units perimeter walls composed of disordered stone with poor-quality mortar and a vast heterogeneity of horizontal structures: some of them consist of wooden floors with chestnut beams of circular cross-section, others are made up of reinforced concrete floors cast on site, still i-beams and flooring blocks and barrel vaults. As for the horizontal covering closures it has been observed that they are all in wood with simple warping of false struts surmounted by a plank placed as a base for the tiles. In the structural units at the ends of the two aggregates, the roofs have a pavilion shape, while in the central area they are at hut or a single pitch. Regarding the walls of the buildings, it has been found that they are in uncertain limestone work, with an adequate mortar/stone material ratio. They have a thickness ranging from a maximum of 70 cm to a minimum of 50 cm, with structural projections on the upper floors of the order of 15%.

The collected data have been introduced in the approach proposed by the authors in [18] to identify for each structural unit the corresponding most probable collapse mechanism toward seismic events: the obtained results are summarized in Table 1.

From Table 1, it can be observed that the most frequent mechanism is the simple partial overturning: it could happen in 13 structural units. While the less frequent mechanism is the global overturning: it could only occur in 3 structural units.

**Table 1.** Local collapse mechanisms emerged from the Qla.

| River Aggregate | | Hill Aggregate | |
|---|---|---|---|
| ID | Local Mechanism | ID | Local Mechanism |
| 01A | Horizontal bending | 01B | Overturning along the openings |
| 05A | Overturning along the openings | 02B | Simple partial overturning |
| 08A | Overturning along the openings | 03B | Simple partial overturning |
| 09A | Overturning along the openings | 04B | Simple partial overturning |
| 10A | Simple partial overturning | 05B | Global overturning |
| 11A | Simple partial overturning | 06B | Overturning along the openings |
| 12A | Horizontal bending | 07B | Simple partial overturning |
| 13A | Horizontal bending | 08B | Simple partial overturning |
| 14A | Horizontal bending | 09B | Simple partial overturning |
| 15A | Simple partial overturning | 10B | Global overturning |
| 16A | Overturning along the openings | 11B | Simple partial overturning |
| 17A | Simple partial overturning | 12B | Simple partial overturning |
| 18A | Horizontal bending | 13B | Horizontal bending |
| 19A | Simple partial overturning | 14B | Overturning along the openings |
| 20A | Horizontal bending | 15B | Global overturning |

### 3.2 Quantitative Level [Qna]

Regarding the quantitative level, the knowledge process of the aggregate has been carried out throughout the digital photogrammetric aerial survey [25]. In particular, it has been used the DJI Phantom 3 Advanced drone with 12 Mpixels camera mounted on a 3-axis gimbal. Through the processing of the images, a metric correct cloud of points has been obtained (error of about 1%), easily navigable and geometrically queryable through a series of tools internal to the cloud management software. From the interrogation of the 3D model obtained, detailed information concerning the geometry of the single individual elements have been obtained (Fig. 10). On the other hand, the mechanical properties of the masonries constituting the various structural units have been deduced directly considering data of the table C8.5.I of circular no. 7/2019 on the basis of masonry type emerged from the knowledge process (Fig. 9).

The collected information have been used to perform a linear kinematic analysis by taking into account the collapse mechanisms emerged from the Qla. For each structural unit composing the aggregates object of investigation, it has been possible to deduced the level of ground acceleration activating the identified mechanisms with reference to the LLS (Lifesaving Limit State) $a_{g,LLS}$ (Fig. 11).

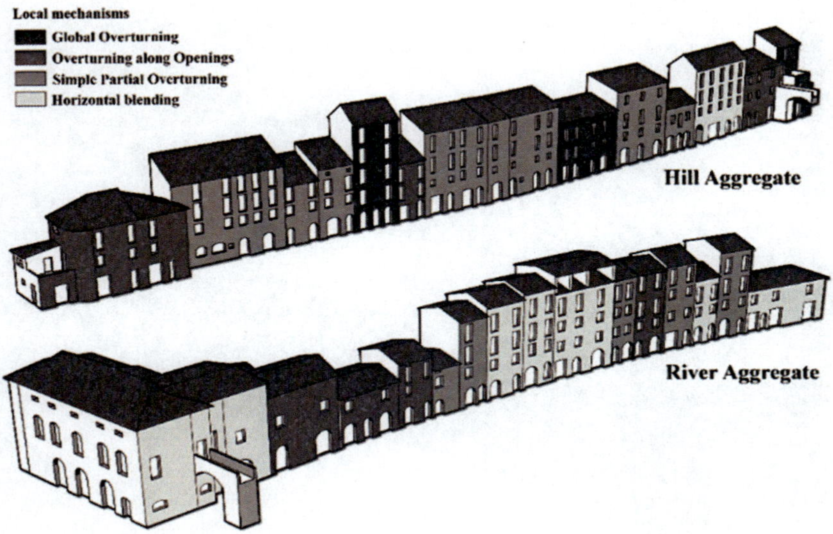


**Fig. 9.** Qualitative level – Local collapse mechanisms identified (M. Saccucci)

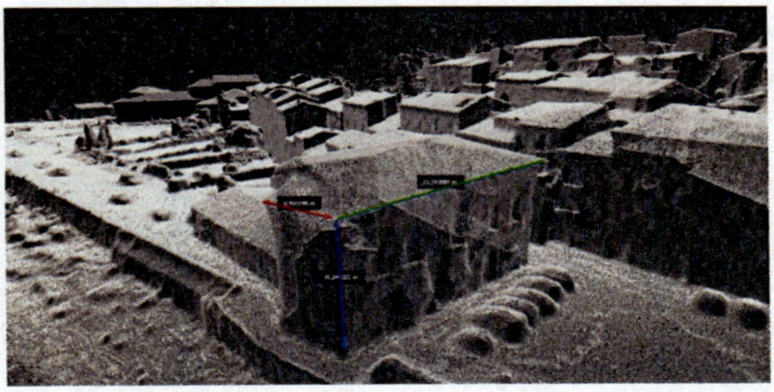

**Fig. 10.** Dense cloud point interrogation by aerial digital photogrammetric survey (M. Saccucci).

Subsequently, in agreement with the indications of national guidelines, it has been possible to assess the level of seismic vulnerability of the structural units by comparing the value of the acceleration activating the mechanisms (capacity) with the expected peak ground acceleration (demand). In Fig. 12 are summarized the obtained outcomes by underlining the cases where the capacity is greater than the demand (YES) and the cases where the capacity is lower than the demand (NO).

Finally, the vulnerability index $I_V$, calculated as the ratio between capacity and demand is reported in Fig. 13 and 14.

As an example, the schemes adopted for the kinematic analysis of the structural units 03B and 10B of the Hill Aggregate are reported. The structural unit 03B consists of 3 floors for a total height of 9 m (Fig. 15). For it the qualitative analysis has returned as the most probable local collapse mechanism the simple partial overturning of the wall of the last level.

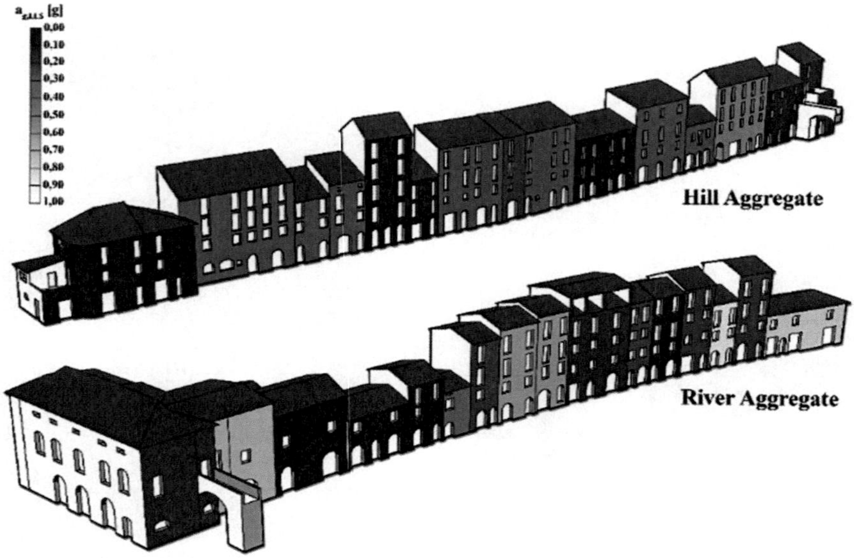


**Fig. 11.** Results deduced from the linear kinematic analysis in terms of ground acceleration activating the identified mechanisms with reference to LLS $a_{g,LLS}$ (Saccucci)

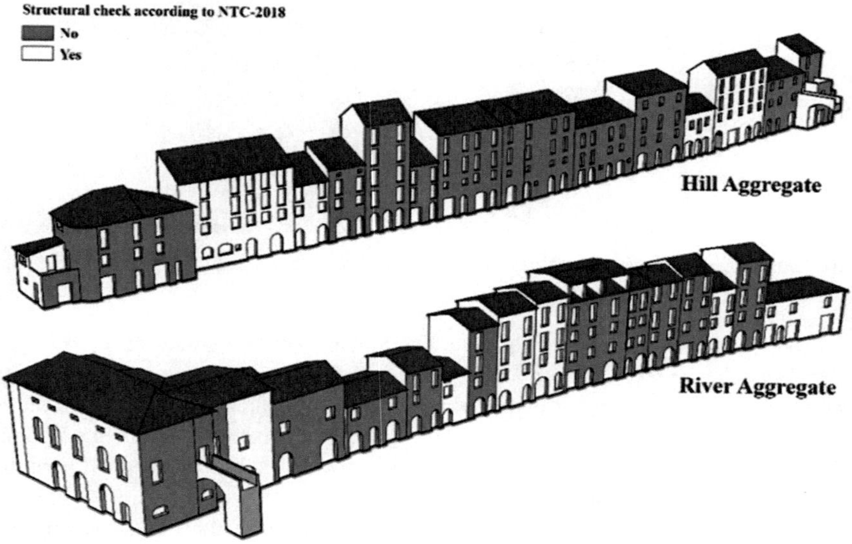


**Fig. 12.** Comparison between the seismic capacity and the seismic demand (M. Saccucci)

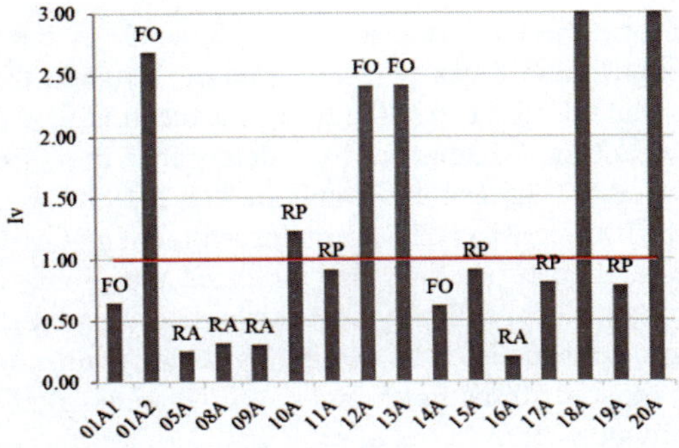


**Fig. 13.** Vulnerability index: River Aggregate (V. Cima).

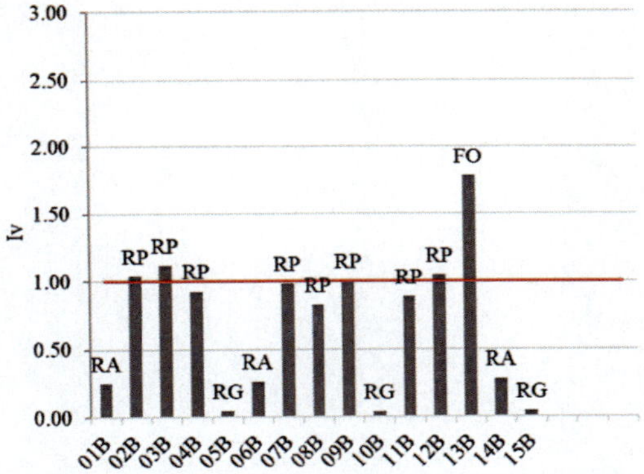


**Fig. 14.** Vulnerability index: Hill Aggregate (V. Cima).

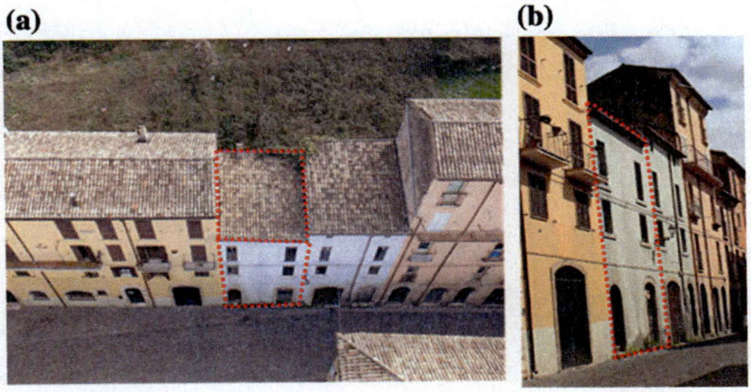


**Fig. 15.** Structural unit 03B: a) aerial photo with drone; b) facade photo (ph. M. Saccucci)

Therefore, to perform the kinematic analysis the facade of the structural unit has been modeled as a single rigid block with a cylindrical hinge at the base of the third level. Due to the fact that the roof is parallel to the façade, on the wall of the third level only the own weight acts (Fig. 16), that has been determined in reference to the specific weight reported in the table C8.5.I of the circular n.7 of 2019.

The structural unit 10 B consists of 3 floors for a total height of 10 m (Fig. 17).

For this unit the qualitative analysis has indicated the global overturning of the facade as the most probable local collapse mechanism. In order to perform the kinematic analysis, the facade has been modeled as a single rigid block with a cylindrical hinge at the base (Fig. 18). In the unit all the floors and coverage are parallel to the facade, thus on the rigid block only the weight forces of each wall act.

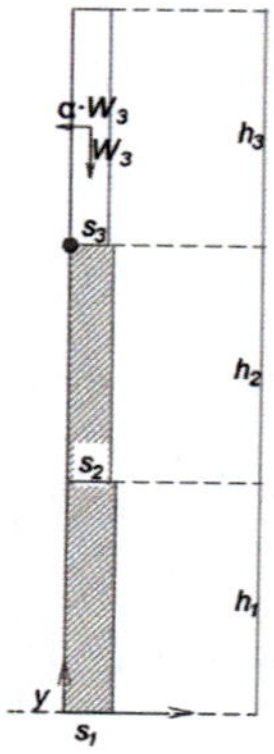


**Fig. 16.** Structural unit 03B: scheme for the partial overturning mechanism analysis (V. Cima).

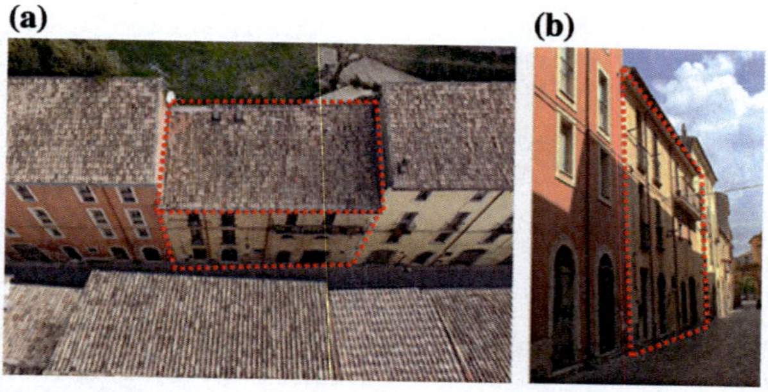


**Fig. 17.** Structural unit 10B: a) aerial photo with drone; b) facade photo (ph. M. Saccucci).

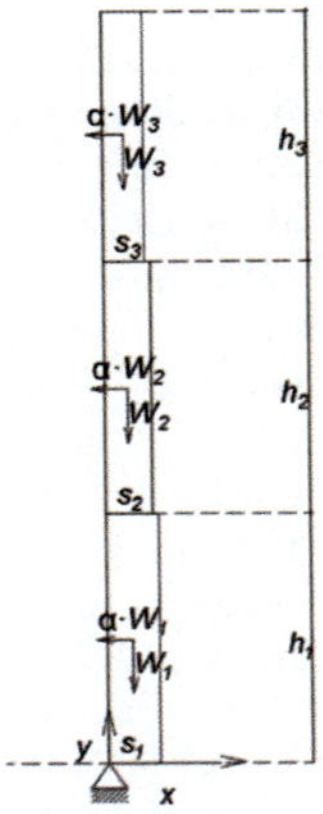


**Fig. 18.** Structural unit 10B: scheme for the global overturning mechanism analysis (V. Cima).

# 4 Conclusions

The assessment of the seismic vulnerability of existing masonry buildings is a topic of particular relevance, mainly for the buildings of historic centers of Italy which are generally arranged in complex aggregates. Indeed, these buildings are often the result of a stratified and non-planned urban development over time and they have non-homogeneous construction features. Therefore, in order to model these buildings, it is necessary a knowledge process whose level of detail can strongly influence the applicability and reliability of the structural analysis methods.

The main goal of this work has been the proposal of a procedure able to link the level of knowledge of existing buildings (and then the set of deducible data) to the methods of structural analysis generally employed for the evaluation of the seismic vulnerability.

A first level is defined on the basis of quick survey activities and/or acquisition from available database (such as the one developed within the project CARTIS (DPC/ReLUIS 2019–2021)): from this level qualitative information about the expected type of collapse mechanism is obtained. On the other hand, the second level is defined on the base of an *as-built* process of knowledge by using advanced instrumental surveys: it allows to obtain quantitative information in terms of seismic acceleration activating the identified local mechanisms.

The proposal of two levels is particularly useful in practical applications. Indeed, since the Qla allows to identify a first quick overview, the Qna, more expensive, can be applied for some of the units on the basis of the results emerged from the previous level.

# References

1. Locati, M., et al.: Italian macroseismic database (DBMI15). National Institute of Geophysics and Volcanology (INGV). https://doi.org/10.6092/INGV.IT-DBMI15
2. Valluzzi, M.R., Munari, M., Modena, C., Cardani, G., Binda, L.: Analisi di vulnerabilità sismica degli aggregati storici: il caso di Castelluccio di Norcia. XII National Congress ANIDIS "Seismic Engineering in Italy", Pisa, pp. 10–14 (2007)

3. Munari, M., Valluzzi, M.R., Cardani, G., Anzani, A., Binda, L., Modena, C.: Seismic vulnerability analyses of masonry aggregate buildings in the historical centre of Sulmona (Italy). In: Proceedings of the 13th International Conference on Structural Faults and Repai (2010)
4. Munari, M., da Porto, F., Valluzzi, M.R., Modena, C.: Valutazione della vulnerabilità sismica degli edifici in aggregato. Il Calcolo delle Strutture Esistenti: Software e Normativa. Studi Esperienze Poposte Conference (2011)
5. Fava, M., Munari, M., da Porto, F., Modena, C.: Valutazione della vulnerabilità sismica locale di edifici esistenti in muratura attraverso analisi cinematica dei meccanismi di collasso e curve di fragilità. In: XVI Convegno ANIDIS L'Ingegneria Sismica in Italia, L'Aquila (2015)
6. Manuale per la compilazione della scheda di 1 livello di rilevamento danno, pronto intervento e agibilità per edifici ordinari nell'emergenza post-sismica (AeDES). Department of Civil Protection, Presidency of the Council of Ministers and others, Rome (2009)
7. OPCM n.3274: Ordinary supplement to the Official Gazette n.72 of 8.05.2003, n. 105 (2003)
8. Aggiornamento delle "Norme Tecniche per le Costruzioni". Ministerial Decree 17 January 2018, Ministry of Infrastructure and Transport, Ordinary supplement to the "Official Gazette" n. 42 of 20 February 2018 - General series (2018)
9. C.S.LL.PP. Circolare n.7 del 21.01.2019, Istruzioni per l'applicazione dell' «Aggiornamento delle "Norme tecniche per le costruzioni"» di cui al DM 17.01.2018 (2019)
10. Binda, L., Cardani, G., Saisi, A., Valluzzi, M.R.: Vulnerability analysis of the historical buildings in seismic area by a multilevel approach. Asian J. Civ. Eng. (Build. Hous.) **7**, 343–357 (2006)
11. Binda, L., Anzani, A., Cardani, G.: Methodologies for the evaluation of seismic vulnerability of complex masonry buildings: case histories in the historic centre of Sulmona. In: Proceedings of 11th International Conference on Structural Repairs and Maintenance of Heritage Architecture, pp. 395–405. Wessex Institute of Technology, Ashurst (2009)
12. D'Ayala, D., Novelli, V.: Seismic Vulnerability Assessment: Masonry Structures. Encyclopedia of Earthquake Engineering. Springer, Heidelberg, pp. 1–20 (2014)
13. Saccucci, M., Cima, V., Grande, E., Imbimbo, M., Pelliccio, A.: Valutazione della vulnerabilità sismica degli aggregati in muratura: il caso di Borgo San Rocco a Sora (Italia). In: XVIII ANIDIS Conference Seismic Engineering in Italy. Pisa University Press, Ascoli Piceno (2019)
14. Giuffré, A.: Sicurezza e conservazione dei centri storici. Il caso Ortigia. Laterza, Bari (1993)
15. Doglioni, F.: Codice di pratica (Linee Guida) per le progettazione degli interventi di riparazione, miglioramento sismico e restauro dei beni architettonici danneggiati dal terremoto umbro-marchigiano. Official bulletin of the Marche Region (2000)
16. Zuccaro, G., Dolce, M., De Gregorio, D., Speranza, E., Moroni, C.: La scheda CARTIS per la caratterizzazione tipologico-strutturale dei comparti urbani costituiti da edifici ordinari. Valutazione dell'esposizione in analisi di rischio sismico. In: Proceedings of the GNGTS (2015)
17. Consiglio Nazionale delle Ricerche (CNR)/Istituto Centrale del Restauro, Raccomandazioni Normal n. 1/88, Alterazioni macroscopiche dei materiali lapidei: Lessico, Rome (1990)
18. Saccucci, M.: Analisi multilivello BIM per la valutazione della vulnerabilità sismica degli edifici in muratura (2020)
19. Remondino, F.: Rilievo e modellazione 3D di siti e architetture complesse. Disegnarecon **4**(8), 90–98 (2011)
20. Borri, A., De Maria, A.: Indice di Qualità Muraria (IQM): correlazione con le caratteristiche meccaniche e livelli di conoscenza. Progettazione sismica **6**(3), 45–63 (2015)
21. Galli, P., Costantino, A., Naso, J.A.: Unmasking the 1349 earthquake source (Southern Italy): paleoseismological and archaeoseismological indications from the Aquae Iuliae fault. J. Struct. Geol. **31**, 128–149 (2009)
22. Guidoboni, E., Ferrari, G., Mariotti, D., Comastri, A., Tarabusi, G., Valensise, G.: Catalogue of strong EQ in Italy (461BC-1997) and Mediterranean Area (760BC-1500) (2007)

23. Alonzi, V.: Terremoto 13 gennaio 1915 - Ore 7,50. La nobile Sora devastata conta le sue vittime. Sora (2015)
24. Branca, C.: Memorie storiche della città di Sora (1847)
25. Pelliccio, A., Marco S., Grande, E.: HT_BIM: parametric modelling for the assessment of risk in historic centers/HT_BIM: La modellazione parametrica per l'analisi del rischio nei centri storici. DISEGNARECON (2017)

# Hazard Risk Analysis. A Worst Case Scenario Approach

Alina-Mihaela Nicuță(✉)

Faculty of Civil Engineering and Building Services, "Gheorghe Asachi" Technical University of Iași, 700050 Iași, Romania
alinanicuta@yahoo.com

**Abstract.** Hazard Risk Analysis in about safety, reduced consequences, customized strategies and perfect action plan in a context subjected to change and risks. Considering the inequitable distribution of hazards, people all around the world are exposed to devastating natural disasters. Several countries restructured their policies, plans and strategies in risk management, invested time, effort and money to reduce the consequences of these phenomena. They started with identification of specificities and vulnerabilities, understanding potential risks and managerial processes. In view of these aspects, the present paper focuses on the analysis of hazard risk, considering a worst case urban scenario for the City of Iași, underlying vulnerabilities and critical points, current action plan, rescue procedures and facilities. The purpose of this study was to ask questions on the urban potential to absorb shocks, resist, adapt and transform and on the population capacity to face and recover physically and mentally from a disaster.

**Keywords:** Hazard risk · Management · Worst case scenario

## 1 Hazard Risk Analysis

### 1.1 Preview on Hazard Risk Categories

In a world subjected to an increased rate of change and danger, hazard risk management must efficiently adapt to global and local tendencies in order to make the process as reliable as possible. International statistics showed that people all around the globe are subjected to significant hazard risks [12, 23]. When it comes to hazards, according to *Daniel Defoe (1660–1731)* the "Fear of danger is ten thousand times more terrifying than danger itself". The main question debated by the present paper refers to the population and authorities' ability and level of preparedness to handle, physically and mentally, a disaster [29]. This inquiry might be reflected by the paper objectives namely: to bring awareness on the natural hazards risks and occurrence according to specific local typologies and degree of severity; to identify local strategies for an efficient disaster management; to increase perception and preparedness to disasters by visualizing on a specific areal worst case scenario. In order to reach the envisaged objectives, has been used the evaluation methodology characteristic to buildings earthquake insurance


A. Rotaru (Ed.): CRIT-RE-BUILT 2019, SSGG, pp. 348–361, 2021.
https://doi.org/10.1007/978-3-030-61118-7_29

degree based on professional expertise performed over buildings and their distribution in a seismic risk class.

For a proper understanding of the issue under discussion it is important to clarify main terms under consideration.

The RISK is the likelihood of an adverse event to take place as a result of a hazard, the probability that an event will occur, producing negative consequences. According to Hollenstein, 1997, Kienholz, 1995 the Risk is a function f of Hazard, Elements at risk, Vulnerability. Figure 1 below details on the three components of Risk function, the vulnerabilities, exposure level and hazards.

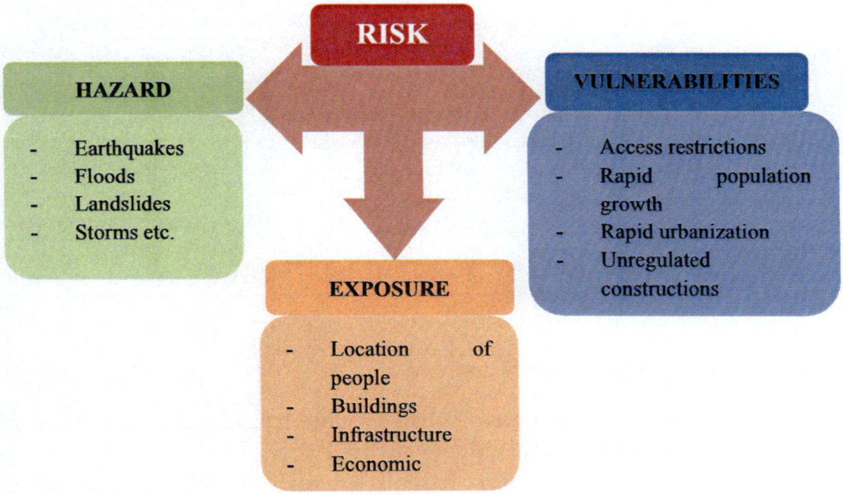


**Fig. 1.** Risk function components

The correlation between a hazard and a vulnerable population, ecosystem or infrastructure creates the risks [5]. If the vulnerability level is high or if natural events become more severe, the risk increases. Elements subjected to risks are: Buildings and facilities, Population, Environment, Production, Economic activities [19].

Manifestation of a risk has associated two types of losses, primary and secondary which are detailed in Table 1, "Looses characteristic to risks" [7, 19].

The HAZARD represents any source or cause of potential damage, harm or adverse effects on something or someone. As observed in Fig. 2 above, "Hazard Typologies", there is a wide diversity of hazards, the two major categories being natural and anthropogenic hazard [25]. The present paper focuses the interest mainly on natural hazards.

**Table 1.** Loses characteristic to risks

| | Human - social | Physical | Economical |
|---|---|---|---|
| Primary | • Fatalities<br>• Injuries<br>• Loss of income or employment<br>• Homelessness | • Ground deformation or quality loss<br>• Structural damage or collapse to buildings and infrastructure<br>• Non-structural damage and damage to contents | • Interruption of business due to building and infrastructure damage<br>• Loss of productive workforce through fatalities, injuries and relief efforts<br>• Capital costs of response and relief |
| Secondary | • Disease<br>• Permanent disability<br>• Psychological impact<br>• Loss of social cohesion<br>• Political actions perceived as inadequate | • Progressive deterioration of damaged buildings and infrastructure which are not repaired | • Insurance industry losses due to weakening of insurance market<br>• Loss of markets and trade opportunities<br>• Loss of confidence by investors, withdrawal of investment<br>• Capital costs of repair |

### 1.2 Management of Emergency Situations

The scientific literature identified several major natural hazards, among these, earthquakes, tsunamis, cyclone, hurricanes, floods etc. [6, 17].

Due to severe manifestations of these hazards, countries like USA or Japan restructured their policies and strategies in hazard management, invested time, effort and money to reduce the consequences, creating a proper management of disasters [23]. European Union prepared also joint strategies and action plans on hazard management [16]. For example Hyogo Framework for Action, "*Managing risks to achieve resilience*", or a ten years plan on *"Building the resilience of nations and communities to disasters"* [2, 3, 11].

These strategies purpose is to contribute to global disaster risk reduction movement and to build a stronger, more systematic and coherent "culture of prevention" in society as part of sustainable development [1, 21].

At national level, Romania implemented different hazard management actions like National Strategy for Emergency Situations Prevention, Decision no. 762/2008 and became a contributory member of International Strategy for Disaster Reduction (ISDR) in order to apply Hyogo Framework for Action [15].

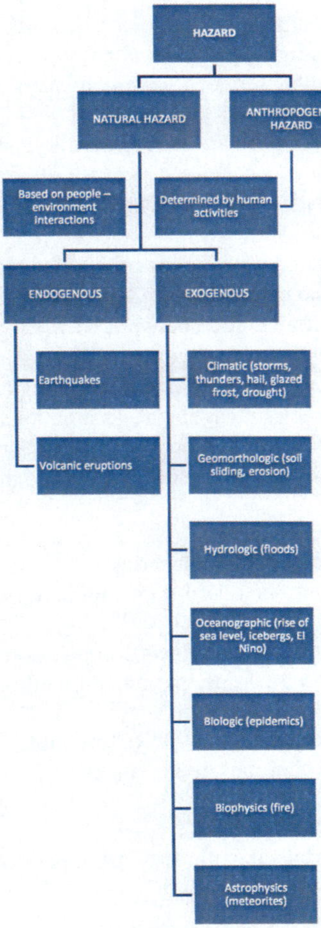


**Fig. 2.** Hazard Typologies

The context of hazard risks manifestation implies the need for efficient interventions and a good management of the situation. Each country has a well-established intervention system and an Organization that complies with these cases. Romania has a National Management System associated with Emergency Situations from 2004, a General Inspectorate for Emergency Situations (IGSU), a medical emergency unit SMURD – Mobile Emergency Service for Resuscitation and Extrication and an emergency phone number 112.

Romania Hazard Risk Management System is composed of temporary and permanent activity structures, including [15]:

- Emergency Committees: national, ministerial, Bucharest and local,

- General Inspectorate for Emergency Situations: transmits the decisions taken by the Government or National Committee to Central authorities and local government,
- Professional community public services for emergency: emergency county inspectorates,
- Emergency Operations Center: permanent or temporary, provides information before or at the time of an emergency,
- Commander action: coordination unit for exceptional events.

The main responsibility related with the planning, management and coordination of information and public education belongs to the General Inspectorate for Emergency Situations. IGSU represents a communication network between population, authorities and organizations qualified for emergency management. This organization has the infrastructure and resources necessary for interventions in emergency cases. It coordinates all the institutions involved in emergency management, as a national contact point and maintains a constant communication with international governmental and non-governmental organizations with responsibilities in the field. Inspectorate mission is [8, 13]:

- Monitor and evaluate emergency cases,
- Population preventive information and education,
- Warning local authorities about the possibility/imminence of an emergency;
- Persons search and rescue,
- Evacuation of persons or property endangered population by providing evacuation measures, installation displaced camps, participation in public transport of people and goods,
- Emergency medical assistance through SMURD modules,
- Provide own and borrowed techniques, materials and equipment for intervention,
- Firefighting,
- Decontamination of the population,
- Neutralizing hazardous materials unexploded during previous military conflicts.

## 2 Worst Case Scenario Approach

The process of planning, implementing, evaluating and adapting strategies, procedures and measures relates to the analysis, reduction and transfer of disaster risks. The aim of this process is to reduce hazards and vulnerability and strengthen the adaptability capacity of individuals, households, communities and state structures [4, 10]. Hazard risk management is a continuous process that involves physical and non-physical measures and takes into account the risk factors within a society. Disaster risk management aims to avoid generation of new risks, improves resilience and contributes to sustainable development [20, 22].

Each hazard manifestation has a certain level, impacting on the recovery level. A hazard arises when an extreme natural event (ex. Storm, earthquake, flooding) strikes a vulnerable society [5]. Whether a natural event becomes a disaster depends on the social, economic, ecologic and politic characteristics of the society in question. Hazard risk management seeks to reduce a society vulnerability to extreme natural events, which can‘t be always prevented, but their impact can be mitigated [14].

**Table 2.** World Natural Catastrophe Losses

| | 2018 | 2017 | 2008–2017 Average values for the last 10 years | 1988–2017 Average values for the last 30 years |
|---|---|---|---|---|
| Number of events | 850 | 740 | 630 | 500 |
| Loses in US dollars | 160,000 | 350,000 | 190,000 | 140,000 |
| Fatalities | 10,400 | 13,000 | 60,000 | 53,000 |

International statistics are abundant in information of the loses associated with disasters, actual information on the number of events, fatalities and economic loses, being detailed in Table 2 [27, 28, 30].

The data found in Table 2 presents a statistic on the number of natural disasters in an evaluative analysis till 2018 by number of events, fatalities and economic loses provided by Munich Re, NatCatSERVICE, March 2019. As observed in the table, the number of hazards taking place at global level is increasing. In spite of this situation, the positive aspect refers to the decrease in the number of fatalities, meaning that the pre-disaster and post-disaster strategies, measures and policies have given results [29]. Statistics provided by the same source, for a number of 850 events, show in Fig. 3 details on natural hazards by type for year 2018 and in Fig. 4 the distribution of these events by continent, for the same year. As observed in the charts, the hydrological and meteorological events have the most significant manifestations, while Asia is the continent mostly subjected to hazards manifestation [26].

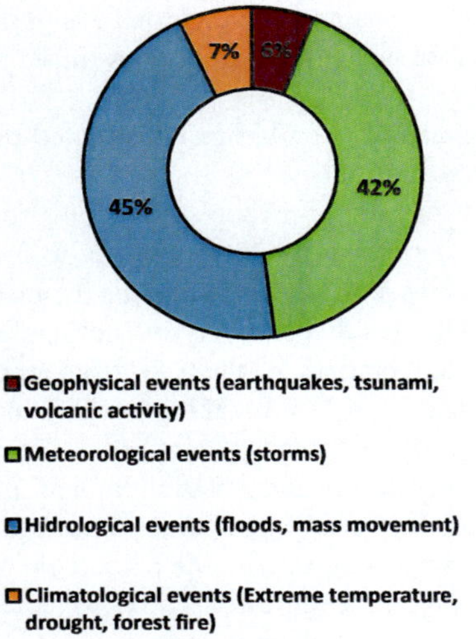


**Fig. 3.** Natural hazards distribution by type for 2018

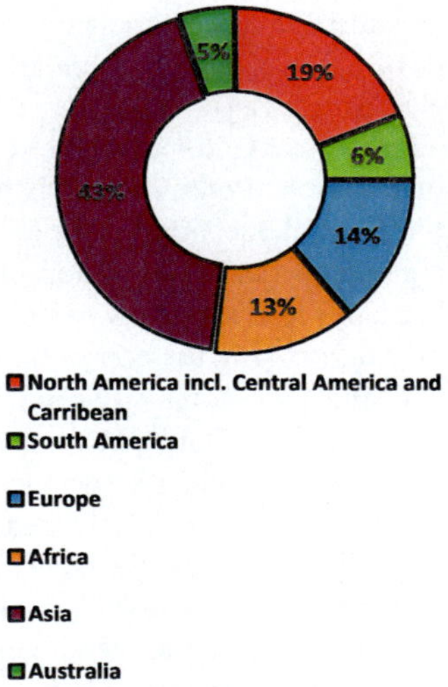


**Fig. 4.** Natural hazards distribution by Continent for 2018

Romania main natural risks are the earthquakes, floods, drought, fires or soil sliding. Among these, the most significant risk is the earthquake. Due to previous experiences in earthquake disasters, Romania seismic activity is under constant supervision by the National Institute of Research and Development for Earth Physics (INFP). In a 5 to 8 ranking of seismic risk, international statistics place Romania at level 6 [9, 24].

The present paper presents a worst case scenario associated with a seismic hazard manifestation for the City of Iași, with particular focus on an area from the city center, Podu Ros Intersection and Bahlui River Bay.

**Fig. 5.** "Red Bullet" marked building.

Romanian legislation system provides a classification of the buildings based on the seismic risk point of view [9].

The endangered buildings are being signalized with a "red bullet", as seen in Fig. 5. Main critical buildings integrate constructions built previous to 1977 that were constructed based on different standards and norms, with little importance to earth-quakes [24].

1977 is considered as reference year in Romania for construction sector since it marked the most significant earthquake registered in our country, with a high number of deaths, casualties and destructions, most of them due to inadequate construction procedures. Following that moment, the legislation in constructions changed, calculation and design of a building are based on new criteria and coefficients in accordance with the seismic activity of our Country.

The City of Iasi still has registered a number of 190 buildings, built between 1870–1963, classified as class I of seismic risk, considered instable and dangerous. A significant number of buildings of this type are situated close to the most important traffic node, Podu Ros Bridge [24].

In Fig. 6 are identified with red color the areas with buildings constructed before 1977, with purple, Podu Ros and Nicolina Intersections.

Podu Ros Intersection has been considered for the paper case study since it is the most significant road intersection in the city, connecting the city center with 5 large neighborhoods leading to city exists. More than this, a fail in accessing this intersection would mean breaking the connection with the city health and safety institutions, since all the hospitals are in the city center. Another characteristic of the City, from strategic point of view, is that Bahlui River crosses a large area of Iasi, breaking the City in half. All connections between the two city parts are made by bridges [23].

A worst case scenario would mean these bridges failure and isolation for half of City population, limited access for emergency units, delay in salvage operations and so on.

Envisioning on the worst case scenario it is important to compile a strategic urban plan for emergency situations and efficient salvage procedures based on weaknesses, considering the strong points, failure points and potential alternatives.

**Fig. 6.** Case Study Areal under surveillance

Main disruptions associated with a worst case scenario include [18, 24]:

- Building collapse,
- Transport infrastructure destruction,
- Human casualties: deceased or wounded,
- Destruction of connections with other cities and with main emergency units,
- Lack of food and water,
- Lack of salvage units and shelters,
- Economic breakdown.

It is vital for certain facilities to be maintained functional in case of a disaster, in order to provide necessary services to the community. These facilities are:

- Police stations,
- Hospitals,
- Fire stations,
- Shelters (ex. Schools, churches).

Worst case scenario is correlated with destruction of main communication systems, bridges and roads. These constructive systems have the role to connect different areas. In their absence, a certain areal remains blocked, potentially without the possibility to provide food and water.

DRDP (Road and Bridges Administration) representatives mention that in case of a natural hazard, they are capable to provide alternative temporary components for bridges and roads, in order to re-establish the connectivity in a short time. Figure 7 presents alternative temporary solutions: temporary bridge and Pontoons Bridge.

**Fig. 7.** Temporary bridge and Pontoons Bridge

In the last years the City developed greatly outside the old City areal. Still, main emergency services, hospitals and police are situated in the old part of the City, leaving the only few private hospitals with reduced capacity, low level of traffic facilities (Arcadia Hospital, Providenta Center) and no emergency services in the new developed part of the city. As previously presented, the two parts of Iasi City are connected by several bridges, along Bahlui River, some of them only with pedestrian traffic, others cover also vehicle traffic. Of these, most significant, from capacity and traffic point of view, are the ones in Podu Ros and Tudor Vladimirescu. It is vital to preserve the functionality of these bridges in case of a hazard, since they provide access to medical assistance and emergency units. Figure 8 presents Bahlui Areal in blue and the road network, with red.

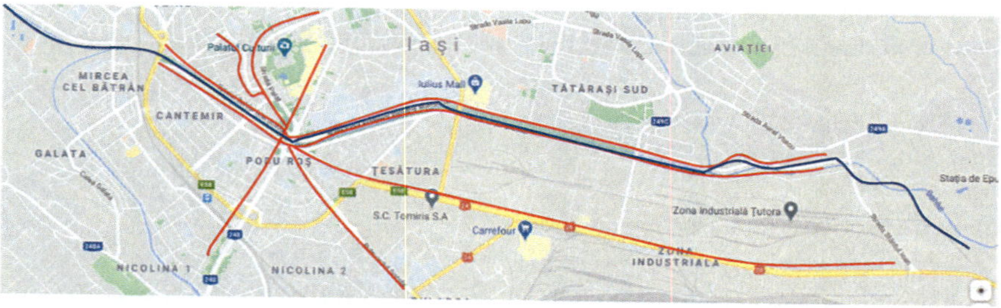


**Fig. 8.** Bahlui Areal and associated road network

Figure 9 and 10 detail on the bridges positioning on Bahlui areal on both sides of Podu Roș Intersection.

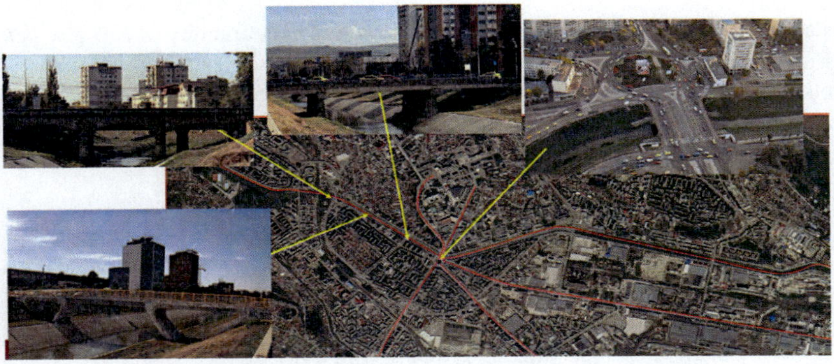

**Fig. 9.** Bahlui Bridges on the right side

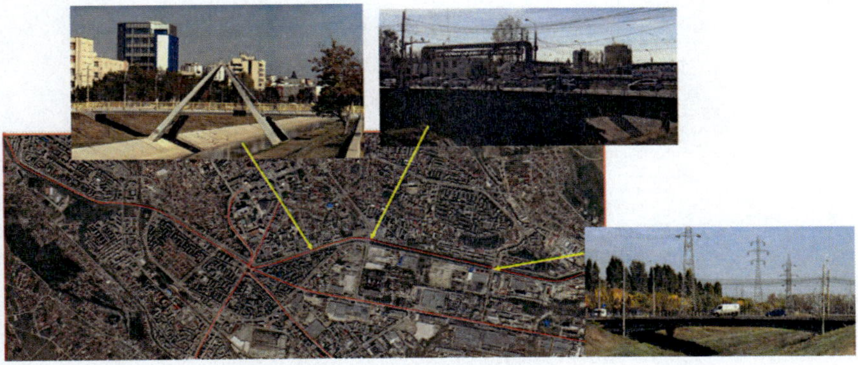

**Fig. 10.** Bahlui Bridges on the left side

Worst case scenario for Iasi City would imply, destruction of transportation infrastructure, consisting in bridges crossing Bahlui River and associated roads and building collapse. Main consequences would be connections breakage, impossibility to access Emergency Centers and salvage operators in due time. More than this, as observed in images above, in the areal close to Bahlui River are situated an extended number of real estates registered by local administration as "dangerous" with significant potential of collapse.

Some of the envisaged solution associated with this worst case scenario refers to:

- Preventive measures: rehabilitations, maintenance works for buildings and bridges,
- Population education on preparing to face hazards,
- Sirens for public alarm,
- Building shelters or activation of old ones.

Table 3 presents a list of old shelters built in the City of Iasi [15] for salvage reasons in case of war or disasters. Currently, for the majority of them were found alternative uses, others than those related to their main attribution and functionality. Also, few people are aware of their positioning and functionality. Actually, every City in Romania has a number of established shelters, their list being accessible on the internet. Still, their functionality is restricted. Complementary, even though the real estate statistics present a significant increase in newly constructed buildings, these are not designed to integrate shelters.

**Table 3.** List of shelters in Iași

| Nr. Crt. | Shelter address |
|---|---|
| 1 | Șos. Păcurari nr.25, BL. 539, SC.A |
| 2 | Șos. Păcurari nr.27, BL. 539, SC. B |
| 3 | Șos. Păcurari nr.29, BL. 539, SC. C |
| 5 | Str. Prof. Anton Șesan nr.8, BL. H4, SC.A |
| 7 | Str. Prof. Anton Șesan nr.10, BL.H5, SC.A |
| 9 | Str. Sf. Lazăr nr.20, BL. K2, SC.A |
| 11 | Str. Sf. Lazăr nr.22, BL.K3, SC.A |
| 12 | Str. Sf. Lazăr nr.24, BL.K4, SC.A |
| 14 | Str. Sf. Lazăr nr.26, BL. K5, SC. A |
| 16 | Str. Sf. Lazăr nr.28, BL. K6, SC.A |
| 17 | Str. Sf. Lazăr nr.47, BL. A5, SC.A |
| 18 | Str. Sf. Lazăr nr.45, BL. A6, SC. A |
| 19 | Str. C. Negri nr. 43, BL.T1, SC. A |
| 20 | Str. C. Negri nr.41, BL.Z1, SC.A |
| 21 | Str. C. Negri nr. 39, BL.Z2, SC.A |
| 22 | Blvd. Independenței nr. NR.24, BL. Y3, SC.A |
| 23 | Blvd. Independenței nr.26, BL.Y4, SC.A |
| 24 | Blvd. Independenței nr. 28, BL.Y5, SC.A |
| 25 | Str. Prof. Paul nr. 9, BL. 338, SC.B |
| 26 | Str. Prof. Paul nr. 11, BL.339, SC. A |
| 27 | Str. Prof. Paul nr. 11-13, BL. 339, SC.A |
| 29 | Str. Prof. Paul nr. 11, BL. 339, SC.B |
| 31 | Str. Pantelimon nr.14, Bl. P. 15, Sc. B |
| 33 | Str. Pantelimon nr. 16, bl. P. 15, sc. C |

Based on these considerations and on the potential risks that threaten this areal would be of great significance to plan alternative shelters in case of a disaster, where people can find refuge and help. Potential shelters can be found in the main important administrative and educational buildings:

- City Hall,
- Schools,
- Churches,
- Innovative shelter structures.

It is important to identify potential units for shelters, integrate them in a City Strategic Plan for Hazards and make them public to the citizens.

## 3 Conclusions

The disturbing statistics on the increase in the rate of disasters worldwide leads us to the conclusion that significant improvements must be brought in the management of hazards. Actions must be taken towards identifying the vulnerabilities, establishing the exposure to risks and potential hazards. Once correlations among these components are established, responsible authorities can elaborate a strategic plan for hazard risk management. In the face of potential dangers each country has established an emergency intervention plan and action. Complementary, at international level, there is a joint intervention strategy and support plan, with application in case of large disasters.

The management of hazard risk is a continuous process that involves physical and psychological measures necessary to avoid generation of new risks, to improve resilience and contribute to sustainable development. Also it seeks to reduce a society vulnerability to extreme events and mitigate the impact. International statistics show an increase in the number of events but a decrease in fatalities meaning that preventive measures and post-disaster actions are fulfilling their purpose.

As European Union member, Romania benefits from the strategy and experience of better developed countries, making efforts to align with their high level of resilience. From a wide diversity of hazards facing Romania, highest risks and vulnerabilities are associated with the seismic activity.

The present paper apart from detailing on the hazard risk management process, presents a potential risk situation in a so called "worst case scenario" for the City of Iași. this scenario can be further extended to any other city in Romania or to national level. The purpose of the case study is to draw attention on vulnerabilities and to make some recommendations on envisaged solutions. Among these solutions can be mentioned preventive measures: rehabilitations, maintenance works for buildings and bridges; population education on preparing to face hazards; sirens for public alarm; building shelters or activation of old ones.

Local administrations have already taken some measures in these directions. Still it is important to further extend and improve them. One topic of discussion might be the existence of shelters. In Iași there is a certain number of shelters, built in certain key points of the city. Unfortunately, in time, these shelters changed their functionality and capacity of use. Even more, the intensive development of real estates in the city doesn‘t take into consideration a planning of alternative shelters in case of disasters, where people can find refuge and help. A potential solution in this direction would be to allocate the attribute of shelter to certain facilities like schools, churches, the City Hall or develop innovative shelter structures.

It is highly important to identify potential units for shelters, integrate them in a City Strategic Plan for Hazards and make them public to the citizens.

## References

1. Alexander, D.: Principles of Emergency Planning and Management. Dunedin Academic Press, Terra Publishing, Dunedin (2002)
2. Bene, C.: Towards a Quantifiable Measure of Resilience, Institute for Development Studies, working paper, vol. 2013, no. 434 (2013)
3. Bene, C., Newsham, A., Davies, M., Ulrichs, M., Godfrey-Wood, R.: Review article: resilience, poverty and development. J. Int. Dev. **26**, 598–623 (2014)
4. Boorman, J., Fajgenbaum, J., Ferhani, H., Bhaskharan, M., Arnold, D., Kohli, H.A.: The centennial resilience index: measuring countries' resilience to shock. Global J. Emerging Market Economies **5**(2), 57–98 (2013)
5. Briguglio, L., Cordina, G., Farrugia, N., Vella, S.: Economic Vulnerability and Resilience: Concepts and Measurements, The World Institute for Development Economics Research (WIDER), Research Paper No. 2008/55, (2008)
6. Bristow, G., Healy, A.: Regional resilience: an agency perspective. Regional Stud. **48**(5), 923–935 (2014)
7. Carter, W.N.: Disaster Management: A Disaster Managers Handbook. Asian Development Bank, Manila (1992)
8. Centre for Research on the Epidemiology of Disasters, Natural Disasters 2018 (2019)
9. Centrul de Cercetare pentru Evaluarea Riscului Seismic (2013). http://www.encipedia.org/articole/proiectare/resurse-utile/harti-de-zonare/harta-de-zonare-seismica-din-p100-1-2013.html. Accessed 30 Sept 2018
10. Cutter, S., Barnes, L., Berry, M., Burton, C., Evans, E., Tate, E., Webb, J.: A place-based model for understanding community resilience to natural disasters. Global Environ. Change **18**, 598–606 (2008)
11. Davoudi, S., Shaw, K., Haider, L.J., Quinlan, A.E., Peterson, G.D., Wilkinson, C., Fünfgeld, H., McEvoy, D., Porter, L.: Resilience: a bridging concept or a dead end? "reframing" resilience: challenges for planning theory and practice interacting traps: resilience assessment of a pasture management system in northern afghanistan urban resilience: what does it mean in planning practice? resilience as a useful concept for climate change adaptation? The Politics of Resilience for Planning: A Cautionary Note, Planning Theory & Practice **13**(2), 299–333 (2012)
12. Facts + Statistics: Global catastrophes (2018). https://www.iii.org/fact-statistic/facts-statistics-global-catastrophes
13. Federal Ministry for Economic Development and Cooperation, Disaster Risk Management. Approach and Contributions of German Development Cooperation, Sprintout Digitaldruk GmbH Publisher. www.bmz.de. Accessed Sept 2019
14. Folke, C.: Resilience: the emergence of a perspective for social-ecological systems analyses. Global Environ. Change **16**(3), 253–267 (2006)
15. General Inspectorate for Emergency Situations – IGSU. https://www.igsu.ro/, https://www.igsu.ro/index.php?pagina=adaposturi_prociv. Accessed Sept 2019

16. Haukkala, H.: The european union as a regional normative hegemon: the case of european neighbourhood policy. In: Whitman, R.G. (ed.) Normative Power Europe: Empirical and Theoretical Perspectives, Palgrave Macmillan, pp. 45–64 (2011). https://reliefweb.int/report/world/natural-disasters-2018
17. Hewitt, K.: Regions of Risk: A Geographical Introduction to Disasters. Addison Wesley Longman, Reading, Massachusetts (1997)
18. Human Development Report 2014: Sustaining Human Progress: Reducing Vulnerabilities and Building Resilience. http://hdr.undp.org/en/content/human-development-report-2014 (2014)
19. Law, S.H., Lim, T.C., Ismail, N.W.: Institutions and economic development: a granger causality analysis of panel data evidence. Econ. Syst. **37**(4), 610–624 (2013)
20. Manyena, S.B., O'Brien, G., O'Keefe, P., Rose, J.: Disaster resilience: a bounce back or bounce forward ability? Local Environ. **16**(5), 417–424 (2011)
21. Martin, R., Sunley, P.: On the notion of regional economic resilience: conceptualization and explanation. J. Econ. Geography **15**(1), 1–42 (2014)
22. Mitchell, T., Harris, K.: Resilience: a risk management approach, Overseas Development Institute (2012). https://www.odi.org/sites/odi.org.uk/files/odi-assets/publications-opinion-files/7552.pdf
23. Nicuta, A.M, Tiganasu, R.: Risk management analysis in a worst case scenario: an urban example. Bulletin of the Transilvania University of Braşov, vol. 11(60), no. 1, Series I: Engineering Sciences (2018)
24. P100-1/2013 Earthquake Design Code. www.mdrap.ro/userfiles/reglementari/Domeniul_I/I_22_P100_1_2013.pdf. Accessed Sept 2019
25. Smith, K.S.: Environmental Hazards: Assessing Risk and Reducing Disaster (3rd edn). Routledge, London (2001)
26. UNISDR, United Nations Office for Disaster Risk Reduction: Global Assessment Report 2013, Geneva (2013). www.unisdr.org/we/inform/publications/33013
27. UNISDR, United Nations Office for Disaster Risk Reduction, Global Assessment Report 2015, Genf (2015). www.preventionweb.net/english/hyogo/gar/2015/en/gar-pdf-GAR2015_en.pdf
28. United Nations Development Programme (UNDP), Time for Global Action (2014). http://www.undp.org/content/undp/en/home/librarypage/corporate/annual-report-2014-2015.html
29. Waugh Jr., W.L.: Living with Hazards, Dealing with Disasters: An Introduction to Emergency Management. M.E.Sharpe, Armonk, New York (2000)
30. World Bank annual report 2014: Main report (2014). http://documents.worldbank.org/curated/en/111781468170952958/Main-report

# The Use of Macro Element Approach for the Seismic Risk Assessment of Brick Masonry Buildings

Marjo Hysenlliu[1] and Huseyin Bilgin[2](✉)

[1] Department of Civil Engineering, Polytechnic University of Tirana, Tirana, Albania
[2] Department of Civil Engineering, Epoka University, Tirana, Albania
hbilgin@epoka.edu.al

**Abstract.** Unreinforced masonry [URM] is the dominant structural type for low to moderate-rise buildings in many Balkan countries including Albania. Its dynamic response is highly inelastic, and generally shows high vulnerability to earthquake loading. In literature, there are several methods available to evaluate the seismic performance of these buildings. The choice of the proper model to use is a matter of paramount importance, as many aspects must be considered in order to reach a good approximation of the structural behavior. Within this context, this paper aims to make a seismic risk assessment by following the equivalent frame approach based on macro-element modeling. Due to the resource and time efficient computations, this approach is becoming more popular among the practitioners and field experts in this area and allows simulating the non-linear behavior of masonry buildings. This method is applied to an old masonry building from the Albanian construction practice that is representative of a typical low-rise residential building. Capacity curves of the investigated building are determined to assess the most probable seismic response of the investigated housing construction in the region under selected ground motions. Finally, seismic risk of the tested structures is evaluated.

**Keywords:** Macro-element approach · TREMURI software · Unreinforced masonry buildings

## 1 Introduction

Selection of accurate and reliable methods for the seismic vulnerability assessment of existing old masonry building stock is seen as a vital step towards the seismic mitigation. The mathematical modelling of the inelastic response of masonry structures can be conducted by detailed nonlinear finite element analyses. Detailed finite element analyses usually require constitutive laws by considering limited tensile strength of masonry. Two- or three-dimensional nonlinear elements are usually adopted for the modeling of unreinforced masonry. Although detailed finite element analysis can capture a deep insight on inelastic response of masonry structures, modeling phase and interpretation of the


A. Rotaru (Ed.): CRIT-RE-BUILT 2019, SSGG, pp. 362–373, 2021.
https://doi.org/10.1007/978-3-030-61118-7_30

analysis results are extremely time consuming. Furthermore, its complexity and some convergence issues during modeling and analysis make this approach not suitable for practical use among the practitioners. Therefore, some simplified analytical models were proposed for masonry structures to overcome the complexity of detailed finite element analysis. In general, these simplified methods adopt equivalent frame elements incorporating nonlinearity, or more complex mechanical sub assemblages, in the modelling and analysis of unreinforced masonry structures. In the case of mixed structures (i.e. confined masonry buildings), the simplified models usually take into account masonry and reinforced concrete elements arranged in series or in parallel, without taking into account for the confining effects.

Conceptual soundness, accuracy and transparency of the mathematical models are the crucial components of an optimal method in seismic vulnerability assessment [1]. According to the recent research in the last two decades, use of existing methods differ from each other in terms precision, assessment scale, computational cost, availability of the data and resources [2]. Although there are a number of methods for the evaluation of the seismic response of existing masonry buildings in the literature, the most optimal choice for seismic evaluation of masonry buildings would be based on simplified nonlinear analysis. Therefore, TREMURI software package [3] based on macro element approach is used in this study.

In this study is aimed to assess the in-plane response of a typical masonry building by means of pushover analysis through macro-element. The use of pushover approach is believed to find a compromise among different necessities to limit the number of required input parameters, the computational cost and convenient versatility to capture building responses. Seismic capacities of the building are calculated for different limit states according to Eurocode 8-3 [4, 9]. Seismic demand estimates are done by following the equal single degree of freedom approach. Finally, seismic performance of the selected building is comparatively assessed under the selected ground motion records.

## 2 Case Study Building

The current case study, presented in Fig. 1, is a typical masonry building constructed by following template designs in 1958, Tirana, Albania. Masonry was commonly used for public and governmental buildings as a low-cost construction method between 1944–1990. Today these buildings are still in use and the main functions are mostly for residential purposes. Typical masonry residential buildings in the Albanian territory were constructed by template designs and relatively low- or mid-rise (two to five story in height). In other words, standard procedures were followed for the design and construction of such buildings. Hence, a considerable number of buildings have the same template designs in different parts of Albania [5]. These structures are composed of two main components, namely load bearing walls and roof diaphragms. The walls are stiff with openings and the diagrams are usually constructed by RC slabs.

Geometric characteristics of the case study were derived from blueprints for building permit and site surveys. It has vertical load bearing elements, which are unreinforced masonry walls. Since the quality of the horizontal load bearing structure is also of great importance in seismically prone areas, horizontal elements above ground floor is reinforced concrete (RC) slabs.

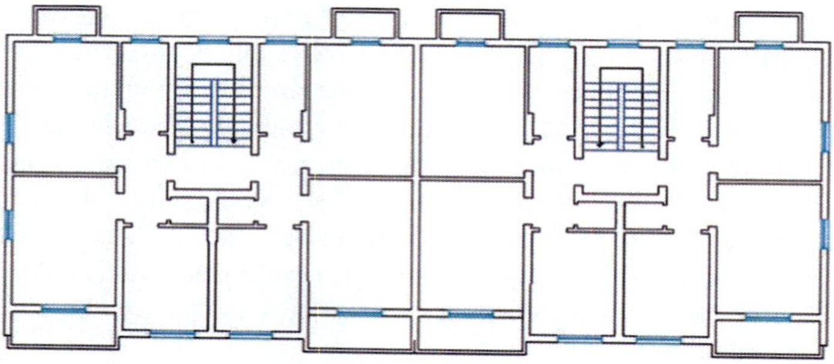

(a) Plan

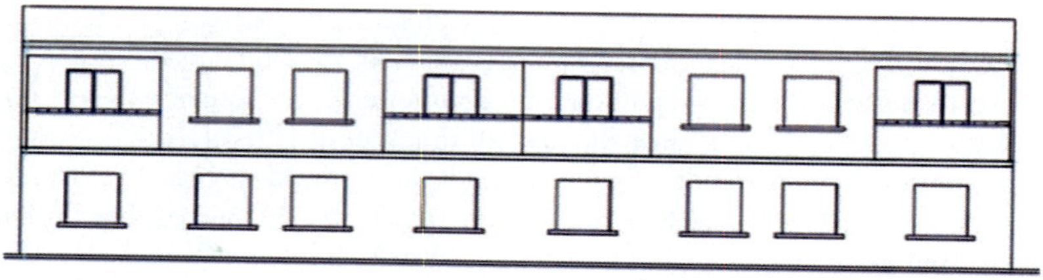

(b) Elevation

**Fig. 1.** Schematic plan and elevation views of the studied building

The building representing this template stock is located at "Sulejman Delvina" street in Tirana, Albania. The building has a total height of 6.00 m, presents a plan are of 27 × 12 $m^2$. Load bearing walls of the building are 25 cm, with red clay bricks M-75 and lime mortar M-25.

To determine the soil profile and the load bearing capacity of the underlying soil, geotechnical/seismological investigations were done on the construction site for the building and the obtained values were adopted for seismic performance analyses.

Soil types and relevant characteristics obtained from seismotectonic measurements were used as input for linear response spectrum calculations given in Table 1 (according to EN 1998-1:2004).

**Table 1.** Soil types in EN 1998-1:2004 [4]

| $V_{s, 30}$ (m/s) | Soil type |
|---|---|
| 209 | C |

# 3 Numerical Modeling

## 3.1 General Workflows

Seismic safety assessment of buildings should be rely on both qualitative (as gathering information about the structure, observation, survey and description of the damage pattern, etc.) and quantitative (as in-situ and laboratory experimental tests, mathematical, etc.) methods that consider the effect of the phenomena on structural response. Any assessment of structural safety is seriously influenced by the uncertainty attached to data (actions, deformations, resistance, etc.), laws, models, assumptions, etc. used in the study, and by the difficulty of representing real phenomena in a precise way.

During the on-site inspections, the damage status of the building was determined by determining the general condition of the buildings, whether there are settlements that can be observed on the ground, and the crack condition in the building.

It was determined that the building was built in accordance with the design projects in terms of member sizes. The structure is geometrically suitable for its project. All dimensions shown in the architectural and design projects of the building have been checked on site. In the seismic performance analysis, the design project was taken as basis together with the obtained test results and the measurements done on the site.

The seismic behavior of the case study building is assessed by deploying pushover analyses using TREMURI software package based on macro-modeling approach. The geometry of the building was defined as given in the blueprints, detailed drawings and the available reports. The roof system was not modelled, however, the corresponding vertical loads were calculated and applied over the respective walls as linear loads. Horizontal diaphragms were assumed properly connected to the masonry walls and the foundation was modeled as elastic based on the findings of the geotechnical investigation done on the building.

Material properties were determined from site investigations and experimental tests. Mechanical characteristics were obtained from the experimental tests with destructive methods according to ASTM C67-09 guidelines [6]. Randomly selected specimens from building were identified to meet the minimum requirements of standards for the number of tests.

Then, the samples were tested to determine the compressive strength of the solid brick units, mortars and the obtained results are adopted for analysis (Table 2).

**Table 2.** Mechanical properties of structural vertical load bearing structures

| $f_k$ [MPa] | $f_{vk}$ [MPa] | $f_{vk0}$ [MPa] | $f_t$ [MPa] | $f_{xk1}$ [MPa] | $f_{xk2}$ [MPa] | E [MPa] | G [MPa] | $G_{fc}$ [Nmm/mm$^2$] | $G_f$ [Nmm/mm$^2$] | v |
|---|---|---|---|---|---|---|---|---|---|---|
| 1.94 | 0.33 | 0.17 | 0.097 | 0.245 | 0.175 | 1940 | 485 | 3.1 | 0.1 | 0.2 |

## 3.2 TREMURI Software

This program, developed under the leadership of Prof. Lagomarsino at the Genoa University, is one of the most common software for the seismic assessment of masonry structures, nowadays. It is based on the equivalent frame-type model approach, following the assumption that the in-plane behavior of masonry panels constituted by a set of one-dimensional spandrel beams and piers, subsequently represented by macro-elements with inelastic response, connected by rigid nodes. While, spandrels are horizontal elements placed in between two vertically aligned openings, piers are vertical elements that carry both dead and imposed loads. This formulation allows the representation of the three principal in-plane failure modes of masonry walls; rocking (bending), shear-sliding and diagonal shear cracking, with a reasonable number of degrees of freedom on the complex response of masonry structures with a limited computational effort.

In this software, the strength criteria governing the in-plane response of spandrel and pier elements is defined according to the Italian Codes for structural design (NTC-2008) [7] and European norms [4]. The flexural behavior, together with bending and compressive failure modes, is based on the beam theory, assuming a rectangular normal stress distribution at the compressed toe. The shear behavior is governed by the diagonal cracking failure criteria as proposed in [8] and adopted by NTC-2008 [7].

Rigid diaphragms were used as orthotropic membrane finite elements (FE), characterized by an equivalent thickness, Young modulus together with the equivalent shear modulus as calculated based on the experimental test results. In the determination of the loads, the provisions in Eurocode 1 and Eurocode 8 were followed. The load bearing capacities of the elements were revealed for the most unfavorable cross section effects obtained from the analyzes. Finally, gravity and live loads equal to 1.00 and 2.00 kN/m$^2$, were assigned to floors, respectively.

Since TREMURI does not allow considering the center of gravity of the structure as control node, pushover analyses were executed under different trial and error combinations to understand the influence of some parameters over the global response of the building. From all the twenty-four different analyses deployed in TREMURI, the most unfavorable one was selected considering the transverse and longitudinal directions.

### 3.3 Dynamic Characteristics and the Lateral Capacity of the Building

The modal analysis was executed using TREMURI and the results were presented for six modes of vibrations. The results of the linear modal analyses were synthesized in Table 3.

**Table 3.** Modal characteristics of the studied building

| Mode | T [s] | $m_x$ [kg] | $M_x$ [%] | $m_y$ [kg] | $M_y$ [%] | $m_z$ [kg] | $M_z$ [%] |
|---|---|---|---|---|---|---|---|
| 1 | 0.118 | 467356 | 90.15 | 0 | 0.00 | 0 | 0.00 |
| 2 | 0.099 | 0 | 0.00 | 46283 | 89.28 | 227 | 0.04 |
| 3 | 0.090 | 2 | 0.00 | 433 | 0.08 | 3 | 0.00 |
| 4 | 0.045 | 50629 | 9.77 | 4 | 0.00 | 13 | 0.00 |
| 5 | 0.033 | 5 | 0.00 | 46456 | 8.96 | 8901 | 1.72 |
| 6 | 0.030 | 24 | 0.00 | 1.431 | 0.28 | 16 | 0.00 |

Upon completing the aforementioned modeling procedure (Fig. 2), the capacity curves of the building were determined by carrying out pushover analysis. Selection of the lateral load pattern has a crucial role on the results. Pushover analysis with invariant load patterns have been used frequently for seismic performance evaluation in several studies. For the scope of the analyses, twelve types of lateral load distributions were applied, and the results are shown in Fig. 3 and 4.

**Fig. 2.** 3D Analytical model of the building

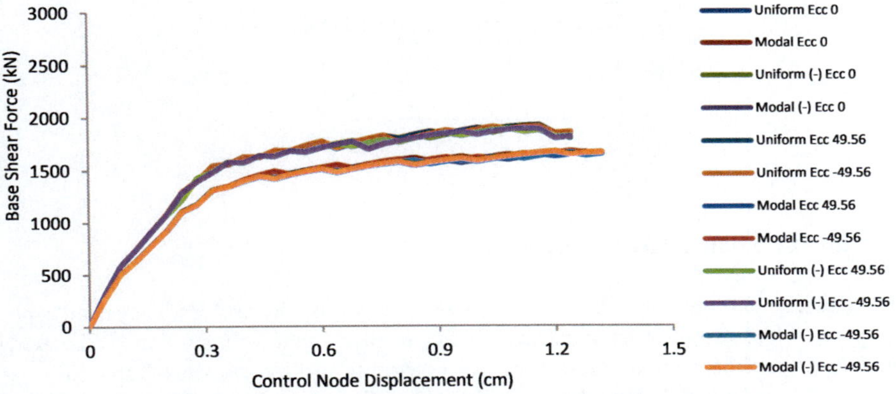


**Fig. 3.** Capacity curves performed by using TREMURI software for different load patterns in the *x*- direction

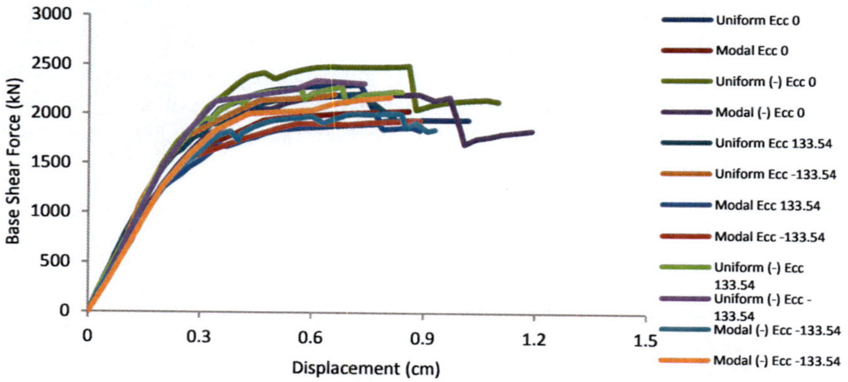


**Fig. 4.** Capacity curves performed by using TREMURI software for different load patterns in the *y*- direction

Damage limit states are quantitative definition of performance levels by a convenient damage indicator capable of representing the seismic performance with appropriate damage thresholds. Three damage limit states levels, i.e., "Limited Damage" (LD), the limit state "Significant Damage" (SD) and the limit state "Near Collapse" (NC) are considered as specified in Eurocode 8 [9]. Consequently, damage limit states of the studied building were calculated according to CEN (2005) [9] and given in Table 4.

**Table 4.** Global displacement capacities (cm) obtained from pushover for the considered limit states under the considering the most unfavorable loading case

| Direction | d (DL) | d (SD) | d (NC) |
|---|---|---|---|
| *x*- | 0.30 | 0.96 | 1.28 |
| *y*- | 0.23 | 0.50 | 0.66 |

## 4 Seismic Risk Evaluation

The displacement capacity values are merely not meaningful themselves. They need to be compared with seismic demand values. Seismic displacement demands were calculated by using a set of 46 ground motions recorded on dense-to-firm soil sites (Table 5).

The "equivalent" SDOF models of the URM building was subjected to ground motion listed in Table 5 to estimate displacement demands. Equivalent SDOF nonlinear response history analyses were carried out using Utility Software for Earthquake Engineering program (USEE) [10]. The "equivalent" SDOF displacement demands were then converted into building displacement demands at the roof level multiplying by the first mode participation factor.

Following displacement demand estimates, seismic performance of the masonry building was evaluated. Statistics of performance evaluation is given in Tables 5 for "Limited Damage", "Significant Damage" and "Near Collapse" limit states (last three columns). In Table 5, the term "exceedance ratio" refers to the number of the building

**Table 5.** Selected ground motion records and Exceedance ratio of ***LD***, ***SD*** and ***NC*** performance levels for the investigated masonry building subjected to the selected ground motions

| No. | Earthquake | $M_w$ | Record and component | Site | PGA (g) | PGV (cm/s) | PGD (cm) | LD | SD | NC |
|---|---|---|---|---|---|---|---|---|---|---|
| 1 | San Fernando 2/9/1971 | 6.6 | LA HOLLYWOOD STOR LOT, 090 | C | 0.21 | 18.93 | 12.42 | 1.00 | 0.00 | 0.00 |
| 2 | San Fernando 2/9/1971 | 6.6 | LA HOLLYWOOD STOR LOT, 180 | C | 0.174 | 14.87 | 6.32 | 1.00 | 0.00 | 0.00 |
| 3 | Friuli, Italy 5/6/1976 | 6.5 | Tolmezzo, 000 | C | 0.351 | 22.03 | 4.11 | 1.00 | 0.00 | 0.00 |
| 4 | Friuli, Italy 5/6/1976 | 6.5 | Tolmezzo, 270 | C | 0.315 | 30.80 | 5.09 | 1.00 | 0.00 | 0.00 |
| 5 | Imperial valley 10/15/1979 | 6.9 | DELTA, 262 | D | 0.238 | 26.00 | 11.99 | 1.00 | 0.00 | 0.00 |
| 6 | Imperial valley 10/15/1979 | 6.9 | DELTA, 352 | D | 0.351 | 33.02 | 19.03 | 1.00 | 0.50 | 0.00 |
| 7 | Imperial valley 10/15/1979 | 5.2 | EL CENTRO ARRAY #11, | D | 0.364 | 34.44 | 16.08 | 1.00 | 0.00 | 0.00 |
| 8 | Imperial valley 10/15/1979 | 5.2 | EL CENTRO ARRAY #11, 230 | D | 0.38 | 42.14 | 18.63 | 1.00 | 0.00 | 0.00 |
| 9 | SuperstitionHills02 11/24/87 | 6.5 | EL CENTRO IMP CO CENTER, 000 | B | 0.358 | 46.36 | 17.53 | 1.00 | 0.00 | 0.00 |
| 10 | SuperstitionHills02 11/24/87 | 6.5 | EL CENTRO IMP CO CENTER, 090 | B | 0.258 | 40.87 | 20.1 | 0.50 | 0.00 | 0.00 |
| 11 | SuperstitionHills02 11/24/87 | 6.5 | POE, 270 | B | 0.446 | 35.80 | 8.82 | 1.00 | 0.00 | 0.00 |
| 12 | SuperstitionHills02 11/24/87 | 6.5 | POE, 360 | B | 0.3 | 32.80 | 11.28 | 1.00 | 0.50 | 0.00 |
| 13 | LOMA PRIETA 10/18/89 | 7.1 | CAPITOLA, 000 | | 0.529 | 35.01 | 9.13 | 1.00 | 0.50 | 0.50 |
| 14 | LOMA PRIETA 10/18/89 | 7.1 | CAPITOLA, 090 | | 0.443 | 29.21 | 5.49 | 1.00 | 0.50 | 0.50 |
| 15 | LOMA PRIETA 10/18/89 | 7.1 | GILROY ARRAY #3, 000 | D | 0.555 | 35.69 | 8.26 | 1.00 | 1.00 | 0.50 |
| 16 | LOMA PRIETA 10/18/89 | 7.1 | GILROY ARRAY #3, 090 | D | 0.367 | 44.67 | 19.33 | 1.00 | 0.50 | 0.00 |

(*continued*)

**Table 5.** (*continued*)

| No. | Earthquake | $M_w$ | Record and component | Site | PGA (g) | PGV (cm/s) | PGD (cm) | LD | SD | NC |
|---|---|---|---|---|---|---|---|---|---|---|
| 17 | CAPE MENDOCINO 04/25/92 | 7.0 | RIO DELL OVERPASS FF, 360 | D | 0.549 | 42.00 | 19.55 | 1.00 | 0.00 | 0.00 |
| 18 | CAPE MENDOCINO 04/25/92 | 7.0 | RIO DELL OVERPASS FF, 270 | D | 0.195 | 10.54 | 7.02 | 1.00 | 0.00 | 0.00 |
| 19 | LANDERS 7/23/92 | 7.3 | COOLWATER, LN | C | 0.283 | 25.64 | 13.71 | 1.00 | 0.00 | 0.00 |
| 20 | LANDERS 7/23/92 | 7.3 | COOLWATER, TR | C | 0.417 | 42.34 | 13.81 | 1.00 | 0.00 | 0.00 |
| 21 | LANDERS 06/28/92 | 7.3 | YERMO FIRE STATION, 270 | D | 0.245 | 51.44 | 43.85 | 0.50 | 0.00 | 0.00 |
| 22 | LANDERS 06/28/92 | 7.3 | YERMO FIRE STATION, 360 | D | 0.152 | 29.71 | 24.63 | 0.00 | 0.00 | 0.00 |
| 23 | NORTHRIDGE EQ 1/17/94 | 6.7 | BEVERLY HILLS - 12520 MULH, 035 | | 0.617 | 40.86 | 8.57 | 1.00 | 0.50 | 0.00 |
| 24 | NORTHRIDGE EQ 1/17/94 | 6.7 | BEVERLY HILLS - 12520 MULH, 125 | | 0.444 | 30.19 | 4.83 | 1.00 | 0.50 | 0.00 |
| 25 | NORTHRIDGE EQ 1/17/94 | 6.7 | BEVERLY HILLS - 14145 MULH, 009 | C | 0.416 | 58.94 | 13.15 | 1.00 | 0.50 | 0.00 |
| 26 | NORTHRIDGE EQ 1/17/94 | 6.7 | BEVERLY HILLS - 14145 MULH, 279 | C | 0.516 | 62.78 | 11.07 | 1.00 | 0.50 | 0.00 |
| 27 | NORTHRIDGE EQ 1/17/94 | 6.7 | CANYON - W LOST CANYON, 000 | D | 0.41 | 43.03 | 11.71 | 1.00 | 0.00 | 0.00 |
| 28 | NORTHRIDGE EQ 1/17/94 | 6.7 | CANYON - W LOST CANYON, 270 | D | 0.482 | 45.38 | 12.54 | 1.00 | 1.00 | 0.50 |
| 29 | KOBE 01/16/95 | 6.9 | NISHI-AKASHI, 000 | D | 0.509 | 37.29 | 9.53 | 1.00 | 0.00 | 0.00 |
| 30 | KOBE 01/16/95 | 6.9 | NISHI-AKASHI, 090 | D | 0.503 | 36.67 | 11.26 | 1.00 | 0.50 | 0.00 |
| 31 | KOBE 01/16/95 | 6.9 | SHIN-OSAKA, 000 | D | 0.243 | 37.86 | 8.55 | 1.00 | 0.00 | 0.00 |
| 32 | KOBE 01/16/95 | 6.9 | SHIN-OSAKA, 090 | D | 0.212 | 27.94 | 7.64 | 0.50 | 0.00 | 0.00 |

(*continued*)

**Table 5.** (*continued*)

| No. | Earthquake | $M_w$ | Record and component | Site | PGA (g) | PGV (cm/s) | PGD (cm) | LD | SD | NC |
|---|---|---|---|---|---|---|---|---|---|---|
| 33 | KOCAELI 08/17/99 | 7.4 | ARCELIK, 000 (KOERI) | C | 0.219 | 17.69 | 13.65 | 1.00 | 0.00 | 0.00 |
| 34 | KOCAELI 08/17/99 | 7.4 | ARCELIK, 090 (KOERI) | C | 0.15 | 39.55 | 35.58 | 0.00 | 0.00 | 0.00 |
| 35 | KOCAELI 08/17/99 | 7.4 | DUZCE, 180 (ERD) | D | 0.312 | 58.88 | 44.13 | 1.00 | 0.00 | 0.00 |
| 36 | KOCAELI 08/17/99 | 7.4 | DUZCE, 270 (ERD) | D | 0.358 | 46.39 | 17.62 | 1.00 | 0.00 | 0.00 |
| 37 | CHI-CHI 09/20/99 | 7.6 | CHY101, E | D | 0.353 | 70.64 | 45.3 | 1.00 | 0.00 | 0.00 |
| 38 | CHI-CHI 09/20/99 | 7.6 | CHY101, N | D | 0.44 | 115.00 | 68.76 | 1.00 | 0.00 | 0.00 |
| 39 | CHI-CHI 09/20/99 | 7.6 | TCU045, E | C | 0.474 | 36.70 | 50.68 | 1.00 | 0.00 | 0.00 |
| 40 | CHI-CHI 09/20/99 | 7.6 | TCU045, N | C | 0.512 | 39.09 | 14.35 | 1.00 | 0.00 | 0.00 |
| 41 | DUZCE 11/12/99 | 7.1 | BOLU, 000 (ERD) | D | 0.728 | 56.49 | 23.07 | 1.00 | 0.50 | 0.00 |
| 42 | DUZCE 11/12/99 | 7.1 | BOLU, 090 (ERD) | D | 0.822 | 62.12 | 13.56 | 1.00 | 1.00 | 0.50 |
| 43 | IRAN_MANJIL 06/20/90 | 7.4 | LONGITUDINAL COMP | – | 0.515 | 43.26 | 14.92 | 1.00 | 1.00 | 1.00 |
| 44 | IRAN_MANJIL 06/20/90 | 7.4 | TRANSVERSE COMP | – | 0.496 | 55.55 | 20.83 | 1.00 | 1.00 | 0.50 |
| 45 | HECTOR MINE OCT 16, 1999 | 7.1 | HEC, 000 | – | 0.266 | 28.58 | 22.54 | 1.00 | 0.00 | 0.00 |
| 46 | HECTOR MINE OCT 16, 1999 | 7.1 | HEC, 090 | – | 0.337 | 41.75 | 13.96 | 1.00 | 0.00 | 0.00 |

cases which do not comply with a given performance level divided by the total number of the cases.

According to many modern guidelines, residential buildings are expected to meet "Life Safety" performance level which corresponds to SD limit states level according to Eurocode 8 under design earthquakes. Among the considered records considered herein, Imperial Valley, Northridge, Düzce and Iran records are extremely destructive with exceedance ratio of SD performance level equal to 1.00.

Average exceedance ratio of considered performance levels is summarized in Table 6. In performance evaluation, for average exceedance ratio, the performance level is considered to be satisfied if the average exceedance ratio is lesser than 0.50. As seen in Table 6, LD performance level is exceeded for both directions. The average exceedance ratio for SD is 0.11 and 0.35 respectively, for both orthogonal directions. Although Collapse Prevention (NC) performance level is not desired for the residential buildings, it is an important criterion for limiting casualties and preventing loss of lives during an earthquake. The average exceedance ratio for NC performance level is 0.09.

**Table 6.** Average exceedance ratio of considered performance levels

| Direction | Limited Damage (LD) | Significant Damage (SD) | Near Collapse (NC) |
|---|---|---|---|
| *x*- | 0.93 | 0.11 | 0.02 |
| *y*- | 0.91 | 0.35 | 0.15 |

## 5 Conclusions

The seismic performance of a typical 2-storey URM building of the Albanian masonry stock has been analyzed. The mechanical properties of the materials used have been determined by experimental tests conducted on the selected building. The capacity of the building was calculated by TREMURI software package that uses macro elements for masonry panels. The seismic demands were estimated under a set of 46 selected ground motions.

Capacity curves highlight the influence of the lateral load pattern on building response. From the capacity curves, the most unfavorable response one was selected for the performance evaluation. According to the analyses results, the building has a higher capacity and resistance in *y*- direction. Except the destructive characteristics of Imperial Valley, Northridge, Düzce and Iran records, it can be said that this structure seems to remain safe under other ground motions.

## References


1. Calvi, G.M., Pinho, R., Magenes, G., Boomer, J.J., Restrepo-Velez, L.F., Crowley, H.: Development of seismic vulnerability assessment methodologies over the past 30 years. ISET J. Earthq. **43**(3), 75–104 (2006)
2. Maio, R.A., Ferreira, T.M., Vicente, R.S.: The use of macro-element approach for the seismic vulnerability assessment of stone masonry buildings. In: 3rd International Conference on Protection of Historical Constructions, Lisbon, Portugal, 12–15 July 2017 (2017)
3. Lagomarsino, S., Penna, A., Galasco, A., Cattari, S.: TREMURI program: an equivalent frame model for the nonlinear seismic analysis of masonry buildings. Eng. Struct. **56**, 1787–1799 (2013)
4. CEN, Eurocode 8: design of structures for earthquake resistance – part I: general rules, seismic actions and rules for buildings (EC8), European Committee for Standardization, Brussels, Belgium (2004)
5. Bilgin, H., Huta, E.: Earthquake performance assessment of low and mid-rise buildings: emphasis on URM buildings in Albania. Earthq. Struct. **14**(6), 599–614 (2018). https://doi.org/10.12989/eas.2018.14.6.599
6. American Association State Highway and Transportation Officials Standard, Designation: C67 – 09. Standard Test Methods for Sampling and Testing Brick and Structural Clay Tile (2008)
7. NTC, Decreto Ministeriale 14/1/2008 – Norme techniche per le costruzioni, Ministry of Infrastructures and Transportations, Gazetta Ufficiale S.O. n. 30 on 04/02/2008 (2008). (in Italian)

8. Turnsek, V., Sheppard, P.: The shear and flexural resistance of masonry walls. In: Proceedings of the International Research Conference on Earthquake Engineering, Japan, pp. 517–573 (1980)
9. CEN, Eurocode 8: design of structures for earthquake resistance – Part III: assessment and retrofitting of buildings (EC8-3), European Committee for Standardization, Brussels, Belgium (2005)
10. Inel, M., Mretz, E., Black, E., Aschheim, M., Abrams, D.: USEE 2001: Utility Software for Earthquake Engineering Report and User's Manual (2001)

# Energy Efficiency. Smart Cities

# Collecting and Reviewing Written Resources that Map the Knowledge Triangle for Transferring Research and Innovation on Sustainable Rehabilitation of the Built Environment in Continuing Education

Carlos Rubio-Bellido(✉) and Francisco Javier Guevara-García

University of Seville, 41012 Seville, Spain
carlosrubio@us.es

**Abstract.** The aim of this study is to map an overview of the education, research and innovation on energy efficiency focused on the rehabilitation of the built environment. The research is organized as follows:

- Introduction and contextualization on energy efficiency.
- The main educational resources are depicted regarding (i) the educational levels on energy efficiency, (ii) the professional attributions, (iii) the role of energy manager and (iv) various examples of existing studies.
- The main research and innovation challenges are presented considering (i) the role of Energy Service Providers (ESPs); (ii) Energy Conservation Measures (ECMs); climate adaption and adaptive comfort; (iv) heating, ventilation and air conditioning systems: adaptive setpoints; (v) mixed mode potential; (vi) Net Zero Energy buildings (NZEB); (vii) climate change and (viii) fuel poverty.
- Conclusions of the educational, research and innovation challenges.
- References including the citations in which the main text is supported.

**Keywords:** Energy efficiency · Built environment · Sustainable rehabilitation

## 1 Introduction

As a result of the oil crisis of the 1970s, concerns about the effects of climate change on the planet have exponentially increased. Currently, global warming and the depletion of non-renewable resources are the main concerns in society. Based on these problems, a greater demand on the improvement of the energy performance has been reflected in several sectors, including building sector. This is due to the bad energy behaviour presented by most of the existing building stock [1–4]. In quantified data, the building sector is responsible for approximately 30% of the energy consumption at a global level [5], generating 40% of pollutant gas emissions in the atmosphere [6, 7]. Given


A. Rotaru (Ed.): CRIT-RE-BUILT 2019, SSGG, pp. 377–392, 2021.
https://doi.org/10.1007/978-3-030-61118-7_31

this circumstance, the European Union has set up the steps required to reach a low carbon economy by 2050 [8]. To achieve this goal, the building sector needs to reduce pollutant gas emissions by 90%, among others. Recently, the Directive 2018/844 [9] has established that European countries need to devise strategies of energy renovation of the existing building stock, with the aim of having efficient buildings before 2050.

In this context, education is a crucial agent for sustainable development. Nowadays, the rapid rise in international students and the commitment of the universities are key points towards a low carbon environment [10]. There has been considerable progress in the incorporation of sustainable development and energy efficiency into the curricula of higher education institutions. However, particularly on competences for sustainable development and on pedagogical approaches, there has been limited research on the connection between how courses are delivered and how they may affect sustainability competences [11]. A combination of pedagogical approaches must be considered to cover all the competences in energy efficiency education.

Regarding the specific case existing buildings refurbishment, the use of alternative or low-energy consumption air-conditioning systems is also a very interesting option; in this case, research is being done to explain how the installation of these systems can affect the building's consumption. [14]; under certain climatic conditions, a free-cooling system can provide comfort with an energy consumption close to zero. Another investigation by Christian, Carsten and Mads [15] explores the possibility of having a zero energy consumption building in Denmark, starting from an existing building and installing different air-conditioning options, all of which are 100% renewable.

The organizations competitiveness needs the improvement of the efficiency and energy savings, as well as, a decrement of greenhouse gas emissions that produce the climate change. On the other hand, the increase in efficiency and energy savings decreases energy dependence.

Europe's future growth depends on greater efficiency in the use of resources and, to a large extent, greater energy efficiency. The Energy efficiency directive can be a valuable part of the basis for growth and job creation in the coming years and would limit European imports of expensive fossil fuels.

## 2 Education on Energy Efficiency

### 2.1 Educational Levels on Energy Efficiency

The application of the educational levels on energy efficiency could be classified in:

(a) Professional training. It is in charge of the training of the technicians in energy efficiency (TEE). In Spain there is a specialty of energy efficiency in the upper modules of professional training.
(b) University training.

- Bachelor's programs. All students of technical degrees have professional attributions for the practice of energy efficiency. However, the degree programs do not incorporate enough energy skills for professional practice, so the currently training offered is incomplete. In recent years there are many students who choose this topic for their final undergraduate project.

- Degree in energy management. In Spain there is no specialized degree in energy management.
- Postgraduate/Master's degree. Nowadays it is the main training implemented. This training is used for the speciality of all technical qualifications. Non-technical graduates (graduated in law, physics, chemistry, environmental sciences, geography, etc.) who studies these masters are enabled as auditor and/or energy manager. On the other hand, in recent years there are many students who choose this theme for their master's degree final project.
- Doctorate program. In recent years there are many students who choose this topic for doctoral thesis PhD.

## 2.2 Professional Attributions

In Spain the professional attributions in energy efficiency are classified into three lines:

Technicians in Energy Efficiency (TEE). It is the technician in charge of the execution, control and maintenance of the energy efficiency measures. Currently it is usually a refrigeration or electrician specializing in energy, but the attributions are not regulated by law.

Energy Auditor (EA). It is in charge of the evaluation of the facilities in the actual state, as well as its diagnosis and proposals for improvement.

Energy Manager (EM). It is in charge of the evaluation, audit, planning, execution control and maintenance of the energy conditions of the buildings, as well as of its proposals for improvement. According to the Royal Decree 56/2016, the minimum requirements will be "to be in possession of any degree or master with knowledge of energy."

## 2.3 The Role of Energy Manager

An energy manager (EM) is a professional who studies, from an integral approach, the use of energy of a building or production process. The competences included are:

- Audit.
- Study and implementation of Energy Conservation Measures (ECMs).
- Control of execution and monitoring.
- Maintenance and control of ECMs.

The academic requirements for the obtaining of the Certificated of specialized and advanced studies on EM are Degree or graduate with knowledge of energy.

**Aims of the studies on Energy Manager:**

- Provide the technical personnel of the curricular/educational training or knowledge necessary to enable them as Energy Managers EM.

- Possibility of inclusion in degree programs of the different engineering.

**Competences developed at advanced form:**

- **G01.** Capacity of search, analysis and selection of the information.
- **G02.** Know and be able to apply to the projects the kinds of elements that show initiative, commitment, enthusiasm and capacity of motivation.
- **G03.** Problem-solving Skills.
- **G04.** Be able to communicate their conclusions with the knowledge and reasons that sustain them, to specialized and not specialized public, in a clear way and with no ambiguities.
- **G05.** Obtain skills in the learning, which allow them to continue studying in an autonomous and self-guided way.

**Targets or specific Competences:**
Main reason: Labor market access. Job offers.

1. Curricular/educational training.
2. To promote the energetic management.
3. Basic principles.
4. Basic knowledge.
5. Resolution of practical cases.
6. Economic, legal and administrative aspects.
7. To propitiate the international homologation:

   - CEM (The Certified Energy Manager® (CEM®)).
   - ISO (International Standardization Organization).

8. Interdisciplinary relationship between the different engineering, especially industrial and building engineering.
9. To encourage an interuniversity and international framework:

   - Share courses/information.
   - Share knowledge and experiences.
   - A common academic commission.

## 2.4 Example of Existing Studies

At present, there is no specific degree of energy management or sustainability. The few studies that there are at the graduate level. In the University of Seville, there is one course of continuous training, and two master:

1. Energy adviser (Continuous training). Type of teaching: Continuous training. No access requirements, 6,00 ECTS, quarterly, non-presential.
   Content: Theory. Project practices. Seminars of companies. Practices in companies.
   Module: Theoretical and practical aspects of energy assessment.

2. Energy manager (Master). Type of teaching: Lifelong training (postgraduate), 60 ECTS, annual, non-presential.
   Content: Theory. Project practices. Laboratory Practices and Practical Work Exhibitions. Seminars of companies. Practices in companies. Travels to companies.
   Module: Energy Management. Market. Tools and tools for energy management. Energy saving measures. Solutions and methodologies. Maintenance. Audit Control. Professional activity. Fine work of the master. Internship in a company/institution.
3. Eco-efficient rehabilitation of buildings and neighborhoods (Master). Type of teaching: Lifelong training (postgraduate), 60 ECTS, annual, non-presential.
   Content: Theory. Project practices.
   Module: Introduction to Sustainability, Eco-efficiency, Economics and Politics Environmental. Environmental Planning and Eco-efficient. Eco-efficient Construction Materials and Products. Building and Environment. Passive Systems. Use and Integration of Renewable Energies in Buildings. Eco-efficient Rehabilitation. Final Master's Project.

# 3 Energy Efficiency Research and Innovation Challenges

## 3.1 The Role of Energy Service Providers (ESPs)

The directive 2012/27/UE and the Royal Decree 56/2016 set certain definitions that affect the Energy Service Providers (ESPs), establishing a wide variety of tasks and actions, which allows the provision by the ESPs of multi energy services with different scopes.

It is necessary to develop the energy services market in order to ensure the availability of both the demand and the supply of energy services. Transparency can contribute to this, through lists of ESPs, as set out in the Directive and the Royal Decree, for which it is advisable to classify them, in coherence with the legal provisions and with existing practice and experience. The classification of the ESPs establishes clarity and confidence in the market about the different typology of existing ESPs, based on their actions. The classification also recognizes the existing singularity with respect to the groups of companies, when providing energy services.

The ESP classification aims at the specialization of the different energy service providers, encouraging them to engage different administrations and companies, with the consequent improvement of the energy efficiency, the energy saving promotion and the reduction of greenhouse gas emissions.

## 3.2 Energy Conservation Measures (ECMs)

The adoption of energy conservation measures (ECMs) constitutes one of the most significant performances considering energy efficiency. From the different elements of buildings, those elements from the envelope mostly contribute to the inefficient energy performance of the existing building stock, due to heat losses or gains occurring through them [12–15].

Thus, adopting efficient ECMs by means of the energy analysis of buildings is fundamental to fulfil the objectives of reducing pollutant gases by 2050. However, the efficiency of adopting ECMs in building envelope is currently a study gap, particularly in warm climatic regions. Most studies has focused on the analysis of the building envelope improvement in cold or mild climatic regions. Some of these studies are as follows: (i) Aksoy and Inalli [16] analysed the influence of passive design parameters, such as the shape factor and the orientation position, in a building located in a cold region of Turkey; (ii) Invidiata et al. [17] analysed the influence of six design strategies of a residential building located in the north of Italy in future scenarios of climate change. These authors studied these strategies from the perspectives of adaptive thermal comfort, evaluation of the life cycle, cost analysis of the life cycle, and the multicriteria decision making, with the aim of selecting the best option for the sustainability improvement of the building; and (iii) Bhikhoo et al. [18] carried out a sensibility analysis in different design aspects of a typical dwelling in Thailand, which is located in the wet-dry tropical climate region (Aw class according to Köppen-Geiger climate classification [19]). Results showed a great influence on the placing of insulating material at the ceiling or the inclusion of balconies in the design.

Moreover, most of these studies focused on public buildings, such as offices or shops: (i) Spyropoulos and Balaras [20] analysed the energy performance of 39 office buildings in Greece, determining the most important aspects; (ii) Rubio-Bellido et al. [21] studied the influence of office buildings on the energy demand in future scenarios, determining that the relationship of the shape and the relationship window-wall can significantly influence the decrease of the energy demand during the design phase of these buildings; (iii) Ge et al. [22] analysed different optimization strategies of energy efficiency, such as the envelope improvement or solar protection, in a research building located in the city of Hangzhóu, in the Cfa climate zone.

### 3.3 Climate Adaption and Adaptive Comfort

Until now, the integration of the climatic adaptability of the users and the low energy consumption has been raised through the adaptive comfort models. These models are set in the international standards EN 15251:2007 [23] and ASHRAE 55-2017 [24], which consider that users can interact with the environment. These models present limitations in warm climates, if the trend to global warming is considered [25, 26].

Their advantage, when compared to models of static comfort, [27] relies on the fact that occupants of naturally ventilated buildings may have a wider comfort range [28] than those who are accustomed to centralized HVAC systems. In addition, current adaptive standards, which are derived from studies carried out by the RP-884 and SCATs projects, prove that indoor temperatures in naturally ventilated spaces are closely related to the external temperatures. In particular, a variety of studies have established the relation between vernacular architecture and its surrounding environment in a way that adaptive comfort plays a crucial role [29]; besides, locally available construction materials are also important [30] and, considering adaptive comfort, even standards for zero energy buildings can be achieved [31].

### 3.4 Heating, Ventilation and Air Conditioning Systems: Adaptive Setpoints

Another aspect to highlight from the existing scientific literature are the setpoint temperatures used in the modellings analysed. The modification of setpoint temperature values has a significant influence on the energy consumption [32]. In this sense, some studies highlighting this aspect are as follows: (i) Spyropoulos and Balaras [33] established setpoint temperatures of 20 °C for heating and 26 °C for cooling in bank branches, according to the national legislation for public buildings in Greece. The results obtained a decrease of the energy consumption for HVAC by 45%; (ii) Hoyt et al. [34] used setpoint temperatures of 18.3 and 27.87 °C for heating and cooling, respectively, in an office building located in 7 different climate zones. By using these setpoint temperatures, a saving between 32 and 73% was achieved; and (iii) Wan et al. [35] studied the impact of climate change on office buildings in subtropical climates and the influence of the setpoint temperatures used. By using setpoint temperatures for cooling higher than 25.5 °C, reductions of the energy demand in different future scenarios were obtained.

Thus, variations of these setpoint temperatures can lead to a variation in amortization periods of the ECMs to be carried out, since the energy consumption varies. However, in most of the studies mentioned above, setpoint temperatures were based on the index Predicted Mean Vote (PMV). In recent years, several research studies have highlighted the importance of using adaptive setpoint temperatures, which can be defined as setpoint temperatures with the aim of keeping the internal operative temperature within the adaptive comfort limits. These research works focus on the application of adaptive comfort models from ASHRAE 55 [36] and EN 15251 [37] in setpoint temperatures, analysing the advantages and limitations presented with respect to the models based on the PMV. Some of these research studies are as follows: (i) Sánchez-García et al. [38] studied the use of adaptive setpoint temperatures in future climate scenarios with the objective of reducing the energy demand in office buildings; (ii) Holmes and Hacker [39] analysed the application of the adaptive thermal comfort approach in different office buildings in United Kingdom, both in current and future scenarios; and (iii) Kramer et al. [40] assigned the lower limit of the comfort area of the model developed by van der Linden et al. [41] for Holland. This model was established in the standard ISSO 74 [42] to the heating setpoint temperature of a museum, which obtained a reduction of the energy consumption by 74%. However, there is a lack of research studies on this field in Spain: (iv) Sánchez-Guevara Sánchez et al. [43] applied the adaptive comfort model from ASHRAE 55-2013 by means of setpoint temperatures that monthly varied, and reductions by 20% in the heating energy demand as well as of 80% in the cooling energy demand were obtained; (v) Barbadilla-Martín et al. [44] compared energy demands of a building with mixed mode by using usual setpoint temperatures as well as setpoint temperatures based on the neutral temperature of a thermal comfort model previously developed in the city of Seville [45]. Usual average setpoint temperatures were 23.5 °C and 22.3 °C for cooling and heating, respectively, whereas the average neutral temperatures were 24 °C and 21 °C for cooling and heating. The results showed reductions by 27.5% and 11.4% in cooling and heating, respectively.

### 3.5 Mixed Mode Potential

Mixed mode buildings have already been widely studied in terms of energy savings [46–48]. However, in terms of thermal comfort, there is still no consensus regarding how mixed mode buildings should be assessed considering natural ventilation and air conditioning separately. There is a similar lack of consensus about whether the adaptive comfort approach is suitable for both modes of operation. Several field studies in mixed mode buildings have been made, although there has been disagreement when some found a unique model for both modes of operation [49, 50] and others did not [51]. Other authors considered that mixed mode buildings should be studied separately for both modes of operation, so developed comfort models for air conditioning and natural ventilation [52–54]. The results were consistent with this, with occupants' actual thermal sensation and acceptance varying when operation modes changed from air conditioning to natural ventilation, or vice versa. But, in general, the adaptive model was more suitable for mixed mode buildings [55]. Furthermore, it is necessary to consider that people might be used to using natural ventilation or air conditioning. In this way, people who tried to adapt their thermal sensation using adaptive comfort strategies more than others were found to switch on the air conditioning less frequently [56]. Since adaptive models might be more suitable for mixed mode buildings, there must be adaptive opportunities. Otherwise, perceptions of overheating could be intensified [57].

### 3.6 Net Zero Energy Buildings

From 2020 new buildings will be NZEB and it will be promoted deep renovations in existing buildings in order to reduce to the maximum their energy consumption [58]. It has been widely worked on the concept of NZEB hitherto and [59] and an international and national standard framework has been generated [60].

In this context, the European Union has established annual thresholds for the use of primary energy consumption and minimum energy production from renewable sources in situ according to climatic zones. Therefore, it is necessary to tack the methodology proposed for the implementation of NZEB taking into account other influential parameters such as the preferences of the users, the indoor environmental quality, the variation of the external conditions in the context of the climate change and socio-economic and cultural factors [61].

### 3.7 Climate Change

Climate change has been profoundly analyzed because of its influence on the construction industry. For this reason, it has become a matter of attention for both governments and research institutions over the last few years. Starting with the foundation of the Intergovernmental Panel on Climate Change (IPCC) in 1988, which has recently published its Fifth assessment report (AR5) [62], there have been numerous studies that focus on climate change, the increase of gas emissions and the scarcity of natural resources. Along this line, sundry prediction models have been generated for various climate scenarios [63]. Most of these models have been developed in the United Kingdom [64], although they have increasingly extended throughout the international framework

[65, 66]. Currently, the IPCC, supported by the United Nations Environment Programme (UNEP) and the World Meteorological Organization (WMO), which is the most widely accepted organization in this matter, envisages multiple emission scenarios for the near future [67]. According to several research projects [68–70], patterns for energy demand and consumption may be altered by climate change; therefore, they should be taken into account when devising schemes for energy supply and should also be carefully considered in order to prevent energy poverty.

### 3.8 Fuel Poverty

Energy poverty, commonly linked with lack of access to energy sources and/or a deficient coverage of energy needs may be quantified in different ways, mainly considering high prices of fuel or energy, low building performance and low level of income. (Thomson and Snell [92]).

The origin, evolution and discussion about the definition of energy poverty can be found in some key publications in this field [71]. For the sake of clarity in the discourse, the definition given by B. Boardman, which is considered a key text in the fuel poverty debate, is adopted in this manuscript [72]. Boardman states that "fuel poverty" can be defined as "the inability to afford adequate warmth because of the inefficiency of the home"; the threshold for affordability is established at 10%, twice the median household expenditure on fuel in 1991 [73]. Since then, despite the 10% threshold being broadly used as a general criteria to evaluate fuel poverty [74–77], many authors have raised their concerns about the real significance of this figure. Methods for measuring energy poverty can be classified into three groups: gathering the impressions about energy services in homes, analysing expenditure and the direct measure of certain energy services.

Several studies have come to light in recent years analysing fuel poverty and their implications in a variety of national contexts, such as Greece [78], Italy [79], Ireland [80, 81], India [82], Japan [83] and the UK [84, 85]. Others have placed focus on their implications for health [86] risks for adults [87] and the death rate [88]. From a perspective of geography [89], these have focused on rural areas [84, 90], the vulnerability of homes [75, 91], and even from a social justice point of view [92].

In recent years, several studies, which are more focused on the development of new indexes to measure fuel poverty, have come to light in order to clarify some issues pertaining to this term, as well as their implications for public policies:

The Building Fuel Poverty Index ($B_{FP}$) [79], which establishes a relationship between building energy performance and the risk of fuel poverty in the Italian context, in order to quantify the number of existing buildings that need economic subsidies.

The Progress out of Poverty Index (PPI) [93] based on a Rwandan case-study, which consists of 10 weighted questions regarding the characteristics of households, whose primary sources can be observed and scored with respect to a poverty line.

The Sustainable Energy Development Index (SEDI) [94], a composite index focused on establishing the sustainability level of both intra- and inter-generational needs and correlated to both the Human Development Index (HDI) and the Energy Development Index (EDI).

The Energy poverty comprehensive evaluation index [95], focused on identifying the characteristics of China's energy poverty both at the national and regional levels, as well as the policy implications of energy poverty alleviation.

Those studies provide a solid scientific base which proves that fuel poverty is present nowadays and can have detrimental effects on occupants and be a burden on the finances of lower-income households [96].

## 4 Conclusions

This study was focused on a comprehensive understanding of the education, research and innovation on energy efficiency with the particular approach of the rehabilitation of the built environment.

Considering the educational resources, various initiatives are now focused on the rehabilitation on the built environment in different educational levels, professional attributions and the appearance of the energy manager as key factor in the future development of education. In this sense, the examples shown an effort but a large room for improvement at educational level.

Regarding the research and innovation challenges, a holistic approach was considered. To accomplish an affordable and sustainable rehabilitation of the built environment, the role of Energy Service Providers (ESPs) es crucial to properly applied Energy Conservation Measures (ECMs). In this sense, there are various important factors in a climate change and fuel poverty context that must be considered in a low energy consumption approach.

This study paves the way for specific educational and research initiatives which contribute to a sustainable and energy efficiency approach for the rehabilitation of the built environment.

## References

1. Horne, R., Hayles, C.: Towards global benchmarking for sustainable homes: an international comparison of the energy performance of housing. J. Hous. Built. Environ. **23**, 119–130 (2008). https://doi.org/10.1007/s10901-008-9105-1
2. Kurtz, F., Monzón, M., López-Mesa, B.: Energy and acoustics related obsolescence of social housing of Spain's post-war in less favoured urban areas. The case of Zaragoza. Inf la Construcción 67:m021 (2015). https://doi.org/10.3989/ic.14.062
3. Lowe, R.: Technical options and strategies for decarbonizing UK housing. Build. Res. Inf. **35**, 412–425 (2007). https://doi.org/10.1080/09613210701238268
4. Park, K., Kim, M.: Energy demand reduction in the residential building sector: a case study of Korea. Energies **10**, 1–11 (2017). https://doi.org/10.3390/en10101506
5. The United Nations Environment Programme: Building Design and Construction: Forging Resource Efficiency and Sustainable, Nairobi, Kenya (2012)
6. European Commission: Directive 2002/91/EC of the European Parliament and of the Council of 16 December 2002 on the energy performance of buildings, Brussels, Belgium (2002)

7. European Union: Directive 2010/31/EU of the European Parliament and of the Council of 19 May 2010 on the Energy Performance of Buildings, Brussels, Belgium (2010)
8. European Commission: A roadmap for moving to a competitive low carbon economy in 2050, Brussels, Belgium (2011)
9. European Union: Directive 2018/844 of the European parliament and of the council of 30 May 2018 amending directive 2010/31/EU on the energy performance of buildings and directive 2012/27/EU on energy efficiency (2018)
10. Shields, R.: The sustainability of international higher education: student mobility and global climate change. J. Clean. Prod. **217**, 594–602 (2019). https://doi.org/10.1016/j.jclepro.2019.01.291
11. Lozano, R., Barreiro-Gen, M., Lozano, F., Sammalisto, K.: Teaching sustainability in European higher education institutions: assessing the connections between competences and pedagogical approaches. Sustainability **11**, 1602 (2019). https://doi.org/10.3390/su11061602
12. De Lieto, V.R., Guattari, C., Evangelisti, L., et al.: Building energy performance analysis: a case study. Energy Build. **87**, 87–94 (2015). https://doi.org/10.1016/j.enbuild.2014.10.080
13. Escorcia, O., García, R., Trebilcock, M., et al.: Envelope improvements for energy efficiency of homes in the South-central Chile. Inf la Construcción **64**, 563–574 (2012). https://doi.org/10.3989/ic.11.143
14. Friedman, C., Becker, N., Erell, E.: Energy retrofit of residential building envelopes in Israel: a cost-benefit analysis. Energy **77**, 183–193 (2014). https://doi.org/10.1016/j.energy.2014.06.019
15. Pacheco, R., Ordóñez, J., Martínez, G.: Energy efficient design of building: a review. Renew. Sustain. Energy Rev. **16**, 3559–3573 (2012). https://doi.org/10.1016/j.rser.2012.03.045
16. Aksoy, U.T., Inalli, M.: Impacts of some building passive design parameters on heating demand for a cold region. Build. Environ. **41**, 1742–1754 (2006). https://doi.org/10.1016/j.buildenv.2005.07.011
17. Invidiata, A., Lavagna, M., Ghisi, E.: Selecting design strategies using multi-criteria decision making to improve the sustainability of buildings. Build. Environ. **139**, 58–68 (2018). https://doi.org/10.1016/j.buildenv.2018.04.041
18. Bhikhoo, N., Hashemi, A., Cruickshank, H.: Improving thermal comfort of low-income housing in Thailand through passive design strategies. Sustain **9**, 1–23 (2017). https://doi.org/10.3390/su9081440
19. Rubel, F., Kottek, M.: Observed and projected climate shifts 1901–2100 depicted by world maps of the Köppen-Geiger climate classification. Meteorol. Z. **19**, 135–141 (2010). https://doi.org/10.1127/0941-2948/2010/0430
20. Spyropoulos, G.N., Balaras, C.A.: Energy consumption and the potential of energy savings in Hellenic office buildings used as bank branches - a case study. Energy Build. **43**, 770–778 (2011). https://doi.org/10.1016/j.enbuild.2010.12.015
21. Rubio-Bellido, C., Pérez-Fargallo, A., Pulido-Arcas, J.A.: Optimization of annual energy demand in office buildings under the influence of climate change in Chile. Energy **114**, 569–585 (2016). https://doi.org/10.1016/j.energy.2016.08.021
22. Ge, J., Wu, J., Chen, S., Wu, J.: Energy efficiency optimization strategies for university research buildings with hot summer and cold winter climate of China based on the adaptive thermal comfort. J. Build. Eng. **18**, 321–330 (2018). https://doi.org/10.1016/j.jobe.2018.03.022
23. CEN EN 15251: Indoor environmental input parameters for design and assessment of energy performance of buildings - addressing indoor air quality, thermal environment, lighting and acoustics. Eur. Commun. Stand. **3**, 1–52 (2007). https://doi.org/10.1520/E2019-03R13.Copyright

24. American Society of Heating R and ACE (ASHRAE): ASHRAE standard 55-2017 thermal environmental conditions for human occupancy. American Society of Heating, Refrigerating and Air Conditioning Engineers, Atlanta GA (2017)
25. Sánchez-García, D., Rubio-Bellido, C., Pulido-Arcas, J.A., et al.: Adaptive comfort models applied to existing dwellings in mediterranean climate considering global warming. Sustainability **10**, 3507 (2018). https://doi.org/10.3390/su10103507
26. Barbosa, R., Vicente, R., Santos, R.: Climate change and thermal comfort in Southern Europe housing: a case study from Lisbon. Build. Environ. **92**, 440–451 (2015). https://doi.org/10.1016/j.buildenv.2015.05.019
27. ISO: ISO 7730: ergonomics of the thermal environment analytical determination and interpretation of thermal comfort using calculation of the PMV and PPD indices and local thermal comfort criteria. Management **3**, 605–615 (2005). https://doi.org/10.1016/j.soildyn.2004.11.005
28. Feriadi, H., Wong, N.H.: Thermal comfort for naturally ventilated houses in Indonesia. Energy Build. **36**, 614–626 (2004)
29. Singh, M.K., Mahapatra, S., Atreya, S.K.: Thermal performance study and evaluation of comfort temperatures in vernacular buildings of North-East India. Build. Environ. **45**, 320–329 (2010). https://doi.org/10.1016/j.buildenv.2009.06.009
30. Desogus, G., Di Benedetto, S., Ricciu, R.: The use of adaptive thermal comfort models to evaluate the summer performance of a mediterranean earth building. Energy Build. **104**, 350–359 (2015). https://doi.org/10.1016/j.enbuild.2015.07.020
31. Attia, S., Carlucci, S.: Impact of different thermal comfort models on zero energy residential buildings in hot climate. Energy Build. **102**, 117–128 (2015). https://doi.org/10.1016/j.enbuild.2015.05.017
32. Ren, Z., Chen, D.: Modelling study of the impact of thermal comfort criteria on housing energy use in Australia. Appl. Energy **210**, 152–166 (2018). https://doi.org/10.1016/j.apenergy.2017.10.110
33. Spyropoulos, G.N., Balaras, C.A.: Energy consumption and the potential of energy savings in Hellenic office buildings used as bank branches - a case study. Energy Build. **43**, 770–778 (2011). https://doi.org/10.1016/j.enbuild.2010.12.015
34. Hoyt, T., Arens, E., Zhang, H.: Extending air temperature setpoints: simulated energy savings and design considerations for new and retrofit buildings. Build. Environ. **88**, 89–96 (2014). https://doi.org/10.1016/j.buildenv.2014.09.010
35. Wan, K.K.W., Li, D.H.W., Lam, J.C.: Assessment of climate change impact on building energy use and mitigation measures in subtropical climates. Energy **36**, 1404–1414 (2011). https://doi.org/10.1016/j.energy.2011.01.033
36. American National Standards Institute/American Society of Heating Refrigerating and Air-Conditioning Engineers (ANSI/ASHRAE): ANSI/ASHRAE Standard 55-2013. Thermal Environmental Conditions for Human Occupancy (2013)
37. European Committee for Standardization EN 15251: Indoor environmental input parameters for design and assessment of energy performance of buildings - addressing indoor air quality, thermal environment, lighting and acoustics, Brussels, Belgium (2007)
38. Sánchez-García, D., Rubio-Bellido, C., Marrero Meléndez, M., et al.: El control adaptativo en instalaciones existentes y su potencial en el contexto del cambio climático. Hábitat Sustentable **7**, 06–17 (2017)
39. Holmes, M.J., Hacker, J.N.: Climate change, thermal comfort and energy: meeting the design challenges of the 21st century. Energy Build. **39**, 802–814 (2007). https://doi.org/10.1016/j.enbuild.2007.02.009

40. Kramer, R.P., Maas, M.P.E., Martens, M.H.J., et al.: Energy conservation in museums using different setpoint strategies: a case study for a state-of-the-art museum using building simulations. Appl. Energy **158**, 446–458 (2015). https://doi.org/10.1016/j.apenergy.2015.08.044
41. van der Linden, A.C., Boerstra, A.C., Raue, A.K., et al.: Adaptive temperature limits: a new guideline in The Netherlands: a new approach for the assessment of building performance with respect to thermal indoor climate. Energy Build. **38**, 8–17 (2006). https://doi.org/10.1016/j.enbuild.2005.02.008
42. Arets MJP: Thermische behaaglijkheid : eisen voor de binnentemperatuur in gebouwen : een nieuwe richtlijn voor thermische behaaglijkheid in (kantoor)gebouwen. ISSO (2004)
43. Sánchez-Guevara Sánchez, C., Mavrogianni, A., Neila González, F.J.: On the minimal thermal habitability conditions in low income dwellings in Spain for a new definition of fuel poverty. Build. Environ. **114**, 344–356 (2017). https://doi.org/10.1016/j.buildenv.2016.12.029
44. Barbadilla-Martín, E., Guadix Martín, J., Salmerón Lissén, J.M., et al.: Assessment of thermal comfort and energy savings in a field study on adaptive comfort with application for mixed mode offices. Energy Build. **167**, 281–289 (2017). https://doi.org/10.1016/j.enbuild.2018.02.033
45. Barbadilla-Martín, E., Salmerón Lissén, J.M., Martín, J.G., et al.: Field study on adaptive thermal comfort in mixed mode office buildings in southwestern area of Spain. Build. Environ. **123** (2017). https://doi.org/10.1016/j.buildenv.2017.06.042
46. Salcido, J.C., Raheem, A.A., Issa, R.R.A.: From simulation to monitoring: evaluating the potential of mixed-mode ventilation (MMV) systems for integrating natural ventilation in office buildings through a comprehensive literature review. Energy Build. **127**, 1008–1018 (2016). https://doi.org/10.1016/j.enbuild.2016.06.054
47. Ezzeldin, S., Rees, S.J.: The potential for office buildings with mixed-mode ventilation and low energy cooling systems in arid climates. Energy Build. **65**, 368–381 (2013). https://doi.org/10.1016/j.enbuild.2013.06.004
48. Chen, J., Augenbroe, G., Song, X.: Evaluating the potential of hybrid ventilation for small to medium sized office buildings with different intelligent controls and uncertainties in US climates. Energy Build. **158**, 1648–1661 (2018). https://doi.org/10.1016/j.enbuild.2017.12.004
49. Barbadilla-Martín, E., Salmerón Lissén, J.M., Guadix Martín, J., et al.: Field study on adaptive thermal comfort in mixed mode office buildings in Southwestern area of Spain. Build. Environ. **123**, 163–175 (2017). https://doi.org/10.1016/j.buildenv.2017.06.042
50. Manu, S., Shukla, Y., Rawal, R., et al.: Field studies of thermal comfort across multiple climate zones for the subcontinent: India Model for Adaptive Comfort (IMAC). Build. Environ. **98**, 55–70 (2016). https://doi.org/10.1016/j.buildenv.2015.12.019
51. Rupp, R.F., de Dear, R., Ghisi, E.: Field study of mixed-mode office buildings in Southern Brazil using an adaptive thermal comfort framework. Energy Build. **158**, 1475–1486 (2018). https://doi.org/10.1016/j.enbuild.2017.11.047
52. Indraganti, M., Ooka, R., Rijal, H.B.: Field investigation of comfort temperature in Indian office buildings: a case of Chennai and Hyderabad. Build. Environ. **65**, 195–214 (2013). https://doi.org/10.1016/j.buildenv.2013.04.007
53. Indraganti, M., Ooka, R., Rijal, H.B., Brager, G.S.: Adaptive model of thermal comfort for offices in hot and humid climates of India. Build. Environ. **74**, 39–53 (2014). https://doi.org/10.1016/j.buildenv.2014.01.002
54. Oropeza-Perez, I., Petzold-Rodriguez, A.H., Bonilla-Lopez, C.: Adaptive thermal comfort in the main Mexican climate conditions with and without passive cooling. Energy Build. **145**, 251–258 (2017). https://doi.org/10.1016/j.enbuild.2017.04.031

55. Luo, M., Cao, B., Damiens, J., et al.: Evaluating thermal comfort in mixed-mode buildings: a field study in a subtropical climate. Build. Environ. **88**, 46–54 (2015). https://doi.org/10.1016/j.buildenv.2014.06.019
56. Kim, J., de Dear, R., Parkinson, T., Candido, C.: Understanding patterns of adaptive comfort behaviour in the Sydney mixed-mode residential context. Energy Build. **141**, 274–283 (2017). https://doi.org/10.1016/j.enbuild.2017.02.061
57. Thomas, L.: Combating overheating: mixed-mode conditioning for workplace comfort. Build. Res. Inf. **45**, 176–194 (2017). https://doi.org/10.1080/09613218.2017.1252617
58. European Commission: COMMISSION RECOMMENDATION (EU) 2016/1318 of 29 July 2016 on guidelines for the promotion of nearly zero-energy buildings and best practices to ensure that, by 2020, all new buildings are nearly zero-energy buildings (2016)
59. D'Agostino, D., Mazzarella, L.: What is a nearly zero energy building? Overview, implementation and comparison of definitions. J. Build. Eng. **21**, 200–212 (2019). https://doi.org/10.1016/j.jobe.2018.10.019
60. European Commission: Directiva (UE) 2018/844 por la que se modifica la Directiva 2010/31/UE relativa a la eficiencia energética de los edificios y la Directiva 2012/27/UE relativa a la eficiencia energética. D Of la Unión Eur 156:75–91 (2018)
61. Attia, S., Eleftheriou, P., Xeni, F., et al.: Overview and future challenges of nearly zero energy buildings (nZEB) design in Southern Europe. Energy Build. **155**, 439–458 (2017). https://doi.org/10.1016/J.ENBUILD.2017.09.043
62. IPCC: Climate Change 2014: Synthesis report. Contribution of working groups I, II and III to the fifth assessment report of the intergovernmental panel on climate change (2014)
63. Jentsch, M.F., Bahaj, A.S., James, P.A.B.: Climate change future proofing of buildings—generation and assessment of building simulation weather files. Energy Build. **40**, 2148–2168 (2008). https://doi.org/10.1016/j.enbuild.2008.06.005
64. Mylona, A.: The use of UKCP09 to produce weather files for building simulation. Build. Serv. Eng. Res. Technol. **33**, 51–62 (2012). https://doi.org/10.1177/0143624411428951
65. Guan, L.: Preparation of future weather data to study the impact of climate change on buildings. Build. Environ. **44**, 793–800 (2009). https://doi.org/10.1016/j.buildenv.2008.05.021
66. Jentsch, M.F., James, P.A.B., Bourikas, L., Bahaj, A.S.: Transforming existing weather data for worldwide locations to enable energy and building performance simulation under future climates. Renew. Energy **55**, 514–524 (2013). https://doi.org/10.1016/j.renene.2012.12.049
67. IPCC Data Distribution Centre. www.ipcc-data.org. Accessed 15 Feb 2016
68. Kalvelage, K., Passe, U., Rabideau, S., Takle, E.S.: Changing climate: the effects on energy demand and human comfort. Energy Build. **76**, 373–380 (2014). https://doi.org/10.1016/j.enbuild.2014.03.009
69. Wang, H., Chen, Q.: Impact of climate change heating and cooling energy use in buildings in the United States. Energy Build. **82**, 428–436 (2014). https://doi.org/10.1016/j.enbuild.2014.07.034
70. Rubio-Bellido, C., Pérez-Fargallo, A., Pulido-Arcas, J.A.: Optimization of annual energy demand in office buildings under the influence of climate change in Chile. Energy **114**, 569–585 (2016). https://doi.org/10.1016/j.energy.2016.08.021
71. Robert, S.C., Koh, L., Marchand, R., et al.: Fuel Poverty. Perspectives from the front line, Sheffield (2012)
72. Boardman, B.: Fuel Poverty: From Cold Homes to Affordable Warmth. Wiley, London (1991)
73. ne Liddell C., Morris, C., McKenzie, P., Rae, G.: Fuel Poverty. 1991 – 2012. Commemorating 21 years of action, policy and research, Belfast, UK (2012)
74. Santamouris, M., Paravantis, J.A., Founda, D., et al.: Financial crisis and energy consumption: a household survey in Greece. Energy Build. **65**, 477–487 (2013). https://doi.org/10.1016/j.enbuild.2013.06.024

75. Walker, R., McKenzie, P., Liddell, C., Morris, C.: Estimating fuel poverty at household level: an integrated approach. Energy Build. **80**, 469–479 (2014). https://doi.org/10.1016/j.enbuild.2014.06.004
76. O'Sullivan, K.C., Howden-Chapman, P.L., Fougere, G.M.: Fuel poverty, policy, and equity in New Zealand: the promise of prepayment metering. Energy Res. Soc. Sci. **7**, 99–107 (2015). https://doi.org/10.1016/j.erss.2015.03.008
77. Legendre, B., Ricci, O.: Measuring fuel poverty in France: which households are the most fuel vulnerable? Energy Econ. **49**, 620–628 (2015). https://doi.org/10.1016/j.eneco.2015.01.022
78. Atsalis, A., Mirasgedis, S., Tourkolias, C., Diakoulaki, D.: Fuel poverty in Greece: quantitative analysis and implications for policy. Energy Build. **131**, 87–98 (2016). https://doi.org/10.1016/j.enbuild.2016.09.025
79. Fabbri, K.: Building and fuel poverty, an index to measure fuel poverty: an Italian case study. Energy **89**, 244–258 (2015). https://doi.org/10.1016/j.energy.2015.07.073
80. Healy, J.D., Clinch, J.P.: Fuel poverty, thermal comfort and occupancy: results of a national household-survey in Ireland. Appl. Energy **73**, 329–343 (2002). https://doi.org/10.1016/S0306-2619(02)00115-0
81. Healy, J.D.: Housing, Fuel poverty and Health: A Pan-European Analysis. Routledge, London (2004)
82. Bhide, A., Monroy, C.R.: Energy poverty: a special focus on energy poverty in India and renewable energy technologies. Renew. Sustain. Energy Rev. **15**, 1057–1066 (2011). https://doi.org/10.1016/j.rser.2010.11.044
83. Okushima, S.: Measuring energy poverty in Japan, 2004–2013. Energy Policy **98**, 557–564 (2016). https://doi.org/10.1016/j.enpol.2016.09.005
84. Roberts, D., Vera-Toscano, E., Phimister, E.: Fuel poverty in the UK: is there a difference between rural and urban areas? Energy Policy **87**, 216–223 (2015). https://doi.org/10.1016/j.enpol.2015.08.034
85. Rosenow, J., Platt, R., Flanagan, B.: Fuel poverty and energy efficiency obligations – a critical assessment of the supplier obligation in the UK. Energy Policy **62**, 1194–1203 (2013). https://doi.org/10.1016/j.enpol.2013.07.103
86. Lacroix, E., Chaton, C.: Fuel poverty as a major determinant of perceived health: the case of France. Public Health **129**, 517–524 (2015). https://doi.org/10.1016/j.puhe.2015.02.007
87. Sharpe, R.A., Thornton, C.R., Nikolaou, V., Osborne, N.J.: Fuel poverty increases risk of mould contamination, regardless of adult risk perception & ventilation in social housing properties. Environ. Int. **79**, 115–129 (2015). https://doi.org/10.1016/j.envint.2015.03.009
88. Teller-Elsberg, J., Sovacool, B., Smith, T., Laine, E.: Fuel poverty, excess winter deaths, and energy costs in Vermont: Burdensome for whom? Energy Policy **90**, 81–91 (2016). https://doi.org/10.1016/j.enpol.2015.12.009
89. Walker, R., Liddell, C., McKenzie, P., Morris, C.: Evaluating fuel poverty policy in Northern Ireland using a geographic approach. Energy Policy **63**, 765–774 (2013). https://doi.org/10.1016/j.enpol.2013.08.047
90. Pereira, M.G., Freitas, M.A.V., da Silva, N.F.: Rural electrification and energy poverty: empirical evidences from Brazil. Renew. Sustain. Energy Rev. **14**, 1229–1240 (2010). https://doi.org/10.1016/j.rser.2009.12.013
91. Walker, R., Liddell, C., McKenzie, P., et al.: Fuel poverty in Northern Ireland: humanizing the plight of vulnerable households. Energy Res. Soc. Sci. **4**, 89–99 (2014). https://doi.org/10.1016/j.erss.2014.10.001
92. Snell, C., Bevan, M., Thomson, H.: Justice, fuel poverty and disabled people in England. Energy Res. Soc. Sci. **10**, 123–132 (2015). https://doi.org/10.1016/j.erss.2015.07.012
93. Desiere, S., Vellema, W., D'Haese, M.: A validity assessment of the Progress out of Poverty Index (PPI)$^{TM}$. Eval. Progr. Plan. **49**, 10–18 (2015). https://doi.org/10.1016/j.evalprogplan.2014.11.002

94. Iddrisu, I., Bhattacharyya, S.C.: Sustainable energy development index: a multi-dimensional indicator for measuring sustainable energy development. Renew. Sustain. Energy Rev. **50**, 513–530 (2015). https://doi.org/10.1016/j.rser.2015.05.032
95. Wang, K., Wang, Y.-X., Li, K., Wei, Y.-M.: Energy poverty in China: an index based comprehensive evaluation. Renew. Sustain. Energy Rev. **47**, 308–323 (2015). https://doi.org/10.1016/j.rser.2015.03.041
96. Bienvenido-Huertas, D., Rubio-Bellido, C., Pérez-Fargallo, A., Pulido-Arcas, J.A.: Energy saving potential in current and future world built environments based on the adaptive comfort approach. J. Clean. Prod. 119306 (2019). https://doi.org/10.1016/j.jclepro.2019.119306

# A Retrospective Analysis on the Thermal Rehabilitation of a School Building - Case Study

Radu-Aurel Pescaru(✉), Laura Dumitrescu, and Irina Baran

"Gheorghe Asachi" Technical University of Iaşi, 67 Bd. D. Mangeron, Iaşi, Romania
radu-aurel.pescaru@tuiasi.ro

**Abstract.** Increasing the energy efficiency of buildings (public or private), is a major objective of the last 20 years. Among public buildings, schools represent a special group due to their specific requirements (indoor climate quality as thermal comfort, air quality, ventilation rate, visual and acoustic comfort).

Among the non-residential buildings, educational buildings cause high energy consumption and often offer poor quality in terms of energy efficiency. Therefore, the energy retrofitting of existing school buildings has been part of the policy of the European Union and each Member State in recent years.

In Romania, most school buildings were built before 1990 and do not meet today's requirements for energy efficiency and sustainability. For this reason, a large part is subject to the thermal refurbishment process. This process started in 2000 and continues until today.

The paper develops a retrospective assessment of the energy efficiency of a school building that was retrofitted more than 10 years ago according to the standards of that time (with additional thermal insulation layers). The energy performance of the building is compared with the actual demand and also the ventilation rate in the building is analysed by an experimental study with a $CO_2$ gas analyser.

**Keywords:** Education buildings · Thermal rehabilitation · Energy efficiency

## 1 Introduction

Nowadays, due to the growth of the world population and the tendency to move to urban areas, and also due to the massive use of more advanced technology, energy consumption and dependence on it are constantly increasing. Environmental (diminishing fossil energy resources, increasing pollutant emissions, climate change, global warming), economic (ability to pay energy bills) and social concerns (energy security) are becoming increasingly important. Thus, energy conservation has become one of the most important issues on both the political and scientific agenda in recent decades.

The built environment consumes a significant amount of energy (about 40% of final energy consumption according to statistics) and is responsible European Union for a large part of greenhouse gas emissions (36% of total GHGz emissions) in Europe. 17% of the existing buildings in Europe and about 12% of the average energy consumption


A. Rotaru (Ed.): CRIT-RE-BUILT 2019, SSGG, pp. 393–403, 2021.
https://doi.org/10.1007/978-3-030-61118-7_32

of non-residential buildings are in educational buildings, which often have poor energy performance quality. These figures could increase as more than a quarter of the building stock of the 2050s is expected to be built in Europe [1].

The constant and large-scale application of determined and efficient measures to reduce energy consumption in buildings is therefore of strategic importance, both at national and international level. Reducing energy consumption and increasing the use of renewable energy in the EU would enable it to comply with the (UNFCCC Kyoto Protocol Paris Agreement Framework Convention Organization United Nations Climate Change) and also meet some of the basic principles of sustainable development.

## 2 Historical Landmarks to the Concept of Energy Efficiency in Buildings

**In Early Industrially Developed Countries.** The continuous technical and technological developments that began at the end of the $18^{th}$ and the beginning of the $19^{th}$ century with the Industrial Revolution, born in United Kingdom, have spread in Europe to the Asian parts of British Empire and in the United States of America and continue today in the post-industrial society. This transition from manual to mechanical production methods led to the fact that, due to the emerging technologies based on energy consumption, more and more energy was needed for the everyday necessities of human life. The paradigm of lifestyle (and housing) has changed dramatically due to intensive urbanization, the development of large industrial areas, population growth and migration to the cities, the limitation of urban land ownership, the construction of tall buildings and the conversion of food production from traditional agriculture (primary agriculture) to the food industry, etc.

As can be seen from Table 1, the focus on technological development and economic achievements has some undesirable negative effects such as the reduction of natural reserves, air, soil and water pollution, $CO_2$ emissions, large-scale climate change, which have been ratified (highlighted) in various documents:

- The *oil crisis* in 1973, produced a sudden increase in energy prices and the impediments in reaching energy resources, forced societies to find different remedies for energy independence.
- In the early 1970s was introduced the idea of Reuse, Recycle, Reduce or 3R, which was also included in the building design and construction practices.
- The *Brundtland Report* (World Commission on Environment and Development-WCED, 1987) called "Our Common Future" introduced the concept of "sustainability". In this document are articulated the importance, need and constituents of sustainable development. According to this report, the three main components of sustainable development are the economic, environmental and socio-cultural sustainability.
- The *Kyoto Protocol*, meant to reduce the onset of global warming by reducing greenhouse gas concentrations in the atmosphere to "a level that would prevent dangerous anthropogenic interference with the climate system" (Article 2), was signed by 160 nations in 1997, and entered into force on 16 February 2005.

**Table 1.** The main steps of rise and development of the energy efficiency in construction are synthesized by T. Harputlugil [2]

| Timeline | Architectural movement | Social events | Effects of social events | Energy efficiency in buildings |
|---|---|---|---|---|
| Before 20th century | Vernacular architecture | Industrial Revolution | Steam power/Electric energy through the grid/Mass production, specialization/Need for new spaces | Passive systems Experience of the past |
| First quarter of 20th century | Modernism (Futurism) | 1st World War/Usage of elevator/New building materials (reinforced concrete, steel)/Invention of air conditioning | First steps of combined heating, cooling and air conditioning | New systems adaptation, Climate independent design |
| Second & third quarter of 20th century | Modernism | 2nd World War/Nuclear energy Space technology/Establishment of European Economic Community (EEC) | Mass production after the world war/The term "optimization"/Basics of solar energy usage/Basics of geothermal energy usage/Basics of heat pump technologies | Systems integration |
| 20th century 1970–1980 | Post-modernism | 1973 Oil crisis | Back to climate based design principles/Passive house principles/Development and adaptation of computer technology | Conservation of energy |
| 20th century 1980–1990 | High-tech | Destruction of ozone layer/1987 Brundtland report/The term "Sustainability"/Developments of personal computers | Global development models Sun houses Passive design principles Environmental consciousness | Sustainability and energy management |
| 20th century 1990–2000 | Deconstruction | Kyoto agreement/Establishment of European Union (EU) Green building certification systems | Environmental consciousness/Passive House standards/Green buildings/Life cycle analysis/Energy based simulation software/Integrated design/Renewable energy sources/Smart buildings | Variety of energy sources Efficiency of renewable energy sources |

(*continued*)

**Table 1.** (*continued*)

| Timeline | Architectural movement | Social events | Effects of social events | Energy efficiency in buildings |
|---|---|---|---|---|
| 21st century 2000–2010 | Sustainability | 2002, 2010 EU Energy directives<br>Energy Performance Building Directives (EPBD) | Green building certification systems/Energy performance certificates/Renewable energy sources/New standards for energy efficiency/Scenarios for climate change | Standards<br>Reduction of emissions<br>Energy efficiency<br>Reducing carbon print |
| | | 2010 EU Energy directive<br>Paris agreement | High performance buildings<br>Net zero and nearly zero buildings | Energy efficiency<br>Clean energy |
| $21^{st}$ century 2020 | | | Positive energy | Energy production |

- On 19 May 2010 the *European Union Energy Performance Directive* (modified the previous form of the 2002 Directive and) introduced the concepts of net zero-emission, and nearly zero-energy have been defined in terms of the standards.

## 3 Methodology and the Aim of the Paper

The analysis of the building-environment relationship is very complex. In order to promote the development of some buildings with low environmental impact or to evaluate the impact of existing buildings, it is necessary to use some indicators for the evaluation of buildings with a quantitative expression.

In this paper we proposed to assess whether the older buildings, which were retrofitted some time ago to reduce heat loss, still meet the newer energy efficiency requirements. The calculation method proposed in Mc 001/1-2006 (Methodology for computing the energy performance of building envelope) was used for this purpose. This is still the official standard currently used in Romania [3].

In that concern the evaluation of the buildings energy efficiency, among other indicators: the unidirectional thermal (heat flow) resistance (R), the adjusted thermal resistance (R′) for each envelope element, and the global coefficient of thermal insulation (G) for the entire building can be used [3, 4].

To estimate the unidirectional thermal resistance, R, for the mean area of each building envelope element is used the flowing equation:

$$R = \frac{1}{\alpha_i} + \sum_{j=1}^{n} \frac{d_j}{\lambda_j} + \frac{1}{\alpha_e}, \left[m^2K/W\right] \quad (1)$$

*with:* $\alpha_i$ *and* $\alpha_e$ *– are indoor and outdoor surface heat transfer coefficients [W/m*$^2$*K];* $\lambda_j$ *– thermal conductivity coefficients of the material from layer, j, [W/m*$^2$*K];* $d_j$ *– the thickness of the layer, j, in [m].*

The assessment of the adjusted thermal resistance for each external building element, R′, is made with the relation:

$$U' = \frac{1}{R'} = \frac{1}{R} + \frac{\sum(\psi l)}{A}, \left[W/m^2K\right] \quad (2)$$

*where:* U′ *- is the adjusted thermal transmittance of each different building envelope element, [W/m*$^2$*K]; R′- adjusted thermal resistance of building element (thermal bridge effects are included) [m*$^2$*K/W]; A,* $A_j$ *- area of the building envelope element with a specific thermal resistance [m*$^2$*]; l - the length of the linear thermal bridge [m],* $\psi$ *- is the linear thermal transmittance,* $\psi = L^{2D} - U_i * l_i$ *[W/mK],* $L^{2D}$ *- is the thermal coupling coefficient obtained from a two-dimensional calculation of the component separating the two environments being considered [W/mK].*

Finally the global coefficient of thermal insulation, G, is calculated with:

$$G = \frac{1}{V}\sum \frac{A_j}{R'_j}\tau_j + 0.34n, \left[W/m^3K\right] \quad (3)$$

*where: "n" is the natural ventilation rate [*$h^{-1}$*], and* $\tau_j \equiv \theta_j = \frac{\theta_i - \theta_u}{\theta_i - \theta_e}$ *depends on the indoor, outdoor air temperatures and also on the air temperature of the unheated spaces[°C]; V is the heated volume.*

**In Romania** - as in most EU member states - similar regulations and standards for quantifying the efficiency of the overall thermal insulation and the thermal bridging effects on the energy balance of buildings exist and are being adopted and implemented. Sometimes progress has not been as effective or obvious here due to economic or technological constraints, but the trend has been a constant increase in thermal efficiency.

Regardless of the type and composition of the building system, the thermal insulation of buildings complied with the standards in force at the time of their construction. Thus, in each construction period, the buildings ensure a level of thermal insulation that meets the standards in force and the technological level specific to that period, even if the materials used are different.

Between 1961 and 1990, most of Romanian buildings were built according to the requirements regarding the thermal protection stipulated in the national technical regulations that were harmonized with those of other countries, respectively the series of STAS 6472 standards. These have been permanently revised from 1961 to 1989.

The specific thermal resistances of envelope elements were established on the basis of the requirement to comply with the minimum conditions of comfort and hygiene and sanitation (Table 2).

Starting with 1985, a higher level of requirements was imposed regarding the reduction of energy consumption by applying the provisions of NP15-84 and later NP15-87, the result being a category of buildings with a thermal protection level increased by about 20% compared to previous one.

**Table 2.** Thermal resistances values for envelope elements (1962–1984)

| Period | Standard | R [$m^2K/W$] | | |
|---|---|---|---|---|
| | | Exterior walls | Terrace and garret floors | Floors over the basement |
| 1962–1968 | STAS 6472-61 | 0.76 | 0.96 | 0.82 |
| 1969–1973 | STAS 6472-68 | 0.80 | 1.02 | 0.82 |
| 1974–1975 | STAS 6472-73 | 0.80 | 1.02 | 0.82 |
| 1976–1984 | STAS 6472-75 | 0.80 | 1.02 | 0.78 |

C107-1997 technical regulations in line with European standards EN ISO has led to further increase by about 20% the level of thermal insulation of buildings designed after 1998.

After 1998 was introduced the adjusted thermal resistances (R′) that take into account the influence of thermal bridges. In order to be able to make a comparison, the values imposed by the previous standards have been reduced with percentages that approximately reflect the presence of the thermal bridges (Table 3).

**Table 3.** Adjusted thermal resistances values for envelope elements (1985–2000)

| Period | Standard | R′ [$m^2K/W$] | | |
|---|---|---|---|---|
| | | Exterior walls | Terrace and garret floors | Floors over basement |
| 1985–1987 | STAS 6472-84 | 0.76 | 0.87 | 0.56 |
| | NP 15 – 84 | 1.20 | 1.55 | 1.08 |
| 1988–1989 | STAS 6472-84 | 0.76 | 0.87 | 0.56 |
| | NP 15 – 87 | 1.20 | 1.55 | 1.08 |
| 1990–1997 | STAS 6472-89 | 1.00 | 1.24 | 0.67 |
| | NP 15 – 87 | 1.20 | 1.55 | 1.08 |
| 1998–2000 | C107/3-1997 | 1.09 | 1.46 | 1.25 |
| | C107/1-1997 | 1.40 | 3.00 | 1.65 |

A new increase in the level of thermal protection was realized for the buildings built after 2010 when the minimum value of the thermal resistances conventionally established under conditions of reduction of energy consumption were revised (Table 4).

The review of the minimum values proposed for thermal resistance $R'_{min}$ (conventionally established under conditions of reduction of energy consumption) was made at successive intervals, taking into account the solutions available on the market and the economic power.

Even if the new buildings are designed and constructed with high levels of energy performance, the great majority of the building stock is represented by older, low energy

**Table 4.** Adjusted thermal resistances values for envelope elements (1998–2011–2019)

| No. | Building element | R′ [$m^2K/W$] after 1.01.1998 | R′ [$m^2K/W$] after 1.01.2011 | R′ [$m^2K/W$] Mc001/2019 |
|---|---|---|---|---|
| 1. | Exterior walls | 1.40 | 1.80 | 2.50 |
| 2. | Exterior carpentry | 0.50 | 0.77 | 0.90 |
| 3. | Floors above the last level, under terraces or garrets | 3.00 | 5.00 | 7.00 |
| 4. | Slabs over unheated basements | 1.65 | 2.90 | 2.90 |
| 5. | Walls adjacent to closed joints | 1.10 | 1.10 | 1.10 |
| 6. | Floors that separate the bottom building from the exterior | 4.50 | 4.50 | 4.50 |
| 7. | Slabs on the ground | 4.50 | 4.50 | 4.50 |
| 8. | Slabs on the inferior side of the heated basements | 4.80 | 4.80 | 4.80 |
| 9. | Exterior walls under the CTS for heated basements | 2.40 | 2.90 | 2.90 |

performance buildings. Educational buildings, which represent 17% of the European non-residential building stock [5] have been predominantly constructed in periods when energy buildings regulation were limited and therefore they are in need of renovation work, having a large energy saving potential.

## 4 Thermal Rehabilitation of a School Building - Case Study

The analysed building belongs to the Construction and Building Services Faculty from Iaşi, constructed in 1973. The load bearing structure is made of shear walls and plane frames (in the façades) of monolith reinforced concrete. The building has a useful area of 3,222.16 $m^2$, a heated volume of 12,185.14 $m^3$, and a floor area on the ground of 786.24 $m^2$.

In 2007, the building was summited to constructive works of thermal rehabilitation, consisting in additional layer of expanded polystyrene (of 5 cm) on the external face of the envelope walls, 10 cm of extruded polystyrene on the roof and top amphitheatre floor, also the old wooden frame windows have been changed with new ones of aluminium frames and thermal insulated glazing system (Fig. 1).

As a result of these constructive measures, the thermal insulation level of the building increased significantly. The values of the corrected thermal resistance of the construction elements, before and after thermal insulation, are presented in the next table (Table 5).

**Fig. 1.** The analysed building: photo and IR image before and after thermal insulation

**Table 5.** The resultant adjusted thermal resistance before and after insulation

| No. | Building element | Area, $[m^2]$ | $R'_{init}$ $[m^2K/W]$ | $R'_{insulat}$ $[m^2K/W]$ |
|---|---|---|---|---|
| 1 | Exterior walls | 1843.71 | 0.651 | 1.412 |
| 2 | Windows | 802.38 | 0.390 | 0.500 |
| 3 | Terrace roof | 653.60 | 0.741 | 2.674 |
| 4 | Amphitheater roof | 181.20 | 1.465 | 1.465 |
| 5 | Ground floor | 786.20 | 2.259 | 2.259 |
| 6 | Amphitheater floor | 48.60 | 0.719 | 1.969 |

Compared to a value of the average thermal resistance calculated for the whole building ($R'_{mean}$) of 0,712 $m^2K/W$, before performing the thermal rehabilitation works (see Table 6), the present value of $R'_{mean} = 1{,}268$ $m^2K/W$ is ~1.8 time grater.

**Table 6.** Some energy efficiency parameters calculated for the studied building

| No. | Building parameter | Symbol | Unit | Initial | Insulated |
|---|---|---|---|---|---|
| 1 | Building envelope area | $A_{env.}$ | $m^2$ | 4315,7 | 4315,7 |
| 2 | Corrected mean thermal resistance | $R_{mean}$ | $m^2K/W$ | 0.71 | 1.27 |
| 3 | Global factor of thermal insulation | $G_1$ | $W/m^3K$ | 0.498 | 0.28 |
| 4 | Annual specific heat consumption | $q_{heat.}$ | $kWh/m^2a$ | 151.34 | 83.9 |
| 5 | Index of building compactness | $A_{env.}/V_{heat.}$ | $m^{-1}$ | 0.35 | 0.35 |
| 6 | Glazed area/Walls area | $A_{gl.}/A_{wall}$ | – | 0.44 | 0.44 |

Using all these constructive measures, it results an energy economy of 42%, the specific secondary energy consumption decreasing from 151.31 kWh/($m^2$.year) for heating to 83.9 kWh/($m^2$.year).

The energy performance certificate of the building, issued in 2014, places the building in energy class B of consumption.

The new windows panels more efficient regarding heat loss are also better sealed against air infiltration. This led to the diminution of the natural ventilation rate and increases the risk of inadequate ventilation and possible build-up of indoor air pollutants, which can produce lethargy and drowsiness to building occupants [6]. To assess this risk, an experimental study using a $CO_2$ gas analyser was conducted between March 26 and March 29 in one of the building's lecture rooms, in use on that period of time. The variation of $CO_2$ measured concentration in the room air is shown in Fig. 2 presented below.

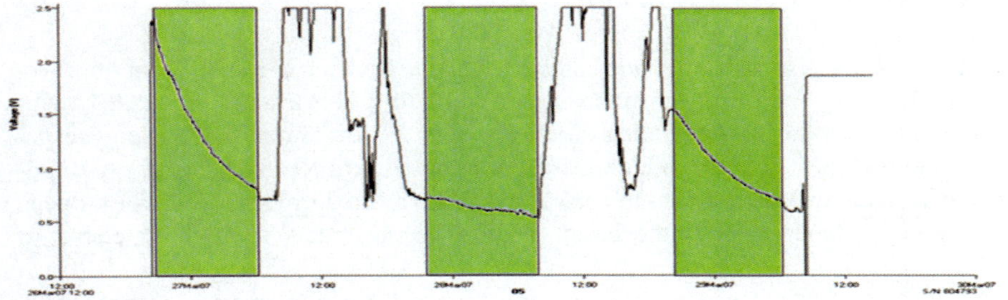

**Fig. 2.** Evolution of $CO_2$ concentration and the selected computing intervals

The values of measured concentrations of $CO_2$ were used for determining the rate of natural ventilation of the room, according to equation:

$$m_L = \frac{\overline{V}}{V} = \frac{\ln C(t_1) - \ln C(t_0)}{t_0 - t_1} \tag{4}$$

*where: V – room volume ($m^3$); $\overline{V}$ – volume of replaced air ($m^3$); $C(t_1)$ – $CO_2$ concentration at the beginning of the measurement (ppm); $C(t_0)$ – $CO_2$ concentration at the end of the measurement (ppm); $\Delta t$ – period of time (hours), $\Delta t = t_0 - t_1$.*

For the calculations were selected three time periods during the night, when the room is empty (not used). The results are presented in tabular form (Table 7).

The values of ventilation rates determined experimentally are much below the minimum requirement of 3 l/(s·person) for teaching and learning spaces, according [7]. For a maximum number of 90 persons and room volume of 432 $m^3$, it results a minimum ventilation rate of 2.25 $h^{-1}$.

The concentrations of $CO_2$ recorded during the three days of study shows an average quantity of 1368 ppm $CO_2$, which means that the indoor air quality is moderate to low (IDA3-4), according to [8]. In addition, using $CO_2$ as an indicator for ventilation, ASHRAE has recommended indoor $CO_2$ concentrations be maintained under 1,000 ppm in schools and 800 ppm in offices (ANSI/ASHRAE Standard 62).

**Table 7.** Values of effective ventilation rate by infiltration corresponding to measured data

| Interval | Date & Hour | Concentration [ppm] | Δt [h] | n $[h^{-1}]$ |
|---|---|---|---|---|
| 1 | 26.03 - $20^{50}$ | 2,485 | 9.12 | 0.12 |
| | 27.03 - $6^{02}$ | 815 | | |
| 2 | 27.03 - $20^{06}$ | 805 | 11.9 | 0.035 |
| | 28.03 - $8^{00}$ | 532 | | |
| 3 | 28.03 - $20^{18}$ | 1548 | 8.00 | 0.10 |
| | 29.03 - $6^{14}$ | 698 | | |

## 5 Discussion and Conclusion

The constructive insulation measures presented above have led to a significant improvement in the energy performance of the analysed building. However, there are still issues that need to be addressed, including those related to the use of renewable energy, especially if the aim is to bring the building as close as possible to Nearly Zero Energy Buildings (nZEB) conditions, which is the new standard set by the European Union - European Commission. Some aspects of the thermal refurbishment strategy are highlighted below.

- The adjusted thermal resistance (R′) obtained for each façade element after the thermal insulation were acceptable for the year 2007 but are unacceptable after 2011 (see Table 4 and Table 5). But sometimes a radical intervention is very troublesome in case of existent buildings.
- In general, educational buildings are characterized by a certain constructive and functional diversity, which requires a specific approach for each individual case, but also has some specific features, such as: large glazed area in the wall, weekly or annual usage program (they are not used during the night, etc.) and occupancy rate, etc.
- The passive constructive energy efficiency measures for buildings (i.e. additional thermal insulation) alone are not strong enough to transform these buildings into nZEB. It is necessary to add active green renewable energy sources, energy generators (such as photovoltaic panels, solar panels or wind turbines) to increase the energy efficiency of the buildings.
- Sustainability indicators are a useful tool to assess the energy efficiency and environmental performance of buildings and to promote design even in case of thermal retrofitting. Concerns about the energy efficiency of buildings in particular sometimes have undesirable effects on the quality of the indoor climate (inadequate air composition, insufficient ventilation, condensation phenomena, etc.). Scientific research concludes that there is a strong correlation between indoor air quality and the learning performance of pupils. Without additional measures, the existing natural ventilation rates could remain below the recommended minimum values.

## References

1. Atanasiu, B.: Implementing nearly zero-energy buildings (nZEB) in Romania – towards a definition and roadmap. Buildings Performance Institute Europe (BPIE), August 2012
2. Harputlugil, T.: Energy efficient building design development: a retrospective approach. In: International Symposium on Energy Efficiency in Buildings, Ankara, Turkey, 13–14 February, pp. 296–303 (2017)
3. Mc 001/1-2006, Methodology for computing the energy performance of building envelope
4. *** SR EN ISO 10211-2:2005 Thermal bridges in building construction – Calculation of heat flows and surface temperatures – Part 2: Linear thermal bridges
5. http://bpie.eu/wp-content/uploads/2015/10/HR_EU_B_under_microscope_study.pdf
6. Canadian Centre for Occupational Health and Safety. Health Effects of $CO_2$ Gas. http://www.ccohs.ca/oshanswers/chemicals/chem_profiles/carbon_dioxide/health_cd.html
7. Building Bulletin 101 – Ventilation of school buildings, Regulations standard design guidance (2006). ISBN 011-2711642
8. EN 13779:2007 - Ventilation for non-residential buildings – Performance requirements for ventilation and room-conditioning systems (2007)

Check for updates

# Life Cycle Cost (LCC) and Sustainability. Proposal of an IFC Structure to Implement LCC During the Design Stage of Buildings

Ignacio Bernardino-Galeana[1(✉)], Carmen Llatas[1], María Victoria Montes[1], Bernardette Soust-Verdaguer[1], Jacinto Canivell[2], and Pedro Meda[2]

[1] University of Seville, 41012 Seville, Spain
ignbergal@alum.es.us
[2] University of Porto, 4200-465 Porto, Portugal

**Abstract.** The role, the economic relevance and the impacts of the construction sector in the world demand, in today's society, increasing requirements in terms of sustainability and efficiency. According to ISO 15686, the LCC (Life Cycle Cost) methodology is a useful technique that allows the evaluation of comparative costs during a specific period of time, taking into account all relevant economic factors, both in terms of initial capital costs and future operating costs. However, the integration of this powerful methodology into building design tools is still poor, limiting the ability of projects to optimize the costs of buildings during their life cycle.

The emergence of IFC (Industry Foundation Classes) technology in the field of architecture, engineering and construction, helped the exchange of information at the design, construction, operation and maintenance stage of a project to be shared independently of the design tools that currently exist, based on a BIM (Building Information Modeling) model that can contain relevant data related to the project/built asset [25]. According to buildingSMART, IFC are the industry's core classes (IfcDoor, IfcWindow, IfcBeam, etc.) that form the basis for globally recognized data exchange and interoperability.

Therefore, this document presents a structured proposal for LCC to implement at IFC. The proposed structure has been developed based on the ISO 15686 standard, which contains the main costs according to the construction phases. With this proposal, it is expected to encourage the implementation of LCC during the design phase of the project in BIM, to achieve more economically efficient buildings.

**Keywords:** Life Cycle Cost (LCC) · IFC · Building data structure · ISO 15686 · Cost optimization · Economic building sustainability · Building Information Modeling (BIM) · Building design tools

## 1 Introduction

Making the best decisions before starting a construction project has become a factor in determining the success or failure of such projects because buildings have become larger


A. Rotaru (Ed.): CRIT-RE-BUILT 2019, SSGG, pp. 404–426, 2021.
https://doi.org/10.1007/978-3-030-61118-7_33

and more complex. Particularly in the life cycle of a construction project, the planning and design phases have an important weight (Kaplan and Norton 2005), in these stages important decisions are made such as the outline of a building and the selection of design alternatives. Furthermore, the preliminary estimate in the initial design phase determines not only the viability of the project but also the total cost of the project (Azman, Abdul-Samad and Ismail).

Decision making in the planning, schematic design, final design phases affects construction costs as well as costs incurred for the maintenance and dismantling of building structures. Therefore, decisions must be made with the life cycle cost (LCC) in mind.

To support decision-making regarding the key elements of a construction project and the selection of design alternatives in the initial design phase of the project, the scope of this study was limited to structuring IFC, including ISO 15686, the LCC methodology in the initial design phase.

This study proposes an IFC import and export method from a BIM model. The criterion was structured and linked to the BIM model, to include LCC data in accordance with ISO 15686. The precision of the prototype was to verify the information on a BIMCollab display applied to a steel construction system, the validity of the method of supporting the shot of project planning and design decisions [30].

## 2 Life Cycle Cost (LCC) Methodology

### 2.1 LCC Definition

Life Cycle Cost (LCC) concept was implemented in a structured way from the 70s, specifically the United States Department of Defense in the area of military aviation [1].

However, most of the methodologies developed by the defense department were oriented to purchasing and logistics processes and exclude the design and production phase. Once the need to apply the LCC methodology in the production design, planning and control processes was recognized, the US National Science Foundation sponsored a conference in 1984. In this conference the research areas were identified and prioritized, receiving the qualifications Highest prioritization areas: economic evaluation in the design phase, life cycle analysis [2].

LCC is defined by ISO 15686 [3] as a “valuable technique that is used for predicting and assessing the cost performance of constructed assets”. The LCC also “allows consistent comparison to be performed between alternatives with different cash flows, Fig. 1 and different time frames” [3].

### 2.2 LCC and LCA

The life cycle assessment (LCA) is an analysis of the possible environmental impacts both in the use of resources and in the consequences of emissions produced during the life cycle of a product or production system.

LCC moves away from the concept of inventory of interpretation of inputs and losses of different elements of the system, directing its interest to predict and evaluate the performance of the costs of the built goods. Life cycle costing is a form of analysis to determine if a project meets customer performance requirements [26].

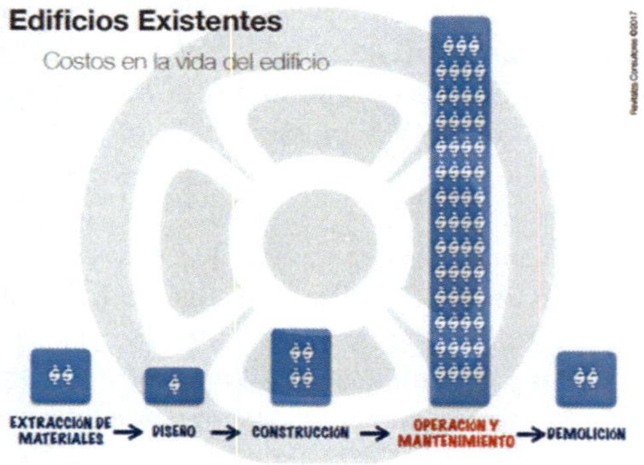


**Fig. 1.** Cost flow, Sustainability workshop, México, 2018

Therefore, the LCC is not an exclusively environmental measurement tool, although it incorporates the concept of "life cycle". It should be remembered that sustainability is more than ecology and, therefore, the LCC can be considered a useful instrument for evaluating the costs that are included in a sustainable project.

## 2.3 LCC and Risk

The LCC has become a main tool in decision-making. [4], projects are not only related to financial concepts, but also to the environmental impact they can produce.

Various techniques are needed to assess risk management to predict the future. This process is determined by parameters or variables such as the discount rate and the Net Present Value (NPV) related to a sensitivity and probability analysis. Two particular techniques related to resource optimization in the LCC construction industry were considered [5]. On the one hand, the discount rate used to convert future values to present values and vice versa and, on the other hand, the NPV is the standard criterion for deciding whether an option can be justified by economic principles.

## 2.4 LCC and Industry Construction Costs

Woodward [6], not only exposes the LCC methodology, but also identifies the costs and benefits based on the goals established by the Royal Institute, identifying the following elements of the LCC: (1) Frontal capital costs; (2) The life of the asset; (3) The discount rate; (4) operation and maintenance costs; (5) Disposal costs; (6) Information and comments and (6) Analysis of uncertainty and sensitivity.

When exploring the most significant variables that have had to attack the construction of the LCC industry, Sterner proposes to include the initial costs (cost of lot, design, construction), operating costs (energy costs, cleaning), maintenance costs (costs annual, substitute costs, contingencies) and recovery costs are also related to previous costs, the discount rate that varies in a proposed sensitivity analysis and is part of the uncertainty exposed by the LCC [37].

It is important to distinguish between the LCC (life cycle cost) and the WLC (cost of a lifetime) in which ISO 15686-5 [3] (reference in LCC) establishes that the LCC study includes: construction, operation, maintenance and demolition, Fig. 2.

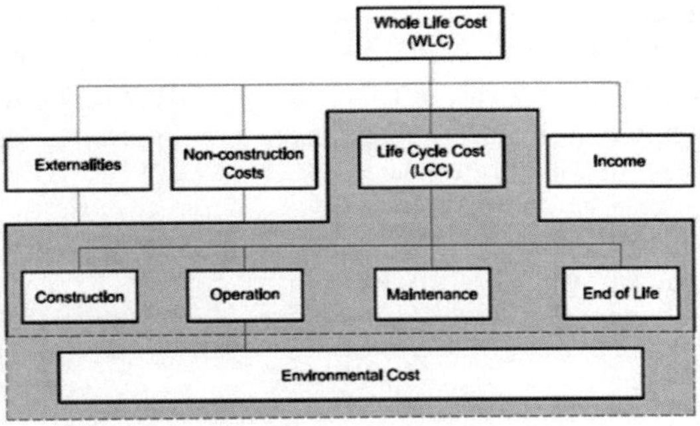


**Fig. 2.** WLC and LCC elements

## 2.5 LCC Utility

In 2007, the European Union (EU) published a report whose objective is to propose a common methodology for the implementation of LCC in EU countries [7] This report was the basis for ISO 15686-5 [3] this report It also intends to connect LCC with LCA. Analysis and application could be directed to evaluating a project based on comprehensive and competitive decision making. On this basis, the plan proposes a series of approaches that are described below:

The use of LCC and stroke as two of the criteria in evaluating a single investment option (for example, the decision to build an asset), where other criteria evaluation could include functionality, aesthetics, construction speed, future investment returns, etc.

The use of LCC and LCA as two of the criteria in the assessment of several alternative investment options (whether constructed entire specific goods or components, materials or assemblies within)

The use of LCC to provide a financial/economic evaluation of the impacts of sustainability that have a monetary value widely accepted and easily calculated;

The use of LCC to provide a financial/economic evaluation of the options identified in an LCA evaluation;

The use of LCC as a means of identifying options with good environmental performance and then carry out an LCC analysis in these options only;

The use of LCC to select cost-effective options; to make a final decision in the light of a process of LCA conducted only for those options.

## 2.6 LCC and Green Buildings

We call green buildings those sustainable constructions that use natural materials, that reduce and optimize their energy and water consumption as much as possible and that integrate into their environment, whether natural or urban, causing the least possible environmental impact [3].

There are different institutions, organizations (US Green Building Council, USGBC) and even independent entities that are responsible for promoting sustainable architecture and construction through the implementation of certification processes that will be in charge of evaluating the degree of sustainability of the buildings. With the voluntary certifications (LEED, EDGE, PEER, WELL, SITES, GRESB, BREEAM), Fig. 3, it is intended to guarantee that green buildings meet a minimum standard of conditions established by each certification body.

**Fig. 3.** LEED in the world, Sustainability workshop, México, 2018

Some of the benefits of having green buildings are: Reduction of energy uses, reduction of water use, and the correct management of waste uses [8].

The purpose of LCC in green buildings is to review initial costs, long-term costs, and possible cost changes, during the decision-making or evaluation process. It also usually includes contributions from other assessments (environmental assessment, design, safety assessment, functionality assessment, compliance assessment) [3].

Current studies based on LCC show the existence of different approaches and contexts that manage and establish an initial point of view for this research [8–14]. LCC-based studies show the growing interest of green buildings against conventional buildings [8].

As can be seen in the following image, carrying out a green project against a conventional project can have an increase of 20–30% of the initial cost, this is due to the use of new technologies to reduce energy costs. The considered range of return on investment in the construction of a green building, (ROI) is between 4–6 years. There are benefits in reducing operational costs, this is due to the economic savings that a green building brings (Fig. 4).

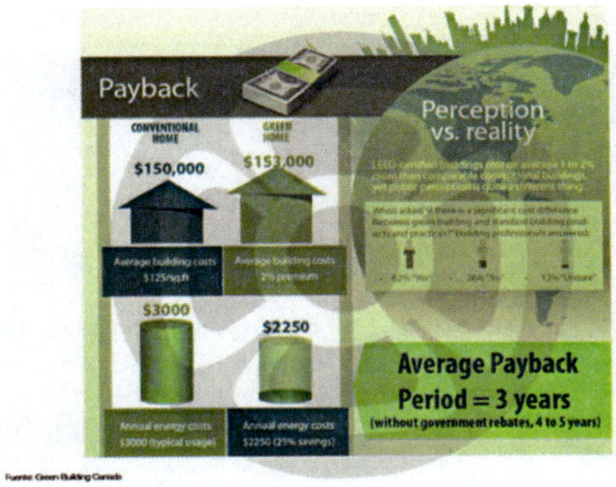


**Fig. 4.** Payback, Sustainability workshop, EU, 2018

## 3 The Industry Foundation Classes (IFC)

The following study wants to publicize the advantages and IFC structure, to relate it to LCC, the application will be made in a case study and verify if there is no data loss during the process.

### 3.1 IFC and Building Information Modelling (BIM)

IFC stands for Industry Foundation Classes, it is a common standard for data exchange in the construction industry and information is shared regardless of the software application used. IFC is an object-oriented file format, which can be used in BIM-based projects [15].

BIM allows that data used throughout the lifecycle of a building remain stored and reused for multiple purposes. According to building SMART [15], the IFC format is the main tool for the implementation of Open BIM, "which represents a universal method for collaboration in the design and construction of buildings based on standard and open workflows." Once exported, the IFC model contains not only the geometry of the building and the building data, but also all information contained in the native BIM files. When native data are exported to an IFC file, data can be transferred between applications. This operation is free and is well documented and allows the use by hundreds of other tools and BIM applications.

In north of Europe, countries such as Denmark have promoted its use for construction projects with public aid. In Finland, the facility management company owned by the state, Senate Properties, now requests the use of software compatible with IFC and BIM in all its projects. In addition, the Norwegian government is mandatory to use BIM IFC projects. In industry, many municipalities, private clients and contractors have already integrated this format in their business.

IFC has its strengths and weaknesses, in accordance with ISO 16739-1 [17] The process involves a loss of data and intelligence of the object. One of the advantages is having an open-format virtual building, which allows contractors to take the first approach to design, Fig. 5.

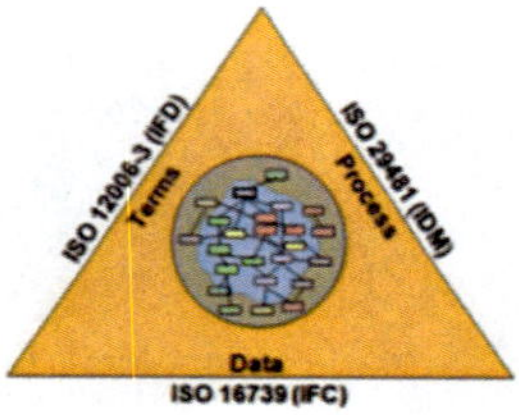


**Fig. 5.** ISO standards. (Source: BuildingSMART international)

Today, the IFC format in building projects is used for visualization and detection of crossings. During the first phase of a project, the design team can merge or reference the discipline of the models independently of the original application using an IFC file, Fig. 6. They are also used to import data from one application to another [16].

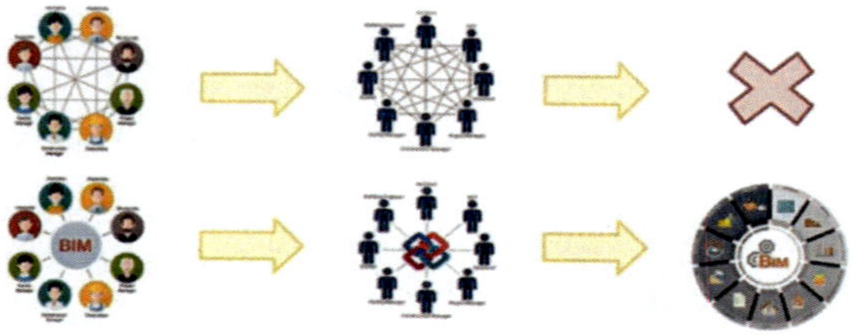


**Fig. 6.** Interoperability and open system. (Source: BuildingSMART international)

## 3.2 IFC Versions

Figure 7 shows the evolution of the IFC way of life, which received several updates and improvements, reflected in different release versions. The most important have been two of them: (1) the IFC2x version released in October 2000 and updated to IFC2x3-TC1 released in July 2007, and (2) the IFC 4 version, released in March 2013 after 6 years of improvements (approximately 1200 improvements and corrections) and documentation.

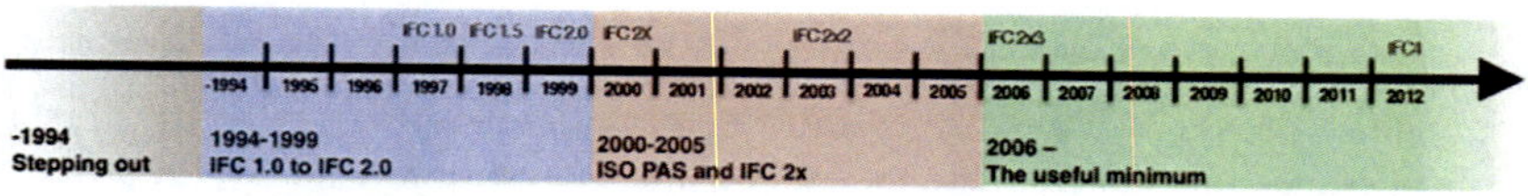


**Fig. 7.** IFC timeline (Source: BuildingSMART international) [15]

### 3.3 Structure IFC

The IFC architecture bases its own structure on:

- Semantics
- Relationships
- Properties

The elements are intended to describe the components of buildings, such as facilities, spaces, areas, furniture, structural elements (pillars, beams, walls, floors, etc.), including the specific properties of each object. Thanks to this subdivision it is possible to associate, to each object, specific greatness such as:

- Shape
- Costs
- Need for maintenance
- Position
- Energy provision
- Connection with other objects
- Security
- Physical and mechanical characteristics

All of this data is generally encoded in one of the three available formats:

- .ifc: predefined file format based on the ISO-STEP standard
- .ifcxml: coding based on XML language
- .ifczip: compressed file of one of these formats, which may also contain attached material such as PDF or images.

It is important to know the type of files when exporting the information, so that the coloration between the figures involved is correct, Fig. 8.

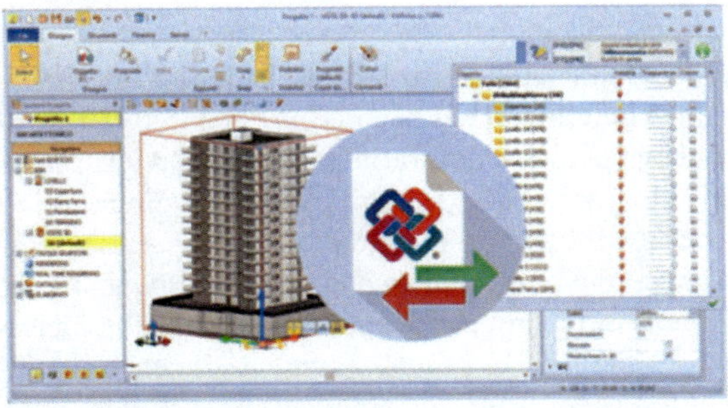

**Fig. 8.** Open BIM (Source: BuildingSMART international)

Model View Definition (MVD) represents subsets of the IFC schema defined by buildingSMART International and groups together certain information useful for specific workflows or uses.

These views are very useful since they simplify the data exchange process and avoid sharing useless or redundant information, following standardized procedures.

The following image shows a schematic that gives an idea of the IFC schema and how a specific MVD, Fig. 9, can simplify it based on useful information for a specific purpose.

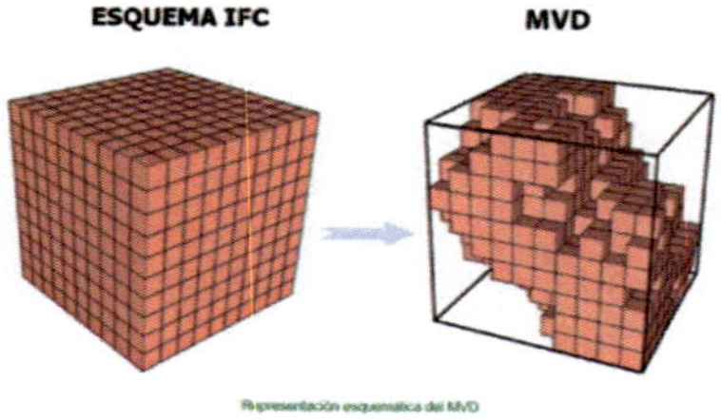


**Fig. 9.** Schematic representation MVD (Source: BuildingSMART international)

In conclusion, considering the image above, we can understand the MVD as a filtered view that collects part (or all) of the information regarding the entire IFC schema.

Imagine then, the different MVDs applied to the federated model, obtaining:

- Structural model
- Architectural model
- MEP model

In this way we can include costs by disciplines in the MVD, which helps us to carry out separate analyzes, understanding that they are part of a construction.

### 3.4 New Trends in IFC

IFC is a data model used to characterize both the building information and geometry along the life cycle of the building. It has become an open ISO standard supported by the industry and promoted by the global non-profit buildingSMART [15], which has created a professional certification program. The first part of the program is called individual rating and aims to standardize and promote the content openBIMtraining, support and accredit training organizations and evaluate and certify individuals, Fig. 10.

According to [18] currently, project planning and execution phases of construction the main uses are compatible IFC. In addition, many designers use this exchange format when working with different software, and architects use solutions such as AECOsim or ArchiCADand engineers using Revitor DDS-Cad. Everyone has their native extensions, but give this option and flexibility to the workflow. The IFC provides geometric compatibility in a common environment and export for analysis in other software. However, one

**Fig. 10.** Exchange of information (Source: Linkedin)

major barrier is that the export of the information in the models is not always correctly translated. For example, IFC Export from the modeling software such as Revit to another software such as ArchiCAD can cause data loss. This lack of interoperability is evidence by [19] as "BIM market coordination problem".

If we follow the BIM methodology, it becomes clear that the IFC can be useful if we know its limitations. To achieve it, user must perform tests to ensure that the IFC format can be used in each specific case [31]. When user start working with the IFC data model, it is warned that the available examples were slight variations each other. For example, the same element in a project had a different structure than in another. At that time seemed a problem of incorrect modeling, but later clarified that the difference was due to the version of IFC in which were developed some models and others (Fig. 11).

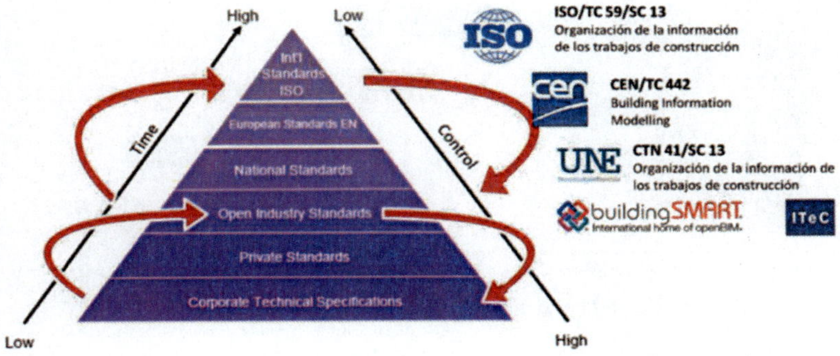


**Fig. 11.** ISO (Source: Linkedin) IFC proposed structure applied to LCC

### 3.5 Current Structure of the IFC

The IFC format is based on a class hierarchy as a tree, starting east of a root node which inherit all other entities directly or indirectly [20]. This root node is called IfcRoot and is a superclass of all other entities. Element unfolds tree hierarchy, beginning by the differentiation of the three main types of entities, IFC includes: (1) object definition, (2) definition of properties and (3) relationships. Each of these nodes acts as abstract definitions of the elements. With this, at every level that the hierarchy can be moved down, new attributes can be added inherited from the parent. Thus, when the leaf nodes are obtained, that is, the products themselves, the full definition is achieved [20].

Figure 12, shows the inheritance tree of a object in IFC. It can be seen, how large the number of levels from the root node to the element. In each of these levels by passing the definition until the product it brings something to it. The IfcRoot inherits the 4 graphic elements of any product: (1) the single global id, (2) reference the owner who created the item, (3) the name and (4) the description. The attributes that make up an entity can be of two types: direct or inverse. Direct attributes contain information stored directly on the instance of that entity, while the reverse (which usually are relations) means that no other entity can have as an attribute to the extinguisher itself [34]. This structure is common to all IFC entities, although the attributes and the relationships can vary. Besides working on different platforms and specific modelers, it is possible to use BIM models in IFC format for making measurements and budgets. Since this is an exportable and importable format by BIM software modeling, calculation or management because it uses a format standard for information exchange, open and internationally accepted code, we will have the freedom to work with software not tied to the platform native creative model for budgeting.

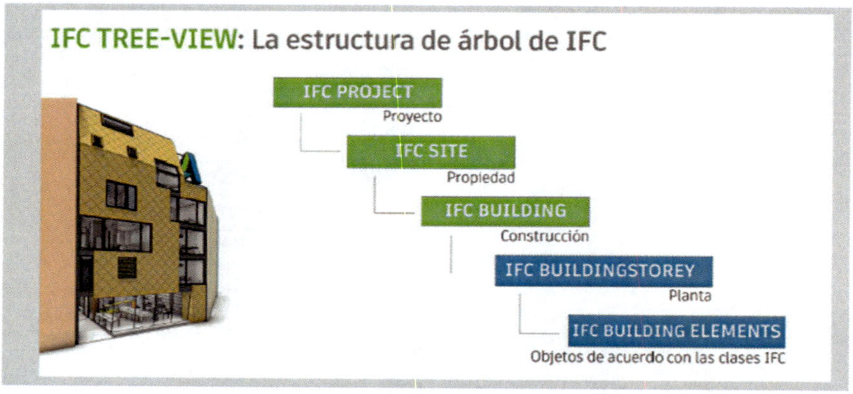


**Fig. 12.** IFC structure (Source: Revit IFC manual [20])

According to [21] an IFC model as in a native pattern, is a graphical database, that can be export from different software modeling, and can be read in the same modeling software. This language IFC, the same way as native models, includes over 600 "classes" the different elements of the building works and civil works. In the IFC models it is defined geometry, materials, properties and relations between structural elements. Due to their complexity not all attributes of the elements of native models can be found in the IFC models, but those for making measurements and budgets are: (1) class or name, (2) material, (3) quantities and (4) phases. Figure 13 shows an EXPRESS-G diagram ofIFC model structure.

An important point in connection with the writing and reading of model elements IFC, is the GUID (Globally Unique Identifier), which uniquely identifies that element, thus making measurements with IFC models can relate each GUID item with one or several

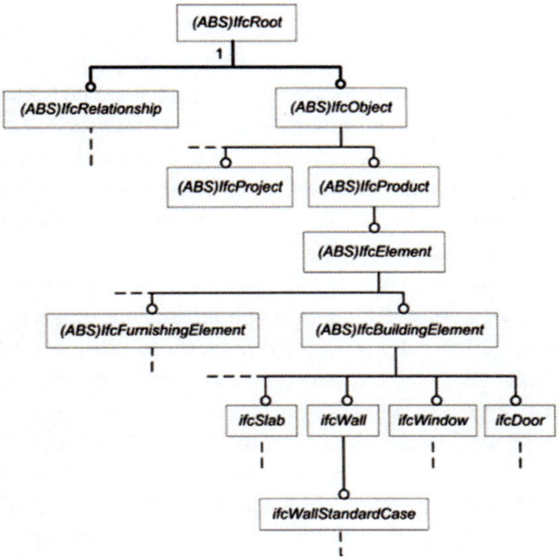


**Fig. 13.** EXPRESS-G diagram of IFC model structure (Source: [22])

budget items. One of the aims of BIM model is the quantities takeoff, measurements and economic control of the project.

In BIM, workflows should be a two-way communication between modeler and responsible measures to resolve such situations as just discussed. The solution adopted in this case is advisable to pass the modeling standards, so that in the future the teams work with that standard adequately the false ceiling in the case of double heights modeled, and the head of measurements has the security that can extract data model with correct placement heights of ceilings, i.e. the data contained in any model are reliable for the use in flow BIM.

Therefore, both BIM standards of a company, as in the BEP, where staying as living documents that are constantly improving. It is unfeasible to make a BIM sit standard for modeling, and think that it will solve all cases and doubts that may arise during modeling [32].

Moreover, BIM Project management recognize several BIM dimensions, related to the working process in building design. The 2D (drawings) includes the traditional plans and drawings (lines, images, renders) [23]. The 3D (model) allows allow the building to be toured [23]. The 4D (time) enables to plan the construction process and the 5D (cost) includes the cost estimation and measurements [23]. Even though there still no consensus to define the sixth dimension, [24] argue that it should be sustainability, due to the relevance of the subject.

### 3.6 IFC and Measurements and Budget of the Construction Project

As in a conventional process measurement, every part of the project will be measured by specialists in each discipline, i.e. the structure calculator provides measurement and budget structure, the facilities, the MEP, etc, or could be only one responsible for measuring and budget. In any case, the profile measurements responsible for budget will have a range of skills in order to achieve the highest performance measurements extraction from the models. The list of requirements and budget measurements responsible for a BIM process are: (1) professional experience on construction project in the relevant discipline; (2) knowledge of the details and construction processes of the work units and (3) the ability to know the designer to consult on the different elements composing a construction detail, system or assembly [27].

To measure the building, and to know to build it, it is sufficient management interface BIM modelers and IFC viewers, and also: (1) to obtain any necessary to correctly display the unit measured view; (2) to get within the model, the necessary tables, including the tables parameters or attributes to order measurements (level, type, class or subclass, material) including in these tables, an organization, segregation parameters, attributes, types, classes or subclasses, as well as calculating the total and partial generated according sub tables; (3) to create parameters or attributes within tables calculated for measurements, e.g.: multiply the section of a profile length to obtain the volume of material; (4) to create parameters or attributes in the model in order to complete the definition, graphic categories of model elements. eg: parameters or reference attributes hinges and hinges number of types of doors, without having to model the hinges to count them; (5) to export tables to Excel; (6) to hand specific measurement software with BIM modelers and (7) measurements in BEP and in the process map.

### 3.7 Measurements in BEP (BIM Execution Plan) and in the Process Map

Like the other project deliverables (plans, reports, specifications, calculations, etc.), the budget will be included as a task in the BEP. In the information flow generation BIM measurements and the budget will develop [36]:

- Input data, i.e., the model with the agreed standards for that model serves to measurements, which corresponds to each phase, and date of receipt, and perhaps other document that could define the project as memories of quality or property requirements.
- Output data, which is the budget itself as document as a spreadsheet, BC3 or format that has been stipulated.
- Responsible extraction of measurements and preparation of the budget and a delivery date or publication.
- A review milestone in each phase, through coordination meetings or checkups.

The above points shall appear in the BEP and/or process map, so that all participants in the process have no doubt the "what", "when", "where" and "how" in the flow BIM. As now, it happens with CAD, measurement may have a unique responsibility. It is often the case that different disciplines (installations, structure, safety and health) do their part of the budget, so each budget in the BEP appears such as a separate task. In that case it will always be necessary responsible for receiving all partial budgets and assemble a single budget, and sheet summary of this, for the "budget of material execution," the "contract budget" and "tender budget".

### 3.8 Structure of Costs to Be Considered in IFC

According to LCC ISO 15686-5 standard [3], the classification of costs can bethe following:

- **Pre-construction costs A0.** Though the theoretical "stages" in which the design process can be divided, construction phase created such as a "strategic definition" where customer is advisable to set the scope of the project, by answering questions such as, the size of the project, the period within which must be operational, the core project requirements and the work teams needed to be developed. This will mark an initial rough idea of the total cost of the project. At this stage, it is also advisable to identify risks, and one of the biggest risks is the gap budget, therefore, factors such as proper evaluation and weighting of the size of the project, deadlines and requirements in the form of qualities and performance of the building will be vital to mitigate this risk. It is also advisable to estimate investment costs.
- **Construction Stage A4–A5.** LCC analysis should consider the costs which are included in the construction phase of the life cycle. Costs should include design and construction, even infrastructure [29].
- **Use Stages B1–B8.** LCC analysis should consider the costs which are included in the use phase of the life cycle. To include operation and maintenance costs of a building. Most of the investment in a real estate asset is not performed during the design phase and construction, in fact, these two concepts usually involve about 30% of the total investment. Most of the investment occurs in operation and maintenance tasks in the building. That is, energy, cleaning, replenishments, security, maintenance and management. This is the building phase where IFC management can produce major savings for the costumer. That is why public and private properties arise not only BIM methodology applied to new projects, but existing buildings, raising the current state models thereof.
- **End of Life stage C1–C5.** LCC analysis must indicate the expenses including the final phase of life (Fig. 14).

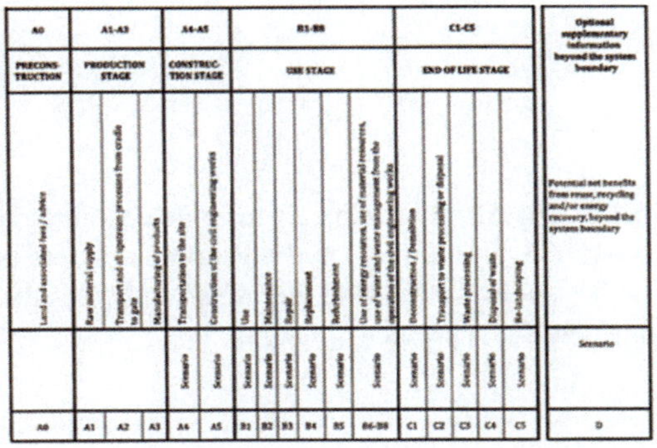


**Fig. 14.** ISO 15686 (Source: [3])

IFC creates an accurate inventory of the elements that make up the real estate asset (spaces, facilities, construction elements) in a single database that is the model, the parameters or attributes and metadata that allow calculating the cost of Operation & Maintenance, producing each of the assets, so that these data can be compiled, sorted and transmitted, and quickly analyzed. One of the advantages of using IFC is that the models are unique and centralized repository of all information of the building. With an updated BIM model building, this analysis is performed accurately and quickly, with the ability to transmit information to conventional formats such as spreadsheets or PDF.

### 3.9 Phases Considered in the IFC Proposed Structure

The phases that have been established in the Implementation Plan developed by BIM Manager for budgeting are:

- Phase 1: Budgeting BIM PBL. Starting from the PBL budget provided by the customer to make a BIM budget for the work to be executed, for it will need to modify some chapters and work units of the original budget, as well as adding new units; due to the expansion of PBL requested by the customer, this new budget will be the budget of the awarded project (hereinafter PA).
- Phase 2: Getting 3D BIM model. Cost discipline primarily part of 3D BIM model developed by the Discipline Corridor. This model contains the BIM objects that make up the linear infrastructure to be constructed. Furthermore, the 3D BIM model works and works existing passage cross drainage along the corridor, is obtained by this cost discipline.
- Phase 3: Getting 5D BIM model. From 3D BIM model will proceed to their interconnection with a tool Budget and measurements in order to obtain a model 5D BIM. For Budget and measurement tools, existing options will be studied to locate the most feasible to implement in a company.
- Phase 4: Initiation of monitoring the work with the 5D Model. It aims to study the process of monitoring work through BIM methodology, identifying work units executed in the first months of work and getting the measurements and proposed work certification.

## 4 Methodology

The methodology carried out aims to develop a life cycle cost integration model, based on the analysis of ISO-15686_5-Buildings and constructed assets_Service_ life planning_Part 5 Cost of the life cycle, within IFC. Figure 15 indicates the link between IFC and the costs to be considered in a project.

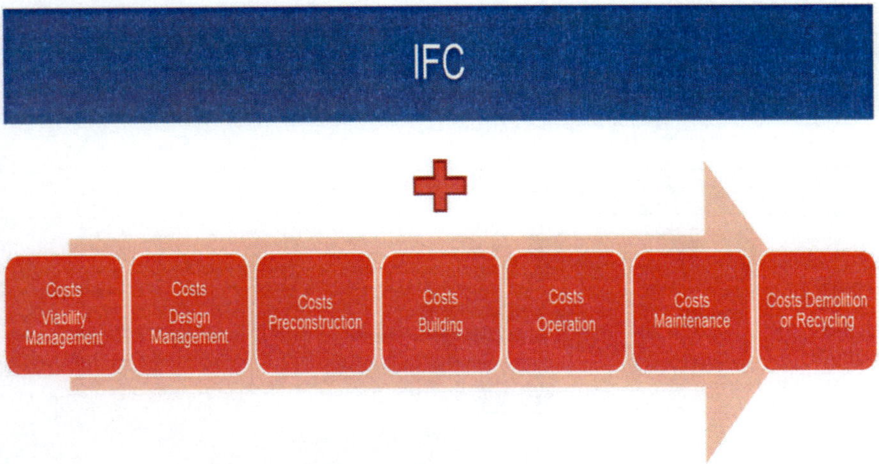


**Fig. 15.** Integration Proposal, IFC and building costs. Own elaboration (2019).

For this, it will be necessary to carry out a cost analysis according to the life cycle of a study typology. The selected study typology will be a residential building, as it is the most developed typology for a new construction, in Andalusia, Spain (according to data from the INE, 2018). Figure 16, shows how the life cycle costs of a home can be implemented in a BIM model. For this, a breakdown of information that influences a project was carried out and the costs that affect each one were successful, based on the ISO-15686 standard

| Model-Level of Detail (Lod) | 1 | 2 | 2 | 3 | | 4 | |
|---|---|---|---|---|---|---|---|
| Information shapers | Architecture | Architecture | Architecture | Installations | | Architecture | |
| | | Structure | Structure | | | Structure | |
| | | MEP | MEP | | | MEP | |
| | | | Specialist (Landscaping, Lighting, Etc) | | | | |
| Information Users | Promotor | Manufacturers | Promotor | Promotor | | Promotor | |
| | Real estate | Products | Real estate | Real estate | | Real estate | |
| | Researchers | Contractors | Researchers | Researchers | | Researchers | |
| | Administration | | Contractors | Contractors | | Contractors | |
| | | | Administration | Users | | Users | |
| | | | | Administration | | Administration | |
| | | | | Facility Manager | | Facility Manager | |
| BIM model |  |  BIM OBJECT |  |  | |  | |
| ISO 15686 LCC | A0 Preconstruction | A1 Product Stage | A5 Construction of the civil engineering works | B1 Use | B2 Maintenance | C1 Demolitions | C3 Recycling |
| Cost | Net present value, TIR | Cost of materials | Design Costs (Professional Fees) | Electricity | Structure | Demolition | |
| | Indirect (Land, permits, facultative address) | | Work execution cost | Water | | Waste Collection | |
| | Income (Sales) | | | Gas | | Reuse / Recycling | |

**Fig. 16.** Proposal for life cycle cost analysis, BIM model, IFC. Original preparation (2020)

IFC files create a construction model based on a predefined structure that constructs the model in a logical way. When saved, the IFC file format sorts the units hierarchically according to their type, this way we can include; project data, costs (project, construction, use, maintenance, demolition, waste) as shown in the following Fig. 17.

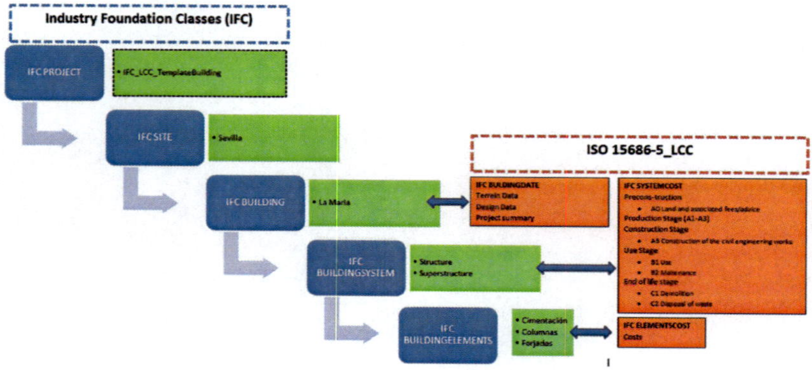


**Fig. 17.** Proposal for life cycle cost analysis, BIM model, IFC. Original preparation (2020)

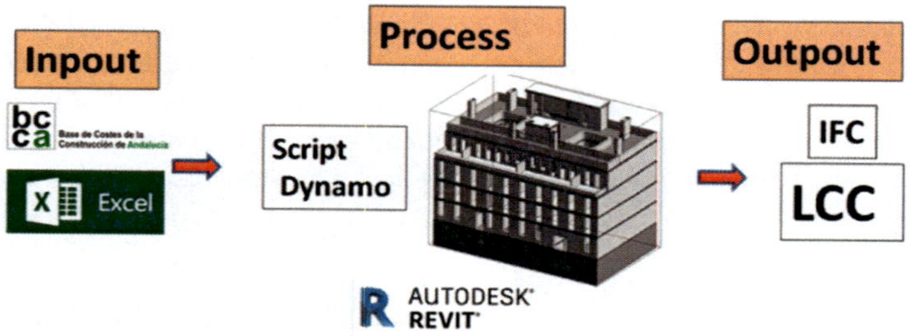


**Fig. 18.** Model proposal. Original preparation (2019)

For the IFC integration of the LCC methodology, a methodological proposal is described, for its instrumental development, as can be seen in the Fig. 18.

It basically consists of the export of IFC tables with cost information, which is intended to allow the exchange of an information model without loss or distortion of data or information. With this information, a script can be proposed in Dynamo, which combines the information from the BCCA cost model to include it in a BIM model.

- First stage: It is to import the tables with the necessary costs to the BIM model.
- Second stage: The script where the IFC and the database will interact is activated.
- Third stage: An IFC file is exported from the BIM model, with LCC data

Using IFC formats allows the transfer of all the information contained in the same model, in addition to coordinating the different models (Architecture, Structure, Installations).

In this proposal, one of the problems that exist; We have prices that are not in the cost base (such as the costs of maintenance and operation of the building, as well as income) which must be calculated according to the recommendations of ISO 15686_5.

## 5 Application Case Study

**Justification of the Choice**
For the LCC analysis, a 3D model made in Revit has been selected, within the ID i project, "Development of a Unified Tool for the Quantification and Reduction of the Environmental, Social and Economic Impact of the Life Cycle of Buildings on BIM Platforms" (Reference BIA2017-84830-R), 2017 State Plan.

It is the steel structure of a multi-family house in Seville, Spain (Fig. 19).

**Fig. 19.** Model BIM. Reference BIA2017-84830-R (2020)

**Fig. 20.** Link Data, original preparation (2020)

**LCC Data Structure Proposal Integrated to IFC**
Step 1: A file must be generated with data, which was obtained from the construction cost base of Andalucia (BCCA), Spain. Once we have made the model, each of the data

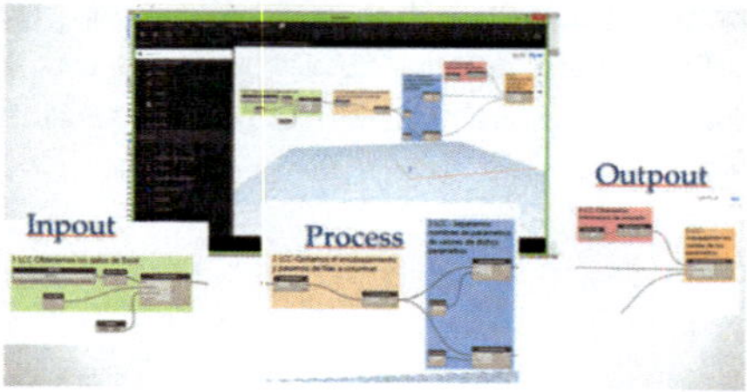


**Fig. 21.** Script dynamo, original preparation (2020)

will be entered, through a Dynamo script, Fig. 21, so that the model houses the updated costs, Fig. 20.
Costs:
Pre construction

- A0 Land and associated rates/advice

Production stage (A1–A3)
Construction stage

- A5 Construction of civil engineering works.

Usage stage

- Use B1
- Maintenance B2

Final stage of life

- C1 demolition
- C2 Waste disposal.

Step 2: Script development for data link of cost information, which helps us automatically enter data into the BIM model.

With Dynamo we will be able to generate a link between the information in the Excel file to the BIM model, Fig. 22.

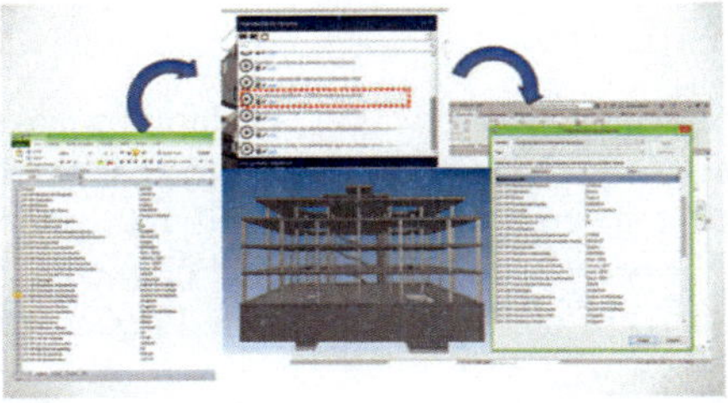

**Fig. 22.** Script dynamo, original preparation (2020)

Step 3: Export the information in the IFC 2x3 version, so you can openly share the data, Fig. 23.

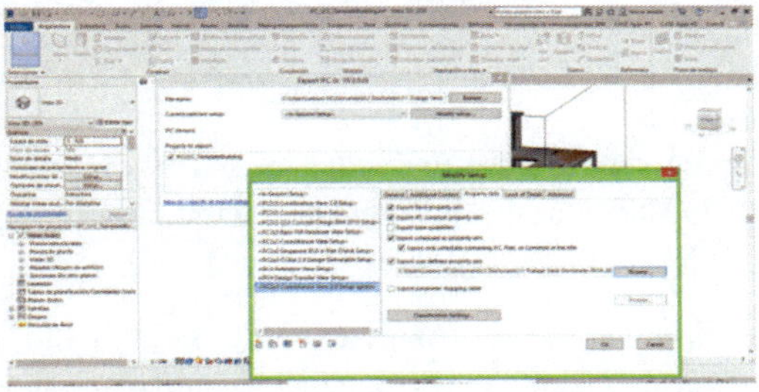

**Fig. 23.** Export, IFC 2x3 version (2020)

Step 4: The IFC information is displayed, in BIMcollab, to verify that the costs are in the exported BIM model. Below is an example of what the result of the execution would look like, export of the information and we verified that the data is not lost and found in our model (Fig. 24).

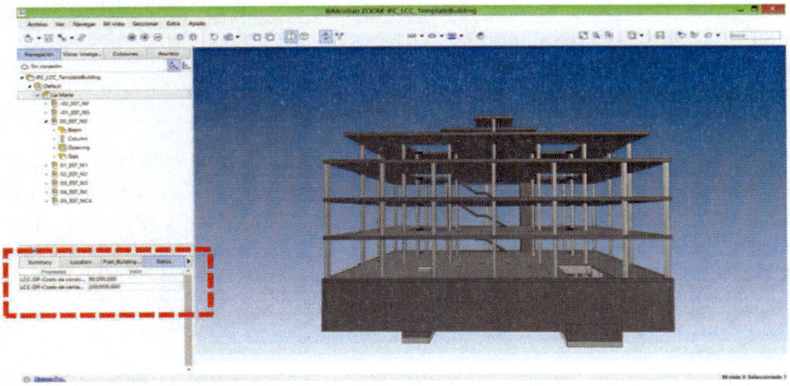

**Fig. 24.** Results visualization, BIMcollab, original preparation (2020)

## 5.1 Results

In reference to our main objective, which was to develop a proposal for an IFC structure to implement LCC, using the ISO 15686 methodology, the results were achieved and visualized on a BIM model, a script has been created and applied to be able to apply to any BIM model that has previously been modeled correctly. This makes possible a quick analysis of which parts of our building we could perfect to realize a sustainable architecture.

Regarding the secondary objective of automating the LCC information process on BIM platforms through the use of parametric software, thanks to the Autodesk Dynamo program and a research path to understand its operation and try to apply it to the LCC, an optimal level has been reached Automation, in which after linking the BIM file with Dynamo, a calculation of less than 30 s is performed, which we consider cannot be compared to that performed manually due to the large time difference used in one and the other, being the first infinitely faster than the second.

Our objective of harmonizing the manual calculation process of LCC by using calculation tables has been achieved in a remarkable way thanks to the excel tables that contain the cost information

In summary, we can assure that the main and secondary objectives proposed at the beginning of the study have been satisfactorily carried out.
Advantage:

- You can evaluate the project in two different ways, a total in which all the elements belonging to the model are taken into account and a comparison between two or more elements, which allows you to choose the one that most interests you. In other words, it allows you to select those groups (or elements) that you are interested in analyzing, so it is not necessary to analyze everything to draw partial conclusions.
- The data that is calculated within the BIM model is exported, it can be linked to our cost file.

Disadvantages:

- It is not updated at the moment, if any characteristic of the model is changed, the calculation would have to be redone.
- It cannot be projected at the same time as analyze.
- The analysis is not intuitive or visual directly on the model.

We can affirm after this study that developing a scrip is a good cost analysis tool for a building since it produces a series of graphs that summarize its LCC behavior. However, the great lack that we observe is that of not being able to use the tool at the same time that we are planning the building, but that we have to carry out the project and once the analysis has been done.

## 5.2 Conclusions

IFC is an emerging technology, created to file sharing, interoperability, including planning and budgeting in the model, which is not currently established for the execution of works. While the various BIM software developers have added a plug to facilitate the work, although not all information is correctly transferred. It is true that there are design and visualization tools that enable really powerful conflict detection and resolution, within the possibilities offered.

Finally, we conclude that the processes and interoperability between software regarding costs, needs further development and should be improved to ensure that the information flow between the model BIM 5D model can be performed satisfactorily.

## References


1. Asiedu, Y., Gu, P.: Product life cycle cost analysis: state of the art review. Int. J. Prod. Res. (1998). https://doi.org/10.1080/002075498193444
2. Fleischer, G.A., Khoshnevis, B.: Incorporating economic impact assessment into computer-aided design. In: Proceedings of the American Institute of Industrial Engineers, Annual Conference Conversion (1986)
3. ISO: ISO 15686-5:2017 - Buildings and constructed assets – Service life planning – Part 5: Life-cycle costing (2017). https://www.iso.org/standard/61148.html
4. García-Erviti, F.: El análisis del coste del ciclo de vida como herramienta para la evaluación económica de la edificación sostenible. Estado de la cuestión. Inf. La Construcción. **67** (2015). http://informesdelaconstruccion.revistas.csic.es/index.php/informesdelaconstruccion/article/view/4015
5. Flanagan, R., Kendell, A., Norman, G., Robinson, G.D.: Life cycle costing and risk management. Constr. Manag. Econ. **5**, S53–S71 (1987). https://doi.org/10.1080/01446193.1987.10462093
6. Woodward, D.G.: Life cycle costing - Theory, information acquisition and application. Int. J. Proj. Manag. **15**, 335–344 (1997). https://doi.org/10.1016/S0263-7863(96)00089-0
7. Langdon, D.: Life Cycle Costing (LCC) as a contribution to sustainable construction: a common methodology. Final Methodol. **18**, 48–54 (2007). https://doi.org/10.2478/v10012-011-0017-8
8. Goh, B.H., Sun, Y.: The development of life-cycle costing for buildings. Build. Res. Inf. **44**, 319–333 (2016). https://doi.org/10.1080/09613218.2014.993566
9. Sayed, S.S.M., Sawant, P.H.: Life-cycle cost and financial analysis of energy components in mass housing projects – a case project in sub-urban India. Int. J. Sustain. Built Environ. **4**, 202–221 (2015). https://doi.org/10.1016/j.ijsbe.2015.07.001
10. Higham, A., Fortune, C., James, H.: Life cycle costing: evaluating its use in UK practice. Struct. Surv. **33**, 73–87 (2015). https://doi.org/10.1108/SS-06-2014-0026
11. Islam, H., Jollands, M., Setunge, S., Ahmed, I., Haque, N.: Life cycle assessment and life cycle cost implications of wall assemblages designs. Energy Build. **84**, 33–45 (2014). https://doi.org/10.1016/j.enbuild.2014.07.041
12. König, H., De Cristofaro, M.L.: Benchmarks for life cycle costs and life cycle assessment of residential buildings Build. Res. Inf. **40**, 1–23 (2012). https://doi.org/10.1080/09613218.2012.702017
13. Moore, T., Morrissey, J.: Lifecycle costing sensitivities for zero energy housing in Melbourne, Australia. Energy Build. **79**, 1–11 (2014). https://doi.org/10.1016/j.enbuild.2014.04.050
14. Tokarik, M.S., Richman, R.C.: Life cycle cost optimization of passive energy efficiency improvements in a Toronto house. Energy Build. **118**, 160–169 (2016). https://doi.org/10.1016/j.enbuild.2016.02.015
15. buildingSMART International: buildingSMART (2019). https://www.buildingsmart.org/
16. Zhang, C., Beetz, J., Weise, M.: Interoperable validation for IFC building models using open standards. J. Inf. Technol. Constr. **20**, 24–39 (2015). https://doi.org/10.18154/RWTH-CONV-213541

17. ISO 16739-1:2018: ISO 16739-1:2018 Preview Industry Foundation Classes (IFC) for data sharing in the construction and facility management industries – Part 1: Data schema. Int. Organ. Stand. (2018)
18. Lilis, G.N., Giannakis, G.I., Rovas, D.V.: Detection and semi-automatic correction of geometric inaccuracies in IFC files. In: 14th Conference of International Building Performance Simulation, BS 2015 (2015)
19. Laakso, M., Kiviniemi, A.: The IFC standard - a review of history, development, and standardization. Electron. J. Inf. Technol. Constr. **17**, 134–161 (2012)
20. Autodesk, Revit IFC manual (2018).
21. https://www.autodesk.co.uk/campaigns/interoperability/ifc-handbook
22. Giannakis, G.I., Lilis, G.N., Garcia, M.A., Kontes, G.D., Valmaseda, C., Rovas, D.V.: A methodology to automatically generate geometry inputs for energy performance simulation from IFC BIM models. In: 14th Conference of International Building Performance Simulation, BS 2015 (2015)
23. Dimyadi, J., Spearpoint, M., Amor, R.: Sharing building information using the IFC data model for FDS fire simulation. In: Fire Safety Science (2008). https://doi.org/10.3801/iafss.fss.9-1329
24. Varios, Spanish journal of BIM 14/01, Spanish J. BIM (2015)
25. Yung, P., Wang, X.: A 6D CAD model for the automatic assessment of building sustainability. Int. J. Adv. Robot. Syst. (2014). https://doi.org/10.5772/58446
26. Van Berlo, L.: Future technology Presentation buildingSMART Summit Beijing. https://www.linkedin.com/pulse/future-technology-presentation-buildingsmart-summit-l%C3%A9on-van-berlo/
27. http://biblus.accasoftware.com/es/ifc-que-es-y-relacion-con-el-bim/
28. Ramirez, D.S.: Desarrollo instrumental de ACV en plataformas BIM: Herramienta paramétrica de visualización de resultados, Trabajo Fin de Master (2018)
29. ISO 21931-2:2019: Sustainability in buildings and civil engineering works — Framework for methods of assessment of the environmental, social and economic performance of construction works as a basis for sustainability assessment — Part 2: Civil engineering works (2019)
30. García-Erviti, F., Armengot-Paradinas, J., Ramírez-Pacheco, G.: Life cycle cost analysis as an economic evaluation tool for sustainable building. Sate of the art, Madrid (2015)
31. OCCS Development Committee Secretariat: About OmniClass. 02/01/2019, de OCCS Development Committee Secretariat Sitio web (2017). http://www.omniclass.org/about/
32. NATSPEC BIM: NATSPEC National BIM Guide - BIM Object/Element Matrix Manual v1.0. 02/01/2019, de NATSPEC BIM Sitio web (2011). https://bim.natspec.org/documents/natspec-national-bim-guide
33. Delany, S.: CLASSIFICATION. 02/01/2019, de NBS BIM Tool Kit Sitio web (2018). https://toolkit.thenbs.com/articles/classification
34. McPartland, R.: What is Uniclass 2015? 02/01/2018, de NBS BIM Tool Kit Sitio web (2017). https://www.thenbs.com/knowledge/what-is-uniclass-2015
35. CSI and CSC: UniFormat™ A Uniform Classification of Construction Systems and Assemblies. 02/01/2018, de Graphisoft Sitio web (2010). https://graphisoft.akamaized.net/cdn/ftp/techsupport/downloads/interoperability/UniFormat_2010m.pdf
36. (2016–2017). BIM Execution Plan v07, New Zealand International Convention Centre. 30/08/2018, de FLETCHER
37. The Construction Specifications Institute and Construction Specifications Canada: MasterFormat® Numbers & Titles. 02/01/2018, de The Edmonton Construction Association (ECA) Sitio web (2016). https://www.edmca.com/media/35207/masterformat-2016.pdf
38. BNA: NL/SfB–Tabellen Inclusief gereviseerde Elementenmethode '91. 03/01/2019, de STABU NL Sitio web (2005). https://www.stabu.nl/mijnstabu/files/NL-SfB_BNA_Boek_2005-ISBN-10%2090-807626-3-6.pdf

# Effects of Container Vehicle Flow on Intersection Capacity in Motorcycle Dominated Cities

Tran Vu Tu(✉) and Nguyen Huynh Tan Tai

Ho Chi Minh City University of Technology and Education, Ho Chi Minh City, Vietnam
{tutv,tainht}@hcmute.edu.vn

**Abstract.** To build a smart city with sustainable infrastructure transport system, studies on traffic flow or traffic management are necessary to reduce traffic congestion, traffic accident, air pollution, etc. as well as to formulate sustainable infrastructure design and maintenance policies. With more than 80% of trips are made by motorbikes, the traffic situation in Vietnam is complicated due to the travel behaviour of motorcycles and the presence of container vehicles in traffic stream. The operation of container vehicle makes the traffic situation worse when its size greatly affects the efficiency of the transportation system. In this paper, the authors investigate the effect of container vehicles on intersection capacity in motorcycle dominated traffic flow. By analysing the data collected at real sites in Binh Duong province of Vietnam, the research estimates the intersection capacity based on observed values of time headway in the moments of non-container vehicle traveling and those of having container vehicles. The research results show that the presence of container vehicles causes reduction in intersection capacity by up to 38.7%. To mimic the current situation as well as to evaluate the effect under various traffic situations, the paper builds a microscopic simulation model to simulate different scenarios. The simulation model concludes the significant effect of container vehicles on intersection capacity once the proportion of container vehicles in traffic flow changes. The research results are useful for practical design of signalized intersection in areas having much container flow as well as building policies leading to sustainable development of transport system in Vietnam.

**Keywords:** Sustainable system · Container vehicle · Traffic · Smart city · Capacity

## 1 Introduction

Since the economy is growing, Vietnam is facing many burden problems related to traffic congestion, pollution, traffic accidents that affect significantly every aspect of society [19–22]. The above problems are very serious in big cities like Ho Chi Minh city, Binh Duong province or Hanoi capital. Many causes were identified as the development of private vehicles, the underdevelopment of infrastructure systems, the presence of container vehicles in mixed traffic flow, motorcycle behavior, etc.…Among them, the operation of container vehicles in mixed traffic flow is the major concern. Although there


A. Rotaru (Ed.): CRIT-RE-BUILT 2019, SSGG, pp. 427–436, 2021.
https://doi.org/10.1007/978-3-030-61118-7_34

have been some studies on the impact of container vehicles on traffic [1, 7, 9, 11, 13–15], its influence in motorcycle dominated traffic flow has not received much attention. Studies on the influence of container vehicles on intersection capacity in the motorcycle environment are needed to understand more about the characteristics of mixed traffic flow, as well as to find solutions for urban traffic planning in developing countries like Vietnam.

## 2 Objective

The paper aims at analysing the effect of container vehicles on intersection capacity in motorcycle dominated traffic flow, including two main parts. The first main part relates to the real investigation into motorcycle dominated traffic to estimate the capacity under consideration of the container vehicle effects. The second main part concerns the simulation analysis in VISSIM [18] to analyze the effect of container vehicles on the intersection capacity under various scenarios.

## 3 Methodology

### 3.1 Capacity Estimation Method

The capacity is defined as the maximum hourly flow of vehicles that can be discharged through the intersection from the lane group under the prevailing traffic, roadway, and signalization conditions [16]. Based on the saturation flow rate, the intersection capacity is calculated by using an adjustment to take the real signal timing into account. The formula is given below.

$$c = (g/C) \cdot s \tag{1}$$

where

$c$ = capacity (*pcu/hour*)
$g$ = Effective green time for the phase (*sec*)
$C$ = Cycle length (*sec*)
$s$ = Saturation flow rate (*pcu/hour)*

The saturation flow rate (see Fig. 1) depends on roadway and traffic conditions and usually be estimated with the time saturation headway [2–6, 8, 10, 12, 17].

$$s = \frac{3600}{h_s} \tag{2}$$

where

$s$ = saturation flow rate (*pcu/hour*);
3600 = number of seconds per hour;
$h_s$= saturation headway ($s$).

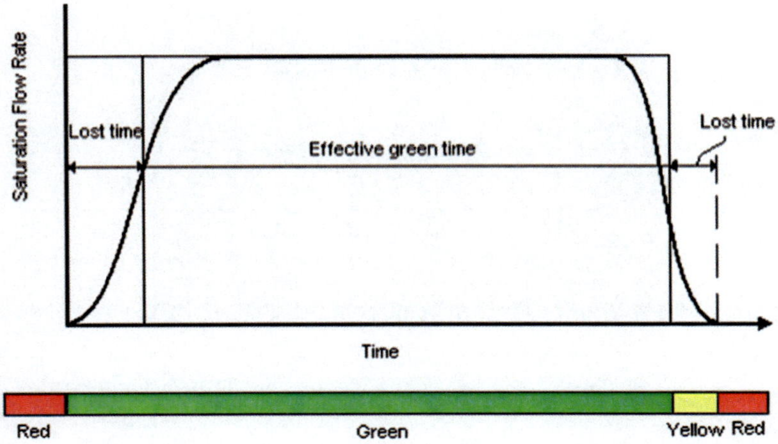


**Fig. 1.** Saturation flow rate at signalized intersections [17]

To investigate the effect of container vehicles on intersection capacity, the research distinguishes the situation of mixed traffic flow into two scenarios. The first one is the scenario of saturation flow with the absence of container vehicle traveling. The second one is that in the presence of container vehicles. The authors measure the saturated headway for each and compare the values of capacity.

### 3.2 Data Collection

The collected data includes geometrical data (the number of lanes, width, length, grade, etc.), control data (traffic signal, lane restriction, traffic signals, priority rules, etc.) and traffic data (traffic flow, traffic proportion, speed, etc.). Through aerial photos and field investigation, all these above information is collected. The research records video tapes and analyzes in the lab to extract information about turning volume, traffic count, vehicle type, traffic proportion, traffic volume, etc. The study site is proposed to be at Di An town of Binh Duong province. This is an intersection connecting DT743 road and DT743B (see Fig. 2). The typical cross-section is shown in Fig. 3.

This intersection is a two-phase signalized intersection. The signal cycle time is 81 s. The signal phases are shown in Fig. 4.

### 3.3 Time Headway Collection

The research measured time headway in two cases at the signalized intersection: the case of having container vehicles traveling and that of the absence of container vehicle. The time headway is then drawn to get the saturated headway. The collection cases are shown in Fig. 5.

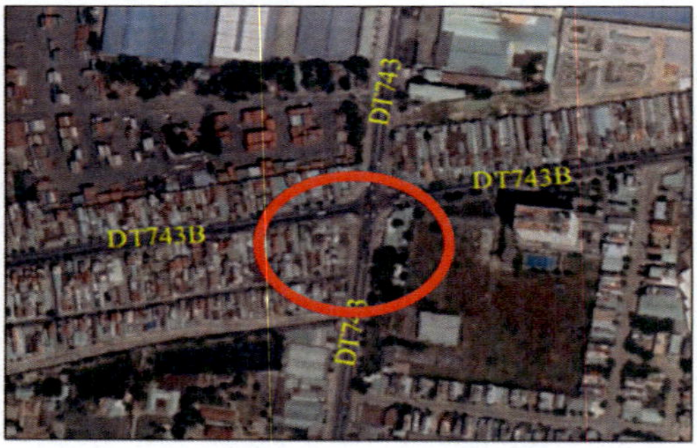


**Fig. 2.** Location of the study site

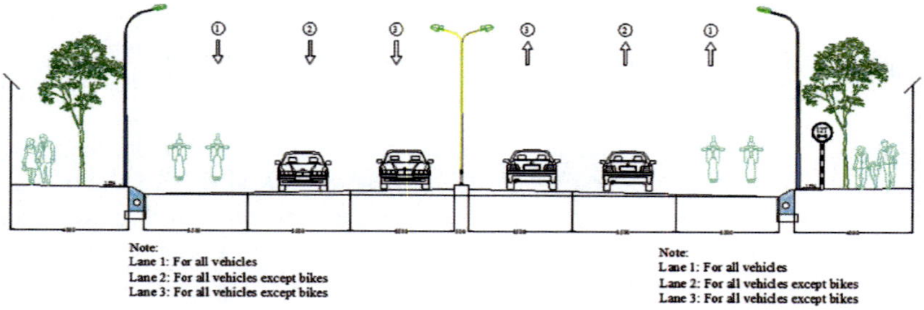


**Fig. 3.** Cross-section of DT743 road

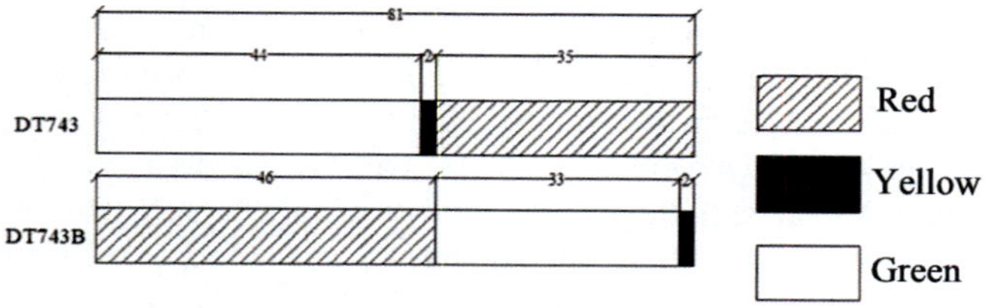


**Fig. 4.** Signal phases at the study intersection.

# 4 Analysis Results

## 4.1 Data Analysis

Based on the recorded videos, the composition of transportation modes traveling at the intersection was calculated. Besides the large number of two-wheel vehicles (nearly 80.6%), the percentage of container vehicle is significant, with the figure of 9.3%. The percentage of each type of transportation modes in the traffic flow at the study site is shown in Fig. 6.

a. Headway measurement in the case of absence of container vehicle

b. Headway measurement in the case of presence of container vehicles

**Fig. 5.** Time headway collection

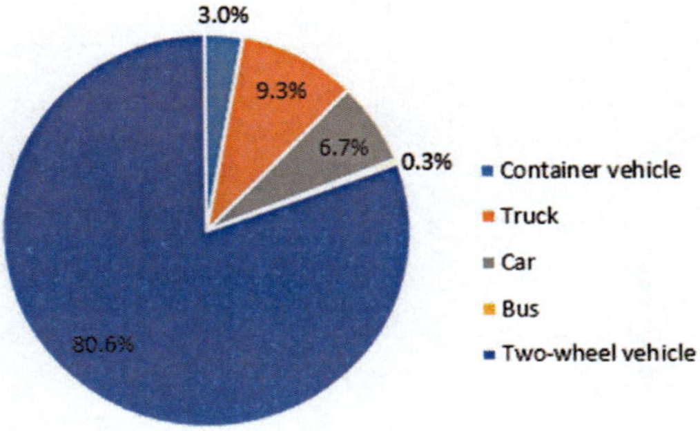


**Fig. 6.** Traffic proportion at the study site

Based on Vietnam standard on Passenger car equivalent unit (PCU) for urban street (TCVN 104-2007), the authors converted all types of vehicle to PCU values to homogenize the heterogeneous traffic flow. The PCU values are determined in Table 1.

**Table 1.** PCU values according to Vietnam standard (TCVN 104-2007).

| Vehicle type | Speed, km/h | | |
|---|---|---|---|
| | Over 60 | 30, 40, 50 | Under 20 |
| Bicycle | 0.50 | 0.30 | 0.20 |
| Motorcycle | 0.50 | 0.25 | 0.15 |
| Car | 1.00 | 1.00 | 1.00 |
| Two-axle truck and small bus | 2.00 | 2.50 | 2.50 |
| Three-axle truck and big bus | 2.50 | 3.00 | 3.50 |
| Container truck | 3.00 | 4.00 | 4.50 |

According to Vietnam Standard (Table 1), the PCU values depend on vehicle types and the travel speed (km/h) on road. The travel speed at the studied roads is designed as 50 km/h. The research uses these PCU values to convert the heterogeneous traffic flow at the studied location. The converted values of traffic flow for each approach of the intersection are compared to determine the typical approach for capacity analysis. Based on the collected data, the research analyzed the traffic flow for each approach. The flow patterns in each direction change over time and differ in each direction. The outline of the approaches with their traffic flow is shown in Fig. 7.

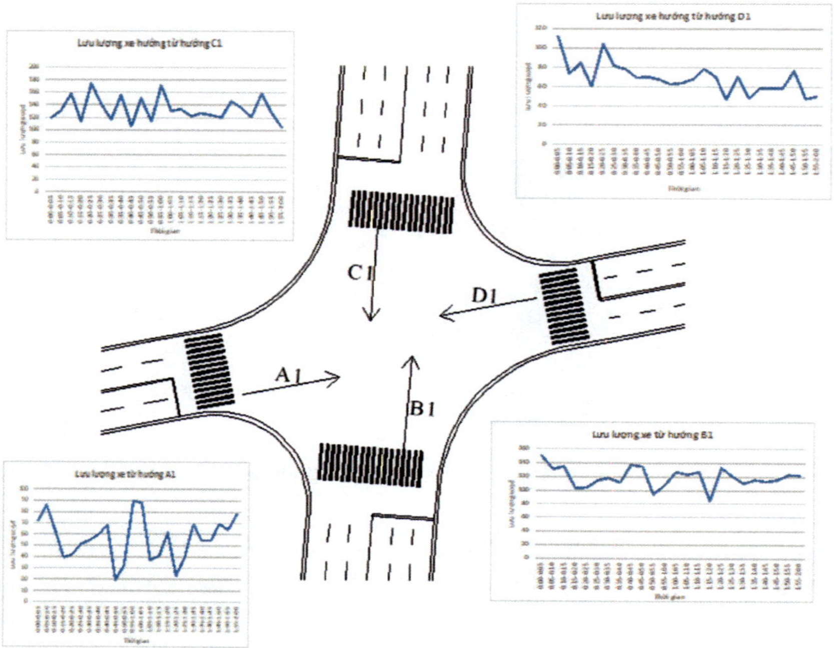


**Fig. 7.** Traffic flow (*pcu/h*) at the intersection

## 4.2 The Effect on Capacity Reduction

Based on the observed values of time headway, the authors determine the saturation flow rate of the typical approaches for each direction (DT743 road and DT743B road). Based on the calculated saturation flow rates, the values of capacity were estimated for the cases of absence and presence of container vehicles. The values of capacity for each approach are calculated and compared as follows:

As shown in Fig. 8, the capacity of typical approach in DT743 road and DT743B road reduces significantly under the current condition of traffic composition due to the influence of container vehicles. As for the approach in DT743 road, the capacity reduces by 30.3% compared to the scenario of no container vehicle traveling. Meanwhile, this figure is about 38.7% for the typical approach in road DT743B. The authors use VISSIM [18] as a tool to simulate the traffic flow at the study intersection. After calibrating the model, the model is validated by using Mean Percentage Error (MPE) as a measure of goodness of fit. The definition is as follows:

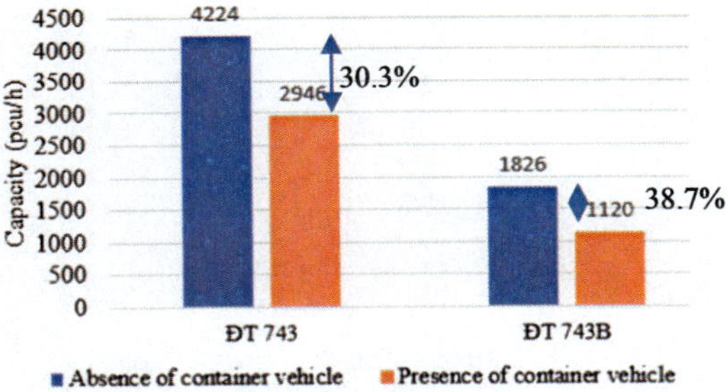


**Fig. 8.** Capacity comparison

$$MPE = \frac{1}{N}\sum_{n=1}^{N}\frac{|O_n - S_n|}{O_n} \tag{3}$$

where,

$S_n$: Simulation values at point $n$
$O_n$: Observation values at point $n$
$N$: The number of observation points

Based on the simulation data and observation data, the traffic flows at the intersection are plotted around the 45-degree line with small relative error (MPE = 1.8%) asserting that the traffic flows in the simulation model and at the real study site are approximately same (see Fig. 9). The validation result is shown in the following figure.

After validating the base scenario, the research simulates traffic scenarios with various values of container vehicle proportion to investigate the effect of container vehicles on intersection capacity. The capacity is indirectly calculated based on the data output from the simulation model. The relationship between intersection capacity and the proportion of container vehicles in traffic stream is shown as in Fig. 10.

As shown in Fig. 10, the number of container vehicles in the flow and the capacity have strong relationship. The proportion of container vehicle increases, the capacity reduces significantly. If the proportion of container vehicles increases by 1%, the capacity of the intersection decreases by around 21 (pcu/h). This is reasonable because of the big size of container vehicle hinders the operation of other modes, especially when container vehicles turning (left or right) at intersections.

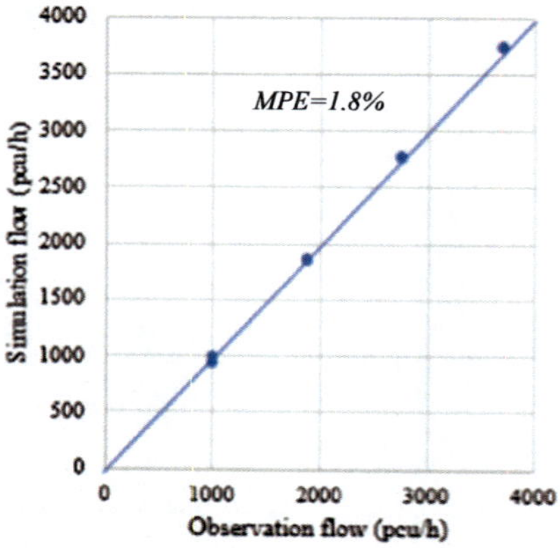


**Fig. 9.** Traffic flow validation

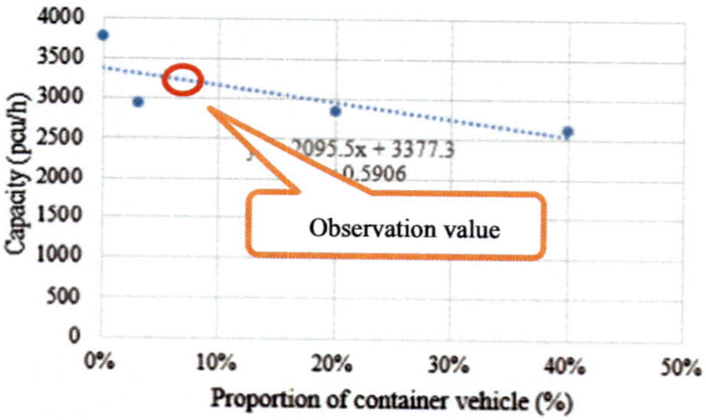


**Fig. 10.** The relationship between capacity and proportion of container vehicles

## 5 Conclusions

The paper presents the effect of container vehicles on intersection capacity in motorcycle dominated traffic flow. The presence of container vehicles in traffic has caused traffic jam and traffic accident in developing countries like Vietnam. By analysing the data collected at real sites in Binh Duong province of Vietnam, the research evaluates the capacity reduction due to the presence of container vehicles. The presence of container vehicles causes reduction in the intersection capacity by up to 38.7%. To evaluate the effect of container vehicles on capacity under various situations, the research builds a microscopic simulation model to mimic the current situation and simulate various traffic scenarios with different values of container vehicle proportion. The simulation model shows that the proportion of container vehicles affects significantly the intersection capacity. If the proportion of container vehicles increases by 1%, the capacity of the intersection decreases by around 21 (pcu/h). Although the study provides good information in planning container transport activities in urban streets, the analysis in this research is simple and has some limitations. Other factors such as traffic composition, the complicated behavior of motorcycles, vehicle sizes, etc. should be considered in capacity estimation process. Moreover, the comparison among methods of capacity estimation should be conducted to show the sensitivity of the results. These should be considered in future research studies.

## References

1. Benekohal, R., Zhao, W.: Delay-based passenger car equivalents for trucks at signalized intersections. Transp. Res. Part A: Policy Pract. **34**(6), 437–457 (2000). Elsevier Science, Limited, Oxford, England
2. Bonneson, J.A., Nevers, B., Pratt, M.P., Bonyani, G.: Influence of area population, number of lanes, and speed limit on saturation flow rate. Transp. Res. Rec. J. Transp. Research Board, TRR 1988, Washington DC (2006)
3. Cohen, S.L.: Application of car-following systems in microscopic time-scan simulation models. J. Transp. Res. Board, TRR 1802, Washington DC (2002)
4. Cohen, S.L.: Application of car-following systems to queue discharge problem at signalized intersections. Transp. Res. Rec. J. Transp. Res. Board (2002). TRR 1802, Washington DC
5. Federal Highway Administration (1989), TRAF-NETSIM User's Manual, FHWA, Washington DC. (Note: This program is currently named TSIS-CORSIM, and the latest version (6.1), 1989, is available from the McTrans Center at the University of Florida, mctrans.ce.ufl.edu. Accessed 21 Nov 2019
6. Joseph, J., Chang, G.-L.: Saturation flow rates and maximum critical lane volumes for planning applications in Maryland. J. Transp. Eng. Am. Soc. Civ. Eng. **131**(12), 946–952 (2005)
7. Kockelman, K., Shabih, R.: Effect of light-duty trucks on the capacity of signalized intersections. J. Transp. Eng. Am. Soc. Civ. Eng. **126**(6), 506–512 (2000)
8. Li, H., Prevedouros, P.: Detailed observations of saturation headway and start-up lost times. J. Transp. Res. Board (2002). TRR 1802, Washington DC
9. Lin, F.-B., Tseng, P.-Y., Su, C.-W.: Variations in queue discharge patterns and their implications in analysis of signalized intersections. Transp. Res. Rec. J. Transp. Res. Board 192–97 (2004). No. 1883. TRB, National Research Council. Washington DC
10. Long, G.: Inter-vehicle spacing and queue characteristics. Transp. Res. Rec. J. Transp. Res. Board 86–96 (2002). No. 1796. TRB, National Research Council. Washington DC
11. Molina Jr., C.J.: Development of passenger car equivalencies for large trucks at signalized intersections. Inst. Transp. Eng. J. **57**(11), 33–37 (1987)
12. Perez-Cartagena, R., Tarko, A.: Calibration of capacity parameters for signalized intersections in Indiana. J. Transp. Eng. Am. Soc. Civ. Eng. **131**(12), 904–911 (2005)
13. Transportation Research Board, Special Report 209: Highway Capacity Manual. Transportation Research Board, National Research Council, Washington DC. Tsao, S.M. & Chu, S.W. (1995), Adjustment factors for heavy vehicles at signalized intersections. J. Transp. Eng. Am. Soc. of Civ. Eng. **121**(2), 150–57 (2000)
14. Washburn, S.S., Cruz-Casas, C.O.: Impact of trucks on arterial LOS and freeway work zone capacity—Part A: impact of trucks on arterial LOS. Final Report. Florida Department of Transportation. Tallahassee, FL, p. 107 (2007)
15. Washburn, Cruz-Casas: Impact of Trucks on Signalized Intersection Capacity, Computer-Aided Civil and Infrastructure Engineering, pp. 1–16 (2010)
16. The theory and concepts of saturation flow rate: https://www.webpages.uidaho.edu/niatt_labmanual/Chapters/signaltimingdesign/theoryandconcepts/CapacityAndSaturationFlowRate.htm. Accessed 21 Nov 2019

17. Saturation flow rate: https://repository.up.ac.za/bitstream/handle/2263/5838/002.pdf%3Bjsessionid%3D4FCD4E6450E063E61E9B02. Accessed 29 Jan 2020
18. VISSIM. http://vision-traffic.ptvgroup.com/en-us/products/ptv-vissim/. Accessed 29 Jan 2020
19. Doan, T.T., Tran, V.T., Sano, K.: A study on motorcycle equivalent unit in urban streets. Asian Transp. Stud. (ATS) **5**(2), 243–255 (2018)
20. Linh, T.T., Sano, K., Amila, J., Tran, V.T.: Two-player game theory based analysis of motorcycle driver's behavior at signalized intersection. Asian Transp. Stud. (ATS) **5**(2), 272–291 (2018)
21. Tran, V.T., Bo, V.T.: Simulation of bus lanes in motorcycle-dominated traffic flow by using agent-based model. Int. J. Traff. Transp. Eng. **7**(2), 232–242 (2017). ISSN 2217-5652
22. Tran, V.T., Sano, K.: Genetic algorithm for optimization in adaptive bus signal priority control. Int. J. Optim. Control: Theor. Appl. **3**(1), 35–43 (2013)
23. Tran, V.T., Nishiuchi, H.: A comparison of motorcycle Ownership and Use in Ho Chi Minh City and Bangkok. In: 4th International Conference on Green Technology and Sustainable Development. IEEE, pp. 825–830 (2018)

# The Effect of Thermal Bridge Junctions Between Pillars and Walls in the Energy Demand of Buildings in Warm Climate

David Bienvenido-Huertas(✉), María Victoria Montes, Carlos Rubio-Bellido, and Jacinto Canivell

University of Seville, 41012 Seville, Spain
jbienvenido@us.es

**Abstract.** Currently, the building stock is energy inefficient. Consequently, the residential sector is one of the main sources emitting Greenhouse Gases, mainly due to the poor thermal performance of envelopes. Thermal bridges are among those envelope elements where heat losses or gains take place. A previous study highlighted the importance of controlling the linear thermal transmittance in junctions, such as slab fronts. However, there is a lack of studies analysing the thermal bridges of pillars and their effect on the energy demand of buildings located in warm climate zones. This study therefore analyses how the linear thermal transmittance of pillars affects the building energy demand. For this purpose, a case study located in Seville was analysed in 3 different climatic scenarios (current, 2050, and 2100). The case study was simulated with 3 different designs of junctions between pillars and walls. The linear thermal transmittance was determined using a two-dimensional simulation, and the energy demand was determined using EnergyPlus. The results of this study confirm the importance of controlling the thermal bridges of pillars and their impact on the energy demand.

**Keywords:** Thermal bridges · Pillars · Junctions · Linear thermal transmittance · Energy demand · Climate change

## 1 Introduction

Existing buildings have a high energy consumption which generates serious effects on the environment [1, 2]. For this reason, the European Union has set a series of goals to achieve a low-carbon economy by 2050 [3]. Among these goals, reduction values of carbon dioxide emissions are established in various sectors. As for the building sector, emissions should be reduced by 90% [3].

To achieve this goal, the main reason of energy consumption in buildings should be addressed: the use of HVAC systems [4, 5]. To reduce this type of energy consumption, different energy saving measures, such as the use of systems with a better performance or the building envelope improvement, could be performed in buildings. Regarding the envelope, controlling its thermal properties is of great importance to reduce energy


A. Rotaru (Ed.): CRIT-RE-BUILT 2019, SSGG, pp. 437–448, 2021.
https://doi.org/10.1007/978-3-030-61118-7_35

consumption [6–8]. Highly effective thermophysical properties for the envelope allow the building energy demand to be significantly reduced [9]. The envelope is generally made up of a series of layers of different materials and thicknesses determining its thermal resistance. However, there are zones in the envelope where the junction between various elements causes a thermal bridge. A thermal bridge is understood as the part of the envelope presenting variations in its thermal resistance due to the junctions of materials with different thermal resistances [10].

These thermal bridges are responsible for generating heat losses in winter and heat gains in summer [11, 12], with variations of up to 30% in the heating demand [13]. The reason is that the thermal bridge can increase up to 35% the thermal transmittance value of a wall [14]. As a result, due to both the lowest thermal resistance and the associated energy losses, the analysis of thermal bridges influences the building energy demand [15, 16] and the determination of energy saving measures that should be implemented.

There are various studies in the scientific literature analysing the effect of certain typologies of thermal bridges in the energy behaviour of buildings, such as reinforced concrete structures [17, 18], curtain walls [19] or ceilings [20]. Nevertheless, there is a lack of studies analysing the effect of improving thermal bridges on the junctions of pillars and façades in corners. Also, there are few studies conducted in warm climates. For this reason, the goal of this study is analysing the energy behaviour of various building solutions considering this type of junctions and applied to a case study located in a warm region. To do this, an actual case study with problems of heat losses in this type of thermal bridge was selected, and the existing design was compared to other two building solutions. The case study was in Seville, which has a Csa class according to Köppen-Geiger climate classification [21]. The linear thermal transmittance was obtained through two-dimensional simulations performed according to ISO 10211:2007 [10], whereas the energy demand was obtained through simulations performed with EnergyPlus.

## 2 Methodology

### 2.1 Case Study and Energy Simulation

To carry out this research, an existing case study in Seville was selected. The building is a single-family dwelling from 2008. After analysing it through infrared thermography, heat losses were found in the thermal bridge of pillars and walls in corners (see Fig. 1 (a)). The analyses were conducted according to EN 13187:1998 [22].

The use of this case study allowed therefore to have a building appropriate to the needs of this research. For its energy simulation, a thermal model was designed in Design-Builder appropriate to the characteristics of the building. As its technical documentation was available, its dimensions and design characteristics could be accurately known. In this regard, one of the most important aspects was to know the thermal properties of the façade (see Table 1).

Regarding the use and internal loads of the building, the usage profile defined by the Spanish Building Technical Code (in Spanish, CTE [23]) was used. This profile is generic and representative of the use of buildings in most building stock designed according to criteria of the CTE. Table 2 indicates the setpoint temperatures defined in the CTE, and Table 3 indicates the percentage contributions of the internal loads. It

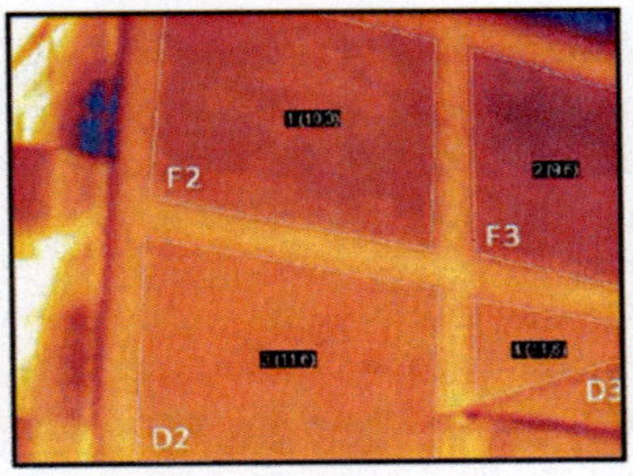

(a)

(b)

**Fig. 1.** Case study of the Research: (a) thermographies conducted in which heat losses can be seen through the thermal bridge between pillars and walls, and (b) 3D scheme of the case study.

**Table 1.** Layer thickness and thermophysical properties of the building façade.

| Layers | Thickness [mm] | Thermal conductivity [W/(mK)] | Thermal resistance [(m$^2$K)/W][a] |
|---|---|---|---|
| Cement mortar | 10 | 0.70 | – |
| Perforated brick | 110 | 0.59 | – |
| Cement mortar | 10 | 1.30 | – |
| PUR insulation | 20 | 0.03 | – |
| Air gap | 40 | – | 0.18 |
| Hollow brick | 50 | 0.44 | – |
| Gypsum plaster | 10 | 0.40 | – |

[a]Thermal resistance obtained from ISO 6946 [27].

is important to note that the maximum load for equipment and lighting is 4.40 W/m$^2$, whereas for the latent and sensitive occupancy are 1.36 and 2.15 W/m$^2$, respectively. As the variable analysed in the energy simulations was the energy demand, it was not necessary to consider special performance requirements for the HVAC system of the case study.

**Table 2.** Setpoint temperatures in the case study.

| Setpoint temperature | Month | Hour | | | | | | |
|---|---|---|---|---|---|---|---|---|
| | | 1–7 | 8 | 9–15 | 16–18 | 19 | 20–23 | 24 |
| Cooling | Jan.–May | – | – | – | – | – | – | – |
| | Jun.–Sep. | 27 | – | – | 25 | 25 | 25 | 27 |
| | Oct.–Dec. | – | – | – | – | – | – | – |
| Heating | Jan.–May | 17 | 20 | 20 | 20 | 20 | 20 | 17 |
| | Jun.–Sep. | – | – | – | – | – | – | – |
| | Oct.–Dec. | 17 | 20 | 20 | 20 | 20 | 20 | 17 |

**Table 3.** Percentage of loads in the case study.

| Load | Hour | | | | | | |
|---|---|---|---|---|---|---|---|
| | 1–7 | 8 | 9–15 | 16–18 | 19 | 20–23 | 24 |
| Occupancy | 100 | 25 | 25 | 50 | 50 | 50 | 100 |
| Occupancy (weekend) | 100 | 100 | 100 | 100 | 100 | 100 | 100 |
| Equipment | 10 | 30 | 30 | 30 | 50 | 100 | 50 |
| Lightning | 10 | 30 | 30 | 30 | 50 | 100 | 50 |

Climate files were obtained through METEONORM. As for the climate files for 2050 and 2100, the climate change scenario of the Intergovernmental Panel on Climate Change (IPCC) was selected [24]. In this sense, the scenario A2 was selected because it is one of the scenarios which has the greatest effects of climate change [24]. It is also the scenario most used in other similar research studies [25, 26].

## 2.2 Determining the Linear Thermal Transmittance

To assess the effect of the thermal bridge, 3 different designs were analysed (see Fig. 2). The first design corresponds to the current design of the building, and the other two designs constitute two proposals to improve the performance of the junction. The advantage of these two new designs is that they could be applied to the building. Therefore, the results also aimed to analyse the actual possibilities of applying the measure. To determine the linear thermal transmittance ($\psi$), two-dimensional simulations were performed in HTFlux. The modelling of the thermal bridges was carried out according to the criteria included in ISO 10211:2007 [10]. In this sense, walls were cut at a distance of three times the thickness of the wall. The surface thermal resistances indicated in ISO 6946 were used as boundary conditions of the simulation [27]: 0.13 $m^2K/W$ for the internal surface thermal resistance and 0.04 $m^2K/W$ for the external surface thermal resistance.

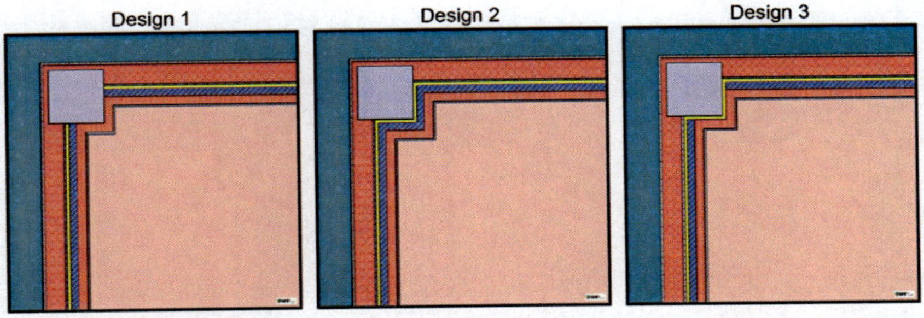


**Fig. 2.** Junction designs between the pillars and walls analysed in this research. Design 1 corresponds to the actual design of the building, and designs 2 and 3 are the improvement proposals.

## 3 Results and Discussion

Firstly, three designs of junction between pillars and walls were simulated to assess the existing differences between the linear thermal transmittance of the current design (design 1) and the two improvement proposals. Figure 3 represents the linear thermal transmittance values obtained by each design, showing that the linear thermal transmittance decreased considerably with the new designs, having the design 2 the lowest linear thermal transmittance. Whereas design 2 obtained a decrease of the linear thermal transmittance of 0.392 W/(mK), the decrease was slightly lower in design 3 (0.344 W/(mK)). Design 2 would be therefore the most recommended option for the decrease of the linear thermal transmittance, without considering other aspects resulting from the execution of the design and the economic cost of the measure.

After determining the design with the greatest decrease of the linear thermal transmittance, the energy saving achieved in the three scenarios considered in this research (current, 2050, and 2100) were studied. Figure 4 represents the monthly values of the heating and cooling energy demand obtained in the simulation process. The tendencies presented by the models simulated in each scenario were similar. However, differences between the two designs were found. The use of design 2 allowed slight decreases to be obtained in the energy demand. These decreases were greater in heating than in cooling.

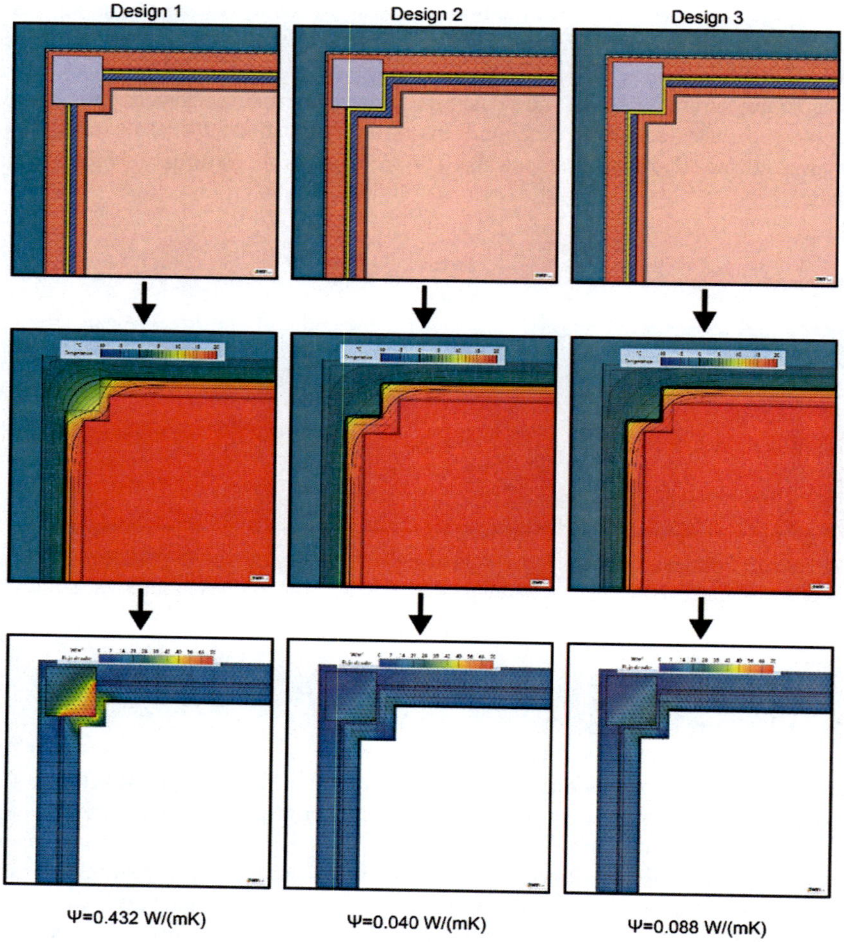


**Fig. 3.** Results of two-dimensional simulation obtained in HTFlux (isotherm and heat flux profiles) and the linear thermal transmittance value obtained.

Tables 4, 5 and 6 indicate the percentage deviations for the different periods considered and the obtained values of energy demand. As can be seen, the effect of design 2 on the monthly energy demand depended on the type of demand. In this sense, the monthly heating demand reached decreases of 298.87 kWh, whereas the monthly cooling saving reached maximum values of only 111.88 kWh.

Regarding future scenarios, the effect of design 2 on energy demands was similar to that of the current scenario. However, the progressive rise of external temperatures generated that the cooling energy demand was more and more high, thus significantly influencing the annual energy demand obtained in future scenarios (see Fig. 5).

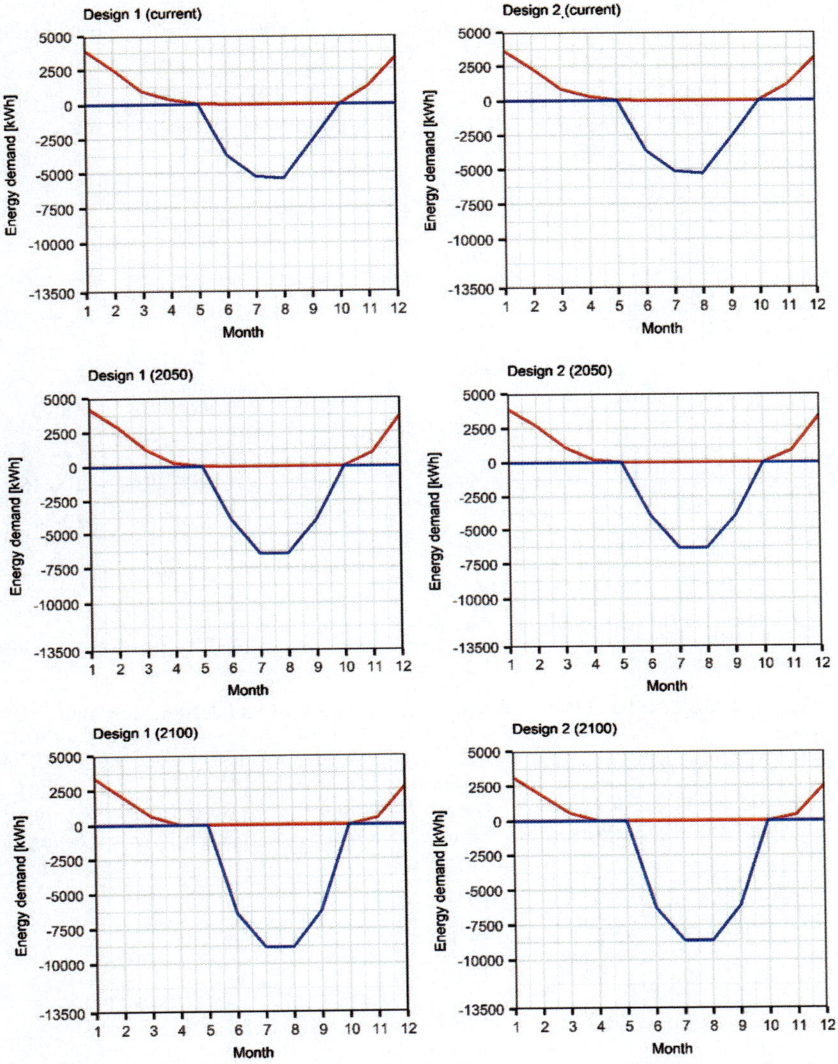


**Fig. 4.** Monthly building energy demand in the different scenarios considered (current, 2050, and 2100) using design 1 and design 2. The cooling energy demand is represented by the blue line, and the heating energy demand by the red line.

The decrease obtained in the total energy demand was low in all scenarios and with a decreasing tendency. In this regard, the decrease achieved in the total annual energy demand was 4.69% in the current scenario, 4.31% in 2050, and 3.64% in 2100. So, the improvement of the thermal bridge achieved low energy savings in the building in the different scenarios considered, with a higher incidence on the heating demand. These savings show the possible low economic profitability of carrying out energy improvement

**Table 4.** Differential of energy demand between design 1 and design 2 for the current period.

| Month | Heating energy demand | | | Cooling energy demand | | |
|---|---|---|---|---|---|---|
| | Design 1 [kWh] | Design 2 [kWh] | Energy saving [kWh] | Design 1 [kWh] | Design 2 [kWh] | Energy saving [kWh] |
| Jan. | 3,927.93 | 3,629.06 | 298.87 | 0 | 0 | – |
| Feb. | 2,500.88 | 2,302.68 | 198.20 | 0 | 0 | – |
| Mar. | 928.75 | 823.52 | 105.23 | 0 | 0 | – |
| Apr. | 344.46 | 282.03 | 62.43 | 0 | 0 | – |
| May. | 37.85 | 27.99 | 9.86 | 0 | 0 | – |
| Jun. | 0 | 0 | – | −3,702.53 | −3,646.33 | 56.20 |
| Jul. | 0 | 0 | – | −5,250.64 | −5,138.76 | 111.88 |
| Aug. | 0 | 0 | – | −5,409.39 | −5,302.93 | 106.46 |
| Sep. | 0 | 0 | – | −2,729.39 | −2,716.05 | 13.34 |
| Oct. | 16.96 | 9.97 | 6.99 | 0 | 0 | – |
| Nov. | 1,260.19 | 1,114.69 | 145.50 | 0 | 0 | – |
| Dec. | 3,417.71 | 3,146.96 | 270.75 | 0 | 0 | – |

**Table 5.** Differential of energy demand between design 1 and design 2 for 2050.

| Month | Heating energy demand | | | Cooling energy demand | | |
|---|---|---|---|---|---|---|
| | Design 1 [kWh] | Design 2 [kWh] | Energy saving [kWh] | Design 1 [kWh] | Design 2 [kWh] | Energy saving [kWh] |
| Jan. | 4,212.87 | 3,902.33 | 310.54 | 0 | 0 | – |
| Feb. | 2,880.90 | 2,661.48 | 219.42 | 0 | 0 | – |
| Mar. | 1,222.32 | 1,098.71 | 123.61 | 0 | 0 | – |
| Apr. | 233.44 | 185.06 | 48.38 | 0 | 0 | – |
| May. | 19.30 | 15.45 | 3.85 | 0 | 0 | – |
| Jun. | 0 | 0 | – | −3,948.27 | −3,902.54 | 45.73 |
| Jul. | 0 | 0 | – | −6,473.60 | −6,338.13 | 135.47 |
| Aug. | 0 | 0 | – | −6,466.29 | −6,336.14 | 130.15 |
| Sep. | 0 | 0 | – | −3,992.82 | −3,945.31 | 47.51 |
| Oct. | 5.59 | 2.58 | 3.01 | 0 | 0 | – |
| Nov. | 976.94 | 855.43 | 121.51 | 0 | 0 | – |
| Dec. | 3,713.64 | 3,430.19 | 283.45 | 0 | 0 | – |

**Table 6.** Differential of energy demand between design 1 and design 2 for 2100.

| Month | Heating energy demand | | | Cooling energy demand | | |
|---|---|---|---|---|---|---|
| | Design 1 [kWh] | Design 2 [kWh] | Energy saving [kWh] | Design 1 [kWh] | Design 2 [kWh] | Energy saving [kWh] |
| Jan. | 3,401.88 | 3,134.75 | 267.13 | 0 | 0 | – |
| Feb. | 2,011.41 | 1,837.36 | 174.05 | 0 | 0 | – |
| Mar. | 623.18 | 550.18 | 73.00 | 0 | 0 | – |
| Apr. | 19.61 | 11.66 | 7.95 | 0 | 0 | – |
| May. | 0.16 | 0.04 | 0.12 | 0 | 0 | – |
| Jun. | 0 | 0 | – | −6,371.87 | −6,255.36 | 116.51 |
| Jul. | 0 | 0 | – | −8,828.43 | −8,621.48 | 206.95 |
| Aug. | 0 | 0 | – | −8,827.21 | −8,629.28 | 197.93 |
| Sep. | 0 | 0 | – | −6,213.47 | −6,113.60 | 99.87 |
| Oct. | 0 | 0 | – | 0 | 0 | – |
| Nov. | 480.56 | 412.31 | 68.25 | 0 | 0 | – |
| Dec. | 2,781.38 | 2,552.72 | 228.66 | 0 | 0 | – |

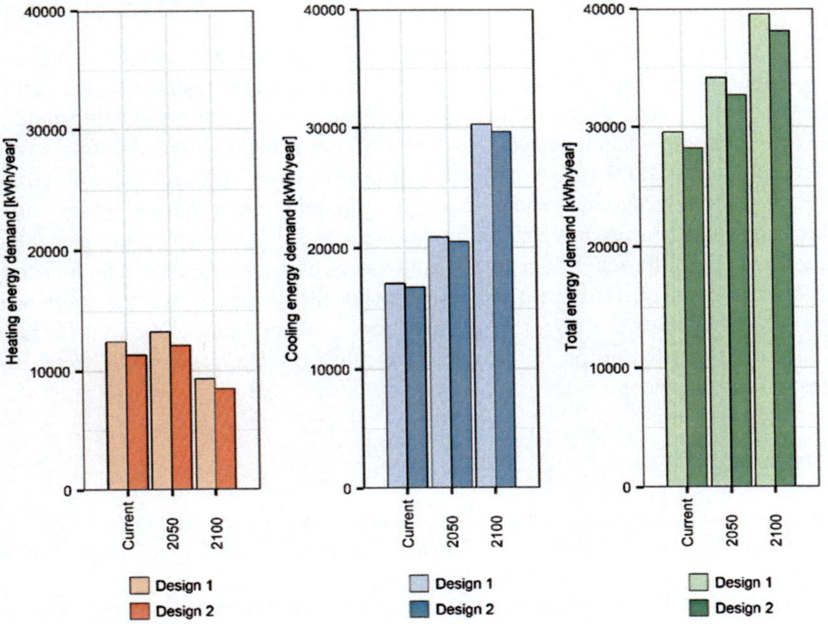


**Fig. 5.** Annual building energy demand in the different scenarios considered (current, 2050, and 2100) using design 1 and design 2.

measures in the existing buildings in warm climates to reduce the linear thermal transmittance in these thermal bridges. These results have a similar trend to that detected in a study of the slab front improvement in the same building [18], where it was detected how the decrease in linear thermal transmittance had a greater effect on the heating energy consumption than on the cooling energy consumption. However, the use of effective designs for these thermal bridges in new buildings are an opportunity due to the low impact produced. In addition, although the savings achieved are low, the effect generated by this low energy saving should be considered at a greater scale as little reductions could significantly affect the decrease of the environmental impact of the sector [28].

## 4 Conclusions

This research aimed at studying the effect of the thermal bridge of the junction of pillars and walls in corners on the energy demand of a case study located in a warm region. The analysis was conducted in a climate context with various scenarios. The results showed that the use of effective designs in the junctions led to significant decreases in the linear thermal transmittance. This decrease achieved decreases in the two types of building energy demand, although the saving in heating is greater than in cooling. As the energy demand with a higher incidence on the buildings of the zone is the cooling energy demand, the annual energy saving obtained is low and with a decreasing tendency in future scenarios.

Nevertheless, the effect generated by this low energy saving at a greater scale could have considerable decrease effects of the environmental impact caused by the building stock. Although the economic profitability of applying these designs in existing buildings is low (due to high payback periods), the use of these designs in new buildings would guarantee a building stock with a more efficient performance. It is also important to note that the results obtained are based on the analysis of a sole type of thermal bridge and in a climate zone. As other research studies have obtained similar energy saving percentages using other types of thermal bridges [18], future steps of this research will be focused on the combined analysis of different improvements of the linear thermal transmittance in buildings located in various regions and the economic assessment of their performance, so an accurate knowledge on the profitability of using effective designs for thermal bridges will be therefore available.

## References

1. European Environment Agency Final energy consumption by sector and fuel
2. European PROJECT iNSPiRe Report, EURAC (European Academy of Bozen/Bolzano) (2015)
3. European Commission: A Roadmap for moving to a competitive low carbon economy in 2050, Brussels, Belgium, pp. 1–15 (2011)
4. Kurekci, N.A.: Determination of optimum insulation thickness for building walls by using heating and cooling degree-day values of all Turkey's provincial centers. Energy Build. **118**, 197–213 (2016). https://doi.org/10.1016/j.enbuild.2016.03.004
5. Vine, E.L., Kazakevicius, E.: Residential energy use in Lithuania: the prospects for energy efficiency. Energy **24**, 591–603 (1999). https://doi.org/10.1016/S0360-5442(99)00013-4

6. Battista, G., Evangelisti, L., Guattari, C., Basilicata, C., de Lieto Vollaro, R.: Buildings energy efficiency: interventions analysis under a smart cities approach. Sustainability **6**, 4694–4705 (2014). https://doi.org/10.3390/su6084694
7. Mortarotti, G., Morganti, M., Cecere, C.: Thermal analysis and energy-efficient solutions to preserve listed building façades: the INA-Casa building heritage. Buildings **7**, 1–22 (2017). https://doi.org/10.3390/buildings7030056
8. Park, K., Kim, M.: Energy demand reduction in the residential building sector: a case study of Korea. Energies **10**, 1–11 (2017). https://doi.org/10.3390/en10101506
9. Goggins, J., Moran, P., Armstrong, A., Hajdukiewicz, M.: Lifecycle environmental and economic performance of nearly zero energy buildings (NZEB) in Ireland. Energy Build. **116**, 622–637 (2016). https://doi.org/10.1016/j.enbuild.2016.01.016
10. International Organization for Standardization ISO 10211:2017 - Thermal bridges in building construction - Heat flows and surface temperatures - Detailed calculations, Geneva, Switzerland (2017)
11. Balaras, C.A., Argiriou, A.A.: Infrared thermography for building diagnostics. Energy Build. **34**, 171–183 (2002). https://doi.org/10.1016/S0378-7788(01)00105-0
12. Sajjadian, S.: Risk identification in the early design stage using thermal simulations—A case study. Sustainability **10**, 262 (2018). https://doi.org/10.3390/su10010262
13. Capozzoli, A., Gorrino, A., Corrado, V.: A building thermal bridges sensitivity analysis. Appl. Energy **107**, 229–243 (2013). https://doi.org/10.1016/j.apenergy.2013.02.045
14. Šadauskiene, J., Ramanauskas, J., Šeduikyte, L., Daukšys, M., Vasylius, A.: A simplified methodology for evaluating the impact of point thermal bridges on the high-energy performance of a Passive House. Sustainability **7**, 16687–16702 (2015). https://doi.org/10.3390/su71215840
15. Aguilar, F., Solano, J.P., Vicente, P.G.: Transient modeling of high-inertial thermal bridges in buildings using the equivalent thermal wall method. Appl. Therm. Eng. **67**, 370–377 (2014). https://doi.org/10.1016/j.applthermaleng.2014.03.058
16. Martin, K., Erkoreka, A., Flores, I., Odriozola, M., Sala, J.M.: Problems in the calculation of thermal bridges in dynamic conditions. Energy Build. **43**, 529–535 (2011). https://doi.org/10.1016/j.enbuild.2010.10.018
17. Ramalho de Freitas, J., Grala da Cunha, E.: Thermal bridges modeling in South Brazil climate: three different approaches. Energy Build. **169**, 271–282 (2018). https://doi.org/10.1016/j.enbuild.2018.03.044
18. Bienvenido-Huertas, D., Quiñones, J.A.F., Moyano, J., Rodríguez-Jiménez, C.E.: Patents analysis of thermal bridges in slab fronts and their effect on energy demand. Energies **11**, 2222 (2018). https://doi.org/10.3390/en11092222
19. Song, J.H., Lim, J.H., Song, S.Y.: Evaluation of alternatives for reducing thermal bridges in metal panel curtain wall systems. Energy Build. **127**, 138–158 (2016). https://doi.org/10.1016/j.enbuild.2016.05.078
20. Ascione, F., Bianco, N., De Rossi, F., Turni, G., Vanoli, G.P.: Different methods for the modelling of thermal bridges into energy simulation programs: comparisons of accuracy for flat heterogeneous roofs in Italian climates. Appl. Energy **97**, 405–418 (2012). https://doi.org/10.1016/j.apenergy.2012.01.022
21. Rubel, F., Kottek, M.: Observed and projected climate shifts 1901–2100 depicted by world maps of the Köppen-Geiger climate classification. Meteorol. Zeitschrift **19**, 135–141 (2010). https://doi.org/10.1127/0941-2948/2010/0430
22. European Committee for Standardization EN 13187:1998. Thermal performance of buildings - Qualitative detection of thermal irregularities in building envelopes - Infrared method (1998)
23. The Government of Spain Royal Decree 314/2006. Approving the Spanish Technical Building Code, Madrid, Spain (2013)

24. Intergovernmental Panel on Climate Change Climate Change 2007: The Physical Science Basis. Contribution of Working Group I to the Fourth Assessment Report of the Intergovernmental Panel on Climate Change (2007). ISBN 9780521880091
25. Bienvenido-Huertas, D., Rubio-Bellido, C., Sánchez-García, D., Moyano, J.: Internal surface condensation risk in façades of Spanish social dwellings. Build. Res. Inf. **47**, 1–20 (2019). https://doi.org/10.1080/09613218.2019.1612729
26. Sánchez-García, D., Bienvenido-Huertas, D., Tristancho-Carvajal, M., Rubio-Bellido, C.: Adaptive comfort control implemented model (ACCIM) for energy consumption predictions in dwellings under current and future climate conditions: a case study located in Spain. Energies **12**, 1498 (2019). https://doi.org/10.3390/en12081498
27. British Standards Building components and building elements - Thermal resistance and thermal transmittance - Calculation method BS EN ISO 6946:2007 (2007)
28. Besagni, G., Borgarello, M.: The determinants of residential energy expenditure in Italy. Energy **165**, 369–386 (2018)

# Tehnico – Economic Analysis to Change the Public Buildings in Energy Efficiency Buildings

Marius-Costel Balan(✉), Vasilică Ciocan, Marina Verdeș, and Emilian Țurcanu

Faculty of Civil Engineering and Building Services, Department of Building Services, "Gheorghe Asachi" Technical University, Iasi, Romania
marius.balan@tuiasi.ro

**Abstract.** The paper presents a study on the use of renewable energy sources to ensure thermal comfort parameters in public high school buildings. The implementation of renewable energy sources is also a real challenge in particular at the level of public buildings. The case study shows a proposed system that ensures the thermal energy for heating and cooling based on heat pump soil to water with vertical loops. Comparative analysis it is made between the proposed system – the heat pump and the base system – natural gas boiler.

**Keywords:** System · Renewable energy · Analysis · Heat pump

## 1 Introduction

One of the effects of technological development of all human societies over the past century is a more pronounced increase in energy consumption, but more pronounced dependence on fossil fuel consumption, especially as oil, natural gas, and coal.

Development and human progress are related to the possibility of ensuring natural resources in general and especially energetic. As technical progress and industrial civilization gain, total energy consumption increased significantly, a trend that manifests itself today.

The need for comfort in administrative buildings today has triggered a variety of energy-consuming devices. Although progress in science and technology has enabled machines with good performance in terms of relatively low energy consumption, however, overall, absolute consumption increases.

Solar energy, wind energy, and biomass are the technologies that have the fastest growth rate to achieve the objectives of the European Union, respectively: Solar energy and wind energy are used to produce electricity, while biomass remains dominant in the thermal energy sector.

Reducing existing building energy consumption impose two main directions:

- to reduce the need for energy through the implementation of energy efficiency measures and


A. Rotaru (Ed.): CRIT-RE-BUILT 2019, SSGG, pp. 449–457, 2021.
https://doi.org/10.1007/978-3-030-61118-7_36

- to offset the remaining building energy needs through the use of renewable energy systems.

It is important to note that building energy efficiency measures should be considered first, as the cost to invest in efficiency measures is approximately half the cost of installing renewable energy generating capacity equal to what the efficiency measures offset. [1]

It is advised that all energy efficiency opportunities are explored and as many are implemented as is feasible before or in conjunction with renewable energy projects for existing buildings.

The present paper focuses only on the opportunities and issues for implementing renewable energy projects for existing buildings.

## 2 Technologies Using Renewable Energy

A modern, clean and renewable heating system of medium-sized buildings is consists of heat pumps, which operate on the principle "inverted refrigerator", takes heat from the outside (earth, air, groundwater) and it cedes through a compressor and evaporator inside the utilization system.

The principle of operation of a heat pump which, a heating agent pumped by a compressor, takes over from the heat of water, soil, or air and it cedes to the underfloor heating system, walls, or at least classic radiators. The natural heat of the earth is the transfer by a system of buried polyethylene pipes, through a fluid, usually a mixture of water with glycol. It is taken into account that the soil at a depth of 1,8–2,0 m has a constant temperature between 8–12 °C, throughout the year, energy consumption is reduced providing efficiency of 4–6 kWh thermal energy at the consumption of 1 kWh electricity.

Modern electric heat pumps obtain three-quarters of the heat required for heating from the environment, the remaining quarter is drawn as electrical power for driving the compressor. Since that electrical power will also finally be converted into heat, it too can be utilized for heating purposes. The coefficient of performance results from the ratio between the transferred heating energy (including the compressor heat generated by the electrical power consumption) and the energy used (power supplied to the equipment), which describes the effectiveness of the heat pump (Fig. 1).

When using power generated from renewables, the total heating energy will have been generated from renewable resources. In such cases, heat pumps represent the only heating system, apart from solar technology, which enables heat to be generated with zero $CO_2$ emissions.

Temperatures and pressures in a cyclic process are commonly illustrated in an "lg p-h diagram". For the basic heat pump process, individual stages – evaporation (1–2), compression (2–3), condensation (3–4), and expansion (4–1) – can be read off as distances (Fig. 3). The coefficient of performance represents the ratio between currently available output and electrical power consumption. The predominant part of the heat transfer to a heating system takes place in the process medium vapor range (circled in red) (Fig. 2).

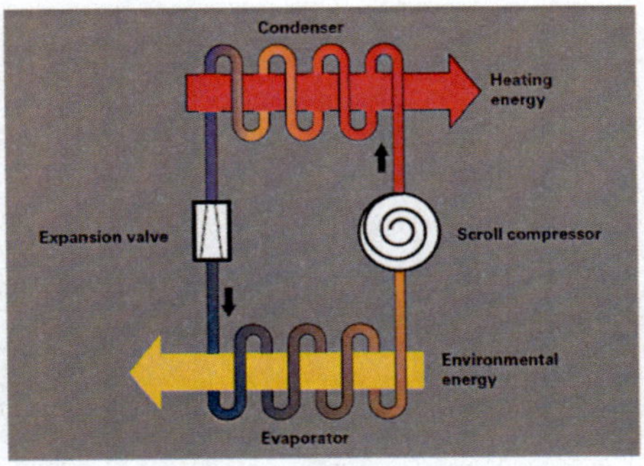


**Fig. 1.** Theoretical principle of heat pumps [12]

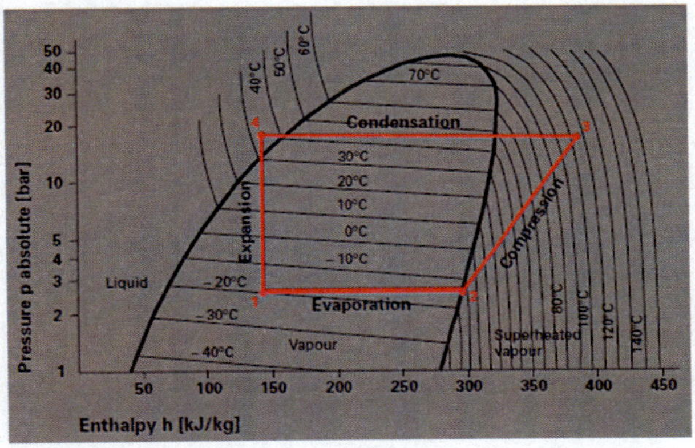


**Fig. 2.** Logarithmical p-h diagram for heat pumps [13]

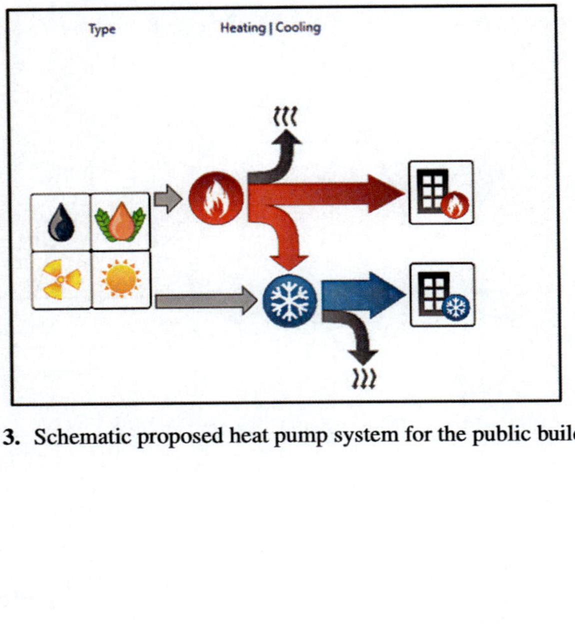


**Fig. 3.** Schematic proposed heat pump system for the public building

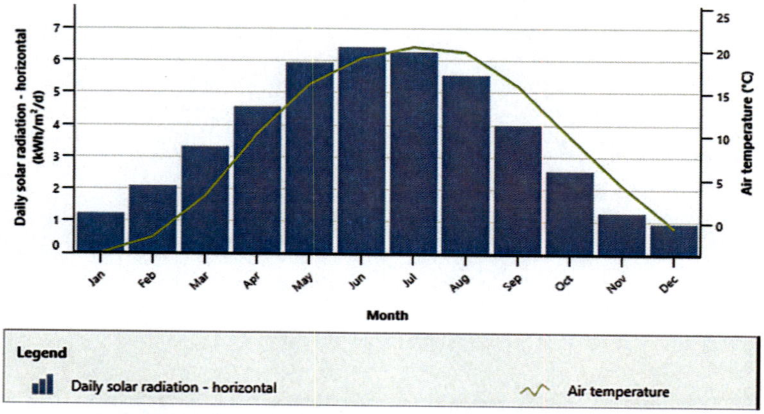


**Fig. 4.** Meteorological data for Iasi city

## 3 Study Case

Simulation of performance, economic, and financial systems to input renewable energy for one public building to ensure thermal comfort (heating and cooling) was performed using the computer program RETScreen. (Figure 3) and was achieved through the performance indicators:

– Load characteristics, (Figs. 5 and 6), calculated as working hypotheses for the base and proposed system;

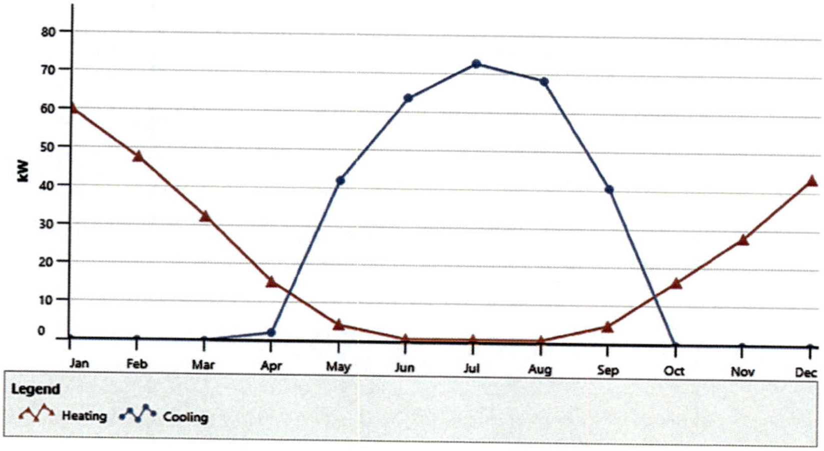


**Fig. 5.** Base system characteristic load

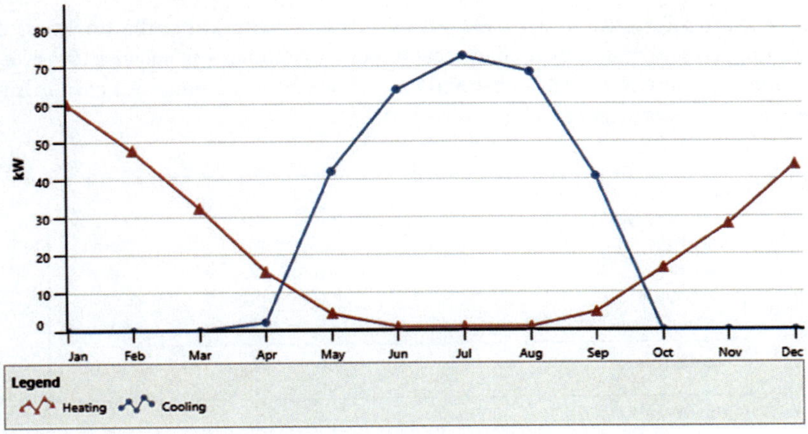


**Fig. 6.** Proposed system characteristic load

- Reduction of greenhouse gases emissions, (Figs. 7 and 8) if we used the proposed system instead of the references system, we make a considerable reduction o GHG emission;

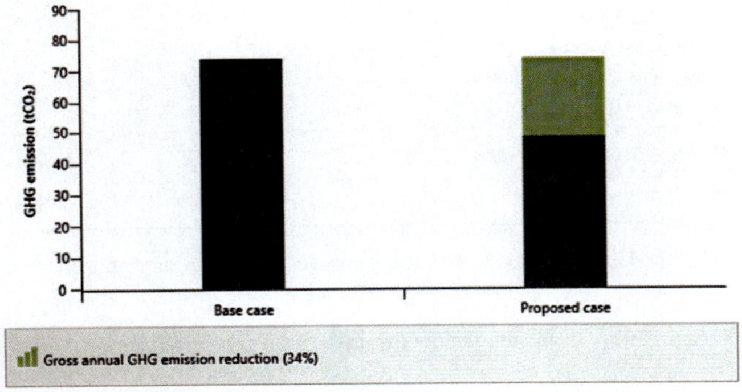


**Fig. 7.** Graphic of GHG emission reduction if the systems proposed are implemented

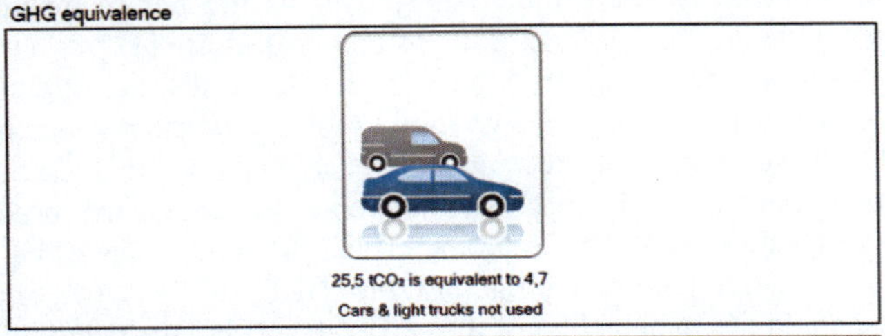


| GHG emission | | |
|---|---|---|
| Base case | 74,1 | $tCO_2$ |
| Proposed case | 48,6 | $tCO_2$ |
| Gross annual GHG emission reduction | 25,5 | $tCO_2$ |

**Fig. 8.** Comparative analysis of GHG emission on the systems

- Fuel cost for references a proposed system, (Fig. 9) shows that the proposed case consuming less more energy than the references case, and this is because the price for natural gas is increasing every year and the efficiency of the proposed case is higher than the references system.

| | | | |
|---|---|---|---|
| Initial costs | | | |
| Feasibility study | -2,7% | ROL | 2.175 |
| Development | -1,1% | ROL | 910 |
| Engineering | -0,74% | ROL | 602 |
| Cooling system | 122% | ROL | -99.084 |
| Balance of system & miscellaneous | -17,3% | ROL | 14.063 |
| Total initial costs | 100% | ROL | -81.334 |
| Yearly cash flows - Year 1 | | | |
| Annual costs and debt payments | | | |
| O&M | | ROL | 0 |
| Fuel cost - proposed case | | ROL | 54.336 |
| Debt payments - 20 yrs | | ROL | 0 |
| Total annual costs | | ROL | 54.336 |
| Annual savings and revenue | | | |
| Fuel cost - base case | | ROL | 69.669 |
| GHG reduction revenue | | ROL | 0 |
| Customer premium income (rebate) | | ROL | 0 |
| Other revenue (cost) | | ROL | 0 |
| Total annual savings and revenue | | ROL | 69.669 |
| Net yearly cash flow - Year 1 | | ROL | 15.333 |

**Fig. 9.** Cost and savings of the proposed case implementation.

- The system price, (Fig. 9) shows the financial cost of the investment for the proposed system.
- The pre-tax internal rate of return (IRR) on equity (%), (Fig. 10) represents the true interest yield provided by the project equity over its life before income tax. It is calculated using the pre-tax yearly cash flows and project life [13].
- The simple payback (year), (Fig. 10) which represents the length of time that it takes for a proposed facility to recoup its own initial cost, out of the revenue or savings it generates. The basic premise of the simple payback method is that the more quickly the cost of an investment can be recovered, the more desirable is the investment [13].
- The Net Present Value (NPV) of the project, (Fig. 10) is the value of all future cash flows. NPV is related to the internal rate of return (IRR). NPV is thus calculated at a time 0 corresponding to the junction of the end of year 0 and the beginning of year 1. Under the NPV method, the present value of all cash inflows is compared against the present value of all cash outflows associated with an investment project. The difference between the present value of these cash flows, called the NPV, determines whether or not the project is generally a financially acceptable investment. Positive NPV values are an indicator of a potentially feasible project [13].

| | | |
|---|---|---|
| Pre-tax IRR - equity | % | Positive |
| Pre-tax MIRR - equity | % | Positive |
| Pre-tax IRR - assets | % | Positive |
| Pre-tax MIRR - assets | % | Positive |
| Simple payback | yr | Immediate |
| Equity payback | yr | Immediate |
| Net Present Value (NPV) | ROL | 262.250 |
| Annual life cycle savings | ROL/yr | 26.699 |
| Benefit-Cost (B-C) ratio | | -2,2 |
| GHG reduction cost | ROL/t$CO_2$ | -1.046 |

**Fig. 10.** Financial parameters of the proposed system.

– The net Benefit-Cost (B-C) ratio, (Fig. 10) is the ratio of the net benefits to the costs of the project. Net benefits represent the present value of annual revenue and savings less annual costs, while the cost is defined as the project equity. The net benefit-cost ratio, similar to the profitability index, leads to the same conclusion as the net present value indicator [13].
– The annual cash flows, (Fig. 11) are the yearly net flows of cash for the project after income tax. It represents the estimated sum of cash that will be paid or received each year during the entire life of the project. The cumulative cash flows, (Fig. 12) represent the net after-tax flows accumulated from year 0 [13].

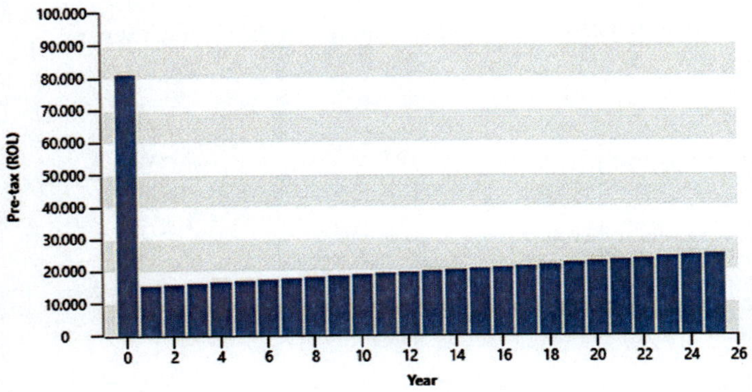


**Fig. 11.** Annual cash flow.

For this calculation the following hypotheses were necessary:

– the meteorological data of the system - the 3rd temperature zone of Romania –Iasi
– the public building is a high school with a surface 5328 $m^2$ well insulated and a regime of 12 h of function; (Fig. 4)
– cost of natural gas for the year 2020 is 0.15 ROL/kWh;

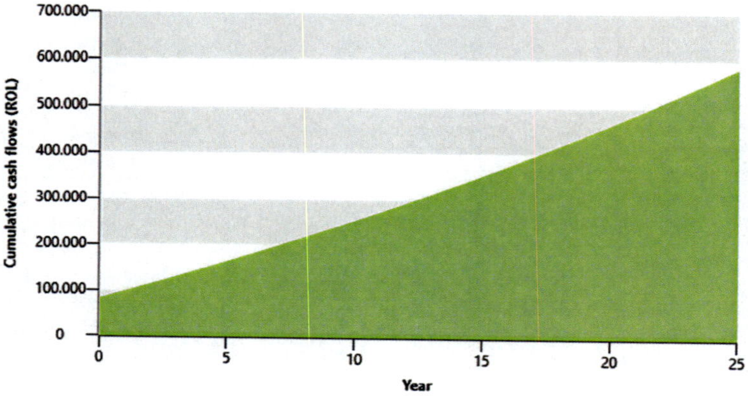


**Fig. 12.** Cumulative cash flow

- cost of electricity for the year 2020 is 0.65 ROL/kWh;
- the inflation rate is 3% considered;
- for reference is using a natural gas plant with an efficiency of 80%.
- load characteristic for the base system (Fig. 5)
- load characteristic for the proposed system (Fig. 6)
- The figures present the simulation results, the economic and financial performance depending on the temperature zone (Table 1).

**Table 1.** Capacity and energy delivered for the proposed system.

| | Technology | Fuel type | Capacity kW | Energy delivered MWh |
|---|---|---|---|---|
| Heating | Heat pump | Electricity | 100 | 196 |
| Cooling | Heat pump | Electricity | 150 | 229 |

## 4 Conclusions

This paper analyzed the performance of economic and financial a heat pump system to ensure the thermal comfort and to increase the energy efficiency for the public high school building.

Analyzing in the light of economic and financial indicators we observed:

- Fuel price for the first system is higher than the base system;
- Reducing greenhouse gas for the proposed system is 34% higher than the base system.

– The payback period for the proposed system is at the same time with the implementation because the performance of the heat pump in heating has a performance coefficient of 4 and in cooling the performance coefficient is 5.

The energy coming from this system is currently a practical endless energy source that can be operated continuously, regardless of weather conditions.

## References


1. Hayter, S.J., Fashrae – Alicen K., Renewable Energy Applications for Existing Buildings. U.S. Department of Energy's National Renewable Energy Laboratory – Integrated Applications Center (2016)
2. Shirazi, A.M., Zomorodian, Z.S., Tahsildoost, M.: Techno-economic BIPV evaluation method in urban areas. Renewable Energy **143**, 1235–1246 (2019). https://doi.org/10.1016/J.RENENE.2019.05.105
3. Masip, X., Cazorla-Marín, A., Montagud-Montalvá, C., Marchante, J., Barceló, F., Corberán, J.M.: Energy and techno-economic assessment of the effect of the coupling between an air source heat pump and the storage tank for sanitary hot water production. Appl. Therm. Eng. **159**, 113853 (2019). https://doi.org/10.1016/J.APPLTHERMALENG.2019.113853
4. Gugul, G.N., Aydinalp Koksal, M., Ugursal, V.I.: Techno-economical analysis of building envelope and renewable energy technology retrofits to single family homes. Energy. Sustain. Dev. **45**, 159–170 (2018). https://doi.org/10.1016/J.ESD.2018.06.006
5. Renaldi, R., Friedrich, D.: Techno-economic analysis of a solar district heating system with seasonal thermal storage in the UK. Appl. Energy **236**, 388–400 (2019). https://doi.org/10.1016/J.APENERGY.2018.11.030
6. Lu, Y., Wang, S., Zhao, Y., Yan, C.: Renewable energy system optimization of low/zero energy buildings using single-objective and multi-objective optimization methods. Energy Buildings **89**, 61–75 (2015). https://doi.org/10.1016/J.ENBUILD.2014.12.032
7. Ozyogurtcu, G., Mobedi, M., Ozerdem, B.: Techno-economic evaluation of a ventilation system assisted with exhaust air heat recovery, electrical heater and solar energy. Energy Buildings **72**, 17–23 (2014). https://doi.org/10.1016/J.ENBUILD.2013.12.025
8. Tagliabue, L.C., Di Giuda, G.M., Villa, V., De Angelis, E., Ciribini, A.L.C.: Techno-economical analysis based on a parametric computational evaluation for decision process on envelope technologies and configurations. Energy Buildings **158**, 736–749 (2018). https://doi.org/10.1016/J.ENBUILD.2017.10.004
9. Ma, W., Xue, X., Liu, G.: Techno-economic evaluation for hybrid renewable energy system: application and merits. Energy **159**, 385–409 (2018)
10. https://doi.org/10.1016/J.ENERGY.2018.06.101
11. Kim, M.-H., Kim, D., Heo, J., Lee, D.-W.: Techno-economic analysis of hybrid renewable energy system with solar district heating for net zero energy community. Energy **187**, 115916 (2019). https://doi.org/10.1016/J.ENERGY.2019.115916
12. Abdin, Z., Mérida, W.: Hybrid energy systems for off-grid power supply and hydrogen production based on renewable energy: A techno-economic analysis. Energy Convers. Manage. **196**, 1068–1079 (2019)
13. https://doi.org/10.1016/J.ENCONMAN.2019.06.068
14. Technical information – heat pumps viessmann
15. http://www.retscreen.net

# Energy Saving Achieved with Adaptive Setpoint Temperatures Based on EN16798-1: Application of the Category III

David Bienvenido-Huertas(✉), Francisco Javier Guevara-García, Daniel Sánchez-García, and Carlos Rubio-Bellido

University of Seville, Av. Reina Mercedes 4A, 41012 Seville, Spain
jbienvenido@us.es

**Abstract.** Climate change is one of the main problems of the society of the 21st century. High emissions of Greenhouse Gases (GHG) are generating more and more extreme living conditions, so GHG emissions should be reduced. Regarding buildings, their high energy consumption is mainly responsible for the contributions of GHG to the atmosphere. Ambitious goals for reducing GHG emissions have therefore been established by 2050. This study analyses the potential of the energy saving achieved using adaptive setpoint temperatures. These setpoint temperatures based on different thermal comfort models as energy conservation measures have been analysed in previous studies. However, the European Committee for Standardization has recently published a new standard. The new European thermal comfort model, the standard EN16798-1, is therefore used in this study. A total of 10 building models in southern Spain were analysed using EnergyPlus. The energy consumption obtained with adaptive setpoint temperatures was compared to that obtained using a static thermal comfort model. The results reflected the potential for the energy saving obtained by using adaptive setpoint temperatures in warm climatic zones.

**Keywords:** Consumption · Adaptive setpoint temperatures · Building · HVAC system · Warm climate

## 1 Introduction

The existing building stock in Europe has a deficient energy behaviour [1, 2]. This behaviour is generated greater greenhouse gas emissions by the existing buildings [3, 4]. To reduce this consumption, a greater implementation of nearly zero energy building (nZEB) is essential. According to the Directive 2010/31/UE, all countries of the European Union must design new buildings or restorations as per the criteria of nZEB from 2020 [4]. Although the designs appropriate to reach nZEB are easy to be implemented in cold climates, their implementation is more complex in warm regions [5].

To ease this transition, many research studies have been developed in recent years to reduce the energy consumption in these regions. It is on this point where the implementation of adaptive thermal comfort models in the use of HVAC systems is reaching


A. Rotaru (Ed.): CRIT-RE-BUILT 2019, SSGG, pp. 458–466, 2021.
https://doi.org/10.1007/978-3-030-61118-7_37

important values of energy saving. Adaptive thermal comfort models are based on the capacity of adaptation of the human being under external variations so that the limits in which the environmental temperature could be adapted to be under thermal comfort conditions. The use of an approach based on this principle leads to important energy savings if they are compared to the tradition criteria of using HVAC systems (i.e., based on the use of static setpoint temperatures). In this regard, some research studies show the savings that could be reached:

- Barbadilla-Martín et al. [6] used as setpoint temperatures the neutral temperatures of a comfort model for mixed-mode buildings. Such model was designed for the climate of Seville. The energy consumptions obtained by using old setpoint and adaptive temperatures were compared (old setpoint temperatures were 22.3 °C and 23.5 °C, whereas adaptive temperatures were 21.5 °C and 24 °C). Both energy savings of 27.5% in cooling and of 11.4% in heating were obtained.
- Sánchez-Guevara Sánchez et al. [7] analysed 3 buildings of social dwellings located in 3 different cities (Avila, Madrid, and Seville). The authors used setpoint temperatures monthly varying by applying the adaptive thermal comfort model, thus achieving an energy saving between 20 and 80% in buildings.
- Sánchez-García et al. [8] analysed the application of daily adaptive setpoint temperatures according in a case study located in the same cities of the study by Sánchez-Guevara Sánchez et al. [7], thus achieving a saving in the energy consumption between 10 and 46% with respect to the recommendations of the Spanish regulation on setpoint temperatures in this type of buildings.

Regarding to adaptive thermal comfort models, there is a wide variety of standards that develop models adapted to each region, such as the EN 15251:2007 standard [9] and the EN 16798-1:2019 standard [10]. The EN 15251:2007 standard was the first European adaptive thermal comfort standard. Nevertheless, this standard has been updated with the EN 16798-1:2019 standard. Such standard sets 3 different categories (I, II, and III), giving recommendations according to the level of expectation of the occupant and the state of the building. Upper and lower limits are established for the operative temperature in each category, and the higher the category, the higher the range of tolerance of comfort limits. For the application of the thermal comfort model, the running mean outdoor temperature ($\theta_{rm}$) (Eq. 1)) must be between 10 and 30 °C. Limits are calculated by using linear correlations according to $\theta_{rm}$ (Eqs. (2 and 3)).

$$\begin{aligned}\theta_{rm} = \big(&\theta_{ext,d-1} + 0.8\theta_{ext,d-2} + 0.6\theta_{ext,d-3} + 0.5\theta_{ext,d-4} \\ &+ 0.4\theta_{ext,d-5} + 0.3\theta_{ext,d-6} + 0.2\theta_{ext,d-7}\big)/3.8\end{aligned} \quad (1)$$

$$Upperlimit(CategoryIII) = 0.33 \cdot \theta_{rm} + 22.8 \quad (2)$$

$$Lowerlimit(CategoryIII) = 0.33 \cdot \theta_{rm} + 13.8 \quad (3)$$

Regarding EN 16798-1:2019, most research studies are based on the thermal comfort limits of category III from EN 15251:2007. The modification leading to the new standard was related to the increase of the range of $\theta_{rm}$ for the lower limit and to the variation of correlations for the calculation of limits. It is therefore possible that the application of the new standard means a variation in the building energy consumption. For this reason, this research applied adaptive setpoint temperatures to a set of 10 different case studies. These case studies were located in Seville as it is one of the main cities on which the previous research studies have been focused to analyse the adaptive models [8, 11]. The use of a regular usage profiles of HVAC systems in the country allowed the viability of using these measures to be determined.

## 2 Methodology

To study the effect caused by adaptive setpoint temperatures in different geometries, 10 different building typologies were modelled (see Fig. 1). These case studies are used as various studies obtained through a positive experience [12–14].

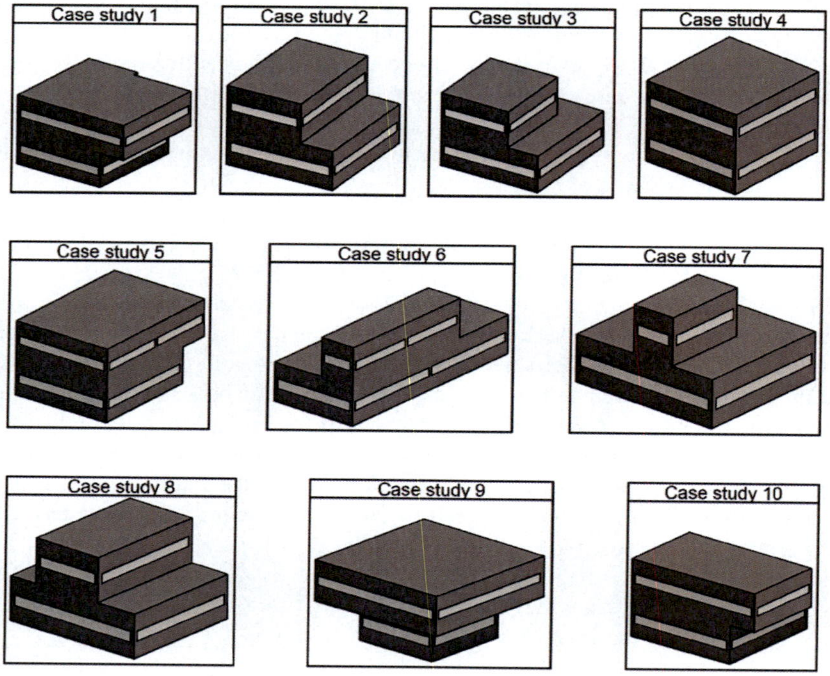


**Fig. 1.** Case studies designed for the research

The case studies were designed with different single-family dwelling typologies. All geometries presented the same number of floors and the glazing percentage was 25% in all façades. Also, the case studies do not have shading elements.

Regarding the thermal properties, designs appropriate to the building stock in Seville were determined. Table 1 includes the thermal properties of the different envelope elements. As HVAC system, a heat pump with a Coefficient of Performance (COP) of 2.10 and with an Energy Efficiency Ratio (EER) of 2.00 was designed in all case studies.

**Table 1.** Thermal properties of the envelope of the case studies.

| Element | Thermal transmittance [W(m$^2$K)] |
|---|---|
| Façade | 0.637 |
| Floor | 0.800 |
| Ceiling | 0.661 |
| Window | 2.761 |

The simulation process of the case studies was performed with EnergyPlus through DesignBuilder. The EnergyPlus weather file of Seville was obtained with METEONORM. Regarding the usage and load profiles of the system, the specifications of the Spanish Building Technical Code (in Spanish, CTE) were considered [15]. In such standard, the residential usage profile typical for Spain is established, including both the load profiles (see Fig. 2) and the setpoint temperatures for these buildings (see Table 2). Regarding the setpoint temperatures, as this study aimed at assessing the energy saving obtained by using adaptive setpoint temperatures according to the category III from EN 16798-1, the static profile of the CTE was adapted to the standard (see Table 2): the setpoint temperature of the adaptive model was calculated according to the variation of the external temperature by using Eqs. (1) and (2), whereas it did not vary in the static model. In the adaptive model, when the adaptive thermal comfort model could not be applied (e.g., when the external temperature was lower than 10 °C), setpoint temperatures were obtained by extending the limit values in the boundary conditions of the adaptive model.

The results obtained in the static model will therefore be useful to have a reference value of the energy consumption associated with the case studies according to the building standard in Spain. Obtaining differences in the energy consumption of the adaptive model shows the possible savings achieved by adaptive models.

## 3 Results and Discussion

To analyse the energy saving, a total of 20 different simulations were performed in EnergyPlus: 10 for the static model and 10 for the adaptive model. Results of monthly and annual energy consumption were obtained by these simulations. Firstly, the results of monthly energy consumption are discussed. As can be seen in Fig. 3 and in Table 3, the use of adaptive setpoint temperatures achieved important savings in the monthly energy consumption. As for the heating energy consumption, such consumption decreased between 0.02 and 1.54 kWh/m$^2$, with a percentage deviation between 44.71 and 100%. As for the

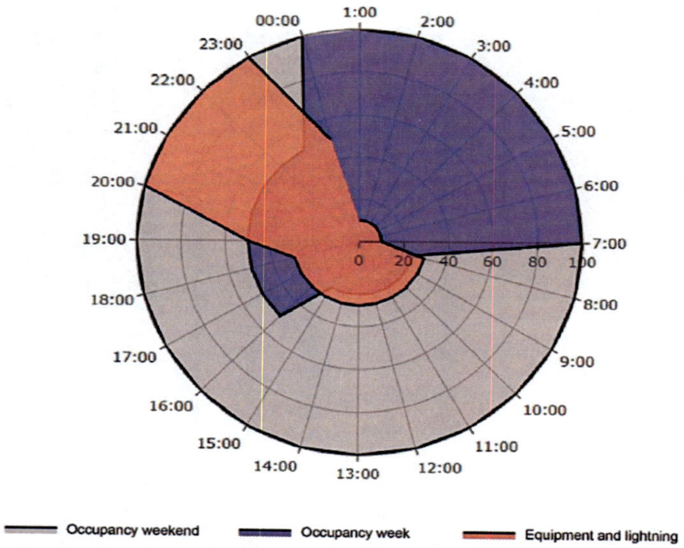


**Fig. 2.** Radar chart with the hourly profiles of equipment, lightning, and occupancy. 100% of equipment and lighting corresponds to an hourly load of 4.4 W/m$^2$ and 100% of occupancy corresponds to an hourly load of 3.23 W/m$^2$.

**Table 2.** Setpoint temperatures used in the static and adaptive models.

| Model | Standard | Limit | Range $\theta_{rm}$ [°C] | Setpoint temperature [°C] | | | | | | | | |
|---|---|---|---|---|---|---|---|---|---|---|---|---|
| | | | | Jan.–May | | | June–Sep. | | | Oct.–Dec. | | |
| | | | | 24–7 | 8–15 | 16–23 | 24–7 | 8–15 | 16–23 | 24–7 | 8–15 | 16–23 |
| Static | CTE | Upper | all | – | – | – | 27 | – | 25 | – | – | – |
| | | Lower | all | 17 | 20 | 20 | – | – | – | 17 | 20 | 20 |
| Adaptive | EN 16798-1 Category III | Upper | (-∞,10) | – | – | – | 26.1 | – | 26.1 | – | – | – |
| | | | [10,30] | – | – | – | (1) | – | (1) | – | – | – |
| | | | (30,∞) | – | – | – | 32.7 | – | 32.7 | – | – | – |
| | | Lower | (-∞,10) | 17.1 | | | – | – | – | 17.1 | | |
| | | | [10,30] | (2) | | | – | – | – | (2) | | |
| | | | (30,∞) | 23.7 | | | – | – | – | 23.7 | | |

(1) $0.33 \cdot \theta_{rm} + 22.8$; (2) $0.33 \cdot \theta_{rm} + 13.8$

cooling energy consumption, the energy saving ranged between 5.27 and 10.31 kWh/m$^2$, with a percentage deviation between 51.74 and 67.78%.

The use of adaptive setpoint temperatures implied a considerable decrease in the cooling energy consumption. In this sense, although the percentage decrease of the energy consumption was lower in cooling than in heating, the low values associated with the heating consumption of the static model caused that the percentage deviation

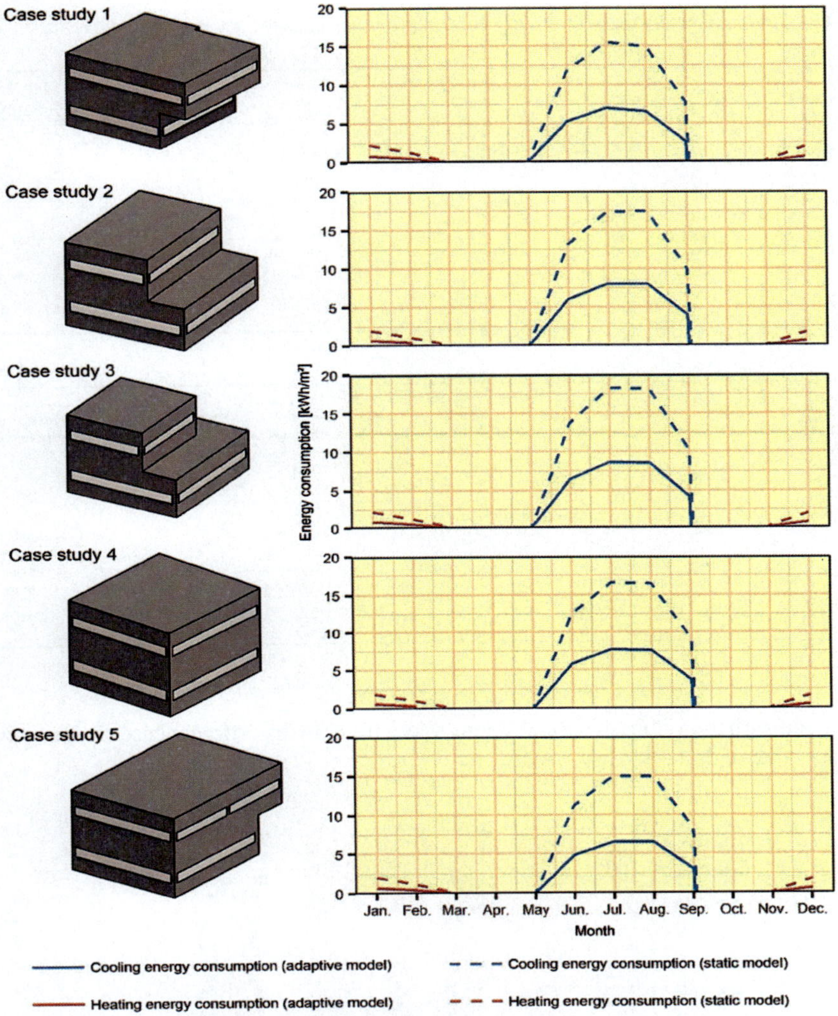


**Fig. 3.** Monthly energy consumption in different case studies

was greater. It is worth stressing that the climate of Seville corresponds to the Csa class of Köppen-Geiger classification [16]: the Csa class is characterised by mild winters and warm summers. Also, the tendency of energy saving was very similar in all case studies and in the different summer months, with an increase in the percentage deviation lower than 20%.

The reduction of energy consumption became more important when analysing data of annual energy consumption. As can be seen in Fig. 4, the energy consumption of the adaptive thermal comfort model was significantly lower than that of the static model. The energy consumption decreased between 31.85 and 34.61 kWh/m$^2$ year. Also, the

**Table 3.** Percentage decrease of energy consumption between the static and adaptive models.

| | Energy saving [kWh/m²year] | | | | | | | | | | | | | | | | | | | |
|---|---|---|---|---|---|---|---|---|---|---|---|---|---|---|---|---|---|---|---|---|
| | Case study 1 | | Case study 2 | | Case study 3 | | Case study 4 | | Case study 5 | | Case study 6 | | Case study 7 | | Case study 8 | | Case study 9 | | Case study 10 | |
| Month | Heating | Cooling | Heating | Cooling | Heating | Cooling | Heating | Cooling | Heating | Cooling | Heating | Cooling | Heating | Cooling | Heating | Cooling | Heating | Cooling | Heating | Cooling |
| Jan. | 1.40 | - | 1.26 | - | 1.33 | - | 1.26 | - | 1.31 | - | 1.38 | - | 1.54 | - | 1.37 | - | 1.48 | - | 1.47 | - |
| Feb. | 0.77 | - | 0.69 | - | 0.71 | - | 0.68 | - | 0.70 | - | 0.77 | - | 0.81 | - | 0.72 | - | 0.82 | - | 0.78 | - |
| Mar. | 0.10 | - | 0.04 | - | 0.05 | - | 0.02 | - | 0.05 | - | 0.11 | - | 0.10 | - | 0.07 | - | 0.13 | - | 0.09 | - |
| Apr. | - | - | - | - | - | - | - | - | - | - | - | - | - | - | - | - | - | - | - | - |
| May | - | - | - | - | - | - | - | - | - | - | - | - | - | - | - | - | - | - | - | - |
| Jun. | - | 6.44 | - | 7.10 | - | 7.27 | - | 6.72 | - | 6.38 | - | 7.21 | - | 7.73 | - | 7.23 | - | 6.50 | - | 7.12 |
| Jul. | - | 8.44 | - | 9.36 | - | 9.57 | - | 8.83 | - | 8.39 | - | 9.35 | - | 10.31 | - | 9.50 | - | 8.58 | - | 9.38 |
| Aug. | - | 8.34 | - | 9.41 | - | 9.59 | - | 8.84 | - | 8.42 | - | 9.43 | - | 10.24 | - | 9.45 | - | 8.51 | - | 9.22 |
| Sep. | - | 5.11 | - | 5.91 | - | 6.07 | - | 5.49 | - | 5.32 | - | 6.18 | - | 6.47 | - | 5.99 | - | 5.27 | - | 5.64 |
| Oct. | - | - | - | - | - | - | - | - | - | - | - | - | - | - | - | - | - | - | - | - |
| Nov. | 0.15 | - | 0.08 | - | 0.11 | - | 0.08 | - | 0.11 | - | 0.15 | - | 0.18 | - | 0.13 | - | 0.21 | - | 0.16 | - |
| Dec. | 1.25 | - | 1.15 | - | 1.23 | - | 1.14 | - | 1.18 | - | 1.23 | - | 1.34 | - | 1.24 | - | 1.29 | - | 1.32 | - |

| | Percentage decrease of energy consumption [%] | | | | | | | | | | | | | | | | | | | |
|---|---|---|---|---|---|---|---|---|---|---|---|---|---|---|---|---|---|---|---|---|
| | Case study 1 | | Case study 2 | | Case study 3 | | Case study 4 | | Case study 5 | | Case study 6 | | Case study 7 | | Case study 8 | | Case study 9 | | Case study 10 | |
| Month | Heating | Cooling | Heating | Cooling | Heating | Cooling | Heating | Cooling | Heating | Cooling | Heating | Cooling | Heating | Cooling | Heating | Cooling | Heating | Cooling | Heating | Cooling |
| Jan. | 62.83 | - | 65.12 | - | 61.13 | - | 66.97 | - | 64.56 | - | 54.91 | - | 51.65 | - | 57.83 | - | 53.93 | - | 58.37 | - |
| Feb. | 59.11 | - | 64.26 | - | 59.38 | - | 67.38 | - | 62.54 | - | 52.48 | - | 49.53 | - | 56.14 | - | 50.32 | - | 55.41 | - |
| Mar. | 92.36 | - | 100.00 | - | 99.05 | - | 100.00 | - | 99.69 | - | 75.99 | - | 74.62 | - | 91.34 | - | 70.15 | - | 82.51 | - |
| Apr. | - | - | - | - | - | - | - | - | - | - | - | - | - | - | - | - | - | - | - | - |
| May | - | - | - | - | - | - | - | - | - | - | - | - | - | - | - | - | - | - | - | - |
| Jun. | - | 55.15 | - | 54.05 | - | 52.91 | - | 53.56 | - | 56.96 | - | 58.20 | - | 52.98 | - | 52.73 | - | 57.10 | - | 51.74 |
| Jul. | - | 54.59 | - | 53.81 | - | 52.46 | - | 53.31 | - | 56.20 | - | 57.68 | - | 53.63 | - | 52.71 | - | 56.44 | - | 51.97 |
| Aug. | - | 55.95 | - | 54.08 | - | 52.85 | - | 53.83 | - | 56.47 | - | 57.86 | - | 54.35 | - | 53.42 | - | 57.72 | - | 53.34 |
| Sep. | - | 65.73 | - | 59.61 | - | 58.96 | - | 59.72 | - | 63.48 | - | 64.20 | - | 61.10 | - | 60.30 | - | 67.78 | - | 62.21 |
| Oct. | - | - | - | - | - | - | - | - | - | - | - | - | - | - | - | - | - | - | - | - |
| Nov. | 73.30 | - | 70.48 | - | 65.08 | - | 76.34 | - | 74.24 | - | 50.65 | - | 44.71 | - | 57.66 | - | 55.64 | - | 62.31 | - |
| Dec. | 63.30 | - | 65.56 | - | 61.69 | - | 67.57 | - | 65.36 | - | 54.30 | - | 49.42 | - | 58.00 | - | 52.43 | - | 58.17 | - |

oscillation with the values of annual saving was similar in the different case studies, with a mean value of 34.61 $kWh/m^2$ year.

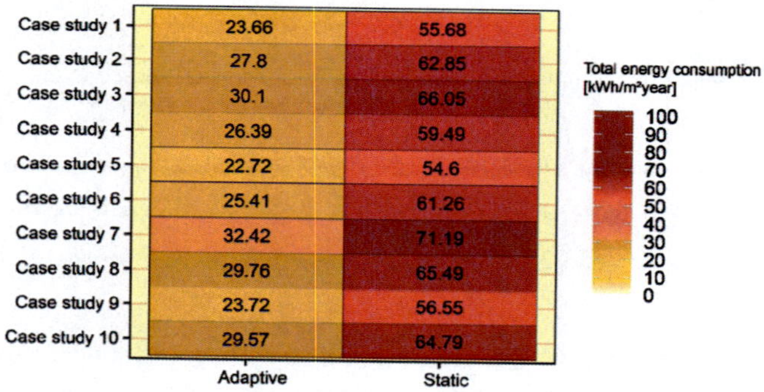


**Fig. 4.** Heat map with the adaptive and static total energy consumption

The use of the adaptive thermal comfort model therefore implied a great decrease in the main consumption in Seville. The obtaining of considerable reductions in cooling consumption is important to be stressed due to the existing difficulties to reduce this type of energy consumption as, to the contrary of heating, significant reductions by improving the thermal properties of the envelope are not achieved. The use of this type of measure could be an opportunity to improve the energy performance of buildings located in warm regions. In addition, the different geometries of the buildings analysed show the potential of application of this type of energy saving measures in buildings.

Regarding the modification obtained with the new limits of the category III from EN 16798-1:2019, it has been confirmed that the modification carried out in this standard does not imply an important variation with respect to the energy saving achieved with EN 15251:2007 in previous research studies, thus stressing the importance and validity of the results obtained in previous research studies as the modifications of the new standard only affect the lower limit which have a low incidence on the energy consumption in warm regions. However, new research works should be conducted in regions with a colder climate.

## 4 Conclusions

The results obtained in this study show the potential of using adaptive setpoint temperatures to reduce the energy consumption of buildings in warm climates. By using the new standard on adaptive thermal comfort (EN 17698-1:2019), saving results similar to those obtained in other research studies with the previous standard (EN 15251:2007) were achieved in warm regions (those zones with a greater potential of use of these systems). These results are also the starting point of the need to establish a clear criterion of the maximum limits of acceptability that could be reached inside dwellings with the implementation of adaptive setpoint temperatures. Further studies will therefore analyse the establishment of possible maximum values to be reached inside dwellings. Also, the modification of the lower limit of EN 17698-1:2019 should be studied in other climate zones to determine the degree of improvement in the energy saving achieved in regions with a greater severity in winter.

## References

1. European Environment Agency Final energy consumption by sector and fuel (2016); Copenhagen, Denmark, 2018
2. European Commission Action Plan for Energy Efficiency: Realising the Potential; Brussels, Belgium, pp. 1–37 (2006)
3. EC Directive 2002/91/EC of the European Parliament and of the Council of 16 December 2002 on the energy performance of buildings; Brussels, Belgium, vol. 1, pp. 65–71. (2002)
4. EU Directive 2010/31/EU of the European Parliament and of the Council of 19 May 2010 on the Energy Perform of Buildings; Brussels, Belgium, 153, pp. 13–35
5. Attia, S., Eleftheriou, P., Xeni, F., Morlot, R., Ménézo, C., Kostopoulos, V., Betsi, M., Kalaitzoglou, I., Pagliano, L., Cellura, M., Almeida, M., Ferreira, M., Baracu, T., Badescu, V., Crutescu, R., Hidalgo-Betanzos, J.M.: Overview and future challenges of nearly zero energy buildings (nZEB) design in Southern Europe. Energy Build. **155**, 439–458 (2017)
6. Barbadilla-Martín, E., Salmerón Lissén, J. M., Martín, J. G., Aparicio-Ruiz, P., Brotas, L.: Field study on adaptive thermal comfort in mixed mode office buildings in southwestern area of Spain. Build. Environ. **123**, (2017) https://doi.org/10.1016/j.buildenv.2017.06.042
7. Sánchez-Guevara Sánchez, C., Mavrogianni, A., Neila González, F.J.: On the minimal thermal habitability conditions in low income dwellings in Spain for a new definition of fuel poverty. Build. Environ. **114**, 344–356 (2017). https://doi.org/10.1016/j.buildenv.2016.12.029

8. Sánchez-García, D., Bienvenido-Huertas, D., Tristancho-Carvajal, M., Rubio-Bellido, C.: Adaptive comfort control implemented model (ACCIM) for energy consumption predictions in dwellings under current and future climate conditions: a case study located in Spain. Energies **12**, 1498 (2019). https://doi.org/10.3390/en12081498
9. EN 15251:2007 Indoor environmental input parameters for design and assessment of energy performance of buildings addressing indoor quality, thermal environment, lighting and acoustics; European Committee for Standardization: Brussels (2007)
10. EN 16798–1:2019 Energy performance of buildings - Ventilation for buildings - Part 1: Indoor environmental input parameters for design and assessment of energy performance of buildings addressing indoor air quality, thermal environment, lighting and acous (2019)
11. Sánchez-García, D., Rubio-Bellido, C., Marrero Meléndez, M., Guevara-García, F.J., Canivell, J.: El control adaptativo en instalaciones existentes y su potencial en el contexto del cambio climático. Hábitat Sustentable **7**, 06–17 (2017)
12. Fernandes, M.S., Rodrigues, E., Gaspar, A.R., Costa, J.J., Gomes, Á.: The impact of thermal transmittance variation on building design in the Mediterranean region. Appl. Energy **239**, 581–597 (2019). https://doi.org/10.1016/j.apenergy.2019.01.239
13. Rodrigues, E., Fernandes, M.S., Gaspar, A.R., Gomes, Á., Costa, J.J.: Thermal transmittance effect on energy consumption of Mediterranean buildings with different thermal mass. Appl. Energy **252**, 113437 (2019)
14. Rossi, M., Rocco, V.M.: External walls design: The role of periodic thermal transmittance and internal areal heat capacity. Energy Build. **68**, 732–740 (2014)
15. The Government of Spain Royal Decree 314/2006. Approving the Spanish Technical Building Code; Madrid, Spain (2013)
16. Rubel, F., Kottek, M.: Observed and projected climate shifts 1901-2100 depicted by world maps of the Köppen-Geiger climate classification. Meteorol. Zeitschrift **19**, 135–141 (2010)

# Supporting Community Participation for Sustainable Smart City Implementation in Indonesian Cities

Sushardjanti Felasari[1](✉) and Muhammad Sani Roychansyah[2]

[1] Universitas Atma Jaya Yogyakarta, Jl. Babarsari 44, Yogyakarta 55281, Indonesia
s.felasari@uajy.ac.id

[2] Universitas Gadjah Mada, Jl. Grafika Utara 2, Bulaksumur, Yogyakarta 55281, Indonesia

**Abstract.** The development of smart and sustainable cities has been a global trend and is growingly implemented in many cities around the world. However, there are still many problems in the implementation of the concept which most rely on in the use of information technology. This paper discusses how community participation can be done to help in implementing the concept of smart city. By using the case of three cities in Indonesia, represent small cities, medium cities, and large cities this paper discusses the level of friendliness of the city from various smart city's dimensions and attributes. Using Important Performance Analysis (IPA), the research found that there are disparities in perceptions between the conditions of existing city services and the expectations of citizens. The disparities show the level of smart city's friendliness which can be used for the Government to determine priorities in implementing dimensions and attributes of smart cities. It also can help to see in what degree the community participation can be supported. The results show that Mobility, Environment and Disaster Management are the most important dimensions of the Smart City that needs more participation from the community in their implementation.

**Keywords:** Community participation · Sustainable smart city · Implementation · Indonesian cities

## 1 Introduction: Is Smart City a Friendly City?

The development of smart and sustainable cities has been a global trend in the last several years. The concept of smart city is present to address complex problems and challenges that arise in urban areas due to the limited resources available. The point of becoming a smart city is that it will increase resilience and improve sustainability of the lives of citizens.

Most of smart city initiatives in world's cities involve 'smartization' of existing cities such as New York, London, and Singapore while several others have been built from the scratch like Songdo in South Korea which apply "ubiquitous computing" to integrate information systems and social systems (U-cities). In a U-City like Songdo, all devices,


A. Rotaru (Ed.): CRIT-RE-BUILT 2019, SSGG, pp. 467–477, 2021.
https://doi.org/10.1007/978-3-030-61118-7_38

components, and services in the city are linked to an information network, mainly through wireless networks [1]. A key objective of a U-city is to provide an integrated environment, in which citizens have access to all types of services, everywhere and any time through ICT devices [2].

However, the implementation of the smart city concepts seems not to be in line with the initial expectations of the community and the reality of its development, which mainly emphasizes on the sophistication of the information technology offered. There is great potential to apply smart technologies to the city challenges but there is also potential issues facing the development of smart cities such as abuse, privacy, sustainability and ethics [3]. Therefore, it is not surprising that in its implementation a smart city sometimes becomes a city that is not friendly to its inhabitants.

Experiences from many cities in the world, which implemented the smart concepts have to consider not only the technology aspects but also the social aspects of the smart city. According to Cohen [4], there are 3 generations of smart cities i.e. smart city 1.0 which is technology driven, smart city 2.0 which is technology enabled and city led, and smart 3.0 which is citizen co-creation models.

By using the case of three cities in Indonesia, this paper starts discussing the level of friendliness of the smart city implementation i.e. small cities, medium cities, and big cities in Indonesia. The level of friendliness of the smart city has been measured based on the perception of its citizens. From the level of friendliness, the dimensions and attributes of the smart cities that need community participation can be highlighted. By knowing in which dimensions or attributes community participation can be involved, it will help the Government city to put the priority in developing the sustainable smart city implementation.

## 2 Learning from Songdo Smart City: Technology vs Social

The idea behind a smart city is that it can use real-time data to improve the livability and sustainability of the city. Songdo City in South Korea has been planned as an aerotropolis and a ubiquitous city (U-City). Since its conception in 2001, South Korea's project to build Songdo smart city has become an international model for urban planning. The Songdo IBD was envisioned as a sustainable, low-carbon, and high-tech utopia[1]. Efficient transport can be seen from the city plan in which offices, parks, hospitals and schools are located nearby. Green concepts have been implemented in providing 40% green spaces and parks, protecting the environment, through environmental energy criteria including water management. About 36,000 people live in the IBD and 90,000 in Songdo City, while more than 20,000 residential units are occupied or under construction. An investment of $35 billion has been ploughed in the city.

As a smart city which is predicted to be a reference for smart city development in other countries, it seems that the construction of Songdo City is not going as expected. The project has been pushed back several times as the number of residents is small compared to its potential. Apart from the quite impressive technology that was built in the city of Songdo, many buildings are still empty.

[1] (https://www.citylab.com/life/2018/06/sleepy-in-songdo-koreas-smartest-city/561374/).

From the point of view of its residents, the hardest thing about living in those eco-friendly master planned utopia is the vibrant community. Lack of interaction such as difficulties to meet the neighborhood creates a feeling of loneliness that have been experienced by many of the residents and it becomes a concern for Songdo's citizens. Technology driven should be combined with social aspects. The acceptance of the end users is one of the important aspects that determines the success of the development of the smart city of Songdo.

## 3 Bottom Up Approach Through Importance-Performance Analysis

There are two main approaches in smart city initiatives i.e. 'top-down' and 'bottom-up' [3]. Top-down approaches focus on technology, efficiency and master planning, integrating data from different systems into a central operations center. Meanwhile bottom-up approaches focus on citizens and how they can use innovative technologies, such as mobile applications, social media, DIY sensors and open data to create solutions to issues that matter to them and enable behavior changes.

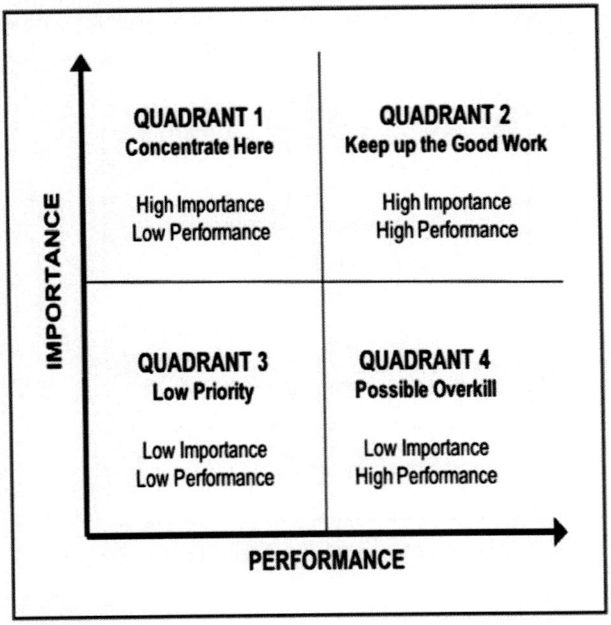


**Fig. 1.** Importance-performance analysis-IPA diagram (adapting from Martilla and James [6])

The smart city concept emphasizes the importance of strategy in choosing priorities, directions, policies, and intelligently utilize technology with automatic computerization (self-configuration, self- healing, self-protection, self-optimization). It also emphasizes the citizen centric approach where people are encouraged to participate actively in the management of the city, so that there is a more dynamic and close interaction between citizens and local government [5].

In choosing priorities of what dimension and attributes of smart city can be implemented for city services in the future, importance-performance analysis introduced by Martilla and James [6] can be used for analyzing the gap between smart city's existing performance (Attribute's Performance) and resident's expectation of smart city in the future (Attribute's Importance) (see Fig. 1).

Using the IPA Diagram in Fig. 1, The City Government is able to give attention to the smart city's attributes in quadrant 1 and quadrant 2 as it reveals the high important attributes for the community and it also measures how the smart city implementation are perceived by the community. In addition, by looking at the dimensions and attributes of smart cities in this quadrant, the support given by the Government related to the way and the form of community participation be involved can be focused.

In the era of communication technology, it is more likely that community participation in urban development can be observed via technology-based media, which shows the involvement of the community. Community involvement has also experienced significant growth, with the help of information technology. Foth et al. (2011) [7] highlighted that there is a clear sense of positioning urban informatics to play a role in the development of the public sphere and community. People can also do engagement in almost all areas and the results can be linked directly to their city.

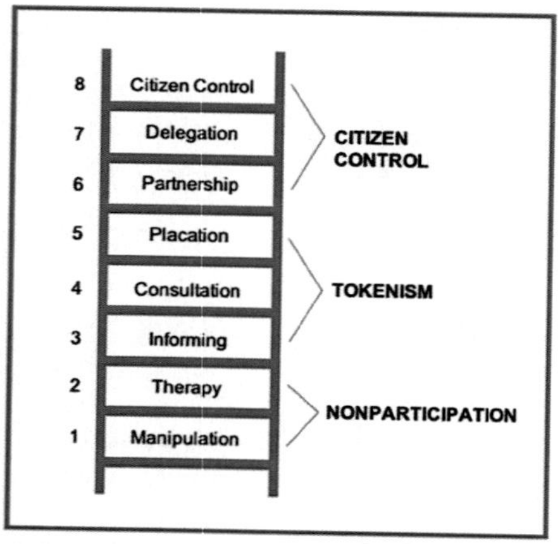


**Fig. 2.** Arnstein's ladder: degree of citizen participation

## 4 Friendliness of Smart City: Community Perception and Government's Priority

According to Arnstein [8] community participation can be classified into 3 groups and are divided into eight levels. The basic level of citizen participation is non-level participation which consists of manipulation (1) and therapy (2). Both are non- participative, which aim is to cure or educate the participants and to achieve public support through public relations. Meanwhile class of tokenism which presented the intermediate level of citizen participation is divided into 3 form of participation i.e. informing (3) - usually one-way flow of information with no channel for feedback, consultation (4) - a legitimate step attitude surveys, neighbourhood meetings, and public enquiries and placation (5) - allows citizens to advice or plan but retains for power holders the right to judge the legitimacy or feasibility of the advice. The advance level of citizen participation is citizens power that consists of partnership (6) - planning and decision-making responsibilities are shared through joint committees, delegated power (7) - citizens have delegated powers to make decisions and to assure accountability of the program to them and citizen control (8) – have bots handle the entire job of planning, policy making and managing a program (see Fig. 2).

Since 2015, the Indonesian government has been working with several organizations and various media to grant awards to smart cities in Indonesia as an appreciation for its implementation. Besides that, The Government of Indonesia in cooperation with a consortium has been trying to map the smart city index for Indonesian cities. National Planning Bureau [9] has also long established a roadmap for policy and strategy of urban development in Indonesia by issuing 3 keywords namely, livable, green, and smart. The award was given to cities in various categories, namely large cities (population > 1.000.000), medium cities (population between 200.000–1.000.000), and small cities (population < 200.000).

A study conducted in 2018 by the authors in 3 cities in Indonesia i.e. Magelang (small city), Yogyakarta (medium city) dan Surabaya (large city) found that there was a gap between perceptions of citizens regarding the existing city services performance and resident's expectation of smart city importance in the future. This is understandable, given that smart cities generally place a high level of information and communication technology in urban services, so that often non-technical aspects of city life that involve humanity are not given enough attention. Therefore, the level of smart city friendliness is measured by the large difference in value between the conditions of the city services the community feel (Performance) with their perceptions of smart cities on the attributes of a smart city (Importance). The performance of the city in recent condition presents how ease to use the attributes of smart city. Meanwhile the importance of the attributes shows the usefulness in the future. The differences between the performance and importance of smart city's attributes can be mapped to evaluate and then to determine the level of priority in smart city development in the future.

By using a Likert Scale with a scale of 1–7, measurements were made on 6 dimensions of Cohen's smart city i.e. Smart Economy, Smart Mobility, Smart Environment, Smart People, Smart Living and Smart Government [10] plus 1 dimension for Indonesian context i.e. Smart Disaster Management [11]. Each then are spelled out to 42 attributes in total (see Fig. 3).

| Dimension | | Attributes |
|---|---|---|
| SMART ECONOMY (SE) | SE-1 | New startups |
| | SE-2 | Internet-based full employment |
| | SE-3 | Innovative in business processsng |
| | SE-4 | ICT-based business license |
| | SE-5 | Online transaction between institutions |
| | SE-6 | Business information availbility |
| SMART MOBILITY (SM) | SM-1 | Clean-energy transport such as bicycle |
| | SM-2 | Innovative and modern parking system |
| | SM-3 | Integrated multi-modal transit system |
| | SM-4 | Real time public transport information services |
| | SM-5 | Smart cards for public transportation |
| | SM_6 | Online transportation apps |
| SMART ENVIRONMENT (SN) | SN-1 | Smart homes |
| | SN-2 | Energy saving house |
| | SN-3 | Public building automation system |
| | SN-4 | Certified green buildings |
| | SN-5 | Solid waste and waste water treatment |
| | SN-6 | Green open space |
| SMART PEOPLE (SP) | SP-1 | Accessible internet at home |
| | SP-2 | Smart phone penetration |
| | SP-3 | Civic engagement activities using ICT |
| | SP-4 | Voter |
| | SP-5 | Number of higher education degree |
| | SP-6 | Creative industry jobs |
| SMART LIVING (SL) | SL-1 | Housing sufficiency |
| | SL-2 | Gini Index |
| | SL-3 | Investment in culture |
| | SL-4 | Smart crime prevention |
| | SL-5 | Accessible health & education services |
| | SL-6 | Life expectancy |
| SMART GOVERNMENT (SG) | SG-1 | Online government services |
| | SG-2 | Electronic benefit payments |
| | SG_3 | WIFI coverage at public space |
| | SG-4 | Sensor coverage |
| | SG-5 | Integrated health and safety services |
| | SG-6 | Open data and apps |
| SMART DISASTER MANAGEMENT (SDM) | SDM-1 | Emergency warning system |
| | SDM-2 | Responsive mitigation planning |
| | SDM-3 | Sensor for monitoring disaster |
| | SDM-4 | Evacuation and rescue training |
| | SDM-5 | Emergency information |
| | SDM-6 | Disaster relief assistance |

**Fig. 3.** Attributes of the smart city measured

Table 1 shows how importance (I) the dimensions of smart cities are according to community perceptions and how they are evaluated in their current performance (P). It can be seen that there are gaps between the importance (I) dan the performance (P).

**Table 1.** The importance and performance of smart city's dimensions based on community perception in Indonesian cities

| No. | Smart city dimensions | Magelang city | | Yogyakarta city | | Surabaya city | |
|---|---|---|---|---|---|---|---|
| | | I | P | I | P | I | P |
| 1. | Smart economy | 5,87 | 4,59 | 6,11 | 4,98 | 6,44 | 4,85 |
| 2. | Smart mobility | 5,92 | 3,65 | 6,42 | 4,77 | 6,57 | 4,36 |
| 3. | Smart environment | 5,91 | 3,75 | 6,31 | 4,15 | 6,62 | 4,18 |
| 4. | Smart people | 6,00 | 5,03 | 6,12 | 4,97 | 6,49 | 4,95 |
| 5. | Smart living | 6,13 | 4,78 | 6,43 | 4,67 | 6,69 | 4,52 |
| 6. | Smart government | 5,96 | 4,42 | 6,44 | 4,57 | 6,64 | 4,76 |
| 7. | Smart disaster management | 6,21 | 4,30 | 6,58 | 4,55 | 6,59 | 4,21 |

I = Importance P = Performance

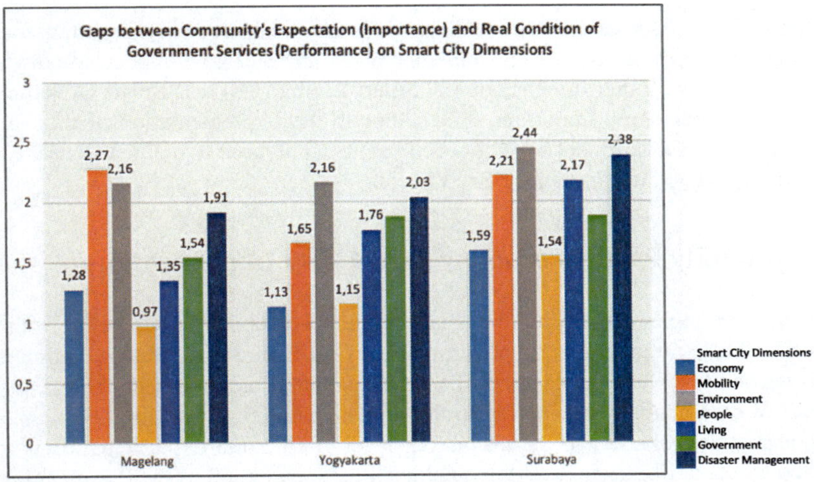


**Fig. 4.** Gaps between community's expectation (Importance) and real condition of city services (Performance)

Figure 4 shows the size of the gap of the 7 dimensions measured. The labels indicate how big the difference between Importance and Performance dimensions of smart cities. The small gaps are shown in the dimensions of Smart Economy (SE) and Smart People (SP). This means that in terms of smart economy and smart people, the performance of the existing city services is fit with the citizen's expectations.

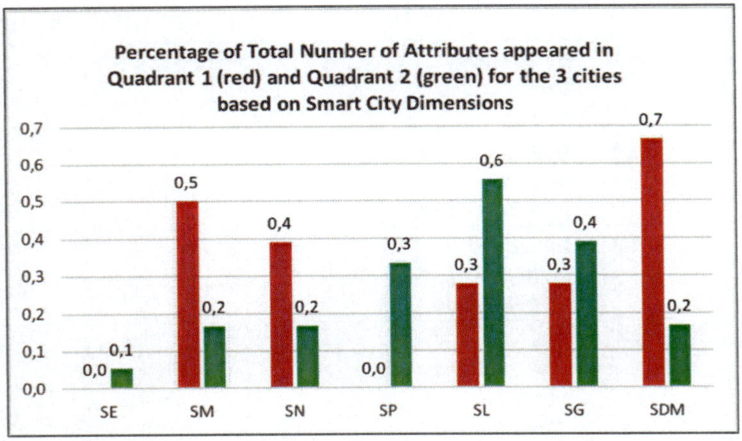


**Fig. 5.** Percentage of total number of attributes based on smart city dimensions

Meanwhile, among 42 attributes in the 7 dimensions, it appears that Smart Mobility (SM), Smart Environment (SN), and Smart Disaster Management (SDM) have the more percentage number of attributes lay on quadrant 1 than the number of attributes lay on quadrant 2. This shows that these three dimensions are considered very important for the community but have not been well implemented by the city government. Meanwhile, for attributes in two other dimensions i.e. Smart Living (SL) and Smart Government (SG) are also considered important. And although the implementation of that Smart Living and Smart Government have not been yet ideal but those two dimensions are still considered good by the community (see Fig. 5).

## 5 Community Participation for Smart City Implementation

The successful implementation of the smart city concept cannot be separated from the role of the stakeholders such as the government, private institutions, and the community. The forms and levels of participation of the stakeholders can be varied. Table 2 shows the level of community participation in the implementation of the smart city concept in Magelang, Yogyakarta and Surabaya. The level of community participation is only indicated by the attributes included in quadrant 1 and quadrant 2 of the IPA diagram. In the shaded dimensions, the eight attributes (SM-2, SM-3, SN2, SN-4, SDM-1, SDM-3, SDM-4 and SDM-5) indicate aspects that need government attentions and can involve the community participation in decisions making.

**Table 2.** Dimension and attributes of smart city and form of community participation

| Dimension | Attributes | | Community participation | | | | | | | |
|---|---|---|---|---|---|---|---|---|---|---|
| | | | Non participation | | Tokenism | | | Citizen power | | |
| | | | M | T | I | C | P | PS | DP | CC |
| Smart Mobility (SM) | SM-1 | Clean-energy transport such as bicycle | | | * | * | * | * | | |
| | **SM-2** | **Innovative and modern parking system** | * | * | | | | | | |
| | **SM-3** | **Integrated multi-modal transit system** | * | * | | | | | | |
| | SM-4 | Real time public transport information services | | | * | * | * | * | | |
| | SM-5 | Smart cards for public transportation | | | * | * | * | | | |

(*continued*)

**Table 2.** (*continued*)

| Dimension | Attributes | | Community participation | | | | | | | |
|---|---|---|---|---|---|---|---|---|---|---|
| | | | Non participation | | Tokenism | | | Citizen power | | |
| | | | M | T | I | C | P | PS | DP | CC |
| Smart Environment (SN) | **SN-2** | **Energy saving house** | * | | | | | | | |
| | **SN-4** | **Certified green buildings** | * | | | | | | | |
| | SN-5 | Solid waste and waste water treatment | | | * | * | | | | |
| | SN-6 | Green open space | | | * | * | * | | | |
| Smart Living (SL) | SL-1 | Housing sufficiency | | | * | * | * | | | |
| | SL-2 | Gini Index | * | * | | | | | | |
| | SL-4 | Smart crime prevention | | | * | * | | | | |
| Smart Government (SG) | SG-3 | Wi-Fi coverage at public space | | | * | * | * | | | |
| | SG-4 | Sensor coverage | * | * | | | | | | |
| | SG-6 | Open data and apps | * | * | * | | | | | |
| Smart Disaster Management (SDM) | **SDM-1** | **Emergency warning system** | * | * | * | | | | | |
| | SDM-2 | Responsive mitigation planning | | | * | * | * | * | * | |
| | **SDM-3** | **Sensor for monitoring disaster** | * | * | * | | | | | |
| | **SDM-4** | **Evacuation and rescue training** | * | * | * | | | | | |
| | **SDM-5** | **Emergency information** | * | * | * | | | | | |

(*continued*)

**Table 2.** (*continued*)

| Dimension | Attributes | | Community participation | | | | | | | |
|---|---|---|---|---|---|---|---|---|---|---|
| | | | Non participation | | Tokenism | | | Citizen power | | |
| | | | M | T | I | C | P | PS | DP | CC |
| | SDM-6 | Disaster relief assistance | | | * | * | * | | | |

M: Manipulation P: Placation
T: Therapy PS: Partnership
I: Informing DP: Delegated Power
C: Consultation CC: Citizen Control

The Mobility (SM), Environment (SN) and Disaster Management (SDM) are considered very important dimensions for the community (High Importance). However, the degree of the citizens participation is quite low for these attributes. On the Mobility dimension (SM), degree of community participation is low at innovative and modern parking systems (SM-2) and integrated multi-modal transit systems (SM-3). In the Environment dimension (SN), attributes such as energy saving house (SN-2) and certified green buildings (SN-4) needs community participation further. Meanwhile, in Smart Disaster Management dimension (SDM), the degree of community participation can be increased in aspects of the emergency warning system (SDM-1), sensors for disaster monitoring (SDM-3), evacuation and rescue training (SDM-4) and emergency information (SDM -5). It can be seen from Table 2 that the degree of citizens participation is mostly only at the lowest level of the Arnstein's ladder of participation. Non-participation level dominates precisely of the involvement whether manipulation, therapy or both. In these circumstances in order to achieve better implementation then support from Government and Public Private company should be improved to increase more participation from the community.

Meanwhile for Smart Living (SL) and Smart Government (SG), the degree of citizen participation is presented better on Table 2. Tokenism has been used to involve the citizens participation in the implementation of smart city. However, more degree of citizen participation can be supported to achieve the advance level. It can be started by partnership with the government and public-private company.

## 6 Conclusion

It can be concluded from the discussion that the level of friendliness of smart cities measure differences in perceptions between the conditions of existing city services and the expectations of citizens. It can help the city government in choosing and determining the priority of which dimensions or attributes of smart cities will be provided or implemented. As a result, it is expected that the implementation of the smart city could be more sustainable as it fits with what the community needs. The level of friendliness also helps to see the degree of community participation in the attributes.

The research found that Mobility, Environment and Disaster Management are dimensions that needs attention from the Indonesian City Government in the implementation of the smart city. Community participation in those dimensions shows the lowest degree of the Arnstein's participation ladder. Supporting the improvement of the citizen participation degree can increase the success rate of smart city implementation as well as its sustainability.

## References

1. Kshetri, N., Alcantara, L.L., Park, Y.: Development of smart city and its adoption and acceptance: the case of new Songdo. Commun. Strat. **96**, 113–1128 (2014)
2. Murgante, B., Borruso, G.: Cities and smartness: a critical analysis of opportunities and risks. In: Murgante, B., et al. (eds.) ICCSA 2013. LNCS, vol. 7973, pp. 630–642. Springer, Heidelberg (2013). https://doi.org/10.1007/978-3-642-39646-5_46
3. Felasari, S., Roychansyah, M.S.: Readiness of incorporating big data to support smart governance of Yogyakarta city. J. Built Environ. Technol. Eng. (JBETE) **2**, 151–158 (2017)
4. Cohen, B.: The 3 Generations of Smart Cities: Inside the development of the technology driven city (2015). https://www.fastcompany.com/3047795/the-3-generations-of-smart-cities. Accessed 27 Sept 2018
5. Roychansyah, M.S., Felasari, S.: Measuring level of friendliness of smart city: a perceptual study. In: IOP Conference Series: Earth and Environment Science (2018)
6. Martilla, J.A., James, J.C.: Importance performance analysis. J. Mark. **41**(1), 77–79 (1977)
7. Foth, M., et al.: From social butterfly to engaged citizen. In: Urban Informatics, Social Media, Ubiquitous Computing and Mobile Technology to Support Citizen Engagement. The MIT Press, London (2011)
8. Arnstein, S.R.: Ladder of citizen participation. J. Am. Inst. Planners **35**(4), 216–224 (1969)
9. Bappenas (National Planning Bureau): Kebijakan dan Strategi Pembangunan Perkotaan Nasional/KSPPN (National Policy and Strategy for Urban Development (2014). http://tataruangpertanahan.com/regulasi/pdf/kepmen/menppn/kepmenppn_65
10. Cohen, B.: What exactly is a smart city? Co.Design (2012). https://www.fastcodesign.com/1680538/waht-exactly-is-a-smart-city
11. Anonym: "Final Report of Jogja Istimewa Smart City Study" (Laporan Akhir Kajian Smart City Jogja Istimewa), Bappeda (Badan Perencanaan Pembangunan Daerah) Kota Yogyakarta dan Pusat Studi Perencanaan Pembangunan Regional (PSPPR), UGM, Yogyakarta (2016)

# Transformation of the Built Environment for the Rehabilitation of Socially Disadvantaged City Districts

# Transformation of the Built Environment for the Rehabilitation of Socially Disadvantaged City Districts

Jiří Boštík(✉), Jan Štefaňák, and Martin Závacký

Faculty of Civil Engineering, Institute of Geotechnics, Brno University of Technology, Veveří 331/95, 602 00 Brno, Czech Republic
Zavacky.m@fce.vutbr.cz

**Abstract.** The article brings an overview of the brownfield topic from a geotechnical point of view. A brownfield defines an non-utilized, dilapidated or ecological damaged property that lost its first function. Demand on the revitalization of such areas, especially within cities, is increasing. However, the revitalization often has to deal with geotechnical aspects as ground investigation, ground improvement or potential settlement of new buildings. Slope stability is also a big issue, especially in mining areas where the remediation of spoil heaps and open-pit excavations should come true. Various underground structures count as specific kinds of brownfields. The potential collapse of the underground structures, mostly unknown, accounts for a risk for existing buildings as well as for new projects. Revitalization of the brownfields is also an opportunity to employ smart city concepts. Geothermal energy is one of the geotechnical contributions to the topic of smart cities. The article illustrates the examined issues from the Czech Republic with an emphasis on Brno city.

**Keywords:** Brownfield · Revitalization · Ground investigation · Spoil heap · Slope stability · Utilization of underground

## 1 Introduction

The socially disadvantaged city district (brownfield) is a real estate (a lot, a building or a site) inadequately used, neglected and potentially contaminated. Brownfield usually contains remains after industrial, agricultural, residential, military or other activities. The usage of those sites cannot be appropriate and effective without undergoing a process of regeneration. The rehabilitation of the brownfields is one of the pre-conditions for the sustainable development of the cities. Numerous policy documents adopted at all levels of public administration support the social importance and the need to revitalize the brownfields. The investments in the brownfield revitalization are usually much more costly for the owners than "greenfield" projects and these investments also entail many risks. The investors must include in the expenditure the costs of removing remains from the past utilization of the site. There is a risk that during the revitalization, they will


A. Rotaru (Ed.): CRIT-RE-BUILT 2019, SSGG, pp. 481–495, 2021.
https://doi.org/10.1007/978-3-030-61118-7_39

encounter unforeseen difficulties which may be linked, for example, to environmental pollution. The implementation is often delayed also due to complicated and unsettled ownership of the sites [1, 2].

European Structural and Investment Funds: "A brownfield is defined as an unused, dilapidated or ecologically damaged property that has lost its original use. Refurbishment and revitalization of such properties for the purpose of new, effective utilization are supported by European Union funds".

The term brownfield was first use in the US in 1992. Whereas the issue of brownfields has been discussed in developed countries since the 1960s, it came to the fore of interest in the Czech Republic thirty years later, in the 1990s, when the Czech economy transitioned from planned to market. The switch to a market economy and the accompanying wave of privatization of state-owned enterprises brought about a rising number of abandoned industrial and agricultural complexes, production facilities and warehouses.

Of course there are other types of socially disadvantaged city districts as defined by their original land use – mining-related areas (old coal mines, brickworks), transport (often railway), military (as a result of army reorganization), agricultural (mainly in rural areas), public amenities (school sites, sports facilities – stadiums, etc.) or residential brownfield sites. More rarely there appear even for example ecclesiastical brownfield sites as well as the paradoxical cases of unfinished new buildings – uncompleted buildings where work was for some reason interrupted and then never resumed. Other specific group include former allotments or individual urban areas (e.g. unused areas that were originally gardens).

As a result of their unique positioning and connection to technical infrastructure brownfield sites often have great development potential and at the same time their rehabilitation is an effective way of restricting the process of suburbanization, that is of reducing the number of greenfield projects that are linked to the annexation of agricultural land.

At the end of 2018, 124 socially disadvantaged city districts, spreading on a total area of 418 ha are registered in the city of Brno, see Fig. 1. Rehabilitation of such inefficient sites is one of the basic conditions for sustainable development, see Fig. 2.

In terms of the mobility and traffic, the rehabilitated socially disadvantaged city district should be easily accessible to its residents and visitors (tourists), traveling in the city should be seamless, comfortable, fast, efficient and also environmentally friendly. The use of the smart city concept results in traffic monitoring, smart traffic management (communication tunnels), smart (underground) parking, traffic sharing, etc. In the rehabilitation of brownfields, it is necessary to use state of the art technologies e.g. utilization of geothermal energy, technical collectors, etc. Brownfields are often found on the outskirts of towns, in difficult geological conditions, so it is often necessary to use special geotechnical methods for their rehabilitation. Regeneration is often linked to the slope stability problems (landfill sites, interaction with present underground objects, old brickworks etc.) [1, 2].

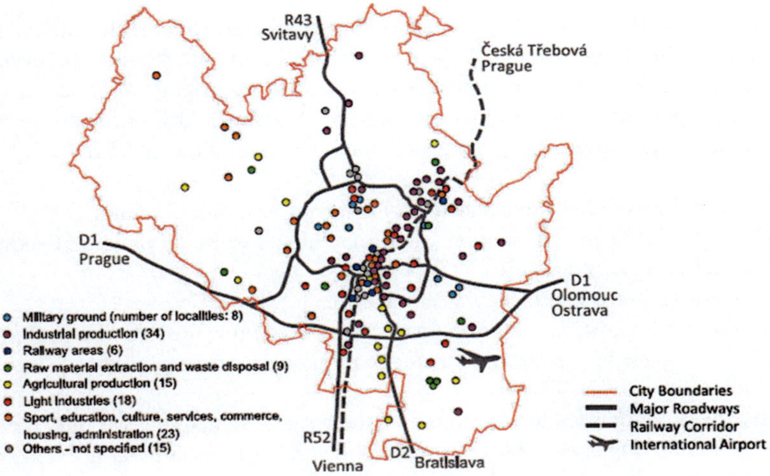


**Fig. 1.** Diagram showing brownfield sites in Brno according to their origin [3]

**Fig. 2.** Shopping centre Brno Vaňkovka – former engineering factory from the 19th century [4]

## 2 Geotechnical Aspects Associated with Rehabilitation of Socially Disadvantaged City Districts

### 2.1 Site Investigation

Site investigation in its widest meaning is the process of relevant site information acquiring, which might affect the construction or performance of a civil engineering project. Such information includes geotechnical, environmental, chemical pollution data etc. On the other hand, the term "ground investigation" is used to cover the detailed work undertaken (excavation, boring, drilling, sampling, testing) to identify the geological and geotechnical characteristics of the site. Ground investigation involves exploring the characteristics of a site as laid down by nature and modified by man and representing the results in a form that can be rapidly understood and assimilated [5].

In civil engineering and building projects the largest element of technical and financial risk usually lies in the ground. This is even more significant for the case of brownfield sites, as the presence of non-natural soils and often occurrence of artificial objects placed into the ground. Site investigation program of brownfield sites must incorporate complementary techniques (e.g. geophysical methods), that allow at least:

- encountering unrecorded subterranean voids (chambers, tunnel, shafts),
- discover remnant of old foundations, that may have historical value and should be investigated by archeologists [6],
- discover ammunitions e.g. from war bombardment [7],
- predict the predispositions of investigated area for future differential settlements, caused by presence of fills, leading to foundation construction problems [8].

If those facts are omitted by investigation, the costs of remedying wrongly designed works, or of mobilizing alternative construction methods, are usually far in excess of the money saved on an inadequately designed site investigation.

## 2.2 Subsoil Improvement

Brownfields are specific in requirement of construction new objects on subsoil, which is significantly affected by human activity – so-called anthropogenic subsoil. As already mentioned, brownfields include territories that are affected by previous human activity, which may result in degradation of subsoil at site:

- Chemical - presence of undesirable substances in the subsoil
- Mechanical:
  - The presence of solid residues that have not been created by nature (e.g. remnants of old foundations, ammunition from war, etc.)
  - Newly deposited material – fill (e.g. spoil heaps – Fig. 3, tailing dams, landfills, etc.). This topic is presented by Vaníček and Vaníček [9] or Watts and Charles [10].

The material could be natural or have been altered artificially prior to deposition. Demolition materials generated during demolition of existing buildings e.g. are found on brownfields sites in built-up areas. Fill is either engineered or non-engineered depending on whether any specific treatment during deposition takes place. As a result, a number of engineering challenges may exist and require different remediation approaches to improve and mitigate problematic behavior. Non-engineering fills may settle variably, have poor bearing capacity and may suffer significant movements due to causes other than the imposed loading [11]. Other problems associated with some wastes include contamination, spontaneous combustion and the emission of gas. The extent to which non-engineered fill will be suitable as foundation material depends largely on its age, composition, uniformity, properties and method by which the material was placed [12].

**Fig. 3.** General view on spoil heap during its construction – filling [9]

The newly deposited materials are characterized by their high compressibility and low bearing capacity. The aim of the geotechnical design is to reduce compressibility and increase the bearing capacity. A number of special foundation technologies exist to achieve this aim. Examples of the suitable methods are:

- Dynamic compaction [13],
- Rapid impact compaction [14, 15],
- Impact rolling [16],
- Vibro stone columns [17, 18],
- Vibro-compaction [13],
- Deep soil mixing [19, 20],
- Reinforcement [21].

These technologies are in focus of research and they are continually improved. Research activities are focused on the theoretical description of individual technologies and on the impact of individual technologies on the environment (e.g. the influence of vibrations). Gaining knowledge about these methods is important for young engineers, because in practice they will increasingly be forced to construct on anthropogenic subsoil.

### 2.3 Geotechnical Structures Using Geothermal Energy

Geothermal energy is one of the oldest energy forms on our planet. The utilization is related to the use of heat flow from the Earth's core. The easiest use is for heating, cooking, bath. However, to reach a sufficient temperature, it is necessary to penetrate deep below the Earth's surface. Therefore, a low temperature (shallow) geothermal energy is usually used. It is also common to use the thermal capacity of the subsoil as a heat accumulator.

Using shallow geothermal energy there is no need to penetrate deep into the earth. The character of heat appliance can help to use even at low temperature. The temperature is often increased by using a heat pump. The use of soil or rock mass for cooling and accumulation can also be included in this category.

Nowadays, there is a common to use a group of geothermal boreholes with heat exchangers drilled in a matrix. The geometric arrangement of such system has a considerable impact on its operation. Specific examples show that, unlike the design, actual boreholes in such system can often be drilled non vertically or even curved. These imperfections then usually deteriorate the thermal properties of the system [22].

Geotechnical equipment, through which shallow geothermal energy is obtained, are usually energy-piles, diaphragm walls and other geotechnical structures (e.g. energy-anchors, energy-tunnel).

**Energy-Piles.** Larger buildings are often based on piles. However, this type of building also usually has specific heating and cooling requirements. To improve the energy budget of the building, geothermal energy can be utilized by energy-piles. The pile is then used in secondary function as a heat exchanger. However, it does not deliver higher temperatures because of the relatively shallow depth of the piles. Therefore, the use of heat accumulation in the subsoil or the use of the groundwater heat often prevails in energy utilization of the piles as for example in project of AZ tower in Brno (Fig. 4).

**Fig. 4.** AZ Tower Brno, height 111 m, total length of large diameter piles - 1800 m [photo R. Holba]

**Diaphragm Walls.** Diaphragm walls are often used to retain the open pit excavation and as underground sealing walls. During the construction, reinforcement is inserted into the groove and then poured with concrete. Inserting pipe winding gives a shallow underground heat exchanger. The diaphragm wall is often used in sealing function, so it is assumed that it will be in contact with groundwater. Thus, heat of the groundwater can be used. The Knightsbridge Palace Hotel in London is a representant of this system. The use of the groundwater heat flowing around a building through a water collector with relatively small thickness of approx. 5 m was considered to employ also in project of the building Palác Trnitá in Brno.

**Energy-Anchors, Energy-Tunnels.** Basically, it is possible to integrate the heat exchanger into any suitable geotechnical element. However, the use in tunnel lining and in the anchors (the energy-anchor can be considered as a reduced geothermal well) is not widespread in the Czech Republic.

## 2.4 Landslides

The problem of the stability of slopes, both natural and cut, has to be faced in many fields of human activity, particularly in civil engineering. Landslides and other slope movements have attracted the attention of man in the same way as other uncontrollable natural phenomena (earthquakes, volcanism, floods) which threaten his life of property. Because of the great damage they cause to farmland, forest stands, communications, engineering structures and buildings, they are also a serious economic problem. The sliding of slopes is often caused by human activity, such as deforestation, bad construction methods etc. When slope stability is disturbed, a number of sliding movements takes place.

Sliding phenomena includes slow, long-term deformations of slopes, which usually occur within a thick zone involving a system of slide surfaces. Those deformation has the characteristic of a viscous movement and it is referred as creep.

The stability problems are closely related to the brownfield sites. There are two main areas related to this issue. One of them is related to the newly man-made deposited slopes (spoil heaps). There is tendency to construct spoil heaps as steep and as high as possible. This tendency brings risk of slope instability, creating landslides with high potential risk to the surrounding area (Fig. 3).

The second area is the exploitation of potentially unstable sites located in built-up localities. Such localities were overlooked in past for their problematic rehabilitation for new building projects. But due to the rising prices of building land, they are now used more often. For example, there is active landslide right in the built-up area of Brno that threatens local cultural heritage Villa Tugendhat (Fig. 5).

Rapid movements of sliding rocks, separated from the underlying stable part of the slope by a plane of separation, are also considered as landslides. On steep rock walls, blocks of solid rocks may become loosened and fall to the foot of the slope. Such a rapid movement is termed a rock fall. There are localities on the north-west of the Czech Republic affected also by this type of rock mass movement (Fig. 6).

Institute of Geotechnics at Brno University of Technology has long tradition with the research and practical application related to landslides. Co-author of one of the first books related to these phenomena [25] prof. Mencl was the former head of the institute.

Series of research projects focused on ground anchors and rock bolts was conducted in last eight years [26, 27] as possibility of landslide or rock fall risk mitigation.

## 2.5 Underground Structures

Cities are facing a problem of limited surface area. The problem is the most significant in densely populated urban districts. However, we can widespread our view to vertical direction and obtain much more options how to on solve the problems with limited surface. If we look upwards, the space is utilized by skyscrapers and various kinds

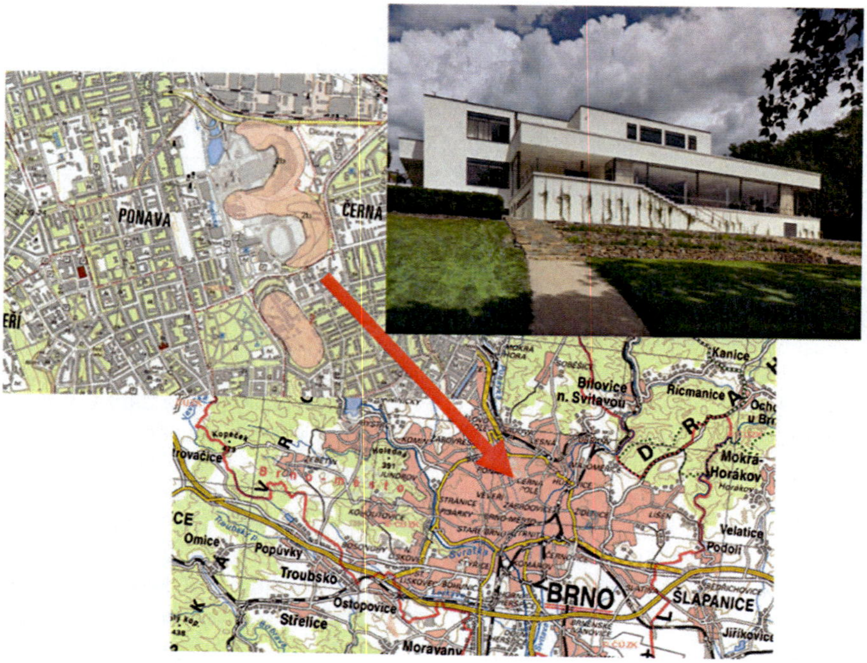


**Fig. 5.** Database of landslides in the Czech Republic – example of active landslide in Brno build-up area [23]

**Fig. 6.** Rock wedges on the road after accidental rock fall in the Bohemian-Saxon Switzerland area [24]

of bridges nowadays. If we look downwards, it is not conspicuous how many options are offering. People have utilized underground space since ancient times typically for storage of food or possessions. Underground has been often used as a shelter, so defense has been also important reason. Transportation became significant issue later. However, transportation also in meaning of water, energy, data, transport etc. Hence, utility lines are often placed under the ground surface.

Underground space utilization has long history in old cities like Paris or London, but underground is extensively used nowadays in every urban area. Tunnels for transit and local road traffic are typical example of current large-scale underground structures. Similar situation is in segment of railway transportation. Good examples are railway by-pass of Innsbruck in Austria, where the city is confined by mountains and new main train station in Berlin, where significant part of the station is subsurface. Local transport is provided by various subway systems in most of capital cities and many other large cities.

**Utility Tunnels (Collectors).** However, underground space is even more often used in small-scale as environment where utility lines are placed separately. Supply of water, electrical energy, heating or sewerage is important for citizens and their comfortable life. State of the art technology of internet, as well as other communication utilities, often uses cables. Fiber optic cables provide the highest speed of data transfer, despite the fact of recent development of wireless technologies. Hence, the cables are still common technology. Increasing demand to capacity of the mentioned utility lines results in construction of more and more lines, what makes chaos in the underground space. Organization of the lines is solution of the problem. The lines can be collected to cable ducts, underground corridors or relatively large utility tunnels, which are alternatives to individual buried lines.

Such a complex topic as underground space, which was briefly introduced, is difficult to include at all. Hence, we will focus on part of utility tunnels, which are in operation in Brno city (Czech Republic).

Densely populated area requires numerous utility lines in order to satisfy demand of citizens and companies. It is logical to organize the lines due to limits of available space, maintenance, cost reduction in case of combined installation, etc. Utility tunnels offer high level of convenience for installation and maintenance of the lines and favorable conditions to enlarge lifetime of cables and pipes. Great advantage of the tunnels is that all service interventions can be done without any traffic restriction or disturbance by noise and dust during excavations on surface. Citizens are then significantly less affected by the works. The other advantage is possibility of easy lines quality monitoring in the utility tunnels. Risk of accident occurrence can be reduced by identification of damaged or degraded sections of lines due to ageing for example. Reconstruction can be planned in advance, directly on problematic places, and an accident is avoided. Supply of media (water, electrical energy, heating, etc.) to citizens is more reliable then. The mentioned approach is following ideas of smart city concept.

One of the problems of underground building construction that arises and must be solved in the case of building a shallow deposited underground structure, is the effect of the driving of the underground structure on the existing built-up area on the surface of then territory. A typical situation is the building of collectors, when it is necessary to guarantee that the deformation in the overburden of the underground structure will not exceed given limit value of the deformation. If it were not the case and the building of the underground structure might in an undesired way influence the existing objects on the surface of the territory, it is necessary to take measures resulting in the limitation of deformations. Among these elimination measures can be included e.g. the location of the underground structure into greater depth, adaptation of sequential excavation, securing the foundations of the threatened objects, cementation of soil surrounding excavation [28], building protective curtains [29]. The last mentioned construction measure has been implemented in the Czech Republic recently in building the network of collector tunnels. An example thereof is the building of the collector tunnel in Starobrněnská Street in Brno or Panenská Street in the capital Prague, where in both cases pillars were built by means of jet grouting. The given solution thus represents the building of another structure element (e.g. a bottoming wall, jet grouting, underground or pile wall) between the underground structure and the foundations of existing object.

However, construction of the utility tunnels is expensive, thus, we need to consider when and where this type of structure is feasible. It depends essentially on amount of involved lines and geological conditions in particular area. On the other hand, high construction costs are balanced by long term lifetime of the tunnels. Hence, planning of the utility tunnels should be done carefully in order to profit from their advantages and do not suffer by their costly construction.

Long-term life of tunnel structures within the infrastructure is expected to a return of higher investment for construction. Structure of tunnel lining can be exposed to negative influences of ground water etc., which cause degradation of used materials during this long period of time. Reduction of bearing capacity is the result, which in the critical case may cause the collapse of whole structure. Thus, it is advisable to know response of the weakened structure in order to identify the negative trends in its behavior early and to proceed with remediation. Consequences of lining degradation can be analyzed, for example, by FEM (Finite Elements Method) modelling of tunnel [30]. Proper choice of a constitutive model is an essential part of any successful application of numerical methods in geotechnical engineering. In most cases, attention is paid to the soil constitutive model. Simpler elastic constitutive models are often used for structural elements, such as tunnel linings, retaining structures etc. These material models however do not involve many aspects of a real structural behavior, such as limited tensile and compressive strength, strain softening and time dependent behavior during service life of a structure. Therefore, advanced material model for a shotcrete lining can be used [31].

**Historical Underground Objects.** As it was mentioned above, many aged underground structures can be found beneath centers of cities with long history. Some of the objects can be known and well maintained, while some of them can be without appropriate maintenance or even completely unknown. Especially unmaintained and unknown underground spaces represent potential risk for society organized in cities. Sudden collapse of structure can cause serious economic or health damages, sometimes even with fatal consequences. Since long history of settlement in Europe, aged underground structures represent problem, which has to be dealt with in many towns and cities.

Brno city in the Czech Republic can be presented as an example where plenty of cellars and corridors are crossing underground of historical center. The cellars and corridors have been built in medieval ages, then many times rebuilt and extended. Many of them have been closed and forgotten. Sometimes the cellars are rediscovered during recent construction activities. Such finding usually makes problem to current project, which have to be adapted in some way, especially when the cellars or corridors have to be preserved due to their historical value. Some projects of historical underground spaces remediation were successful and Brno city open them for tourists as museums, for example Labyrinth underneath "Zelný trh" (Vegetable Market) and Ossuary underneath the Church of St. James. This is the second largest ossuary in Europe (according to the number of skeletal remains.

When known historical cellars or corridors could be affected by new construction project, geotechnical monitoring should be installed to observe potential changes. If negative trend of e.g. deformations exceed certain acceptable limit, some prepared safety measurements have to be done in order to avoid serious damages. Sensors of the geotechnical monitoring can be modified to allow continuous measurement with on-line data transfer. Then experts can have access to the data in real time and make prompt safety analysis or decisions. Hence, employment of on-line geotechnical monitoring is reasonable and also support ideas of smart city concept.

## 3 Conclusions

- There are many definitions of socially disadvantaged city districts – brownfields in the world, even in the Czech Republic, the definition varies by institution.
  For example:

- Ministry of the Environment of the Czech Republic: Brownfields are all land areas that have been fundamentally disturbed by human activity and cannot be further used effectively or may potentially pose a threat to the environment.
- Ministry of Regional Development of the Czech Republic: Brownfields are all land plots and properties in urban areas that have lost their original purpose and are underused. Such properties are economically and physically detrimental to their surroundings and to themselves etc. [3, 4].

- Utilization of brownfields has been discussed in developed countries since the 1960s. It came to the fore of interest in the Czech Republic thirty years later, in the 1990s, when the Czech economy transitioned to a market economy. The pioneer of brownfields redevelopment in the Czech Republic is Prof. I. Vaníček.
  At present, rehabilitation of socially disadvantaged city districts is supported by various Czech and European financial programs, e.g. Operational Programme Enterprise and Innovation for Competitiveness, Operational Programme Environment etc. [3, 4].
- The whole process of the brownfields redevelopment can be divided into the following steps:

- site location identification,
- first phase of investigation,
- preliminary economic analysis,
- project and completion of site remediation

- design and completion of construction of new development (including foundation engineering, reuse of old foundations) [3].

- There is a significant role of the geotechnical engineering for revitalization of socially disadvantaged areas, especially from the view of sustainable construction, from the view which is very sensitive for all society at this time. The team of Brno University of Technology (BUT) in the RE-BUILT project is focused on the revitalization of newly deposited materials from a geotechnical perspective. This problem is in more detail described in Chap. 8 "Earth structures in Environmental Engineering" [9]. The importance of this issue is illustrated by the establishment of the TC307 Sustainability Technical Commission at the International Society for Soil Mechanics and Geotechnical Engineering [31]. The big issue that needs to be considered and solved during remediation of big areas of man-made fill (which will be used as brownfield sites for building project more and more intensively in future) is stability of the fills [32, 33]. The problems which relate to structures, are following:

- Free fall filling is more often a manner of spoil heap construction.
- Deposited material has high porosity (macro-porosity) between individual clods
- Individual clods are sensitive to weathering [34].
- Properties are time dependent, not only as function of weathering but also as the result of water saturation, depth of deposition etc. [9].
- Another geotechnical problem is associated with filling of tailing dams. During sedimentation in free water the gravitational force is decreased by uplift and therefore settling particles are deposited under very loose state with high porosity. This loose state has very low shear strength and high susceptibility to structural collapse (especially during dynamic loading). The biggest and heaviest particles sediment close to the main dam and create there the zone with highest permeability. As the result, final body of settled waste is anisotropic and nonhomogeneous, susceptible to differential settlement.

Changes in the shape of man-made fill created by free filling has to be often done during their rehabilitation and transformation to building sites. It may be necessary to use variety of retaining structures to ensure stability. Such structures require, due to their dimensions, incorporation of anchoring elements. The team of BUT has also wide experience with anchors design and testing based on research carried out at the university [35–37].

- The very important question from geotechnical point of view is: How to prepare brownfield site for new reuse? The role of the geotechnical engineer is to design appropriate method for improvement, reinforcement or treatment subsoil at brownfield site. It means, that geotechnical engineer needs to have theoretical and practical knowledge. The knowledge, which is transferred to students at universities, should be based on knowledge obtained from literature, own research and professional activities.
- Next topic, which is important for this site, is ground investigation. One of the simple but very quick and useful technique for determination of space differences in density of

fills is the group of penetration methods (CPT, SPT, DHP etc.). The research program, aimed partially on the evaluation of density of backfill by dynamic penetration probe, was done by the BUT research team [38, 39]. The results of research are incorporated into the lectures of RE-BULIT teaching programme.

- Collectors (Utility tunnels) could be constructed in order to supply energy and other utility lines to both industrial and residential areas. Some of the industrial areas may become brownfields. It is necessary to deal with the existing collectors when reuse of the area is planned. Furthermore, construction of new collectors either in brownfield areas can be useful in order to systematical organizing of utility lines. Especially when the area will be used for e.g. administrative or residential purposes and ideas of smart city concept can be implemented also by this way.
  Historical and various other types of existing underground spaces are often challenging during construction of new projects in cities. Useless underground openings could be also considered as subsurface brownfields. They can be removed or adapted to some new purposes.
  Thus, it is useful for civil engineers to have overview within the mentioned underground spaces subtopic when they deal with remediation of brownfields. Issue of construction of new collectors and rules of maintenance for existing structures should be discussed. Furthermore, interaction of new surface projects and existing underground structures is important where geotechnical monitoring should be applied. The noted issues can be presented by practical examples from Brno city.
- The use of the smart city concept results in smart traffic management (communication tunnels), smart (underground) parking etc. It is necessary to state of the art technologies such as energy structures, during the transformation of the built environment for the rehabilitation of socially disadvantaged city districts.

## References


1. Úvod I BROWNFIELDY CZ. http://www.brownfieldy.eu. Accessed 14 May 2019
2. Investment opportunities – Statutory city of Brno. https://www2.brno.cz/index.php?lan=en&nav01=20608&nav02=20621&nav03=22317. Accessed 14 May 2019
3. Brno brownfields 2015. https://www.brno.cz/fileadmin/user_upload/Podnikatel/Brownfields/mmb_brownfields_EN_2015.pdf. Accessed 22 Jan 2020
4. Obchodní centrum Galerie Vaňkovka. https://www.k4.cz/galerie-vankovka/t1064. Accessed 22 Jan 2020
5. Sarsby, R.: Environmental geotechnics, 2nd edn. ICE Publishing, London (2013)
6. Piga, C., Piroddi, L., Pompianu, E., Ranieri, G., Stocco, S., Trogu, A.: Integrated geophysical and aerial sensing methods for archaeology: a case history in the punic site of Villamar (Sardinia, Italy). Remote Sens. **6**(11), 10986–11012 (2014)
7. Frýbová, P., Drápalová, R., Vlček, P., Lubojacký, O.: Předběžný IF a GF průzkum pro návrh GT a HG průzkumných prací na lokalitě Červený kopec: Závěrečná zpráva. VENTIMIGLIA s. r. o., Brno (2013)
8. Mei, Y., Hu, C.M., Wang, X.Y.: Experimental research on deep collapsible loess foundation treatment by dynamic compaction under super high fill. Appl. Mech. Mater. **256–259**, 129–138 (2013)

9. Vaníček, I., Vaníček, M.: Earth Structures in Transport, Water and Environmental Engineering, 1st edn. Springer, The Netherlands (2008). https://doi.org/10.1007/978-1-4020-3964-5
10. Watts, K., Charles, A.: Building on Fill: Geotechnical Aspects, 3rd edn. BRE Press, Watfords (2015)
11. Charles, J.A., Watts, K.S.: Building on Fill: Geotechnical Aspects, 2nd edn. BRE Press, London (2001)
12. Burland, J., Chapman, T., Skinner, H., Brown, M.: ICE Manual of Geotechnical Engineering: Volume II Geotechnical Design, Construction and Verification, 1st edn. ICE Publishing, London (2012)
13. Kolymbas, D., Fellin, W.: Compaction of Soils, Granulates and Powders (Advances in Geotechnical Engineering and Tunneling), 1st edn. CRC Press, Boca Raton (2000)
14. Serridge, C.J., Synac O.: Application of the Rapid Impact Compaction (RIC) technique for risk mitigation in problematic soils. In: 10th Congress of the International Association for Engineering Geology and the Environment, IAEG 2006, p. 294. The Geological Society of London, London (2006)
15. Adam, D., Adam, C., Falkner, F.-J., Paulmichl, I.: Vibration emission induced by Rapid Impact Compaction. In: Proceedings of the 8th International Conference on Structural Dynamics, EURODYN 2011, Leuven, Belgium, pp. 914–921 (2011)
16. Waddell, P.J., Moyle, R.A., Whiteley, R.J.: Geotechnical verification of impact compaction. WIT Trans. Ecol. Environ. **141**, 73–85 (2010)
17. Kirsch, F.: Vibro stone column installation and its effect on ground improvement. In: Proceedings of the Numerical Modelling of Construction Processes in Geotechnical Engineering for Urban Environment, Bochum, Germany, 23–24 March 2006, pp. 115–124. Taylor and Francis, London (2006)
18. Castro, J., Karstunen, M., Sivasithamparam, N.: Influence of stone column installation on settlement reduction. Comput. Geotech. **59**, 87–97 (2014)
19. Madhyannapu, R.S., Puppala, A.J., Nazarian, S., Yuan, D.: Quality assessment and quality control of deep soil mixing construction for stabilizing expansive subsoils. J. Geotech. Geoenviron. Eng. **136**(1), 119–128 (2010)
20. Sukontasukkul, P., Jamsawang, P.: Use of steel and polypropylene fibers to improve flexural performance of deep soil-cement column. Constr. Build. Mater. **29**(1), 201–205 (2012)
21. Koerner, R.M.: Designing with Geosynthetics, 5th edn. Prentice Hall, Upper Saddle River (2005)
22. Leiter, A.: Geothermal Energy – Influence of the Borehole Geometry (Doctoral Thesis). Brno University of Technology, Brno (2017)
23. Svahové nestability. https://mapy.geology.cz/svahove_nestability/. Accessed 22 Jan 2020
24. iDNES.CZ. https://www.idnes.cz/zpravy/domaci/ze-silnice-v-hrensku-zmizi-spadla-skala.A120903_163652_usti-zpravy_alh. Accessed 22 Jan 2020
25. Záruba, Q., Mencl, V.: Landslides and Their Control: Development in Geotechnical Engineering, 1st edn. Elsevier, Amsterdam (1982)
26. Chalmovský, J., et al.: Statistical-numerical analysis for pullout tests of ground anchors. Baltic J. Road Bridge Eng. **12**(3), 145–153 (2017)
27. Štefaňák, J., Miča, L.: Response surface method analysis of the ultimate capacity of intelligent composite anchoring element. In: Proceedings of the 16th Danube - European Conference on Geotechnical Engineering, 7–9 June 2018, pp. 785–790. Ernst & Sohn, Berlin (2018)
28. Boštík, J.: Stabilita mělce uložených podzemních děl. In: Geotechnika 2006, pp. 225–230. ORGware, Stupava (2006)
29. Miča, L. Boštík, J.: A study of the influence of protective curtains during the excavation of an underground structure in the urban area (2005, unpublished)

30. Závacký, M., Chalmovský, J., Štefaňák, J., Miča, L., Bílek, P.: Modelling of tunnel lining degradation. In: Road and Rail Infrastructure V, Proceedings of the Conference CETRA 2018, pp. 1499–1505. University of Zagreb, Zagreb (2018)
31. ISSMGE – TC307 Sustainability. https://www.issmge.org/committees/technical-committees/impact-on-society/sustainability. Accessed 20 Jan 2020
32. Igwe, O., Chukwu, C.: Slope stability analysis of mine waste dumps at a mine site in Southeastern Nigeria. Bull. Eng. Geol. Env. **78**(4), 2503–2517 (2018). https://doi.org/10.1007/s10064-018-1304-8
33. Rybalchenko, S., Verhovov, K., Kudriavtcev, S., Kudryavtsev, S., Yoo, C., Zhussupbekov, A.: Landslides-flows on rock dumps of coal mining enterprises. In: MATEC Web of Conferences, vol. 265. EDP Sciences, Les Ulis Cedex (2019)
34. Novianti, V., Marrs, R.H., Choesin, D.N., Iskandar, D.T., Suprayogo, D.: Natural regeneration on land degraded by coal mining in a tropical climate: lessons for ecological restoration from Indonesia. Land Degrad. Dev. **29**(11), 4050–4060 (2018)
35. Chalmovský, J., Štefaňák, J., Miča, L.: An innovative approach to the design of ground anchors. In: 15th International SGEM GeoConference on Science and Technologies in Geology, Exploration and Mining, vol. 1.2, pp. 25–35. SGEM World Science, Albena (2015)
36. Štefaňák, J., Miča, L., Chalmovský, J., Leiter, A., Tichý, P.: Full-scale testing of ground anchors in Neogene Clay. Procedia Eng. **172**, 1129–1136 (2017)
37. Štefaňák, J., Miča, L.: Design parameters assessment for ground anchors in tertiery fluvial clays. Eng. Struct. Technol. **3**(6), 106–113 (2014)
38. Leiter, A., Štefaňák, J., Miča, L.: Development of automatic datalogger system for heavy dynamic penetrometer. In: 18th International SGEM GeoConference on Science and Technologies in Geology, Exploration and Mining, vol. 1.3, pp. 287–294. SGEM World Science, Albena (2018)
39. Štefaňák, J., Miča, L.: Full-scale compaction testing of recycled demolition materials. In: 17th International SGEM GeoConference on Science and Technologies in Geology, Exploration and Mining, vol. 17.41, pp. 99–106. SGEM World Science, Albena (2017)

# Mitigation of the Impact of Landslides in Built-up Areas

Jan Štefaňák(✉)

Faculty of Civil Engineering, Institute of Geotechnics, Brno University of Technology, Veveri 331/95, 602 00 Brno, Czech Republic
stefanak.j@fce.vutbr.cz

**Abstract.** Paper brings a review of creep slope movement phenomena, its damage to structures, and the applied stabilization techniques in urban locations. The holistic approach to such locations based on proper ground investigation and design of preventive and remedial measures is presented on three case studies in Brno, Czech Republic. The need to respect the given geological conditions is emphasized. The related non-structural preventive measures (preventing infiltration, not undercutting the slope toe), and structural remedial measures (underpinning, anchored pile walls) application in the specific condition are described. The process of numerical analysis of strengthening pile retaining wall supporting the cut constructed in the lower part of the sliding body is presented. The importance of residual soil strength parameters for analysis of landslides is emphasized also.

**Keywords:** Landslide · Remedial measures · FEM · Ground anchors · Residual strength

## 1 Introduction

The problem of the stability of slopes, both natural and excavated, has to be faced in many fields of human activity, particularly in civil engineering. The sliding of slopes is not uncommonly caused by human activity, such as deforestation, bad construction methods, etc. When slope stability is disturbed, a number of sliding movements take place. Sliding phenomena include slow, long-term deformations of slopes, which usually occur within a thick zone involving a system of slide surfaces. Those deformations possess the characteristics of a vicious movement and are referred to as creep [1]. Since the 1940's it has been understood that earth acting on the structures the sliding body may significantly differ from those in stable slopes. Design codes (e.g. SIA 267) reflect this knowledge and recommend application of landslide pressures for construction in permanent landslides. The possible effect of boundary proximity may include increased pressures on the existing structures. This becomes particularly important when a building in this area has to be replaced by a new one. It raises the question about the magnitude of earth pressures, that should be used in the design of both the excavation and the structure. Understanding the mechanism of the downhill pressure transfer within the


A. Rotaru (Ed.): CRIT-RE-BUILT 2019, SSGG, pp. 496–505, 2021.
https://doi.org/10.1007/978-3-030-61118-7_40

stable part of the slope, below the landslide, is critical for finding the answers to these questions. Intuitively, there should be a transition zone, within which high landslide pressures subside to the regular (at rest) earth pressure in the stable slopes [2]. The creep earth pressure is to be considered in such cases [3]. Continuing urbanization brings construction into landslide-prone areas. There are two active landslides just near the city center in Brno, The Czech Republic.

## 2 Landslides in the Build-up Area of Brno

Brno is the capitol of the South Moravian Region of the Czech Republic located 140 km north from Vienna. There are two active landslides just near the city center in Brno, The Czech Republic (see Fig. 1).

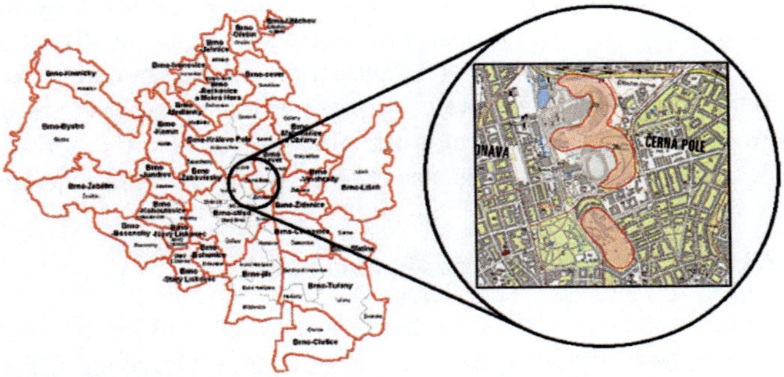


**Fig. 1.** City plan of Brno with the detail of two active landslides (the pink shaded areas in the right part of figure) in close vicinity of the city Center (based on [4] and www.geology.cz, last accessed 2019/10/10)

There are at least three interesting sites located directly in these landslides, or in its close proximity, which worth to be mentioned. The first of them is a typical brownfield site [4] – "Za Lužánkami" football stadium. The second object located in the landslide zone is the Villa Tugendhat. The third site, where the problems with those landslides are faced, is located under the slope, i.e. in the area of the lower part of sliding mass (see Fig. 2).

In this third case, the problems are associated with the construction of parking lot for the hospital. The objectives of the paper are 1/to emphasize the need for a proper evaluation of findings from the initial desk study and walk-over survey aimed at the geotechnical character and geological history of the city areas prone to creep sliding movements. This step should precede any decision on the further development of such areas. Omitting this step can lead to unnecessary significant costs for repairing possible damages, 2/to present available non-intrusive geophysical survey methods and their application in the ground investigation of landside areas, 3/to point out typical damages of buildings located in a different part of the sliding body (i.e. in the crown and near the toe). Examples of structural and non-structural remedial measures of these damages are

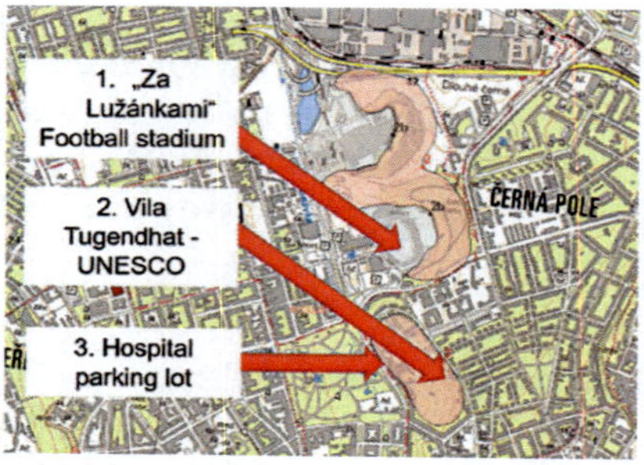


**Fig. 2.** Localization of the discussed cases in the active landslide located in the "Černá pole" city district (adapted from www.geology.cz, last accessed 2019/10/10)

also discussed. 4/The last objective is to present possibilities of design and assessment of remedial measures by numerical methods, together with an identification of some important points related to this approach (e.g. the use of residual strength for calculation).

The contribution tries to bring awareness of geotechnical issues closer also to professionals from other fields involved in urban planning. Hence, the holistic approach based on the description of case studies was chosen to achieve these objectives.

## 2.1 Case Study No. 1: Za Lužánkami Football Stadium

This stadium, built in the period 1949–1953, was reconstructed, in the 1960s and 70s with the building of new stands to give a capacity of 50 000 making it the largest in the country. After the revolution in 1989, the whole sports facility was acquired by a private company. The stadium began to fall into disappear, and in 2001 no longer met the requirements of the competition at the highest level. Hence, the football club had to move to another stadium. From that time the stadium has been deteriorating [4]. Interpreted results from the geophysical survey [5] in Fig. 3 show the borders of partial mass movements (white dashed line). It is obvious that the construction of the stadium itself was a significant factor that initiated landslides in this area, as the toe of the slope was undercut.

**Fig. 3.** Interpretation of the 2011 geophysical survey results (right); the photograph of the shallow landslides on the opposite site of the hill and the mitigation measures from 1993. (taken from [5]).

The stability problems already have roots in history. It can be seen in the right picture in Fig. 3 the view on the hill above the "Za Lužánkami" stadium from the other side than the stadium is located (the year 1993). As there was brickwork for decades, the soil near the toe of the hill was mined for the bricks production. This, of course, initiated landslides. The brickwork owner has an effort to secure the facility against landslide. There was a subtle gabion gravity wall constructed near the toe of the hill. The effect of such a structure on the stabilization of the whole landslide is very questionable if any. During the construction of the shopping center on the place of former brickworks, the anchored earth retaining structures were built-up on this side of the hill [5]. As was mentioned before, the stadium is out of the focus of developers currently. Hence, no other remedial measures to stabilizing the landslide on the stadium side have been constructed up to now.

## 2.2 Case Study No. 2: UNESCO World Cultural Heritage Building Villa Tugendhat

Villa Tugendhat is one of the pioneering prototypes of modern architecture in Europe. It was designed by the German architect Ludwig Mies van der Rohe and built in 1929–1930. It is the only example of modern architecture in the Czech Republic inscribed in the list of UNESCO World Cultural Heritage sites. The Villa Tugendhat underwent renovation and restoration work between 2010 and 2012. Both, the structure and the adjoining gardens, were restored to their original appearance following the completion of the Villa in 1930. The interiors have been equipped with exact replicas of the original furnishings [6]. The renovation started after the alarming state of the load-bearing structure of the building with many cracks appearing (see Fig. 4). Those cracks were initiated by the creep movement of soil mass, as the building is founded directly on the crown of the sliding body.

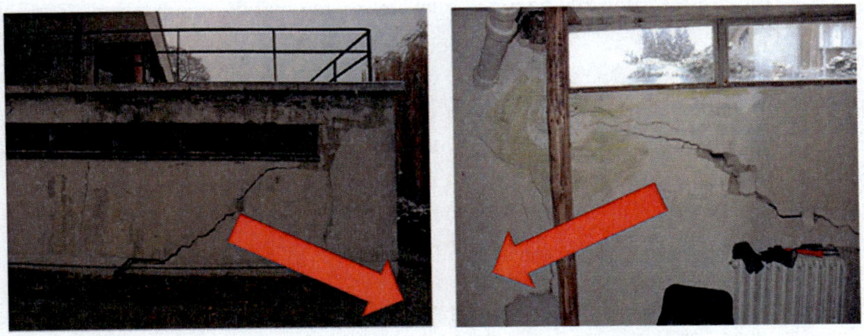

**Fig. 4.** Direction of the soil mass movements and resulting cracks through the walls: view from the exterior (left) and from the interior (right) (taken from http://www.geology.cz/app/pasport/viewdbs.pl?db=3&map=24-32-25, last accessed 2019/10/10).

As the building is an invaluable part of history, much effort was made both into the investigation and consequent restoration activities. The geophysical investigation was conducted, where the GPR (Ground penetrating radar) measurements, magnetic profiling, and electrical conductivity techniques were incorporated [7]. The Geophysical investigation reveals, that the surface layers under the villa consist of loess and loess clay with a thickness of up to 5 m. These loess layers are largely saturated by water and damaged by water seepage in the area of the villa. There is a layer of up to 2 m thick of sandy-gravel material, probably an old Pleistocene terrace, under the loess layer. Groundwater can infiltrate the area of the villa and its surroundings through this sandy-gravel terraces. The subsoil consists of Badenian calcareous clays, occasionally sandy clays. These are covered only by eluvio-deluvial quaternary sediments and fillings. The spread foundations of the villa are in loess and partly in the underlying neogene clays. The boundaries of the partial sliding masses are marked by dashed lines in the Fig. 5. The crown of one of the landslides bodies is located directly under the villa. The picture results from the interpretation of the combination of the electrical methods and GPR [7].

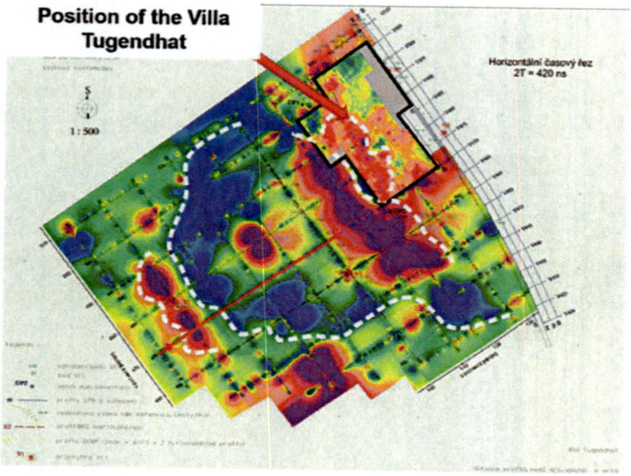


**Fig. 5.** Localization of the landslide boarders intersecting the building foundations (taken from [7]).

The complexity of movements is best documented by measurements conducted directly on the foundation structures. The translation movement is also accompanied by the rotation of the footings. The largest movements clearly correlate with anomalous soil weakening. Intensive subsoil damage reaches a depth of 5 to 6 m below the basement floor of the villa in most cases. However, it goes even up to a depth of 8 m in some cases [7]. Remedial measures consisted of transferring the stress from the current shallow foundations to the deeper sound strata below the sliding surfaces. Where existing foundations are inadequate, they can be improved by underpinning, which involves either providing new foundations or, more usually, extending the existing foundations downwards to reach stiffer or more stable ground. Underpinning has a variety of applications, including 1/Increasing the bearing resistance of foundations, prior to refurbishment work, and 2/Preventing foundations being undermined by nearby excavations [8]. The most common use is as a remedy for damage caused by foundation movement. Hence, the mini-piling underpinning was designed as a remedial measure for the Vila Tugendhat. In this method of underpinning, loads are transferred directly from the structure to the piles, either by needles or cantilevered pile caps, or by placing the piles directly through the existing footing.

## 2.3 Case Study No. 3: Cut for the Hospital Parking Lot

The third site, where the problems with above-discussed landslide was faced, is located under the slope, i.e. in the area of lower part of sliding mass. In this particular case, the problems are connected with the disputable construction process of parking lot for the hospital.

**Description of the Structure**

In 2010 the investor decided to construct parking lot behind the buildings of his real estate. The retaining structure was designed in the form of pile wall. Piles, structures made of piles, and pile-like structures are usually useful structural elements, to support deep excavations and cuts in slopes, and to retain creeping or sliding slopes. Depending on the static system and the dimensions, the structural elements transfer forces mainly by shear ("dowel") or mainly by bending ("beam") to the ground. In numerous cases, they are particularly effective in combination with other structural measures like (pre-stressed) anchors and/or drainage systems [3]. The discussed pile-wall was designed as a contiguous wall. A contiguous pile wall (the piles were installed with a gap between individual piles) consists of bored cast-in-place piles installed along the line of the wall. The size of the gap is chosen on the basis of the site dimensions, and also the particular ground conditions, but is typically between 50 mm and 150 mm (in this particular case, the distance was 450 mm). It was also designed as an embedded retaining wall. (Embedded wall penetrates into the ground at its base, to gain lateral support from the passive resistance of the ground in front of the lower part of the wall). Hence, the passive earth resistance was used as a sole resistance to provide stability for the wall. The designer originally decided to build up the wall without any anchoring or other sources of lateral support for stability, despite the well-known fact that the structure is located in/under the body of the active landslide. The proposal was to monitor the deformations of the wall and to add the eventual anchors later. The deformation starts to increase immediately after the structure was built-up and the cumulative horizontal displacement measured at the pile heads was 180 mm in 2016.

**Geotechnical Model Development**

The investor decided to evaluate the current state of retaining structure. Consequently, the remedial measure should be designed to stop the progress of deformations of the wall. The GPR (Ground-penetrating radar) measurements, magnetic profiling and electrical conductivity techniques were incorporated in the geophysical survey conducted in the target area by Frolka and Hubatka [9]. The position of the boundaries of the partial sliding masses is depicted in the Fig. 6 together with the position of evaluated structures.

The georadar measurements reveal the presence of a number of sliding surfaces and landslides masses in the area. These sliding bodies are located in the upper parts of the weathered Badenian clay and claystone complex. The active slip surfaces are located at a depth of 3 to 6 m. Next picture (Fig. 7) which shows a geological interpretation of

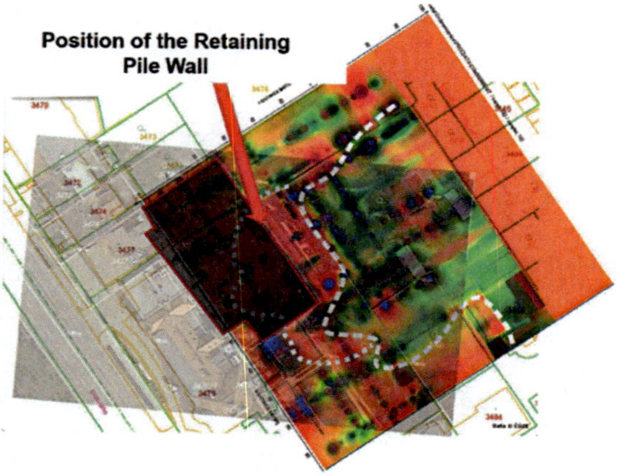


**Fig. 6.** Localization of the landslide boarders intersected by the retaining wall (adapted from [9] and Google maps).

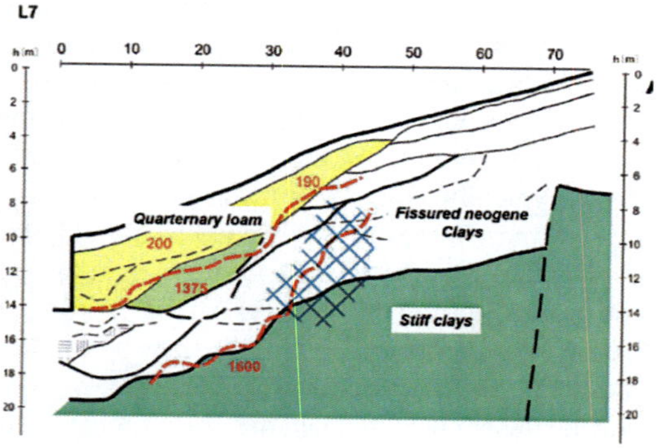


**Fig. 7.** Geological interpretation of down the slope cross-section through the slope based on the geophysical survey (adapted from [9]).

the studied area is based on a combination of all the results of the geophysical survey method that have been implemented in the survey conducted by Frolka and Hubatka [9].

Hence, it can be concluded the area of interest is located in a complex of Badenian clays, which are covered with loess. Coverage thickness ranges from 2 to 5 m. The base of the disturbed Baden complex lies at a depth of about 8 to 10 m and is more or less parallel to the existing terrain. There are also displayed results of seismic measurements (wave velocities) in the picture (see Fig. 7). The velocities in the cover layers range between 110 and 310 m/s. The main refractive interface runs at a depth of 3–5 m. The value of the refracted wave velocities of 1375 m/s indicates a relatively strongly damaged rock environment. Stronger (or intact) rocks can be expected only from depths of 8–10 m, according to the results of seismic tomography. The velocity refracted seismic waves increases to about 1600 m/s in those strata [9].

Because of the complexity of the problem, the FEM analysis was conducted. Numerical methods are extremely powerful and have the potential for dealing with most of the likely scenarios that may arise in urban development. However, there are several issues. The problems involve both, the ground and various structural elements, so understanding of both structural and ground behavior is necessary. The knowledge of the numerical method being used, and the constitutive models used to represent the various soil strata and structural members is necessary as well, etc. Hence, the FEM model geometry was based on the geological interpretation of geophysical measurements. The Mohr-Coulomb constitutive model was adopted, and the calculations were performed using the Plaxis 2D software [10].

The question arises in this moment, what shear strength parameters shall be used in constitutive models. Skempton [11] concluded that the drained post-peak drop in strength observed in overconsolidated clays is due to an increase in water content due to dilation, and due to the orientation of clay particles parallel to the direction of shearing [12] In normally consolidated clays, the post-peak drop in strength is due entirely to the orientation of particles parallel to the direction of shearing. Large shear displacements in one direction are required to orient the clay particles parallel to the direction of shear and obtain a residual strength condition in both normally consolidated and overconsolidated clays. Such displacements are usually achieved during the land sliding process. Hence, the residual shear strength shall be used in landslides calculations (Table 1).

**Table 1.** Values of soil parameters for FEM analyses.

| Material | $\gamma_{unsat}$ [kN/m$^3$] | $\gamma_{sat}$ [kN/m$^3$] | c′ [kPa] | ϕ′ [°] |
|---|---|---|---|---|
| Quaternary loam | 20,0 | 20,0 | 0 | 18 |
| Stiff clays | 21,0 | 21,0 | 20 | 15 |
| Fissured Neogene clays | 21,0 | 21,0 | 0 | 12* |

*Residual strength

It was proposed to add 24 ps of ground anchors with the tendon made of 6pcs of steel strands 15.7/1770 as a remedial measure [9]. The bonds of anchors have to be placed in the competent stiff clay strata under the presumed shear surface. The calculation was conducted in these stages:

1. Initial stress state calculation using gravity loading procedure
2. Installation of pile walls
3. Activation of anchors
4. Excavation
5. Determination of factor of safety via fi-c reduction technique

There are two snapshots from the Plaxis 2D software showing the stage 1. and the stage 4. after activation of the ground anchors and after excavation was finished (see Fig. 8). The boundaries of lithological units are derived from the geological interpretation of the geophysical survey (compare with Fig. 7).

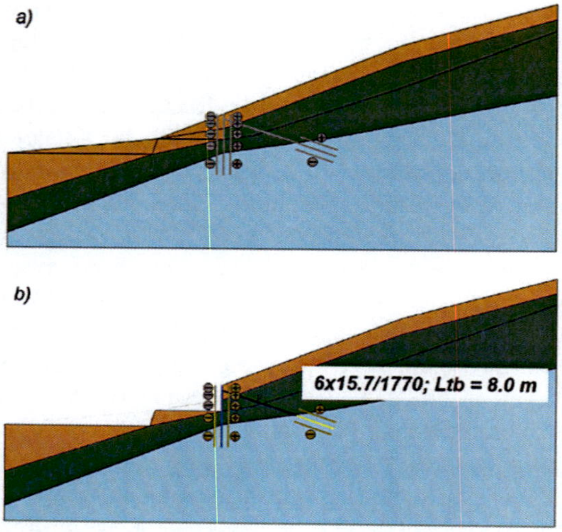


**Fig. 8.** Geotechnical model used for FEM calculation. A) The initial phase; b) The final phase after activation of the anchors and after excavation (adapted from [9]).

### Results of FEM Calculation

From the results (see Fig. 9), it can be seen, that the critical slip surface is located in the fissured clay strata, where the residual shear strength was considered. Even after all the additional supporting force from anchor was applied, the final factor of safety was still about FoS = 1.2 Hence, the remedial measures are able to minimize movement of the wall, and probably stabilize the landslide temporarily. However, the complete termination of the landslide movement can hardly be assured. It should be emphasized, that this landslide was probably reactivated, directly by the excavation of the toe of the slope.

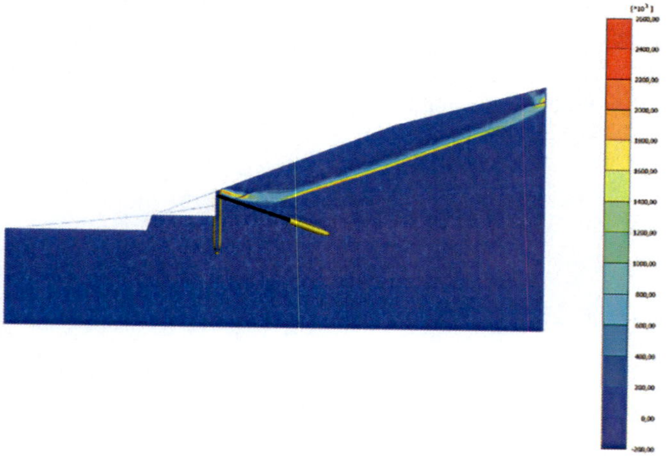

**Fig. 9.** FEM analysis results – total deviatoric stress indicating position of the critical slip surface identified by the fi-c reduction technique.

## 3 Summary

Recommendations for construction in permanent landslide bodies are available and the mechanisms relatively good understood [2]. The remedial measures for mitigation of the translation movement of footings (accompanied by the rotation of the footings) were demonstrated in the example of the Villa Tugendhat case. It is also appropriate to combine such structural measures with the prevention of the water infiltration into the shear surface. The underpinning was designed as a remedy for damage caused by foundation movement. It extends the existing foundations downwards to reach the stiffer and more stable ground.

Piles, structures made of piles, and pile-like structures are usually useful structural elements, to support deep excavations and cuts in slopes, and to retain creeping or sliding slopes. In numerous cases, they are particularly effective in combination with other structural measures like (pre-stressed) anchors and/or drainage systems [3]. The numerical analysis of the pile-wall retaining structure that undercut the toe of the landslide body was presented. The importance of the choice of appropriate shear strength for simulation was depicted. As the large shear displacements in one direction are usually achieved during the land sliding process, the residual shear strength shall be used in landslides calculations. The best is to avoid the undercut of the toe of land-sliding soil body, or the proper effective structural measures like pre-stressed anchors (The bonds of pre-stressed anchors have to be placed in the competent stiff clay strata under the presumed shear surface.) and/or drainage systems must be carefully designed. Otherwise, additional mitigation measures can be very expensive with questionable results.

## References

1. Wyllie, D.C., Mah, C.W., Hoek, E.: Rock Slope Engineering: Civil and Mining, 4th edn. Spon Press, New York (2004)
2. Puzrin, A.M.: Landslide influence zone - a hidden hazard, 27th Prague Geotechnical Lecture. In: Prague Geotechnical Days 2019 Geotechnical Design for Potentially Unstable Ground, pp. 4–6. SG Geotechnika a. s., Charles University, Faculty of Science, Czech and Slovak Society for Soil Mechanics and Geotechnical Engineering, Prag (2019)
3. Adam, D.: Piles as retaining structures in slopes - case histories. In: Prague Geotechnical Days 2017 - Resume: Piles as Retaining Structures, pp. 3–17. Czech Academy of Sciences, Prague, Czech Republic (2017)
4. BRNO Brownfields 2015: City of Brno - City Strategy Office, Brno, Czech Republic (2015)
5. Hruška, J.: Lužánky_Boby tenisové kurty Sprint. Geofyzika a. s., Brno, Czech Republic (2011a)
6. Villa Tugendhat in Brno is wonderful example of early modern architecture. https://www.tugendhat.eu/. Accessed 10 Oct 2019
7. Hruška, J., Hubatka, F., Valter, V.: Vila Tugendhat. Geofyzika, a. s., Brno, Czech Republic (2011b)
8. Charles, J.A.: Geotechnics for Building Professionals, 1st edn. IHS BRE Press, UK (2005)
9. Frolka, J., Hubatka, F.: Nemocnice Delta. Kolejconsult & servis, Brno (2011)
10. PLAXIS 2D: Reference Manual. Version 2016, Build 8122. Plaxis, Delft (2016)
11. Skempton, A.W.: First-time slides in overconsolidated clays. Geotechnique **20**(4), 320–324 (1970)
12. Budhu, M.: Soil Mechanics and Foundations, 3rd edn. Wiley, New York (2011)

# Rehabilitation vs Demolition Methodology to Compare the Waste Generated in Alternative Scenarios of Building Elements in BIM During the Design Stage

Rocio Quiñones[1](✉), Carmen Llatas[2], Maria Victoria Montes[3], and Isidro Cortés[1]

[1] Department of Graphical Expression and Building Engineering, University of Seville, 41012 Seville, Spain
mquinones@us.es
[2] Department of Building Construction I, University of Seville, 41012 Seville, Spain
[3] Department of Building Construction II, University of Seville, 41012 Seville, Spain

**Abstract.** The construction sector is one of the highest waste generators in many countries. Several studies evidence the close relationship between the decisions taken during the design phase and the reduction of the construction and demolition wastes (CDW). Moreover, many studies show the advantages of Rehabilitation versus Demolition. However, a main barrier to apply waste minimization strategies in projects is the lack of information included in the design tools themselves. The present paper aims to describe a methodological framework based on Quantification and Reduction Models of Construction Wastes, used during design stages of buildings and integrated into a building's design methodology such as Building Information Modeling (BIM). The method is conceived to guide designers, contribute to measure and predict CDW. A case study is also provided. CDW of two alternative scenarios are obtained and compared: the demolition versus the refurbishment of a roof. It also contributes to develop a sustainability simulation of a building by obtaining the construction wastes from its BIM model.

**Keywords:** Sustainability · Construction and demolition waste · Building Information Modelling · Building design

## 1 Introduction

The construction industry is essential to realize the built environment and a key sector in the sustainable development of the world production model. In Europe, although construction currently generates 9% of Gross Domestic Product and provides 18 million jobs [1], it is also responsible for 50% of energy consumption, 50% of materials extracted, 33% of water consumption and 33% of the generated waste [2]. Due to the increase of the cost of construction projects and the environmental awareness [3], the construction industry is under an immense pressure to become a more efficient industry in the use


A. Rotaru (Ed.): CRIT-RE-BUILT 2019, SSGG, pp. 506–518, 2021.
https://doi.org/10.1007/978-3-030-61118-7_41

of resources and in the minimisation of generated waste. Specifically, it is considered crucial to pay more attention to the end of the useful life of buildings because demolition activities represent more than 50% of the total waste production of the construction industry [4].

Directive 2008/98/EC establishes the principle "who pollutes pays" and the "expanded responsibility of the producer of pollution." It includes recycling and recovery targets that must be achieved by 2020: 50% of certain household waste and other similar waste, and 70% of waste from building construction and demolition (CDW). It also requires Member States to adopt waste management plans and prevention programs to achieve these objectives. In this way, the European Commission aims to help the construction industry to be more competitive, resource efficient and sustainable. Its approach to waste management is based on the "waste hierarchy" that indicates an order of priority when configuring waste policy and its management at an operational level: prevention (zero waste), reuse (once waste is generated), recycling, recovery and, as last and least desirable option, dumping in landfills and incineration without energy recovery [5].

The emergence of new technologies in the Architecture, Engineering and Construction (AEC), based on collaborative strategies, such as Building Information Modelling (BIM), is considered an opportunity to address the problem of construction and demolition wastes (CDW) from early design stages, to obtain more efficient results [6]. Several researchers [7, 8] point out that the use of the BIM platform can help minimise CDW. However, BIM remains a novel methodology that has not developed its full potential in the industry, both at the level of research and practitioners [9]. Therefore, there is a clear need to accelerate the development and standardisation of BIM tools for CDW and thus provide broad coverage in life cycle information flows of construction projects [10].

Hence, this work aims to establish a methodological framework based on two validated Quantification and Reduction of Construction and Demolition Waste Models [11, 12], to be used during building design stages and integrated into the BIM methodology. The proposed method is intended to help designers in the early prediction and measurement of CDW. At the same time, a case study is provided where the CDW of two alternative scenarios modelled in BIM are calculated and compared: demolition versus rehabilitation of a flat roof.

## 2 Reduction of Construction Waste

While CDW occur during construction execution activities, more and more voices agree that they are caused by decisions and activities throughout the whole building life cycle. There is a consensus in scientific literature that project design, material acquisition, construction, renovation and demolition contribute directly or indirectly to the generation of wastes [4, 9, 13, 14].

Other recent works [8, 9, 12, 13, 15–17] reveal the importance of reducing CDW in building planning and design stage as a first step to minimise waste by eliminating the root causes of its generation. In addition, it is claimed that inappropriate design decisions, lack of constructive details and unexpected design changes can lead to an increase of up to 33% in the volume of construction waste [18].

New production models such as "lean construction" [8, 19], "zero waste" [20] or the circular economy [21] also agree on the importance of reducing CDW at their source instead of recycling and recovering them, while they innovate through the involvement of people and the development of a more efficient work from the beginning of the design process. The influence that stakeholder's project engagement exerts on the reduction of CDW is considered crucial [8, 13], since CDW generated by late information, customer's last-minute requirements, bad distribution and slow revision of plans could be avoided by improving their coordination and communication during the design stage.

However, practitioners and policy makers are still far from implementing CDW minimization measures at the design stage and continue to consider them secondary about the traditional objectives of the project, that is, time, cost and quality [20]. Some authors [7] detect a lack of effective tools that allow accurate estimation of CDW. In conclusion, for the minimization of CDW, quantitative prediction methods and tools are considered essential by practitioners, since they can provide them with fundamental data to assess the actual volume, type, time and place of generation.

## 3 BIM Technology and Construction Waste

The great advances have forced the construction industry, considered one of the most resistant to change, to use new methods that allow it to survive. In recent years, new technologies are emerging in the international sector of Architecture, Engineering and Construction (AEC). This is the case of the BIM methodology. Since it first appeared in 1992 [22], the BIM concept is evolving from a 3D modelling technology to a comprehensive tool and project management process [3, 7, 23, 24], which allows a better exchange of information among the participants in construction projects, during the entire life cycle of the building, for precise decision making and a better delivery process for the final product.

The international ISO Standard 2010 defines BIM as "a shared digital representation of the physical and functional characteristics of any built object (…) that constitutes a reliable basis for decision making". Many authors estimate that BIM is a set of processes [25–27]. For Bazjanac [28], BIM is "the act of creating a construction information model". Eastman [24] defines BIM as "a modelling technology and the processes associated to produce, communicate and analyse construction models". The Building SMART International (BSI) association, which promotes the use of BIM worldwide, defines it as "a business process for the generation and use of construction data, to design, build and operate the building during its life cycle". Although most BIM processes are assisted by informatic tools, the scientific community agrees that it is a new working method that links new processes assisted by an important technological component [29].

Due to the strong commitment of the physical resources of construction projects and their unique and irreversible nature, it is too expensive, practically unfeasible, to try different design and construction schemes before a project is built without using new technologies. In this context, BIM has been revealed as a powerful technology and a versatile tool that helps predict and understand possible improvements and changes in a building before they are adopted on site [13, 30].

Therefore, BIM is considered a viable technology to address the performance problems of construction projects before having to make a large investment of time or money [31]. There is a consensus on the potential of BIM to minimise CDW during design scenarios. Many authors consider that BIM is a useful tool in the early stages of a project to identify, evaluate and effectively reduce the generation of CDW throughout building life cycle [3, 6, 7, 14, 17, 18, 32].

It is widely recognised that integrated building design based on BIM can improve the accuracy and quality of the design, thus reducing errors, repetitions, inefficiencies in communication and lack of coordination amongst the project participants and unexpected changes in design, which they are frequent causes of CDW generation [13, 32]. BIM mechanisms, such as alternative design simulations, design reviews, details, collision detection, digital prefabrication and construction activity planning can reduce the amount of construction waste by 4.3–15.2% [3, 6, 8, 13, 14, 18].

However, the most critical authors [3, 30, 33] warn that BIM is not a panacea. In any case, they consider that the implementation of BIM and the improvement of interoperability between different computer programmes are very necessary to effectively address waste minimisation, proposing a way forward rather than adding mere rhetoric. For this purpose, more research and implementation are required making this functionality a reality.

Recent studies [14, 24, 31, 34], which have mapped the global research on BIM, reveal that CDW research is not a specific topic among BIM research areas. Main issues identified focused on the general environmental assessment of buildings, especially energy efficiency [35, 36]. Likewise, existing BIM and CDW studies have been developed mainly in the management of CDW during the execution of construction works, being considered as the stage in which wastes are generated [3].

Similarly, the actual implementation of BIM in the construction industry for the minimization of CDW is also relatively low, as emerged from latest studies [10, 13]. While stakeholders in the construction industry seem to be aware of the benefits of BIM to eliminate a large number of causes of waste from its beginning to its completion in a construction project, especially at the design stage, evidences show that the use of BIM to reduce them is not yet an extended practice.

Now, none of the world's leading BIM software products include waste minimization and prediction functionality. Nevertheless, it is evident from the reviewed literature that there is a promising demand for the use of BIM to reduce the generation of waste in construction.

## 4 Methods and Tools for Minimizing Construction Waste Integrated in BIM

Although they are still limited, some specific methods and tools assisted by BIM have been developed to minimize and manage the amount of waste from construction, renovation and demolition projects.

Won et al. [18] developed a method to quantify how many CDW would be avoided with the use of BIM through a design validation process on the Autodesk Revit platform and demonstrated through two case studies that it could be a possible solution to eliminate the main causes of CDW. Furthermore, considering the possibility of identifying errors without BIM, using the traditional approach based on drawings, the amount of CDW was analysed and compared with the CDW avoided through the use of BIM. In addition, they claimed that with the use of BIM they could minimize the construction waste generated not only during the design stage, but also during the building construction. For example, waste generated during the construction phase could also be minimized by maximizing the reuse rate by selecting appropriate cutting lengths, areas and volumes of materials based on the precise amount obtained by BIM. A limitation of their study is that the volume of construction waste generated by formworks and packaging materials was not considered. They justify these methodological aspects arguing that, in South Korea, formwork information is not commonly included in BIM models, and the percentage of packaging waste is very insignificant, representing only 0.02% of total construction waste.

Porwal and Hewage [37] proposed a BIM-based method to reduce steel waste in reinforced concrete structures by selecting appropriate bar lengths and considering available cutting lengths. BIM was used to simulate the architectural and structural design requirements and to compare the results in order to make the necessary changes in the design, to reduce and reuse the reinforcement bar waste. In addition, BIM was used as a database to communicate project information between various design teams (architectural, structural, mechanical, electrical and plumbing). The limitation of this method is derived from reducing only a specific type of construction waste, steel waste.

Mercader et al. [38] led an investigation on the quantification of CDW that was expected to be generated during the implementation of a building project through a BIM model. They determined the amount of waste generated during the execution of the structure, total volume of concrete waste and total weight in kg of steel waste, as well as the total amount of material resources consumed. Previously, Mercader et al. [39] had already demonstrated the possibility of automatically quantifying the waste generated in a model developed in BIM during earthworks and foundation process. They compared the results with the manual procedure, and these results varied in decimals of cubic meters; variations that were irrelevant to the magnitudes of the project considered.

Bilal et al. [8] defined an integrated descriptive database of construction materials as a first step of a construction waste simulation tool to be implemented in BIM. In addition, they coined the term BIM Waste Analysis (BWA) to capture the entire BIM process, to predict and design construction waste. The BWA consists of four key steps: (i) analysis of the construction model, (ii) waste prediction, (iii) waste visualization and (iv) waste minimization. Moreover, it is considered that, in a construction waste simulation tool, BWA must be integrated. First, the waste simulation tool would read BIM models and estimate the types and quantities of waste that arise from the different elements used in construction. Subsequently, the tool would visualize the production of waste from the construction elements and highlights those that produce massive amounts of waste. Finally, designers could further investigate these elements to be able to use different materials and strategies with lower waste production through the specification of smart materials.

Recently, Akinade [4] developed a BIM tool (BIM Waste) for construction waste analysis at the design stage, integrated into Autodesk Revit, based on hybrid models of Artificial Intelligence (AI), due to its ability to store and process large amounts of data. AI models were incorporated into Autodesk Revit as a complement to allow the prediction of construction waste from building designs. System performance was assessed by a test plan and two case studies. Results show that the tool works well and predicts construction waste according to the types of waste, types of elements and levels of construction. The study provides a clear direction on how construction waste (CW) management strategies could be integrated into a BIM platform to streamline CW analysis. Notwithstanding, the author recognizes as a limitation of the study that it was conducted in the context of the United Kingdom, so the findings have a bias in AI to conceive the environmental potential of BIM at the end of the construction life cycle.

Cheng and Ma [7] designed a BIM application programming interface (API) for the estimation and planning of demolition and reuse waste that provides alerts to contractors ahead of time in project planning. This API can extract material and volume information through BIM models in the Autodesk Revit platform and integrate it for detailed waste estimation and planning. This system can not only serve as an automated, fast and accurate tool for estimating waste before demolition or reuse, but also serves as a tool to calculate the waste disposal rate for different waste facilities and pick-up truck requirements. This estimation and planning system have been validated in a 47-story residential building in Hong Kong. Some of the limitations of this tool are that it focuses more on waste management than on minimizing the root causes that generate waste.

Therefore, it can be concluded from the literature review that there is a lack of tools and decision-making methods for minimizing CDW during the design phase of the building.

# 5 Methodology

Through a case study, the different scenarios of the methodology based on the integration in the BIM methodology of the Quantification and Reduction Models of Construction and Demolition Wastes [11, 12] are established. A native BIM program is used and the enrichment of BIM objects through attributes (see Fig. 1).

The case study chosen is that of a flat roof covering 25 $m^2$ under two different scenarios; in Stage 1, the demolition of the roof and its subsequent reconstruction are established, while in Stage 2 the rehabilitation of the flat roof is proposed.

## 5.1 Stage 1: Data Collection for the Quantification and Waste Reduction Model

It is essential to start from reliable and validated data on the quantification and reduction of waste generated in the construction processes. Therefore, the first step is to collect data on the production of waste generated in the following activities: demolition, construction and rehabilitation of a flat roof in the validated Quantification and Reduction of CDW Models [11, 12], (see Table 1 and 2). These models develop a systematic structure of the construction process, a waste classification system and factor-based analytical expressions.

## 5.2 Stage 2: Development of CDW Attribute Database in BIM

Then, in the BIM environment chosen to model and provide CDW attributes to the case study - the Allplan program -, the BIM database is created covering the types of wastes to be evaluated (see Fig. 2).

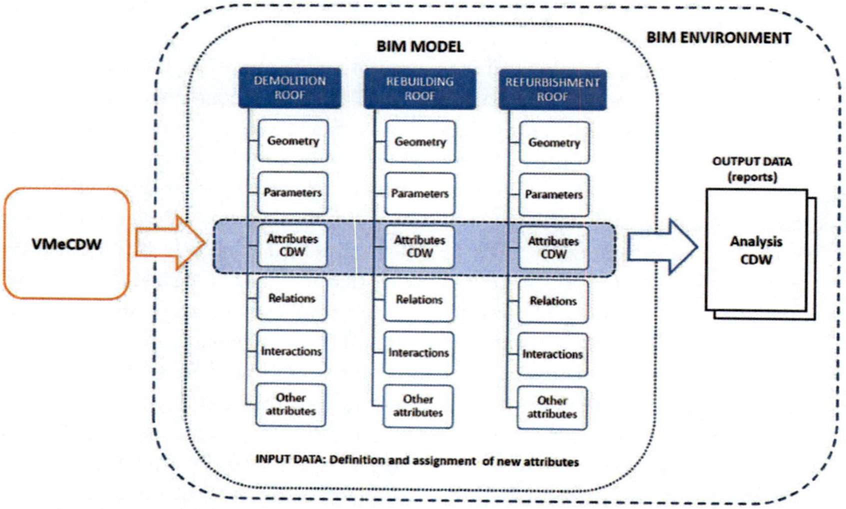


**Fig. 1.** Methodology scheme based on Quantification and Reduction Models of Construction Wastes [11, 12] integrated into Building Information Modeling.

### 5.3 Modelling Architectural Elements in BIM and Attribute Assignment

The roofs are modelled, for the different scenarios of the case study, with the specific tool "Roof", thus obtaining quantifiable BIM objects. Then, with the order "Assignment of attributes to any element", and using the previously created database, the types of wastes corresponding to each case are assigned and the corresponding values of each type of waste are entered by a manual and systematized procedure (Fig. 3). This Allplan tool allows to define any property or attribute, provide it with value and link it to any object so that data can be analysed and evaluated later through other tools.

### 5.4 Analysis and Discussion of Results

To analyse and evaluate the CDW of each of the proposed scenarios, a report that identifies the different constructive elements and the stages to which they are subjected is designed. This report allows the comparative study of results (Fig. 4 and Table 3). From this analysis it is concluded that Rehabilitating/refurbishing the roof (Stage 2) would reduce up to 99% of the CDW waste generated in the case that the roof was demolished and rebuilt (Stage 1).

**Table 1.** CDW expected in building/site work elements in Stage 1 (Source: data obtained from Llatas, 2011; and Llatas and Osmani, 2016)

| INPUT DATA | | | | OUTPUT DATA | | |
|---|---|---|---|---|---|---|
| **STEP** | **Quantity** | **Unit** | **BUILDING ELEMENT/ SITE WORK** | **EWL CODE** | **Type of waste** | **Volume (m3)** |
| 1 | 1 | m2 | Demolition of floor (Thick: 25 cm) and flat roof | 17 09 04 | Mixed CDW | 0.520000 |
| | | | | | **TOTAL** | **0.520000** |
| 2 | 1 | m2 | Rebuilding of concrete floor (Thick: 25 cm, ceramic small vault) | 15 01 02 | Plastic | 0.000554 |
| | | | | 15 01 03 | Wood packaging | 0.004619 |
| | | | | 15 01 06 | Mixed packaging | 0.000052 |
| | | | | 17 01 01 | Concrete | 0.001901 |
| | | | | 17 01 03 | Ceramics | 0.005960 |
| | | | | 17 02 01 | Wood | 0.000170 |
| | | | | 17 04 05 | Iron | 0.000022 |
| | | | | 17 09 04 | Mixed CDW | 0.000080 |
| | | | | | **TOTAL** | **0.013358** |
| | 1 | m2 | Rebuilding of flat floor | 15 01 01 | Cardboard | 0.000244 |
| | | | | 15 01 02 | Plastic | 0.004627 |
| | | | | 15 01 03 | Wood packaging | 0.038560 |
| | | | | 15 01 04 | Metallic | 0.017330 |
| | | | | 15 01 06 | Mixed packaging | 0.000608 |
| | | | | 10 13 99 | Lime | 0.000032 |
| | | | | 17 01 01 | Concrete | 0.003148 |
| | | | | 17 01 02 | Bricks | 0.000595 |
| | | | | 17 01 03 | Ceramics | 0.004967 |
| | | | | 17 03 01 | Bituminous | 0.000096 |
| | | | | 17 06 04 | Insulation | 0.000440 |
| | | | | 17 09 04 | Mixed CDW | 0.000088 |
| | | | | 17 05 04 | Soil and stones | 0.000793 |
| | | | | | **TOTAL** | **0.071528** |

**Table 2.** CDW expected in building/site work elements in Scenario 2 (Source: data obtained from Llatas, 2011; and Llatas and Osmani, 2016)

| INPUT DATA | | | | OUTPUT DATA | | |
|---|---|---|---|---|---|---|
| **STEP** | **Quantity** | **Unit** | **BUILDING ELEMENT/ SITE WORK** | **EWL CODE** | **Type of waste** | **Volume (m3)** |
| 1 | 1 | m2 | Refurbishment of a flat roof (waterproofing, thermal insulation thick:40 mm, non-adhered pavement) | 15 01 02 | Plastic | 0.000497 |
| | | | | 15 01 03 | Wood packaging | 0.001954 |
| | | | | 15 01 06 | Mixed packaging | 0.000026 |
| | | | | 17 01 01 | Concrete | 0.001009 |
| | | | | 17 03 01 | Bituminous | 0.000020 |
| | | | | 17 06 04 | Insulation | 0.000440 |
| | | | | 17 09 04 | Mixed CDW | 0.000004 |
| | | | | | **TOTAL** | **0.003950** |

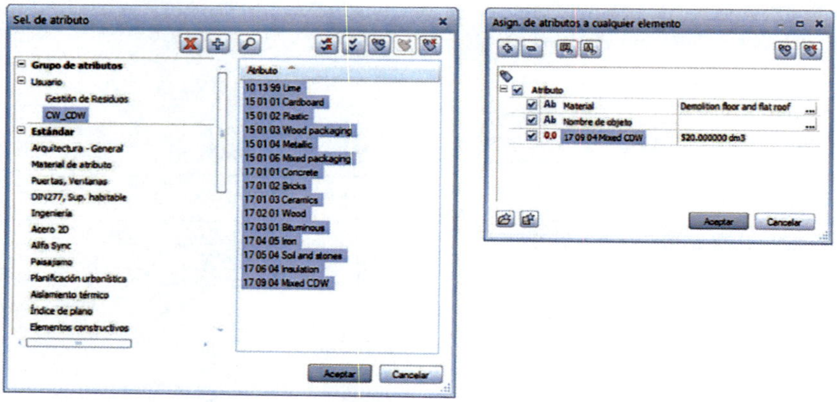


**Fig. 2.** CDW database: definition of CDW attributes.

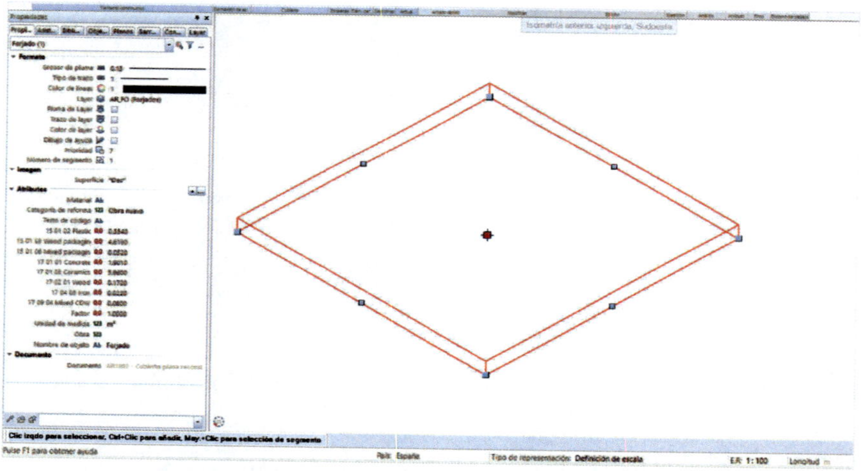

**Fig. 3.** CDW database created in Allplan: assignment of CDW attributes.

In addition, another advantage of Stage 2 is that it facilitates the selective separation of wastes and its subsequent removal, promoting their recovery. While in stage 1 most of the waste is made up of a fraction of mixed materials, making it difficult to recover (Table 3 and Fig. 5).

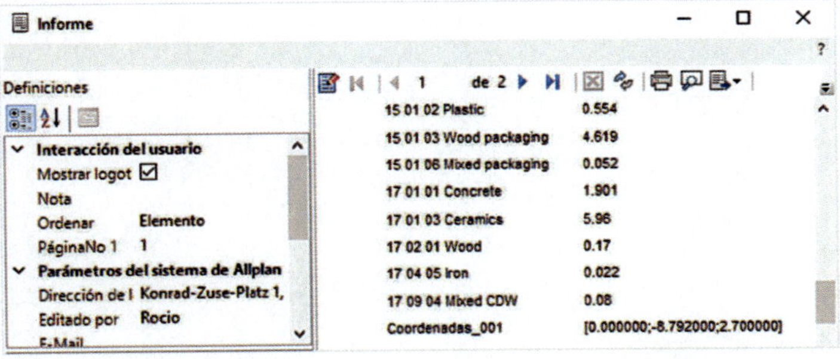


**Fig. 4.** Analysis of results. Custom reports in Allplan.

**Table 3.** CDW. Comparative study of results.

| | Type of waste (European Waste List code) | Scenario 1 (Stages 1&2) | Scenario 2 |
|---|---|---|---|
| **CDW in building: Flat Roof 25 m²** | 10 13 99 Lime | 0,000800 | 0 |
| | 15 01 01 Cardboard | 0,006100 | 0 |
| | 15 01 02 Plastic | 0,129525 | 0,012425 |
| | 15 01 03 Wood packaging | 1,079475 | 0,048850 |
| | 15 01 04 Metallic | 0,433250 | 0 |
| | 15 01 06 Mixed packaging | 0,016500 | 0,000650 |
| | 17 01 01 Concrete | 0,126225 | 0,025225 |
| | 17 01 02 Bricks | 0,014875 | 0 |
| | 17 01 03 Ceramics | 0,273175 | 0 |
| | 17 02 01 Wood | 0,004250 | 0 |
| | 17 03 01 Bituminous | 0,002400 | 0,000500 |
| | 17 04 05 Iron | 0,000550 | 0 |
| | 17 05 04 Soil and stones | 0,019825 | 0 |
| | 17 06 04 Insulation | 0,011000 | 0,011000 |
| | 17 09 04 Mixed CDW | 13,004200 | 0,000100 |
| | **CDW (m³)** | **15,122150** | **0,098750** |

One of the difficulties we have encountered, from an operational point of view, is that BIM programs developed for architectural modelling, consider a maximum of 4 decimal; this fact has conditioned the units we have worked with. In the reference quantification model the values are expressed in $m^3$. This problem has been overcome with the use of the $dm^3$ unit in the early stages of the methodology. Finally, in the reports the results are expressed in $m^3$.

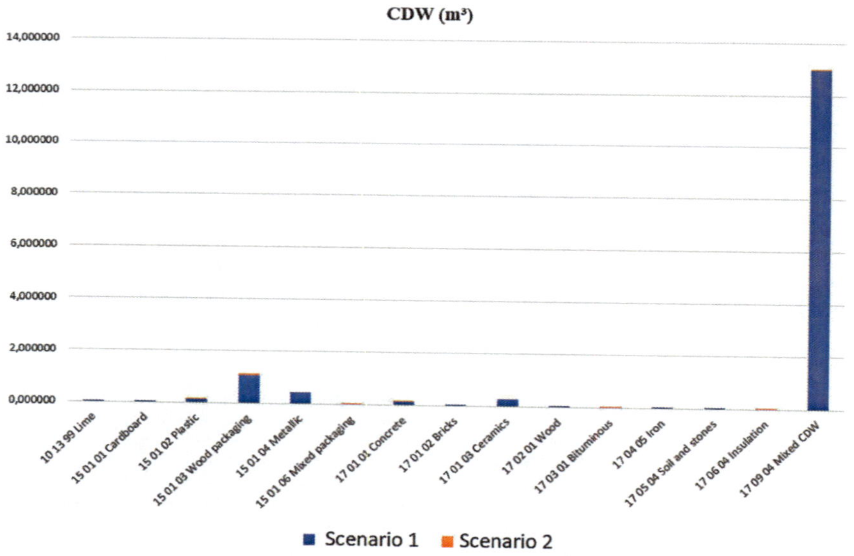


**Fig. 5.** Comparative analysis between Scenario 1 & Scenario 2.

## 6 Conclusions

The results reveal that the Waste Quantification and Reduction Model can be implemented in BIM environments. The proposed methodology and the information flows generated, through the indexed database, allow us to know the waste produced by the construction systems defined in the early scenarios of the design, helping in decision making. However, the following limitations must be considered:

- The attributes related to the quantification of wastes do not exist by default in the database of BIM modellers.
- The use of more than four decimals imposes the need for the use of units that are expressed with fewer decimals.

The digitalization of the design process using BIM objects with attributes allows a real response to the growing need to include environmental considerations in the projects, such as the production of wastes at the beginning of the design process. For example, the case study demonstrates with quantitative values how a Rehabilitation Stage reduces the wastes of a Demolition and Reconstruction Stage up to 99%.

The study forms a basis for future research on the optimization of the integration of CDW in BIM. The proposed methodology presents a high potential since all data can be centralized in a single database and customize the information that you want to evaluate.

## References

1. European Commission, Construction in EU (2019). http://ec.europa.eu/growth/sectors/construction/index_en.htm. Accessed 02 Nov 2019

2. European Commission: The European construction sector: a global partner (2016). http://ec.europa.eu/DocsRoom/documents/15866/attachments/1/translations/en/renditions/native. Accessed 01 June 2020
3. Lu, W., Webster, C., Chen, K., Zhang, X., Chen, X.: Computational building information modeling for construction waste management: moving from rhetoric to reality. Renew. Sustain. Energy Rev. **68**, 587–595 (2017)
4. Akinade, O.O., Oyedele, L.O., Omoteso, K., Ajayi, S.O., Bilal, M., Owolabi, H.A., Alaka, H.A., Ayris, L., Looney, J.H.: BIM-based deconstruction tool: Towards essential functionalities. Int. J. Sustain. Built Environ. (2017). In Press. http://dx.doi.org/10.1016/j.ijsbe.2017.01.002. Accessed 02 May 2017
5. Directive 2008/98/EC on Waste (Waste Framework Directive). http://ec.europa.eu/environment/waste/framework/. Accessed 01 June 2020
6. Won, J., Cheng, J.C.P.: Identifying potential opportunities of building information modeling for construction and demolition waste management and minimization. Autom. Constr. **79**, 3–18 (2017)
7. Cheng, J.C.P., Ma, L.Y.H.: A BIM-based system for demolition and renovation waste estimation and planning. Waste Manag **33**, 1539–1551 (2013)
8. Salgin, B., Arroyo, P., Ballard, G.: Exploring the relationship between lean design methods and C&D waste reduction: three case studies of hospital projects in California. Revista Ingeniería de Construcción **31**(3), 191–200 (2016)
9. Ajayi, S.O., Oyedele, L.O., Akinade, O.O., Bilal, M., Alaka, H.A., Owolabi, H.A.: Optimising material procurement for construction waste minimization: an exploration of success factors. Sustain. Mater. Technol. **11**, 38–46 (2017)
10. Bilal, M., Oyedele, L., Qadir, J., Munir, K., Akinade, O.O., Ajayi, S.O., Alaka, H.A., Owolabi, H.A.: Analysis of critical features and evaluation of BIM software: towards a plug-in for construction waste minimization using big data. Int. J. Sustain. Build. Technol. Urban Dev. **6**, 211–228 (2015)
11. Llatas, C.: A model for quantifying construction waste in projects according to the European waste list. Waste Manag **31**, 1261–1276 (2011)
12. Llatas, C., Osmani, M.: Development and validation of a building design waste reduction model. Waste Manag **56**, 318–336 (2016)
13. Domingo, N., Su, M., Egbelakin, T.: Use of Building Information Modelling in Construction Waste Source Reduction. RICS COBRA AUBEA, Sydney (2015)
14. Liu, Z., Osmani, M., Demian, P., Baldwin, A.: A BIM-aided construction waste minimization framework. Autom. Constr. **59**, 1–23 (2015)
15. Bilal, M., Oyedele, L., Munir, K., Ajayi, S.O., Akinade, O.O., Owolabi, H.A., Alaka, H.A.: The application of web of data technologies in building materials information modelling for construction waste analytics. Sustain. Mater. Technol. **11**, 28–37 (2017)
16. Gómez-Soberón, J.M., Saldaña-Márquez, H., Gámez-García, D.C., Gómez-Soberón, M.C., Arredondo-Rea, S.P., Corral-Higuera, R.: A comparative study of indoor pavements waste generation during construction through simulation tool. Int. J. Sustain. Energy Dev. **5**(1), 243–251 (2016)
17. Rajendran, P., Gomez, C.P.: Implementing BIM for waste minimization in the construction industry: a literature review. In: 2nd International Conference on Management, Proceedings, Kuala Lumpur, Malaysia (2012)
18. Won, J., Cheng, J.C.P., Lee, G.: Quantification of construction waste prevented by BIM-based design validation: case studies in South Korea. Waste Manag **49**, 170–180 (2016)
19. Anerao, S.D., Deshmukh, S.S.: Waste minimization by lean construction technology. Int. Res. J. Eng. Technol. **03**(08), 1703–1707 (2016)

20. Veleva, V., Bodkin, G., Todorova, S.: The need for better measurement and employee engagement to advance a circular economy: lessons from biogen's 'zero waste' journey. J. Clean. Prod. **154**, 517–529 (2017)
21. Adams, K.T., Osmani, M., Thorpe, T., Thornback, J.: Circular economy in construction: current awareness, challenges and enablers. Proc. Inst. Civ. Eng. - Waste Resour. Manag. **170**(1), 15–24 (2017). http://dx.doi.org/10.1680/jwarm.16.00011. Accessed 05 May 2017
22. Van Nederveen, G.A., Tolman, F.P.: Modelling multiple views on buildings. Autom. Constr. **1**(3), 215–224 (1992)
23. Cao, D., Li, H., Wang, G.: Impacts of building information modeling (BIM) implementation on design and construction performance: a resource dependence theory perspective. Front. Eng. Manag. **4**(1), 20–34 (2017). http://dx.doi.org/10.15302/J-FEM-2017010
24. Eastman, C., Teicholz, P., Sacks, R., Liston, K.: BIM Handbook: A Guide to Building Information Modeling for Owners, Managers, Designers, Engineers and Contractors. Wiley, New Jersey (2011). https://doi.org/10.1002/9780470261309
25. Azhar, S., Hein, M., Sketo, B.: Building information modeling (BIM): benefits, risks and challenges. In: Proceedings of the 44th ASC Annual Conference, Auburn, Alabama (2008)
26. BSI: BuildingSMART. http://www.buildingsmart.org/. Accessed 01 June 2020
27. NIBS: United States national building information modeling standard. Version 1-Part 1: overview, principles and methodologies (2007)
28. Bazjanac, V., Maile, T.: IFC HVAC interface to EnergyPlus- a case of expanded interoperability for energy simulation. SimBuild, IBPSA-USA National Conference Boulder. CO, pp. 1–7 (2004)
29. Eastman, C., Teicholz, P., Sacks, R., Liston, K.: The BIM Handbook, 1st edn. Wiley, New Jersey (2007)
30. Young Jr., N.W., Jones, S., Bernstein, H.: The Business Value of BIM: Getting Building Information Modeling to the Bottom Line. McGraw-Hill Construction, New York (2009)
31. Santos, R., Costa, A.A., Grilo, A.: Bibliometric analysis and review of building information modeling literature published between 2005 and 2015. Autom. Constr. **80**, 118–136 (2017)
32. Cheng, J.C.P., Won, J., Das, M.: Construction and demolition waste management using BIM technology. In: 23rd Annual Conference of the International Group for Lean Construction, Proceedings, Perth, Australia, pp. 381–390 (July 2015). http://www.iglc.net. Accessed 01 June 2020
33. Mandujano, M.G., Alarcón, L.F., Kunz, J., Mourgues, C.: Identifying waste in virtual design and construction practice from a lean thinking perspective: a meta-analysis of the literature. Revista de la Construcción **15**(3), 107–118 (2016)
34. Yalcinkaya, M., Singh, V.: Patterns and trends in building information modeling (BIM) research: a latent semantic analysis. Autom. Constr. **59**, 68–80 (2015)
35. Nguyen, T.H., Shehab, T., Gao, Z.: Evaluating sustainability of architectural designs using building information modeling. Open Constr. Build. Technol. J. **4**, 1–8 (2010)
36. Oduyemi, O., Okoroh, M.I., Fajana, O.S.: The application and barriers of BIM in sustainable building design. J. Facil. Manag. **15**(1), 15–34 (2017)
37. Porwal, A., Hewage, K.: Building information modeling–based analysis to minimize waste rate of structural reinforcement. J. Constr. Eng. Manag. **138**(8), 943–954 (2012)
38. Mercader, P., Cózar, E., Ramírez de Arellano, A.: Sistema BIM de cuantificación automática de los CDW, actas del I ENCACS, pp. 499–508, Argentina (2015)
39. Mercader, P., Cózar, E., Ramírez de Arellano, A.: Automatic quantification for waste management with BIM models. World SB14 Barcelona (2014)

# Revitalization of Brownfield Sites – Example of the Solution of the Geotechnical Issues

Jan Štefaňák(✉)

Faculty of Civil Engineering, Institute of Geotechnics, Brno University of Technology, Veveri 331/95, 602 00 Brno, Czech Republic
stefanak.j@fce.vutbr.cz

**Abstract.** There are many obstacles to the regeneration of brownfield sites, and many of them are associated with geotechnical hazards. Evaluation of the stability of waste-soil landfill after its extension linked with the processes related to the rehabilitation of the former factory brownfield site in Brno, Czech Republic is reported in this paper. The geotechnical hazard coming from the uncertainties in the density and strength of the waste soil material and in the position of borders between natural slope and the placed waste material are discussed. The use of the Digital Model of Relief (DMR) for the purpose of geomorphology evaluation is presented. The planned extension will not affect the stability of landfill, based on the results of the study. The management of the brownfield sites by the city of Brno is outlined as well in the paper.

**Keywords:** Brownfield · Stability · Landfill · Revitalization

## 1 Specifics of Brownfield Sites

Today's society is based on materialist principles, emphasis is placed on high profitability while at the same time reducing costs. This principle of the functioning of the system is essentially correct, as the result should be greater efficiency in the use of resources, avoiding waste. Land is one of the most valuable resources. Its efficient use is a condition for the effective functioning of society. However, it is often easier for an investor to place a project on a greenfield site for generally known reasons. Of course, finances are essential in this matter. In greenfields, there is a lower overall cost of the implementation of the project. However, undeveloped land, including good quality agricultural land, is not a renewable resource and is limited. This is the reason, why it is good to focus on using brownfield sites for the implementation of new projects. It is brownfield sites, frequently extensive territory in the built-up areas of cities, which have great development potential. Although there are many obstacles to the regeneration of brownfield sites, it is highly desirable that these be overcome, either by private investors or the public sector [1]. Although the term brownfield is widely used [2, 3], it has no universally recognized definition. It has sometimes been regarded as synonymous with contaminated land. More commonly it has been understood as signifying the opposite of greenfield in planning


A. Rotaru (Ed.): CRIT-RE-BUILT 2019, SSGG, pp. 519–526, 2021.
https://doi.org/10.1007/978-3-030-61118-7_42

terms, where greenfield is taken to mean: "land that has not been previously developed." Brownfield land should be regarded as "any land that has been previously developed, particularly where it has been occupied by buildings or by other types of a permanent structure and the associated infrastructure." Charles (2005) says: "The term Brownfield has been widely adopted to describe previously developed land. It represents a broader concept than either derelict land or contaminated land since a brownfield site is not necessarily either derelict or contaminated" [4]. There can be many geotechnical hazards associated with rehabilitation of brownfield sites related to the potential contamination [5], poor load carrying capacity and stability issues [6], differential settlements [7] etc.

The first objective of the paper is to briefly outline the process of brownfield site management in Brno, Czech Republic. The database of such sites and the condition for inclusion sites in the database will be described briefly. Paper continues with the description of the process of the preliminary evaluation of the stability of soil landfill extension (The extension is planned as a part of former factory brownfield site revitalization project.). The main objective is to point out the related geotechnical hazards coming from additional loading and from the uncertainties in the density of stored waste soil material and in a position of borders between the natural slope and the body of landfill.

## 1.1 Management of Brownfield Sites in Brno

Brno is the capitol of the South Moravian Region of the Czech Republic located 140 km north from Vienna. Records of brownfield sites in Brno originally served [1] mainly as the basis for creating an analytical territorial analysis, and the Master Zoning Plan for Brno. Nonetheless, the issue of brownfield sites goes far beyond urban planning.

### Database of Brownfield Sites and Urban Planning

An emphasis on the use and renewal of brownfield sites is often present also in Brno's significant conceptual documents. For this reason, the Office of Development Sites of Brno city keeps records on brownfield sites, and not only from the perspective of urban planning but especially because of the requirement for the effective management of the city's property and in the broader context of the development of Brno City [1]. Next picture (Fig. 1) shows a map where the intensity of crime is highlighted. The more saturated sectors are, the more crimes occur in the given district. The higher number of crimes occurs in those districts where the brownfield sites are located.

Hence, the brownfields issue should not be perceived as only being about the regeneration or repairing of dilapidated sites and buildings. A positive contribution can also be made even by the temporary use of abandoned premises and buildings, which can prevent further dilapidation and degradation.

### Aerial Parameters of Brownfield Sites in Brno

A condition for inclusion in the Brownfield database is, that a locality is run down, insufficiently and inappropriately used and is somewhere where further development is uncertain. The database covered localities with an area above 0.5 ha and a utilization rate of predominantly up to 30%. (Primarily in the context of the city center this value is rather high since there we can find localities (various nooks, corners, courtyards, unfinished buildings, empty plots, etc.) which have the nature of brownfield sites but are small in

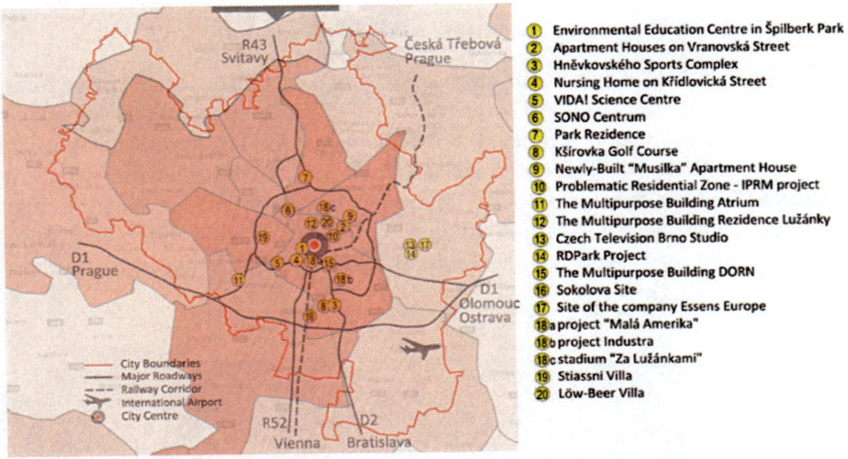


**Fig. 1.** Selected Brownfield sites revitalization projects (based on [1] and www.mapakriminality.cz, last accessed 2019/10/10).

area. These, even if rather small, spaces (from only a few dozen square meters), have a negative impact on perceptions of the city, (not just those of the inhabitants or regular visitors but on tourists as well). For this reason, especially in a city environment, these mini-brownfield localities are becoming the focus of interest) Based on these conditions, a total of 128 sites covering a total area of approximately 378 ha were defined [8].

## 2 The Former Ergon Factory Brownfield Site

Ergon Brownfield site is located in the Former Ergon factory complex in north-east city districts Maloměřice and Obřany. Construction of the factory began in the late 1980s. The buildings were originally intended for the Research Institute of Agricultural and Construction Mechanization. The project was sold to the company Ergon, which produced prosthetic implants (an artificial device that replaces a missing body part). The factory was completed and equipped by the production facilities in 1989. The political and social changes led to the end of the project. From that time the area has been deteriorating. Some recent deteriorating processes at the site and it's wider vicinity are illustrated in Fig. 2.

**Fig. 2.** Natural deteriorating processes on the ERGON site during 1996–2012 (modified after http://gis.brno.cz/mapa/assets/local/documents/rozvojove_lokality/geonika/1910_typ_povrchu.pdf)

### 2.1 Revitalization Program of Ergon Brownfield Site

The developer has the plan to build up a residential area, that will be located in the former Ergon factory complex in Brno. All buildings in the 6.5-ha complex will be demolished, before the start of construction process. Project construction will take place in several stages. The neighborhood will incorporate an ideal mix of individual and mass housing, including facilities for services. Classical apartment buildings will surround the central square, commercial space will be located on the ground floor. Villa apartments provide more intimate housing, where every three floors will be divided between 2 or at most 3 apartments (see Fig. 3).

**Fig. 3.** Visualisation of the developer project for the new residential area in the place of Ergon Factory Brownfield site (https://www.trikaya.cz/projekty/ctvrt-pod-hady)

Until all the legal and business issues will be solved, they found another very interesting and socially beneficial use of this ruin. They lend the dilapidated area to the firefighters and rescuers to practice managing of the specific catastrophic events. The practicing of managing the simulated explosion of bomb in parking House took place there. Such activity can be seen as an example of beneficial temporary use of brownfield sites as they are, as was mentioned in the introduction.

### 2.2 Ergon Brownfield Site Revitalization – Related Stability Issues

There will arise a big amount of waste soil, that will be excavated during the construction of a new residential area.

The investor looks for the possibility of store part of this waste soil directly in his area of interest. The reason is, that permanent storing of waste soil on landfill brings additional costs to the project. They identified kind of trench or let's say depression next to the road that connects the locality with the town. They intended to store the waste soil here and use it for filling this depression. The depression is located in the upper part of the slope. This slope has a very unusual shape, which seems not to be the result of natural processes. The preliminary desk study was performed, including archive documents [9]. The results of this study identified the origin of the unusual slope. The mass of soil was excavated during the construction processes related to the building of the original building of ERGON factory in 1989. It was stored directly on the surface of the slope located just at the gate to the factory. The soil was probably placed there with only minimal compaction from the trucks bringing it onto the landfill. The investor demanded to evaluate the current stability of the locality. The next demand was to evaluate the stability of the locality after placing new waste soil on the top of the current slope. This has to be done in two cross-sections.

There hazards were identified, that need attention:

- Additional loading placed on the crest of the slope is something, that generally can trigger a landslide, furthermore
- The relative density and consequently the strength of the waste soil placed here in 1989 is unknown, (Hence, very conservative values must be taken into account)
- The position of the border between the original slope terrain and the added waste material is unknown.

### 2.3 Preliminary Evaluation of Stability of Soil Landfill Extension Planned During the Ergon Factory Brownfield Site Revitalization

The indicative measurement of the slope inclination was done first during the walk-over survey of the locality. The measured values were higher than those deduced from the results of the desk study. Hence, the additional support tool was used for control. called the Digital Model of Relief (DMR). Next picture (see Fig. 4) shows the 3D visualization of terrain with the position of desired cross-section.

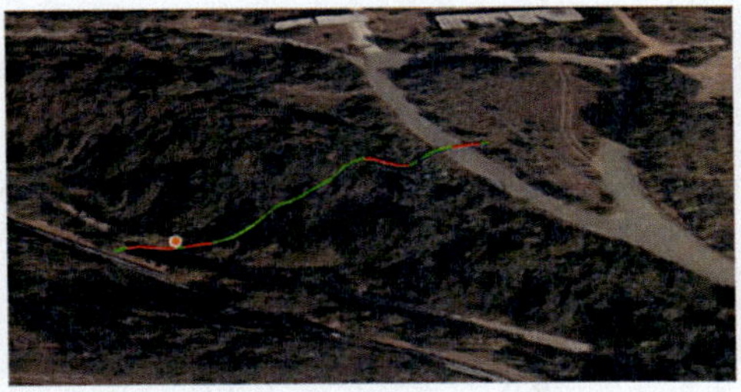

**Fig. 4.** Digital model of relief with the position of the cross-section for the construction of geotechnical model (based on the https://ags.cuzk.cz/dmr/)

There is a relatively easy way how to use the *txt files with data exported from the DMR application for drawing the geometry of the slope in AutoCAD. The function is needed to be written (CONCATENATE), that corresponds to the AutoCAD command for drawing a polyline. The comparison of the surface shape based on the on DMR and the originally presumed is shown in the Fig. 5.

It can be seen, that the shape and the position of the border between the natural slope and between the waste material placed on it differ.

To calculate the factor of safety, a slip surface is assumed and one or more equations of static equilibrium are used to calculate the stresses and factor of safety for each surface

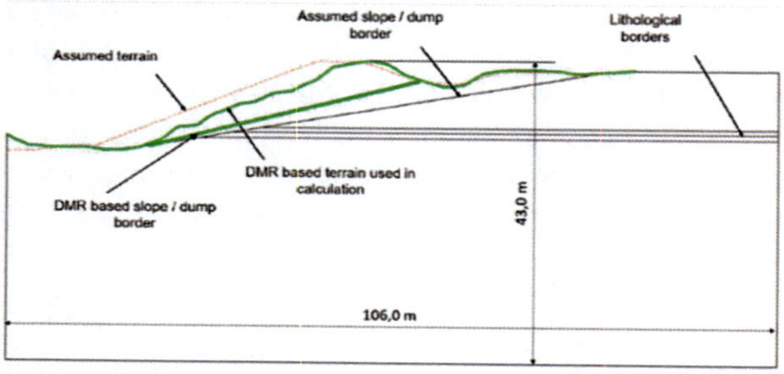


**Fig. 5.** Refinement of the initial model geometry based on the data from DMR application (https://ags.cuzk.cz/dmr/)

assumed. The term slip surface is used here to refer to an assumed surface along which sliding or rupture might occur. The factor of safety is assumed to be the same at all points along the slip surface. Thus, the value represents an average or overall value for the assumed slip surface. If a failure occurs, the shear stress would be equal to the shear strength at all points along the failure surface. The assumption that the factor of safety is constant would be valid. If the slope is stable, the factor of safety probably varies along the slip surface. A number of slip surfaces must be assumed to find the slip surface that produces a minimum factor of safety. The surface with the minimum factor of safety is termed the critical slip surface [10]. Very conservative values of input parameters were used in a calculation, as can be seen from the Table 1, due to the lack of the more precise values based on laboratory tests in this preliminary phase of design.

**Table 1.** Values of soil parameters for stability calculations.

| Material | $\gamma_{unsat}$ [kN/m$^3$] | $\gamma_{sat}$ [kN/m$^3$] | c′ [kPa] | ϕ′ [°] |
|---|---|---|---|---|
| (Y) fill | 18,0 | 18,0 | 5 | 25 |
| (F6) clay-immediate plasticity | 20,0 | 20,0 | 10 | 25 |
| (G3-GF) silty gravel | 19,0 | 19,0 | 0 | 40 |
| (F4-CS) clayey sand | 19,0 | 19,0 | 12 | 27 |
| (S4-SM) silty sand | 18,0 | 18,0 | 5 | 30 |

In the Fig. 6, the geotechnical model of the calculated slope is shown. The body of additional material, that is planned to be placed behind the crest of the slope, is shown also. The program *Slope Stability* from the family of GEO5 programs was used for calculation. The program can be used to perform slope stability analysis (embankments, earth cuts, anchored retaining structures, MSE walls, etc.). The slip surface can be considered as circular (Bishop, Fellenius/Petterson, Janbu, Morgenstern-Price or Spencer methods) or polygonal (Sarma, Janbu, Morgenstern-Price or Spencer methods).

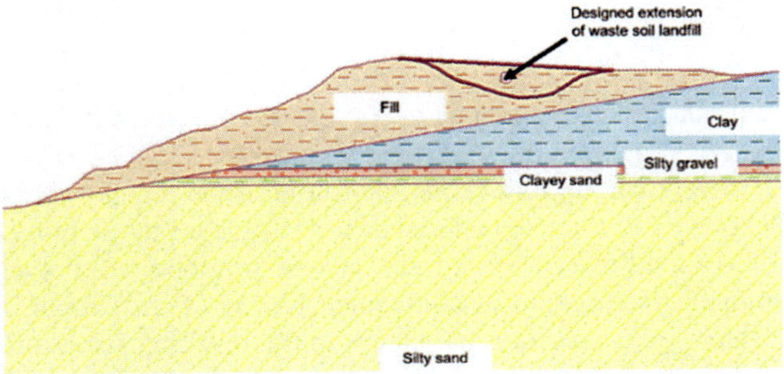


**Fig. 6.** Geotechnical model for evaluation of the stability – second phase with the additional waste material

## 3 Results

The shape of the critical slip surface according to Bishop's theory [11] is shown here. The current state without added material was evaluated as same as the state with newly deposited soil added. The verification was done according to the factor of safety:

$$F_p/F_a = \text{Factor of safety} \quad (1)$$

Optimization procedure was employed to locate a slip surface with the lowest value of the factor of safety. The resulting shapes of the critical slip surface after the optimization and the values of the factors of safety (for the current state and after placing of newly deposited waste material) are shown in the Fig. 7.

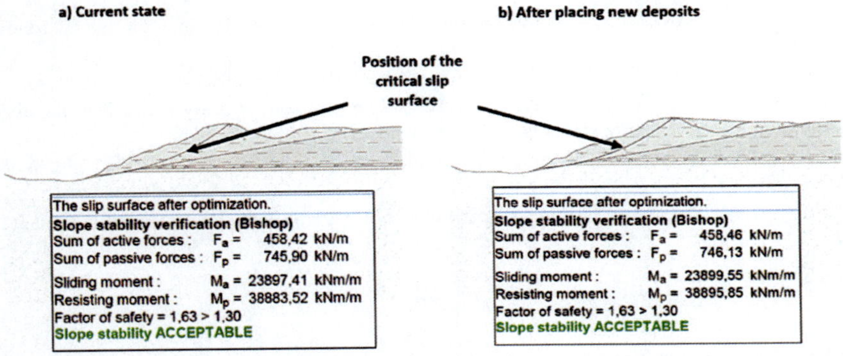


**Fig. 7.** Snapshot from the software with the resulting position of the critical slip surfaces, resulting forces, moments and factors of safety for both evaluated states: (a) current state, (b) after placing new deposits.

## 4 Conclusions

The results of preliminary study aimed on the evaluation of soil landfill extension show, that the factor of safety after landfill extension remains the same, as without this additional loading. Hence, it can be stated, that the filling of the trench located behind the crown of landfill body by new waste material doesn't influence the stability of the slope. Hence, the additional load placed behind the crest of the slope will not trigger sliding processes. Additional investigation in consequent phases of the design is, of course, inevitable, for reducing the uncertainty inherent in the resulting factor of safety. Such investigation should be aimed on the evaluation of the relative density (and consequently the soil strength parameters) of the waste soil placed on the natural slope in 90's. The position of the border between the original slope terrain and the additionally placed waste material should be investigated also.

## References

1. BRNO Brownfields 2015. City of Brno - City Strategy Office, Brno, Czech Republic (2015)
2. Rafson, H.J., Rafson, R.N.: Brownfields: Redeveloping Environmentally Distressed Properties. McGraw-Hill, New York (1999)
3. Adams, D., Watkins, C.: Greenfields, Brownfields, and Housing Development. Blackwell Science, Oxford (2002)
4. Charles, J.A.: Geotechnics for Building Professionals, 1st edn. IHS BRE Press, Watford (2005)
5. Sarsby, R.W.: Environmental Geotechnics, 2nd edn. ICE Publishing, London (2013)
6. Burland, J., Chapman, T., Skinner, H., Brown, M.: ICE Manual of Geotechnical Engineering: Volume II Geotechnical Design, Construction and Verification. ICE Publishing, London (2012)
7. Charles, J.A., Watts, K.S.: Building on Fill: Geotechnical Aspects, 2nd edn. BRE, London (2001)
8. BRNO Brownfields 2013. City of Brno, Brno, Czech Republic (2013)
9. Štefaňák, J.: Brno, ulice Pod Hády: předběžná studie posuzení stability svahu. Vysoké učení technické v Brně, Fakulta stavební, AdMaS, Brno, 20 s. (2019)
10. Duncan, J.M., Wright, S.G., Brandon, T.L.: Soil Strength and Slope Stability, 2nd edn. Wiley, Hoboken (2014)
11. Bishop, A.W.: The use of the slip circle in the stability analysis of slopes. Geotechnique **5**(1), 7–17 (1955)

# Conservation, Rehabilitation and Integration of Cultural Heritage

# Conservation, Rehabilitation and Integration of Cultural Heritage

Chavdar Kolev(✉)

Higher School of Transport "Todor Kableshkov", 158, Geo Milev Street, Sofia, Bulgaria
ch_kolev@abv.bg

**Abstract.** The report presents a study of world literature, charters, recommendations and regulatory standards in the area of restoration of architectural monuments. We have studied the knowledge gained from the time of Schliemann to the present. There is also an analysis of a number of publications from Bulgaria, as well as publications by the author in the same field. Emphasis is placed on geotechnical technologies to restore the foundations of architectural monuments. The practical experience of restoration activities in a number of countries is also extensively presented. This study provides guidance for a series of lectures systematically developed on the same topic.

**Keywords:** Restoration · Monuments · Charter · Micro-piles · Churches · Stone

## 1 Introduction

The preservation, restoration, and integration of cultural and historical heritage is nowadays a result of modern archeology development, which began in the mid-eighteenth century with the excavations of Pompeii and especially of the city of Troy - the remarkable Schliemann's discovery in 1871. The latter was followed soon by excavations of Mycenaean civilization in Peloponnese. However, even at Troy excavations, Schliemann was sharply criticized for the ruthless methods of working, which revealed the intended purpose, but at the same time destroyed many other artifacts of later eras such as the explosions when work was starting up. At that time archeology was already being studied at universities and a number of working rules were taken into consideration. Nearly 150 years after Schliemann, archeological approaches, as well as architecture, urban planning, and construction, have developed. A number of legal norms have been adopted in the civilized world, which regulate the activities for the preservation, restoration, and integration of cultural heritage in modern cities.

The celebrated fathers of the restoration and integration of the cultural and historical architectural heritage of the XIX century are Eugene Viollet-le-Duc in France and William Morris in England etc.

The modern city development is a continuation of traditions inherited from the past. Architectural monuments are precisely the starting points of urban planning, which determine the specific image and spirit of a city. They define the concept that has dominated


A. Rotaru (Ed.): CRIT-RE-BUILT 2019, SSGG, pp. 529–540, 2021.
https://doi.org/10.1007/978-3-030-61118-7_43

urban development throughout all ages. Preservation and restoration of architectural monuments nowadays are subject to specific principles and requirements, which are not always regulated by law but have established global requirements for the protection of cultural heritage. In addition to the general principles, there are too many architectural and construction rules in the theory and practice of restoration related to the design, technology, materials, and organization of works. It is these problems that are subject to the current study and in particular - their condition today.

The topic is too broad and therefore within this study will be considered only the starting points and rules for the restoration of architectural monuments and modern technologies to strengthen their foundations, especially micro-piles technology.

# 2 A Brief Literary Overview and Current State of the Problem

## 2.1 The Actual State of the Problem in Europe and Worldwide

Guided by the topic of the study, nine contemporary papers have been selected to give an initial idea of the problem in Europe and worldwide. Based on the information from these reports, knowledge will be methodically expanded and refined until a sufficiently clear picture is obtained due to reasoned opinions and opportunities for putting the theory into practice. The first document we launch is the European Cultural Heritage Strategy for the 21st Century, adopted by the Council of Europe in 2017. This strategy is detailed in Bold and Pickard [1].

The Strategy [1] provides a comprehensive introduction to the history of European integration in the field of protection and restoration of cultural heritage and sets out the principles and working methods. These principles have been systematically presented in the development of projects of different scales.

The second document under consideration is also of fundamental importance, giving the European Union policy guidance on the protection of cultural heritage. This is the "Opinion of the European Committee of the Regions - Towards an integrated approach to cultural heritage for Europe" [2], compiled and published in 2015. The 48 points recorded in the document deplete the guidelines and recommendations for the development of this activity.

The third document we are paying attention to in this study is the final report [3] from the implementation of STORM, an interesting research project done by 20 participants from 7 European countries within the Union's Horizon 2020 research and innovation program, continued from 2016 to 2018. The study focuses on the effects of natural disasters, which are among the most serious threats to cultural sites.

Fourth in place is the Dutch engineer Van Balen [4]. This report focuses on current issues and new requirements for preservation and maintenance of new monuments, discussing in detail the risk assessment, monitoring and accountability activities of officials engaged in these activities. It is through these factors assessment that form the proper documentation of monuments and compliance with current information requirements.

Important methodological issues for students' education in the field of protection and restoration of cultural monuments are also presented in detail in the report of the Egyptian architect Embaby [5]. His paper aims to propose an educational methodology for dealing with heritage conservation projects: "adaptive reuse of historical buildings" in design studios of architecture and interior design programs, by promoting a design philosophy which supports the integrated approaches of revitalizing heritage values of the traditional communities and creates new activities based on conservation principles, sense of historic buildings and its cultural context, and continuous with local communities needs.

Another aspect of cultural heritage conservation and restoration approaches are the Proceedings [6]. Reports in the compendium reflect the Chinese approach to the issue - from philosophical through historical and up to narrowly professional technical point of view. There are comparisons of the Chinese approach with the European and others. There are reports from British and Japanese professionals that further enrich the range of opinions and concepts.

A multidisciplinary approach for choosing the best opportunity to preserve and restore cultural heritage in developing countries is given in a report by two Dutch scientists - engineers van Egmond and Erkelens [7]. The report provides useful schematics and examples of the necessary steps in the recovery process. Useful Case Studies are delivered.

Another non-European approach to the protection of cultural heritage is shared in the report by Egyptian scholars Elnokaly and Elseragy [8]. The experience of restoring historic buildings in cities and combining them with new constructions is presented. Examples from Cairo, Tunisia, and others are showing up.

In the end, there is Spiekermann's article [9]. The old cities of Aleppo in the Syrian Arab Republic, Shibam in Hadramaut/Yemen, and Sibiu (former Hermannstadt) in Romania are threatened by physical, social, and economic decay.

## 2.2 Authoritative Sources of Geotechnical Knowledge of Micro-Piles Technology

An early publication of the American Association of State Highway and Transportation Officials (AASHTO) from 1988 is the Manuel on Subsurface Investigations [11]. It outlines all the traditional laboratory and field methods for geotechnical exploration, valid to this day. It also shows the zoning of geotechnical characteristics in different states. A greater number of later American published techniques for the analysis of micro-piles and geotechnics, in general, have been developed on the basis of this practical guide.

The core of the modern micro-pile analysis is presented by Bruce et al. [12]. The steps of the computational procedure, the restrictive conditions of the computational model, the required volume of geotechnical study, a table with the average values of the main soil varieties are outlined there. Taking these data into consideration, the analytical formula and its variations for the carrying capacity of micro-piles are regulated.

Micropiles are regulated in Europe through general guidelines in Eurocode 7 and in standard EN 14199, but these documents are to be developed in the coming years.

Very useful is Ischebeck [13]. Ischebeck is the author of the modern type of rigid anchors and micro-piles and he shares several practical examples, with which he demonstrates a convincing combination of experimental results and theoretical analysis in order to successfully solve micro-pile problematic.

The variety of micro-pile applications is very well presented in Lizzi's review [14]. The author shows illustrations of foundations reinforcement of masonry with micro-piles, reinforcement of inclined towers like the one in Pisa, strengthening of landslides and deep excavations, tunnel vaults, and more. The presented report shows the empirical approach of sizing through the measured deformations of completed analogous structures.

## 2.3 More Interesting Bulgarian Contributions in This Area

A new chapter in our research opens the book of Ivanova [10]. At the beginning of the book, we've got a remarkable foreword on the architecture, history, and issues of the same building by two remarkable Turkish architects. She also pointed out that the most significant damages in architectural monuments come from the soil base. If their soil base turns out to be weak over time to withstand the load. the building or monument may tilt or crack.

In addition to the most recent publications on the subject of the study presented above, we have included a very valuable classical knowledge of the reinforcement and restoration of building structures of the past. They are written by some prominent Bulgarian scholars and can largely serve as textbooks in this discipline. These are the monographs by Ignatiev et al. [15] and Venkov's [16]. Thereafter the author gives useful information about the history of construction and restoration of "St. Stefan", the Bulgarian church in Istanbul, built entirely of steel.

Large-scale restoration works on historic monuments and infrastructure in the beautiful town of Balchik, located on a huge landslide at the steep coast of the Black Sea, are subjects of two papers [17] and [18]:

- The paper [17] presents in detail the geological features of Balchik and the evolution of deep landslides there. The role of hydrodynamic groundwater pressure and marine abrasion of the landslide has been emphasized. The implemented and designed reinforcement measures are shown with diagrams and details. The comprehensiveness and originality of the project measures for the permanent stabilization of the area are being emphasized.
- The paper [18] details the experience of the first large drainage gallery in Bulgaria to strengthen deep landslides. The adopted tunnel technology with TBM, which was first implemented for similar purposes, is also highlighted. The shared experience contributes to the further technological development of tunnel construction and reinforcement activities. The effect of the drainage gallery really proved decisive for stabilizing the huge landslide under the town of Balchik and for preserving the cultural monuments.

The huge practical and theoretical experience gained from the strengthening the 70 m deep landslide under the town of Balchik and the subsequent local fortification measures for the cultural monuments in the city are presented in detail and analyzed in Kolev's monograph [19]. There are a few papers by the same author dedicated to examples of the practice for the strengthening of church foundations in Bulgaria. Such paper is [20]. An interesting technical problem in the restoration of an old church has been studied and described historically in [21]. An original geotechnical approach for stabilizing the landslide under a church in Balchik by means of diaphragm walls and anchors is described in [22]. The proposed reinforcement technology is complex, but ensures the independence of the building from the development of landslides around it. One more paper [23] is dedicated to the analysis of geotechnical sustainability and strengthening the church of St. Stefan in Istanbul. The undertaken strengthening of the soil under the church is theoretically substantiated in the same report.

### 2.4 Brief Information About the Bulgarian Experience in Strengthening Architectural Monuments Through the Technology of Micro-Piles

The restoration of architectural monument buildings is the topic of the paper [24]. The successful restoration of the building is due to very well executed micro-piles under the foundations. The architectural qualities of the fortified monument are completely preserved; all the fortification elements and activities have remained discreet.

## 3 Purpose and Scope of the Study

The study focuses on two issues:

- The principles of the restoration of the architectural monuments, regulated in the charters;
- The technical capabilities of micro-piles for the restoration of architectural monuments such as churches, fortresses, towers, etc.

On the way to these goals, research will be limited to:

- General methodological and legal framework;
- Problems in the field of construction and geotechnics;

The study is intended to benefit both students and scientists, mostly engineers, architects, builders, investors, and municipal executives. It has a comparatively narrower professional profile, and so, from numerous problems, it excludes topics devoted to:

- Town planning and architectural details;
- Building materials for non-structural elements;
- Installations in buildings;
- Energy efficiency;
- Ecology.

Excluded topics are also important, but they require additional time, space, and effort to elaborate on the timelines set out in this study.

### 3.1 Principles of the Restoration of the Architectural Monuments, Regulated in the Charters

The main sections of the cited Strategy [1] are grouped as Technical Co-operation and Consultancy Program, consisting of:

- Monuments and sites;
- Historic towns;
- Territories.

Each section of the Strategy structure has a uniform structure of statements, which includes the following items:

- A brief summary of past achievements;
- Aims and approaches of the programs;
- Methodology;
- The taken actions;
- Summary of principal results.

On the other hand, the expected results can be also uniformly grouped: political, institutional, social, economic and territorial development, knowledge and education, site significance.

It is noteworthy that the accumulation of knowledge and educational goals are an integral part of cultural heritage preserving and restoring process, along with other indisputable effects of this activity. It is along this path that our research evolves to contribute to all other applicable science.

There are many European regulations within the subject; also there are national codes for the restoration works. There are many good practices of cultural heritage management and conservation.

A good example of international cooperation is mentioned respective the above cities: Aleppo, Shibam, and Sibiu. As Aleppo and Shibam are listed on UNESCO's World Heritage List, and Sibiu applied for this status in 2006, there is international interest and commitment to preserve these cities and to support them in finding ways for a viable future [9]. Public Authorities, NGOs, and inhabitants face challenges in preserving historical buildings and social and cultural structures that have been growing over hundreds, even thousands of years, in all their colorful variety. At the same time, they must create an adequate environment for living and working in these places. Physical preservation alone could turn the old cities into huge museums; hence, a comprehensive approach of preservation and development must focus on improving the living conditions of the inhabitants as well as sustaining the economic viability of the locations. At the same time, it must focus on making these historical sites better places to live while building on given physical and social structures turns out to be the most appropriate way of preserving urban heritage. The German Agency for Development Cooperation (GTZ) supports those cities, as well as others, in this effort. In addition to the information in the report, according to me, it is interesting to have a new discussion these days because Aleppo was virtually destroyed last year by the war in Syria.

### 3.2 Classical Technological Approaches for Restoration of Architectural Monuments

The classic technologies for restoration and strengthening of architectural monuments remain the most widely used worldwide. The strict requirement for protection of the original monument and the unstable original construction are the main prerequisites for this. Building materials that are as close as possible to the original ones should be used, the construction should not be overloaded, the general architectural image should be preserved, etc.

## 4 New Tasks Ahead of the Study

The new tasks this study poses are due to the contemporary problems that have arisen and are reduced to two aspects:

- Technogenic causes of deformation of old cultural monuments;
- Technical capabilities of micro-piles for the restoration of architectural monuments such as churches, fortresses, towers, etc.

### 4.1 Technogenic Causes of Deformation of Old Cultural Monuments

**Impact of Vehicle Vibrations and Construction Mechanization on the Construction of Old Cultural Monuments.** These are very common reasons for deformation nowadays when the number of cars in cities exceeded all expectations, and the construction of huge new buildings significantly changed the groundwater levels in the respective areas. The large difference in the foundation depth of new and existing buildings has in many cases caused unwanted deformations and loss of stability. The passage of heavy-duty trucks near old buildings has caused the cracking and demolition of masonry structures built over a hundred years ago.

There is a clear example of damage to a monument from new roads in Istanbul. St. Stephen's Church, as well as all the other large buildings in Istanbul's Fener district, were built on long timber piles because the soil was very weak. A newly built highway and dredging of the bay have destabilized the shore and cracks in the church building over the past 30 years. Recently, this church was completely restored and regained its architectural splendor. The work on the restoration and assessment of the artistic value of the monument are presented in detail in the book [10]. The cut ground base under the church was reinforced with a belt of injections up to a depth of 20 m using Jet Grouting technology. In this way, all deformations were stopped and the details of the building superstructure were restored. The technical decision for St. Stephan is presented in [23]. The problem is not decided by codes and standards, nor is there a specific methodology for it. So far, it finds a number of successful practical answers and realizations in our country and around the world. The hypothesis proposed in [23] is theoretically justified entirely with the methods of Soil Mechanics and in accordance with practical experiences. Several functional relationships between soil characteristics and structural elements have been sought, suggesting optimum cement solution parameters. The role of groundwater and their dynamics is also taken into account.

We have new research and design of vertical vibration protection of the Central Military Club building in Sofia - national architectural monument. The building is cracked due to the prolonged operation of vibrating rollers around it. The first results are already available, but a long series of additional analyzes are yet to be done and new dependencies, valuable for the practice and theory of barriers against vibration and earthquake effects on buildings, will be also identified.

**Change of Groundwater Level in the Area Around the Building.** There was an original technical solution, presented in [20] for cultural heritage protection in Veliko Tarnovo City in Bulgaria. The pressurized steel water pipe under the street had a breakdown. As a result of the water pressure, a stone retaining wall collapsed down to the original apse of a famous church - a national monument. The restoration works are combined with the creation of conditions for new archaeological excavations and the creation of a gallery for exhibiting the historic layers below the street. There was constructed a complex bridging underground structure in difficult technological conditions. The bridging structure simultaneously ensures street traffic, retains ground and landslide pressure from the slope and serves as a facade and gallery. The applied technology did not damage the archeological layers.

Another interesting case was in the Pazardzhik City, where a 180-year-old basilica was fortified. There were cracks appearing and evolving one hundred and forty years after the construction of the arches and walls of the Cathedral of the Holy Virgin in Pazardzhik that threaten it from collapse [21]. While getting familiar with the archival materials and the initial survey of the area, there were other features revealed in the geological profile of the foundation - swollen Pliocene clays and the improvement of the surrounding area are the reasons for the permanent decrease of the groundwater, which led to the rotting of the wooden piles under the temple. The construction of sewerage and lining of the nearby irrigation canal mainly caused the subsequent emergency condition of the building. The forced lowering of the groundwater level has led to the rotting of the timber piles under the building and the appearance of cracks in the walls. The examination of cracking shed additional light on the deformation mechanism of the building and pointed to additional measures to strengthen the foundation with stripes and in the transverse direction. The negative effect of the excessively long reinforcement process was noted. The risk of collapse was evaluated and recommendations were made to improve reinforcement. The performed analysis can be directly used in future planning and restoration works.

### 4.2 Technical Capabilities of Micro-Piles for the Restoration of Architectural Monuments Such as Churches, Fortresses, Towers, etc.

One of the most commonly used modern technologies for strengthening the base in such cases is that of micro-piles. The implementation of micro-piles under the foundations is an effective means of strengthening the structure and at the same time does not violate the architectural form. An opportunity is created for the permanent restoration of the monument.

Micro-piles' practice does not have a very long history, so it started at about 30–40 years ago. In its essence, this technology can be likened to a symbiosis between two construction means - one of the pile foundations and one of the injections. After all, current piles can best resemble rigid injection anchors, placed vertically or sum-up vertically on the ground beneath building foundations.

The technology of micro-piles and their application in the construction and renovation of buildings is developing faster than their theoretical analysis; the theory is catching up with practice. This is not an isolated case in Geotechnics and there are mainly two reasons for it:

- Soil is not a homogeneous and isotropic environment;
- The injection is a process that is difficult to control to achieve desired shapes and parameters of a healthy environment. That is why, at this stage, the analysis approaches are mostly empirical and need constant experiments. This is also evident from the brief overview of several key publications for the design and execution of micro-piles.

Micro-piles are still under development and need to be studied because they are very convenient for use in restoration work on buildings, bridges and other monuments. The answers to these new research tasks will be done through the own studies of current project participants and with references to the published world experience.

The limit states for micro-piles are further analyzed in [12]. Table criteria are given to evaluate the durability of this structure type. Recommendations of [12] lay in the Micropile Design and Construction document issued by the US National Highway Institute. This document is by far the most complete guide to the design and execution of micropiles in all aspects. In addition to analysis and design features, there is guidance on materials, machinery, technology, valuation, contracting, etc.

A good example of the great potential of micropilots in strengthening architectural monuments is the case of the cracked building of the Military Club in Sofia ten years ago - a national cultural monument. The building of the Military Club [24] is one of the most beautiful architectural monuments within the Bulgarian capital of Sofia. The building is a brick structure with three floors and a basement. Foundations are made out of massive stone blocks on the clay. The building is located in the heel of a slight slope, near the highest point of the city. During the period from 2006 to 2008, a tunnel for the Sofia subway route was built under the street at depth of about 20 m. A deep shaft control between tunnel and ground, surrounded by diaphragm walls was constructed about 10 m aside, in front of the building. The ground is multi-layered, including sand aquifer above in silt and plastic clay. During the construction of diaphragm walls of the trench, a draining effect into the soil quickly formed a very steep seepage curve between the building and the shaft. The rapid decline and great amplitudes of the groundwater level resulted in an adverse settlement of the foundation under the front facade and arches on the facade of the building are cracking, fissures are dissolving, and a real danger for the stability of the structure occurred. After the investigations, injection with micro piles is performed along the contour of the building facade. The possibilities of technology helped to achieve a high-pressure injection into the soil under the building and to be slightly lifted foundation and to close the cracks in the arches and columns.

There are many examples in all Europe for the great potential of micro-piles for the restoration of building structures. A residential building in Sofia was built 50 years ago and It possesses four floors and two entrances. The building had dangerous large deformations caused by its inclination. Foundations are strips in the depth of 2.50 m each. The reason for the uneven subsidence is swelling silt thicker than 6 m below the building. The first cracks appeared more than ten years ago and reinforced concrete elements have gradually shown cracks 5 to 10 cm of separation. Prior to the restoration of the upper structure, the ground was stabilized and foundations were reinforced by a system of inclined, vertical micro-piles to mitigate swelling of clay and stop further deformation of the building. These were micro-piles injections, installed using the pressure of the grout. The plan for their deployment is done according to the measured subsidence and tilting of foundations. The injection was performed in two phases to achieve higher injection pressure. The lengths of the piles were 4 m each. Control measurements were performed one year after repairs and showed no new subsidence. The restoration of the roof structure is also completed.

The results of ongoing research on these issues are finding their place in the lectures and reports within the Project RE-BUILD.

## 5 Conclusion

The activities for the restoration of the cultural monuments have a history of more than 150 years. The accumulated knowledge and practical experience from them has led to strict regulations in a number of charters and codes in Europe and around the world. All documents emphasize the basic principles of approach to cultural heritage: protection of the monument in all circumstances; restoration of the monument subject to preservation of the original.

There are many events and factors that can damage architectural monuments. Nowadays, new causes have appeared, such as car vibrations, deep excavations for modern buildings and much more. At the same time, construction technologies are developing very rapidly, offering us new means for strengthening and restoring architectural monuments. Micro-pilots are just this type of new construction technology, which has a variety of possibilities and is subject to future improvement and wider application in practice.

The restoration traditions in different countries have explored, highlighted, and compared so we achieve knowledge crossing and a new level of understanding of the restoration works.

## References

1. Bold, J., Pickard, R.: An integrated approach to cultural heritage. The Council of Europe's Technical Co-operation and Consultancy Programme (2017). ISBN 978-92-871-8529-7
2. Opinion of the European Committee of the Regions - Towards an integrated approach to cultural heritage for Europe. EU, Committee of the Regions, EDUC-V/046, 111th plenary session, 16–17 April 2015
3. Current practice for management and conservation of cultural heritage. Part D1.1: Safeguarding Cultural Heritage through Technical and Organizational Resources Management - project of the European Union's Horizon 2020 research and innovation programme under grant agreement n°700191 (2016)

4. Van Balen, K.: Challenges that preventive conservation poses to the cultural heritage documentation field. In: 26th International CIPA Symposium, International Archives of the Photogrammetry, Remote Sensing and Spatial Information Sciences, vol. XLII-2/W5, Ottawa, Canada, 28 August–01 September 2017 (2017)
5. Embaby, M.: Heritage conservation and architectural education: 'An educational methodology for design studio'. HBRC J. ISSN: (Print) 1687-4048. https://www.tandfonline.com/loi/thbr20
6. International principles and local practices of cultural heritage conservation. In: Conference Proceedings of the National Heritage Center of Tsinghua University and ICOMOS, Beijing, China, 5–6 May 2014 (2014)
7. Van Egmond, E., Erkelens, P.: Reaching best practice in rehabilitation of culture heritage buildings in developing countries. In: Proceedings, CIB World Building Congress, Toronto (2004)
8. Elnokaly, A., Elseragy, A.: Sustainable heritage development: learning from urban conservation of heritage projects in Non Western contexts. Eur. J. Sustain. Dev. **2**, 31–54 (2013). ISSN 2239-5938
9. Spiekermann, M.: The sustainability of urban heritage preservation - The Case of Aleppo. Inter-American Development Bank, Discussion paper no. IDB-DP-125 (2010)
10. Ivanova, B.: The Bulgarian architectural complex of the Golden Horn, Paradigm, Sofia (2018). ISBN 9789543263547
11. Manuel on Subsurface Investigations. American Association of State Highway and Transportation Officials (AASHTO) (1988)
12. Bruce, D., Cadden, A., Sabatini, P.: Practical advice for foundation design – micro-piles for structural support. Engineering (2005). https://doi.org/10.1061/40777(156)14
13. Ischebeck: Micro-pile foundations for vertical and horizontal load. Design examples and load test results. In: 12th International Workshop on Micropiles, Krakow, Poland, 11–14 June 2014 (2014)
14. Lizzi, F.: Micro-piles: past, present… and future. In: GEOTECH-YEAR 2000 Conference, Bangkok, Thailand (2000)
15. Venkov, V., Ignatiev, N., Nedelchev, V.: Restoration and Reinforcement of Massive Structures of Buildings. Technica, Sofia (1988)
16. Venkov, V.: Strengthening of Flat Foundations. Technica, Sofia (1964)
17. Kolev, C., et al.: Stabilization of a landslide in the town of Balchik. In: International Society for Soil Mechanics & Foundation Engineering. European Regional Sub – Committee on Stabilization of Landslides, pp. 67–78. Bogazici University, Istanbul (1988)
18. Kolev, C.: La galerie de drainage comme la solution generale contre les glissements profonds de la ville de Balchik. In: Bell (ed.) Comptes Rendus du Sixieme Symposium International, Glissements de terrain, Landslides, стр. 765–766, Balkema, Rotterdam, 10–14 Fevrier 1992 (1991). ISBN 90 54 10 032 X
19. Kolev, C.: Comprehensive Systems for Geo-Protection. Technica, Sofia (2007). ISBN 978-954-03-0677-3
20. Kolev, C.: Restoration of the facilities behind the Church of the Holy 40 Martyrs. In: Scientific Conference on Construction Problems, pp. 26–27, UACG, Sofia, 22 January 2004. Construction City (2004)
21. Kolev, C.: Anthropogenic Deformation Causes of the Cathedral of the Assumption of the Blessed Virgin Mary in Pazardzhik. Monuments, Restoration, Museums, pp. 38–44, 2–3 2005. Arch&Art, Sofia (2005). ISBN 773399
22. Kolev, C., Tzonev, A.: The consolidation of the Church in Balchik. In: Bell (ed.) Proceedings of the 6th International Symposium, Landslides, Balkema, Rotterdam, pp. 765–766, 10–14 February 1992 (1991). ISBN 90 54 10 032 X

23. Kolev, C.: Hypothesis for improving the shear strength of weak soils when founding with suspended piles. In: Proceedings from the Scientific Conference with International Participation VSU 2007, vol. II, pp. 163–168, Sofia, 15–16 May 2007 (2007)
24. Kolev, C.: Restoration of the military club building in Sofia, by micro-pile injections. In: Proceedings of Seventh International Conference on Case Histories in Geotechnical Engineering, Chicago, IL, USA, 29 April–4 May 2013 (2013). Paper no. 3.02C

# Particularities of Moldavian Wood Heritage

Ioana Olteanu-Donţov(✉), Marian-Gabriel Vartolomei, Mihaela Movilă, and Oana-Mihaela Banu

"Gheorghe Asachi" Technical University of Iasi, 67 Prof.dr.doc. D.Mangeron Street, 700050 Iaşi, Romania
ioana.olteanu@tuiasi.ro

**Abstract.** Wood represents one of the traditional materials used in construction. Along the years unique wood constructions were built all around the world, some of which are now included in the international wood heritage. However, several challenges appeared along the year when searching for rehabilitation solution, but now international restrictions are given in order to protect and consolidate structures made of wood. The article presents some of the fundamental rehabilitation principles according to ICOMOS used in the rehabilitation and conservation process, as well as a synthesis of possible decay factors common to wood structures.

The second part of the paper focuses on the Romanian wood heritage, shortly presenting some of its particularities and a case study. The case study makes a damage inventory on the church from Draguseni and proposes adequate rehabilitation solutions in accordance with current restrictions for heritage structures.

**Keywords:** Wood heritage · Wood decay · Wood church

## 1 Introduction

During the last decades, structural vulnerability assessment and rehabilitation techniques have been intensively developed, refined and adapted to modern resources [2]. Research on timber structures is still under development and the results obtained until now are not as impressive as those on concrete and steel. One of the main reasons for this is the variability of timber properties which is much higher than for other building materials. Furthermore, the timber material properties depend on the entire load and moisture history of the structure [19, 20].

Timber, together with stone, are the oldest building materials. Wood capacity to withstand both tension and compression made it the most complete one, before steel became available. Timber heritage is vastly spread and dates back to millennia ago (Fig. 1), but by being very sensitive to lack of maintenance, it also became the most vulnerable part of historical buildings, often suffering severe degradation which require major interventions [4].

The words "wooden built heritage" refer to all types of wooden buildings and other wooden structures that have cultural significance or are parts of historic places, and include temporary, movable and evolving structures [7]. The value of a monument is


A. Rotaru (Ed.): CRIT-RE-BUILT 2019, SSGG, pp. 541–552, 2021.
https://doi.org/10.1007/978-3-030-61118-7_44

**Fig. 1.** Oldest wooden building, Horyuji, Japan, dates from around 600 a.d. [14]

given by its aesthetic, anthropological, archaeological, cultural, historical, scientific and technological heritage values [2]. Special attention should be paid to the interaction of wood elements with other construction materials in mixed structures.

Vulnerability assessment and building rehabilitation solutions are regulated, at international level by the International Council on Monuments and Sites (ICOMOS). ICOMOS is a network of experts that benefits from the interdisciplinary exchange of its members, from different fields which works for the conservation and protection of cultural heritage places. The members of ICOMOS contribute to improving the preservation of heritage, the standards and the techniques for each type of cultural heritage property [23]. Some of the fundamental principles for heritage buildings which are suitable also for historic timber structures are presented below [16, 17].

The *authenticity principle* is a wide concept which extends to structural level, aside from its historic, aesthetic value and social significance. An authentic property expresses its cultural values in a truthful and credible way through a variety of attributes such as its form, materials, functions, management system, location, spirit etc. [17]. The load bearing structures of monumental buildings, carry also a technological quality. These structures are precious witnesses of the efforts and inventiveness of mankind in search for more advanced techniques in the application of materials, idealization of new shapes and steadier configurations, taking the best profit from the available materials. *The conservation* of heritage buildings includes other levels of intervention such as preservation and reconstruction [17]. The *restoration* is defines as the act or process by which the shape, characteristics, as well as the character of a building are reconstructed with precision, respecting the characteristics of the initial period, by removing the interventions that occurred over time and reconstructing the missing elements [17]. The *consolidation* is the reconstruction or renewal of a construction part, in order to obtain an increase in the structure's bearing capacity, which includes the ductility or strength improvement of individual elements, or by introducing new ones, in order to improve the structure behavior to horizontal loads. The *rehabilitation* process consists in the strengthening of a damaged building, with the purpose of restoring it to the same level of operation as before degradation [1, 22]. Rehabilitation is, in many cases, associated with interventions on historical urban sites and, special attention should be given to these cases [10]. Engineers involved in these actions have to be updated on modern rehabilitation techniques, especially on those accepted in case of monuments, taking the cultural dimension of conservation to the same level as the technical one.

ICOMOS committee for timber structures gave, in 2017, a list of Principles for restoration and conservation of wooden built heritage, partially presented below [18]:

- recognize the importance of timber structures from all periods as part of the cultural heritage of the world;
- take into account the great diversity of historic timber structures;
- take into account the various species and qualities of wood used to build them;
- recognize that wooden heritage provides evidence of the skills of craft workers and builders and their traditional, cultural and ancestral knowledge;
- recognize the vulnerability of structures wholly or partially in timber due to material decay and degradation in varying environmental and climatic conditions caused by humidity fluctuation, light, fungal/insect attacks, wear and tear, fire and other disasters;
- recognize the increasing scarcity of historic timber structures due to vulnerability, misuse and the loss of skills and knowledge of traditional design and construction technology;
- take into account the great variety of actions and treatments required for the preservation and conservation of these heritage resources;
- take into account the excellent behavior of wood structures in withstanding seismic forces;
- recognize the relevance of community participation in protection of the wooden heritage, its relation with social and environmental transformations, and its role in sustainable development.

According to the present guidelines for intervention, three very important recommendations should always be present [16]. The intervention works should be recognizable, reversible and always follow the principle of minimum intervention. These are widely accepted principles, and are the core of modern conservation approach. The fact that the interventions should be recognizable is one more way of keeping their intrinsic authenticity, once it should be clear what is truly original from what is added or changed, keeping invasion to a minimum, avoiding hurting the aesthetical value of the original structure. The reversibility of an intervention allows future substitution with more suitable or advanced solutions without sacrificing original fabric or leaving strong traces of past interventions [16].

The fact that historical structures have resisted and performed their functions until the present times should be taken into account before any intervention and the idea of improvement should be preferred to total compliance with the building codes [16].

Research should continue in order to increase the possible rehabilitation solutions to be used by the designer in case of historical buildings which exist in a variety of materials, shapes and complexities. Regarding wood heritage, the lack of specialized workers available represents a real problem in order to fulfill any rehabilitation project. Carpenter knowledge and experience can be of great value in conservation, and measures should be taken for the revival and teaching of traditional techniques. At the same time, investigation on the effectiveness and structural behavior of those techniques will allow its resurrection in future conservation works [21].

## 2 Wood Decay Factors

Wood is a traditional construction material, used for centuries, due to the fact that is easy to find, even though, having an organic composition is naturally predisposed to aging and deterioration caused by environmental or biological factors.

The natural decay factors category include: the oxygen and pollutants, which are reactive factors leading to chemical, physical and biological process; temperature, light and humidity, which are activation factors ensuring the necessary energy for the chemical process [8]. The long term exposure of wood to solar radiation and extreme weather conditions, lead to several decay processes. The ultraviolet rays deteriorate the material. The chemical decomposition process under the influence of light can be observed by the changes of natural wood color and the development of very fine cracks in the cell walls or between them. Also, the heat caused by direct or diffused solar radiation, can lead to an asymmetrical heat gradient on the thickness of the pieces, causing additional tensions in the wood. Heat negatively influences wood by: dimensional changes, increasing inter-molecular and interatomic distances (weakening of chemical bonds), and reducing physical characteristics [6].

Another important enemy for the wood monuments is humidity variation. Humidity also intensifies the chemical and photochemical processes that leads to color changing, material vulnerability and decrease of mechanical strength. The variation of air relative humidity has multiple effects on the wood from the historical monuments and cultural assets, as: contraction and swelling; dimensional instability; moisture content variation; instability of adhesives and consolidation agents; self-oxidation processes; aging of the material, through combinations of water vapor, oxygen and air pollutants; biological degradation [5]. Certain limits of temperature and humidity, favors the action of biological factors that find adequate conditions to develop.

The degradation caused by biological agents is produced by: blue and mold fungi, rotten fungi and xylophages insect [6]. Fungi decay are spread by microscopic airborne spores, needing a minimum humidity of 22%. Color, size and appearance of the mycelium, the strands and the fruit body lead to the identification of the fungal species causing the decay. Wet rot is the most common decay cause. There are two types: those that can attack the cellulose, but not the lignin are called brown rots and those capable of attacking both cellulose and lignin, white rots, which leave the affected wood fibrous and pale in color [6]. Serpula Lacrymans is a particular species of brown rot, which causes larger damage and grows very fast. Can reach 10 mm in only one day. Wood properties decrease with its weight. Specialists consider that a decrease of 10% in the weight of the wood structure indicate that the mechanical properties can be affected by 80% [20].

With respect to the insects that can affect the wood, two categories are identify depending on its humidity level: low humidity (25%–30%) or dry wood (below 10%) [8]. Wood infestation with insects can occur in the following stages: in the forest, where is always accompanied by fungi attack establishing a symbiosis relationships; in the warehouse; or by using wet wood which presents incipient forms of insect attack. Beside humidity, temperature is another important factor for insects' growth, a range of 10 to 40 °C being ideal [5]. Depending on the insect species, different temperature are favorable. Research showed that all wood component can serve as food for insect development. Wood attacked by insects (in association or not with the attack of fungi) leads to the physical-mechanical strength below the healthy one, with an intensely brown-red color, and a brittle, dusty and flowy behavior [5, 20].

The human factor is also important for the wood monuments deterioration, by ignorance, negligence, defective management and, not least, through public institutions which do not comply with their responsibilities to protect their assets [19]. The restoration process is slower than the degradation one, hence preventive measures of conservation should be considered to protect the wood heritage. These measures can include: the improvement of microclimatic conditions in a museum and their permanent control (humidity, temperature, illumination, exposure, visiting conditions etc.); ensuring proper storage, as well as the proper handling, movement and packing of cultural goods; organization of the system of inventory and secure of cultural goods (registration, verification, security against thefts, fires etc.); applying some procedures for wood preservation and slowing and/or stabilizing the degradation processes (biological protection, chemical protection, design protection) [19].

As a result of the degradation factors, the wood, undergoes irreversible changes in time: deformation, color degradation, cracking, exfoliation, molding, rotting and mechanical strength weakening due to insect attack [8]. All this leads to changes which makes the wood monument more and more fragile and vulnerable to accentuated degradation and, in the situation of non-intervention on long term, the disappearance of the monument can occur.

## 3 Romanian Wood Heritage

Not many people know that in Romania are about 1,500 old wooden churches, built before the 20th century, which impress with their beauty and legends created around them [3]. The country is famous for its Cultural and Wood Heritage which has from houses to tall village churches in Maramures, in the North part of the country. Also, there are Popular technique "machineries" as wind and watermills and beautiful Orthodox iconostases – a wall of icons and religious paintings, separating the nave from the sanctuary in a church; iconostasis also refers to a portable icon stand that can be placed anywhere within a church. Several Village Museums are placed around the country in order to recreate and the show the traditional houses and its customs, which represent a point of interest for the tourists [11].

Founded by Royal Decree in 1936, and covering around 15 hectares on the shores of Lake Herăstrău, the Village Museum from Bucharest, Fig. 2 is one of the greatest outdoor museums in Europe [13]. Here are more than 300 original houses, farmsteads, windmills, watermills and churches from all of Romania's historic regions. Every exhibit has a plaque showing exactly from where it was brought. The majority of houses date from the mid XIXth century, but there are some, which date from XVIIIth. The highlight of the museum is probably the steep belfry of the wooden Maramureş church, complete with exquisite but faded icons. It was created in 1936 by Dimitrie Gusti, Victor Ion Popa, and Henri H. Stahl [3, 12]. The museum is organized as a village, with clusters of houses belonging to different areas of the country. They are completely furnished, have gardens and winding paths around so a visit here is like going back in time, visiting real old time rural areas [9].

**Fig. 2.** Village Museum from Bucharest, Romania [15]

Local researchers started to investigate the influence of the environment on the conservation-restoration of wood heritage by comparing the damage state of houses from the Village Museum placed in Bucharest with the houses from another museum placed near Sibiu, where the air quality, temperature and moisture differ.

A core from the Romanian wood heritage is the Maramures region, located in the northern part of the Romanian Carpathian Mountains. In 1900, the forest covered 90% of the Maramures area, which is over 10000 km$^2$. Today there are over 280 villages [3]. The abundance of timber resources combined with the skill and craftsmanship of the local people, brought the area the name of "land of wood". The region is famous for the large number of well-preserved wooden churches, approximately 94 built before 1850. The wood churches developed due to the restrictions imposed from 1278 onward, for the Orthodox churches to be built from durable materials [3]. Generally, the walls are built of blockwork (squared logs laid horizontally) with intricate joints, cantilevered out in places to form brackets or consoles supporting the eaves. The churches are quite small and dark inside and painted with Biblical scenes, byzantine style. The most characteristic features are the tall tower bell above the porch entrance and the massive roof that seems to dwarf the main body of the church. Some of them have been listed by the UNESCO as a World Heritage Site in 1999, for their religious architecture and timber construction traditions [9].

Maramures churches are considered to be the most elaborate structures of the Romanian wooden heritage. The majority of them are built in the XVIIIth century, on the highest ground of the village, to escape seasonal mud. They are founded upon huge blocks of wood rather than stone [3]. Maramures churches stand out due the carpenters' talent showed up in a display of beauty and creativity. These churches were built without any help of iron nails, and still manage to reach impressive heights. In fact, here, is considered to be the highest wooden building in Europe - the church of Surdesti, reaching 72 m (236 ft). The Maramures wooden churches harmoniously combine the major architectural elements of Eastern and Western Europe, creating a symbiosis of the Byzantine plan and the Gothic forms [3].

The churches structure is remarkable, as every element has a double role, constructive and decorative. The sizing denotes a sure sense of proportions; the inner space is conceived at a human scale, while resting imposing. The double roof, the horizontal lines of the registers, are new elements that have considerably changed the architecture of the façades [15]. Along with the bell tower, the roof challenges the robustness and stability of the base. All these elements converge towards the impressive monumentality of these churches, considered to be true wooden cathedrals.

Almost 100 old wooden churches from Maramures, survived for hundreds of years, bringing to the present the talent and imagination of the local artisans who built them in the unique 'Gothic style of Maramures' [9]. The wooden church of Barsana, Fig. 3, was built as a monastery church in 1720, and features some impressive interior frescoes with baroque and rococo influences painted by Toader Hodor in 1806. Originally, part of a monastic complex closed in 1791, the church was moved to its present location at the beginning of the XIXth century [3]. The church from Barsana is one of the oldest in the region, which can be easily seen in the way the color of the wood washed out throughout the years, so from the distance it looks almost white.

**Fig. 3.** Barsana church, Romanian wood heritage [12]

## 4 Visual Inspection on Damages and Rehabilitation Recommendation for a Wooden Church

In the XVIIth century, in Moldavia, oak logs were commonly used in construction joined in the "hinged tail" of the churches. This technique involved the placement of one or two massive oak logs on the stone base above which the walls of the interior are made which are also from oak, carved on the edges, fastened and joined together [3]. Same technique was applied to the studied church placed in Draguseni village.

In an old document which presents churches from Moldavia before XIXth century, the Draguseni church appears to be built between 1760 and 1780, by Iordache Lazu patriarch. The church was built on site, according to local tradition from oak logs. The current church is the second one, the original one, made from fir wood was destroyed by a fire in 1936 [9].

## 4.1 Previous Interventions

Along the years, the church suffered several interventions, from which only the most significant ones are presented. At the end of the XIXth century and beginning of the XXth, the porch was added, the roof was covered with metal sheet, the lower beams were replaced with stones, the walls were reinforced with steel plates and the 3 wooden walls that made up the western facade were replaced with a single stone wall (Fig. 4). These data are part of an expertise of Horia TEODORU made in 1929 [3].

**Fig. 4.** Draguseni church state in 1929 according to H. Teodoru

A second intervention was made in the 1930s after Horia TEODORU's expertise report and consisted in changing the roof and covering it with shingles, replacing the stone walls with wooden logs, restoring the porch and adding the bell tower above it, restoring the western façade by changing the stone wall with 3 wooden walls.

The third intervention was in the 1960s and consisted of the replacement the shingles from the roof. The most recent intervention was in 2014–2015 and consisted of insulating with mineral wool the porch and other walls, installing wooden shutters at the bell windows and restoring of the shingle from the roof covering, Fig. 5.

**Fig. 5.** Draguseni church, actual state (personal archive)

## 4.2 Church Description

The wooden church in Drăgușeni is made of oak beams, carved on the edges, joined in the system “double locks”. It has a quarry stone base, which is supported by two massive oak footings, the first one is rectangular, and the second profiled, with a rounded retraction forming a double footing system. The building has a ring of twisted wooden belt on the outside. The roof is made of shingles, with a slight break in the slope in order to connect it with the porch covering. The walls rest on large consoles, with cuts.

The monument has a rectangular plan, with a length of 16.06 m, a width of 5.74 m and the height of about 7 m. Above the porch is the bell tower. The altar is lowered by 0.5 m from the longitudinal walls, creating a traditional worship annex. A window is placed in the entrance area, made by an X-shaped cutout, provided with steel bar. The original window of the altar is placed in a rectangular frame, fixed in wooden nails. The vaulting system of the roof rests on wood beams being continued by cantilever beams. The age of the church are sustained by the entrance door which has a particular framing, Fig. 6. Inside, the church is divided into four rooms, specific for orthodox churches. Between the porch and the narthex is a wall used to enter the church. The narthex has three faces with a small window and covered by a semi cylindrical dome. The nave is rectangular with two windows separated by a wall from the narthex. Above the nave is a spherical dome rested on an octagonal contour smaller on the west side. The framing of the entrance door presents a series of extremely valuable decorative elements: representations of fantastic animals, plant motifs, being appreciated by specialists, especially for their uniqueness compared to other churches from Suceava area [3, 9, 15]. The painting are of more recent date (1877), but do not have aesthetical value for specialists to be included in the national heritage.

## 4.3 Damage Assessment and Recommendations

Several damages were noticed in the visual inspections, as well as inappropriate rehabilitation works which reduce the monument value. Foundation deterioration is presented

**Fig. 6.** Draguseni church, inside view (personal archive)

in Fig. 7. It can be seen the unfavorable effect of humidity on the foundation oak logs on the north and north-east façade, on the left photo and the different settlement at the foundation level due to water infiltration in the photo from the right side.

**Fig. 7.** Foundation damages due to water infiltration and humidity (personal archive)

The different settlement affected also the oak log structure by relative rotation of the logs. The secondary effect of this is the local crushing on the logs. The water infiltration due to poor drainage system and the missing downpipes affected also the connection between the porch and the church. Other damage causes observed were the aging of the binder between the foundation stones and their decay due to insects. Also, the painting become darker due to the smoke from the burned candles. The effect of all causes could have been significantly reduced if periodical maintenance work had been performed.

The heritage church value was unfortunately reduced due to some interventions performed by locals without taking into consideration the existing restrictions regarding rehabilitation solutions that can be used in case of monuments. Among this works, the most visible ones are: the porch thermal insulation with mineral wool; the use of wooden blinds at the bell windows and mounting of plank on the narthex ceiling;

Generally, the church is in good shape, but some rehabilitation works should be performed in order to stop further deterioration. In order to rehabilitate the structural system two possibilities can be adopted. The cheaper one consist in executing a good drainage system to collect and transport the pluvial water away from the church foundation. This can be done by pouring sidewalks throughout the perimeter of the church in order not to allow rainwater to infiltrate at the foundation. To create a rainwater sewerage channel and directing them to a collecting basin or to the sewerage system of the village. To remove the materials and works that are not part of the initial structure of the monument, even if this means reducing human comfort during religious ceremonies.

A more expensive rehabilitation process, includes the works previously presented, as well as the followings: replacement of the affected logs due to the decomposition of the material or insects; removing the relative rotation of the logs; restoring the foundation and cleaning the paintings. In order to perform this works, it is necessary to dismantle log by log the structure and to replace the damaged or destroyed elements, after which the structure is reassembled taking into account the exact order in which it was dismantled. This method removes the relative rotations caused by different settlements.

## 5 Conclusions

The identity of a nation is given also by its cultural and architectural heritage. Even though historical events, poor maintenance, biological and chemical agents or even human interventions may lead to degradation, current practice focuses a lot on finding solution to limit heritage decay.

Wood, as a construction material is a traditional one, with which experienced carpenters manage to give as a remarkable number of wooded structures which withstood the time challenges and are now included in the wooded heritage. The rehabilitation process, for wood structure in particular, is regulated at international level by ICOMOS.

Diverse decay factors need to be considered when analysis the degradations of a wood structural system. From the air, pollution, light, humidity to different species of fungi and insects and even human actions all contribute to the current state of a wood monuments. The article made a short review of the major degradation factors.

The practical part of the article presents the Romanian wood heritage. A detailed analysis regarding damage causes and rehabilitation solution are performed on Draguseni church. Here, beside degradations due to poor maintenance works, inappropriate rehabilitation solutions were encountered. Among the recommended urgent solution is realizing a drainage system to collect the rain water and direct it to the sewage system in order to avoid further settlement of the foundation.

## References


1. 1997 law privind ratificarea Convenţiei pentru protecţia patrimoniului arhitectural al Europei, adoptată la Granada la 3 octombrie 1985
2. Anwar, M.: Practical techniques for restoration of architectural formation elements in historical buildings. World J. Eng. Technol. **7**, 193–207 (2019)
3. Babos, A.: Tracing a Sacred Building Tradition – Wooden Churches, Carpenters and Founders in Maramures until the turn of the 18th Century. Department of Architectural History, School of Architecture, Lund Institute of Technology, Lund University, Sweden (2004)
4. Barreira Pinto, L.F.: Inventory of Repair and Strengthening Methods – Timber, Master Thesis for Advanced Masters in Structural Analysis of Monuments and Historical Constructions, Erasmus Mundus Programme (2008)

5. Bostenaru Dan, M.: Timber frame historic structures and the local seismic culture—an argumentation. In: Bostenaru Dan, M., Armas, I., Goretti, A. (eds.) Earthquake Hazard Impact and Urban Planning. EH, pp. 213–230. Springer, Dordrecht (2014). https://doi.org/10.1007/978-94-007-7981-5_11
6. Brancoa, J.M., Neves, L.A.C.: Robustness of timber structures in seismic areas. Constr. Build. Mater. **33**, 3099–3105 (2011)
7. Branco, J.M., Tomasi, R.: Analysis and strengthening of timber floors and roofs. In: Costa, A., Guedes, J., Varum, H. (eds.) Structural Rehabilitation of Old Buildings. Building Pathology and Rehabilitation, vol. 2, pp. 235–258. Springer, Heidelberg (2014). https://doi.org/10.1007/978-3-642-39686-1_8
8. Cavalli, A., Cibecchini, D., Togni, M., Sousa, H.S.: A review on the mechanical properties of aged wood and salvaged timber. Constr. Build. Mater. **114**, 681–687 (2016)
9. Constantinescu, B.: Romanian Architectural Wooden Cultural Heritage – The Present Status – A Survey (2005). http://citeseerx.ist.psu.edu/viewdoc/download?doi=10.1.1.495.39&rep=rep1&type=pdf
10. Drdácky, M., Mlázovsky, V., Ruzicka, P.: Lessons from Historic Carpentry, European Research on Cultural Heritage: State-of-the-Art Studies. ARCCHIP ITAM, Prague (2006)
11. http://romaniatourism.com/world-heritage-sites.html
12. https://romaniatourstore.com/blog/the-truth-behind-the-wooden-churches-of-maramures/
13. https://www.nuntatraditionala.ro/locatie-sedinta-foto/muzeul-satului-dimitrie-gusti/
14. https://www.timetravelturtle.com/horyuji-japan-oldest-wooden-building/
15. https://www.uncover-romania.com/attractions/unesco-heritage-romania/wooden-churches-of-maramures/
16. ICOMOS: International Charter for the Conservation and Restoration of Monuments and Sites (The Venice Charter) (1964)
17. ISCARSAH Committee (ICOMOS): Principles for the Preservation of Historic Timber Structures adopted by ICOMOS General Assembly, Mexico City (1999)
18. ISCARSAH Committee: Principles for the conservation of wooden built heritage (2017)
19. Kohler, J.: Reliability of Timber Structures. Institute of Structural Engineering, Swiss Federal Institute of Technology, Zurich (2007)
20. Riggio, M., Sandak, J., Franke, S.: Application of imaging techniques for detection of defects, damage and decay in timber structures on-site. Constr. Build. Mater. **101**, 1241–1252 (2015)
21. SC Manisa Construct SRL: Manual privind tehnologia consolidarii, restaurarii si protectiei biologice la structurile de lemn ale monumentelor istorice, Bucuresti (1997)
22. UNESCO: Convention Concerning the Protection of the World Cultural and Natural Heritage, Paris (1972)
23. www.icomos.org

# The Rehabilitation Process of an Emblematic Historic Building of Iasi County

Oana-Mihaela Banu[1(✉)], Sebastian Cusutură[2], Ioana Olteanu-Donţov[1], and Mihaela Movilă[1]

[1] "Gheorghe Asachi" Technical University of Iasi, 67 Prof.dr.doc. D.Mangeron Bvd., Iaşi 700050, Romania
oana-mihaela.banu@academic.tuiasi.ro

[2] SC CONSTRUCTURAL SRL, 20b Laţcu Vodă St., Iaşi, Romania

**Abstract.** In the last decades, the modern society exhibited an increasing interest towards the necessity of preserving the national identity also through its built cultural heritage. The numerous historic buildings from Iasi County, most of them being badly affected by the environmental conditions and the improper maintenance, drawn the attention of the competent authorities upon their major responsibility to undertake wide restoration works.

One of the emblematic architecture monument of Iasi is a central palace which is a valuable historic landmark for the local cultural heritage of the city. The sumptuous palace is placed in the historic core of Iasi, in the Union Square, and during its lifetime, it served different purposes, such as hotel, bank and political party office. Unfortunately, the communism period dramatically changed its destiny. The upstairs rooms of the three-storey palace have been transformed into social apartments. Nowadays, only the ground floor hosts an art gallery.

At the moment, the palace is severely damaged, being classified into the first seismic risk category. The town hall, the actual owner of the palace, has initiated an important rehabilitation and consolidation process of it, partly with European funds. The paper focuses on the main rehabilitation and consolidation proposed solutions. It also presents the conclusions drawn by the team of experts and describes the necessary intervention measures for the structural performance improvement.

**Keywords:** Historic landmark · Historic masonry buildings · Structural damage · Rehabilitation works · Structural safety

## 1 Introduction

### 1.1 The Importance of Historic Buildings' Conservation

The historic buildings are part of a country's history and represent the cultural and architectural heritage for the present generations, but mostly for the future ones. Thereby, within the actual context of sustainable development, there is a growing awareness of


A. Rotaru (Ed.): CRIT-RE-BUILT 2019, SSGG, pp. 553–567, 2021.
https://doi.org/10.1007/978-3-030-61118-7_45

modern society to preserve the universal cultural and natural heritage for the future generations [1].

In the post-war period, the significant impact on the built environment, the mass destructions and substantial structural damages of the historic monuments from the entire Europe, prompted the necessity of developing common international theories, methodologies and techniques regarding the protection, conservation and restoration of the built cultural heritage. The International Council on Monuments and Sites (ICOMOS) is the first professional association founded in 1965 in Warsaw as a result of the "Venice Charter for the Conservation and Restoration of Monuments and Sites" from 1964 [2], which offers advice to UNESCO on World Heritage Sites [3].

The main goal of restoring the historic buildings is to maintain and perpetuate the historical landmark of a city. There are both cultural and practical reasons for a community to promote and sustain the preservation, restoration and rehabilitation of its historic buildings. They have the great potential of offering valuable benefits for the touristic development of a country and especially, of a city. Any historic building has the particularity of providing the visitors with impressive and attractive memories through its singularity in terms of architectural design and craftsmanship.

The rehabilitation process of the existing buildings also presents the advantage of a positive environmental impact through the natural resources conservation and the natural land saving. The historic buildings are usually made of durable construction materials which are resistant to premature deterioration. So, the preserved natural materials can be also valuable assets to the cultural heritage of a city, since they could have been supplied from sources that no longer exist [4]. Another great benefit is provided by the elimination of waste disposal necessity and of air pollution which can be caused by the demolition of damaged buildings in the historic center of a city.

Historic buildings restoration involves major responsibilities and great specific care. Sometimes, their structural integrity can be impaired by the lack of maintenance and repairing works. The rehabilitation process demands to maintain as much of the initial structural system as possible, beside the original architecture. There are cases when the structural integrity is so much affected that a new system of structural elements has to be designed and its integration to strengthen the existing one without altering the building's architectural features is the real challenge. For the safety, security and operational reasons, the rehabilitation process requires a rigorous conformity to the national building codes and regulations. Preservation experts' assistance and a thorough documentation of the building's historical background are also needed [5].

## 1.2 The Conservation Principles of Historic Buildings

The Romanian historic buildings had an unprosperous fate mainly during the communism period. Some of them have been improperly used, others have been transformed to serve various inappropriate purposes, and other ones have been abandoned to be simply destroyed by the time passing. The seismic events of high intensities, the difficult foundation soils, the diminished materials' properties, the structures massiveness and the lack of specific norms and regulation for addressing the adequate maintenance and repairing methods, also contributed to the actual damage state of the historic buildings. But still, Romania boasts a quite impressive architectural heritage. Iasi is one of the cities hosting the most iconic historic buildings from Romania and it was officially declared Historical capital of Romania, in December 2018. Among the beautiful landmark sites of Iasi, the Palace of Culture, the Metropolitan Cathedral, the National Theater, Golia Monastery, the Palace of the University, the Military School, and the Great Synagogue are only a few of them deserving to be outlined (Fig. 1).

**Fig. 1.** A collage of the most beautiful landmark sites of Iasi (Bogdan Muraru/creative commons.org) [6].

Most of the historic buildings are made of masonry. They usually have a basement, a reduced number of storeys and an attic floor. The load bearing walls and the foundations are massive stone works and sometimes the footings are made of natural stones.

In our country, the earthquake hazard is high and the structural rehabilitation under these circumstances represents a great challenge for the civil engineers due to the various structural configurations, materials and technologies, more or less common, which have been used along the years for the construction of the historic buildings [7]. The historic masonry buildings could not have been designed in accordance to the currently used building codes and nowadays they do not have an adequate seismic capacity [8]. Consequently, an appropriate conservation process is needed to preserve the cultural, aesthetic and affective value of a historic building, without impairing its architecture, its character-defining features and finishes [4].

The modern intervention works applied in case of historic buildings include:

- *rehabilitation*: the process of reconstruction or upgrading of a damaged building with the aim of restoring its initial operational and safety level;
- *consolidation*: the process of increasing the strength and/or stiffness of the structural members or system with the aim of improving the structural performances for any potential exceptional event (e.g. occurrence of a significant seismic event);
- *restoration*: the process of accurate reconstruction of character-defining shapes, features and finishes of a damaged building, as it was initially constructed, paying great care to the historical background and by eliminating any previous interventions [9].

The time evolution of structural behaviour in accordance with the intervention measures is illustrated in the next figure (Fig. 2). It shows that the lifetime of a historic building and the safety level are highly correlated with the maintenance works which have the important role of early damage prevention [7].

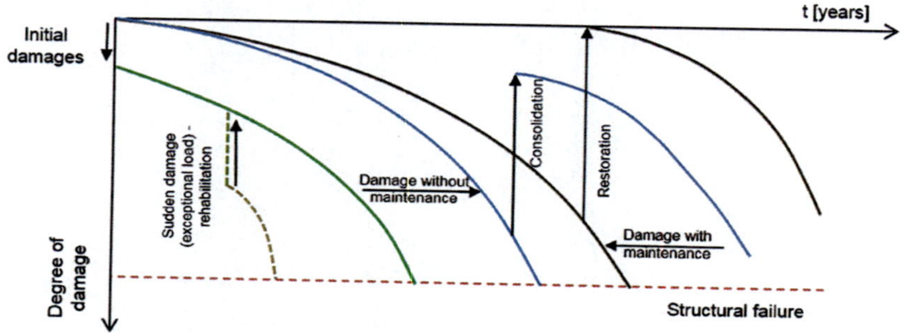


**Fig. 2.** The evolution of structural damage in accordance with the intervention measures [9].

### 1.3 Damage Assessment Methodologies of Historic Masonry Buildings

According to the current Romanian methodology which regulates the risk assessment and the intervention measures for historic buildings, MP 025-04-"Methodologies of risk evaluation and strengthening solutions for heritage buildings" [10], these intervention processes and works need to satisfy the following criteria:

- *efficiency*: the intervention works need to be efficient and this feature have to be demonstrated by qualitative and quantitative assessments;
- *compatibility*: the intervention works need to be compatible with the original structural system and the used materials must have chemical, mechanical and architectural characteristics identical to the initial ones;
- *durability*: the intervention works need to be performed using techniques and materials having an assessed durability similar to the other existing materials;
- *reversibility*: the intervention works have to be as much as possible reversible, so that an utilized solution to be easily removed when a future and maybe a better one will be adopted.

The implementation of all the previous mentioned aspects requires strong collaboration of experts with advanced knowledge in structural engineering, geotechnical engineering, architecture, history, archaeology and materials science [11]. The seismic safety assessment is even a more complex issue due to the fact that the historic monuments cannot comply with the damage assessment and the consolidation design rules of newly constructed buildings [12]. The intervention works may range from simple reparation works, when the cultural value of the existing building can be preserved, to individual or overall consolidation works, when the structural safety level is badly affected [13].

The structural performance assessment of a historic masonry building is based on: the identification of the construction details, the analysis of the historical background and an accurate survey of the actual damage state. Within this context, the needed information are: the structure's geometry; the masonry pattern and texture; the structural members' constructive details and cross sections; the physical, chemical and mechanical properties of the constituent materials; the masonry characterization as a composite material; the location of the structure under consideration and the characteristic seismic and climatic actions; the neighbouring buildings; the foundation soil properties [14].

# 2 Case Study

## 2.1 Historical Background

One of the emblematic architecture monument (class II – according to [15]) of Iasi is a central palace constructed between 1913 and 1914 by a Jewish family. The dome has been completed in 1915 and at its opening, an art exhibition took place inside it. The architect of the Art Nouveau style palace was Herman Clejan. This is a valuable historic landmark for the local cultural heritage of the city (class B – according to [15]), being listed by the Romanian Ministry of Culture and National Patrimony in the National Register of Historic Monuments [15]. The sumptuous palace is placed in the historic center of Iasi and during its lifetime it served different purposes, such as hotel, store, jewelry, bank and political party office. Unfortunately, during the communism period, the upstairs rooms of the three-story palace have been transformed into social apartments and have been progressively damaged due to the lack of maintenance and repairing works. Only the ground floor functioned as an art gallery for the Romanian Union of Plastic Artists – Iasi Branch, for many years, being somehow preserved [16].

Down the ages, the palace resisted three crucial events: the big flood from 1932, when numerous old buildings from Iasi have been partially or completely destroyed, the bombing of Iasi county during the World War II and the 2nd most powerful earthquake recorded in Romania in the 20th century, the 1977 Vrancea earthquake, when most of the buildings that survived the flood and the bombing events, finally collapsed [16].

The advanced damage state of the central palace determined the town hall, the actual owner of the palace, to evacuate the inhabitants and to start a wide rehabilitation, restoration and consolidation project of the building. The newly rehabilitated palace will mainly serve for cultural purposes and will host various art exhibitions [16] (Fig. 3).

**Fig. 3.** The central palace in 1930 [17] and 2009 (Argenna/commons.wikimedia.org) [18].

## 2.2 Building Geometry and Structural Configuration

The entire built-up area sums up 585.70 $m^2$ and the building has 3 stories, a ground floor and a basement, with different heights and compartmentalization. The maximum height of the building is of 16.51 m at the eave level, 21.77 m at the ridge level and 27.71 m at the dome level (see Fig. 4).

The foundation is made of natural stone bound with a lime-sand mortar and the superstructure is consists of structural masonry walls, made of different units of burnt clay bricks, solid and cored bricks, bound with a lime-cement mortar, and reinforced concrete (RC) slabs. The RC slabs transfer the loads directly to the masonry walls or indirectly by means of a system of RC beams. The strip footings beneath the walls have the thickness ranging from 0.75 m to 1.95 m, and the foundation depth is of 1.60 m from the basement slab level. The superstructure walls' thickness varies from 0.90 m to 1.80 m. The reduced thickness of the slabs, of 8 cm to 10 cm, cannot provide a satisfactory contribution to the lateral resistance of the structure [16].

The basement of 2.85 m height includes a technical room, a winery, a kitchen and bathrooms. The ground floor has a height of 5.56 m and large open spaces with big windows, and it lately hosted a bookstore and an art gallery. The upper floors have different heights, 4.3 m, 3.85 m, and 3.6 m respectively, with load bearing masonry walls and reinforced concrete slabs. The wooden roof structure and the dome were severely damaged and even reconstructed along the palace lifetime, mainly after the 1940 and 1977 Romanian earthquakes, and lately due to the partially destroyed metal roof covering [16].

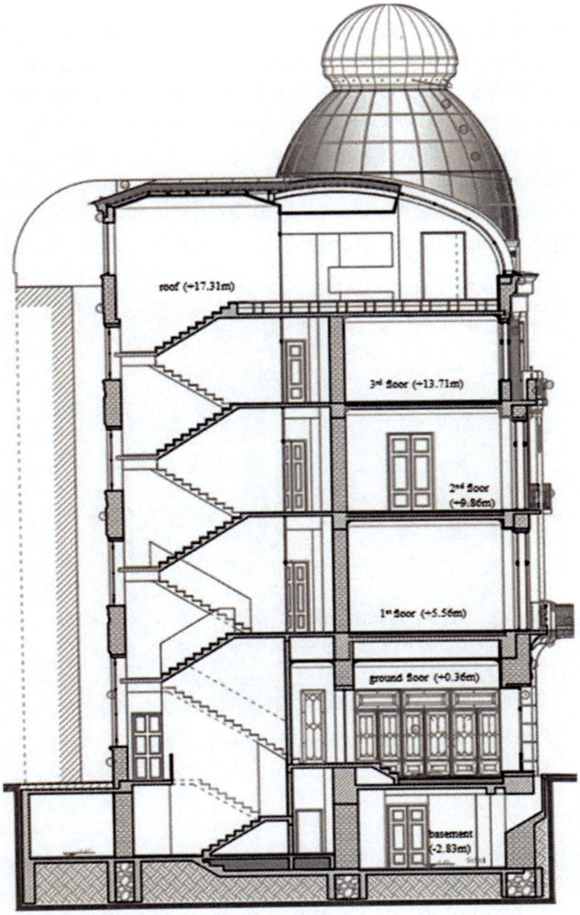


**Fig. 4.** Building transverse cross-section [16]

## 2.3 Building Damage Assessment

According to the Romanian design provisions for earthquake resistant structures [12], the building is placed in an area with moderate seismicity, with the design peak ground acceleration (PGA) of 0.25 g (2.452 $m/s^2$) and the corner period ($T_C$) of 0.7 s. It is ranked in the second importance class, being a cultural heritage building, and in the first class of seismic risk [16, 19], meaning that it is supposed to probably collapse or to be significantly damaged in case of a future earthquake larger than or similar to the 1977 Vrancea earthquake, and urgent structural interventions are essential [20].

The past intervention measures applied to the palace, especially after the two strong Romanian earthquakes from 1940 and 1977, and after the bombing during World War II, could not be performed in accordance with any standardized methodology of heritage buildings' restoration and strengthening [2, 10]. These interventions haven't been at least preceded by an adequate damage evaluation process and haven't been executed conforming to a rehabilitation project. Unfortunately, these measures contributed to a progression of the damage state. They included: the exterior connection of the dome to the palace façade by means of some steel strips; the cracks repairing with inadequate filling mortar; local replacements of the masonry units with new types of bricks and mortar; plaster, coating and architectural finishes improper replacement; local concrete underpinning of the footings; improper repairing of the wooden roof structure and of the steel covering [16]. The next figure illustrates the palace after the World War II (Fig. 5).

**Fig. 5.** The central palace after the World War II [16]

A recent expert survey of the damages [16] revealed some important issues that will be hereinafter mentioned.

The foundation soil is a high compressible silty clay which actually has an increased moisture content. The level of the underground water is very close to the footings. The foundation soil suffered along the years differential settlements. The super-structure cracks have been initiated from the foundation structure (see Fig. 6).

The open ground floor with some vertical RC elements and the various compartmentalization of the interior masonry walls form the upper stories do not provide the vertical regularity and that increases the seismic vulnerability of the palace [21]. The RC slabs of reduced thickness cannot behave as diaphragms to properly withstand horizontal earthquake forces. The masonry walls are not homogeneous structures since different types of masonry units have been used or replaced along the years: solid bricks and cored bricks (see Fig. 7). The utilized lime-cement mortar has reduced mechanical properties and the overall strength and stiffness of the masonry walls has been also impaired by the construction of some chimneys inside their structure.

The plaster exfoliation caused by the environmental conditions determined the corrosion of the reinforcement, as Fig. 8 illustrates, and the severe damage of the masonry elements and of the exterior architectural finishes, as shown in Fig. 9.

**Fig. 6.** Cracks developed in the concrete cover and in the elements of the super-structure [16]

**Fig. 7.** Plaster exfoliation revealing the non-homogeneity of the masonry walls [16]

**Fig. 8.** Plaster efflorescence and exfoliation followed by steel reinforcement corrosion

The integrity of the wooden roof structure and of the dome are badly affected by the water leaks and by the lack of maintenance works, which are so necessary in case of wood structural elements and of metal coverings (see Fig. 11).

**Fig. 9.** The severe damage of the masonry elements and of the architectural finishes [16]

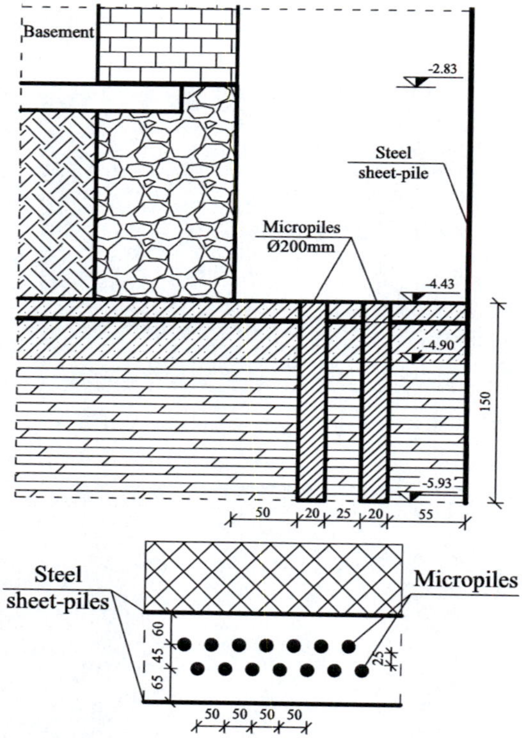


**Fig. 10.** a) The proposed soil stabilization solution; b) micropiles arrangement [16]

## 2.4 Structural Rehabilitation Solutions

The underlying soil stabilization aiming to improve its strength and to increase the resistance to softening by water, can be achieved by jet grouting process. It consists of creating some micropiles under the existing foundation level, which have to be embedded in the good foundation soil (Fig. 10a). They need to be executed on the exterior contour

**Fig. 11.** a) The dome metal covering badly affected by the environmental conditions [16]; b) current rehabilitation stage

of the building, on two rows at a specified distance, so that to form in plane equilateral triangles (Fig. 10b). This process has to be very carefully executed so that to avoid any additional settlement or any local failure of the existing footings. The previous strengthening of the stone footings by cement injection is highly recommended [22].

The infrastructure consolidation is needed to increase the strength, stability and stiffness of the affected footings due to the differential settlements. This can be performed by using a framing system which consists of encasing the existing natural stone footings in a new reinforced concrete system with inferior and superior lateral ring beams vertically connected like the Vierendeel trusses and transversally connected together, as showed in Fig. 12 [23, 24]. The compound effect between the new RC system and the old foundations can be ensured by using some steel rods and grout or a special adhesive. The use of this kind of framing system has the advantage of enhancing the moisture transfer from the masonry walls. The construction of some additional transverse reinforced concrete beams is indicated for an improved load transfer at the foundation level and to ensure a better behaviour to any possible differential settlement.

The water proofing of the foundation and of the basement walls is also required because of the high level of the underground water and of the increased moisture content of the foundation soil [16].

In case of super-structure masonry walls, serious decay of the mortar joints and of the bricks have been observed on the façade of the palace. The repointing of the eroded joints, the mortar or grout injection of the occurred cracks and local repairs or even partial replacement of the masonry units which are deteriorated beyond repair are necessary. It is of utmost importance to previously determine the physical, chemical and mechanical properties of the existing materials and to use new ones with similar size, colour, texture and durability, as much as possible. The cleaning process has to be carefully done and only in justified circumstances so that not to induce more harm to the masonry [25].

Due to the overall weakening of the structural system of the building, partial concreting or the construction of some new reinforced concrete elements may be needed, such as: tie-columns, ring beams, lintels, corner area bindings, vertical reinforced concrete strips. The spatial rigidity improvement can be achieved by using a cross-tying system

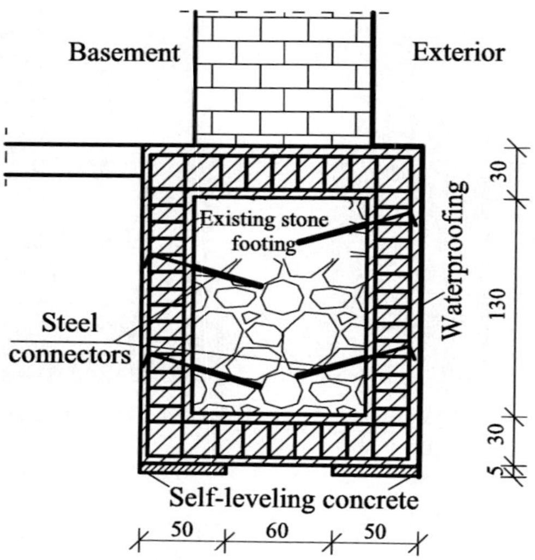


**Fig. 12.** Strengthening solution of the stone footings [16]

and by inserting some vertical RC elements (lamellar cross-section of 15–20 cm thickness) inside the structure of the load bearing masonry walls on the inner face, as shown in Fig. 13. At the same time, the chimneys embedded in the walls need to be removed and their structure to be reconstructed in order to improve their load bearing capacity. The use of a ring beam at the roof level is also needed to further enhance the structural stability. All these intervention measures have to be meticulously done so that not to impair much more the initial architectural features of the palace.

In order to gain a better behaviour to lateral loads, a reinforced topping concrete, with thickness in the range of 5–10 cm, can be placed on the RC slabs. Before pouring the topping concrete it is very important to obtain a perfect clean and smooth surface. This can require a previous grout injection of the existing cracks or the repairing of the concrete segregations, if they will be discovered.

The interior and exterior decorative plasterworks and architectural finishes require a thorough restoration and reconstruction process, paying careful attention to the selection of the used materials and techniques in order to preserve the original Art-Nouveau style of the palace.

Unfortunately, the severely damaged structure of the roof and of the dome cannot be saved. The entire wooden structure is intended to be replaced by a steel structure and the dome to be covered with a glass cladding system. The destination of the obtained space inside the dome is that of an art gallery, as the initial owner desired.

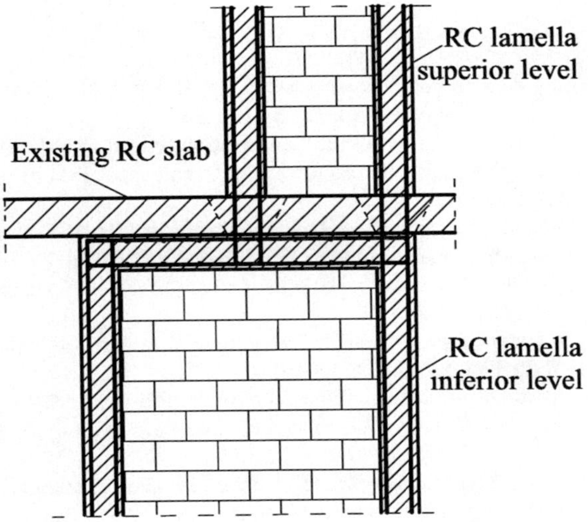


**Fig. 13.** Strengthening solution of the masonry walls with additional RC elements [16]

## 3 Conclusions

The intervention measures needed to restore the historic buildings or the architecture monuments are very complex and have to be carefully selected and performed. They can range from simple repairs or restoration processes to the consolidation or the rehabilitation of the affected structural members or of the entire structural system when the safety level of the considered building is significantly decreased due to the suffered damages.

The central palace from Iasi, located in the historic centre of the city, was exposed along the years to major events, such as floods, earthquakes, the city bombing during World War II, but also to various environmental conditions, and the lack of maintenance works or the wrongly applied intervention measures progressively contributed to its actual advanced damage state.

Wide consolidation and restoration works are now undertaken aiming to strengthen the entire structural system of the palace and to improve its stability in the event of a strong seismic action, but not only. Also the foundation soil necessitates stabilization for the improvement of its strength and to increase its resistance to softening by water, taking into consideration that the ground water level is very close to the footings. All these interventions are performed in accordance with the currently standardized norms and methodologies related to the conservation, restoration and strengthening of heritage buildings.

The aim of the local authorities is to restore the historic status of this beautiful architecture monument to become again a valuable landmark for the local cultural heritage of Iasi city. A media library, some art exhibition galleries and some cultural centres of the city will be hosted by the newly rehabilitated palace.

## References


1. Avrami, E.: Making historic preservation sustainable. J. Am. Plan. Assoc. **82**(2), 104–112 (2016). https://doi.org/10.1080/01944363.2015.1126196
2. Venice-Charter: The Venice Charter for the Conservation and Restoration of Monuments and Sites. The second international congress of architects and specialists of historic buildings, Venice, Italy (1964)
3. ICOMOS Hrsg.: The international council on monuments and sites. Heritage at Risk. World Report 2014–2015 on Monuments and Sites in Danger, pp. 8–10 (2016)
4. LNCS Homepage. https://www.cammconstruction.com/blog/benefits-historic-building-restoration/. Accessed 12 Oct 2019
5. Subramaniam, S.R.: A review on repair and rehabilitation of heritage buildings. Int. Res. J. Eng. Technol. (IRJET) **3**(4), 1330–1336 (2016)
6. LNCS Homepage. https://commons.wikimedia.org/wiki/File:Ia%C5%9Fi_Colaj.jpg, Bogdan Muraru [CC BY-SA 4.0]. https://creativecommons.org/licenses/by-sa/4.0. Accessed 22 Oct 2019
7. Pleşu, R.: Contributions Regarding the Rehabilitation Methods of Mansonry Structures. Ph.D. Thesis, Iasi, Romania (2013). (in Romanian)
8. Vicente, R., et al.: Seismic vulnerability and risk assessment of historic masonry buildings. In: Costa, A., Guedes, J., Varum, H. (eds.) Structural Rehabilitation of Old Buildings. Building Pathology and Rehabilitation, vol. 2, pp. 307–348. Springer, Heidelberg (2014). https://doi.org/10.1007/978-3-642-39686-1_11
9. Soveja, L.: Structural Restoration of Orthodox Cultural Buildings. "Matei-Teiu Botez" Academic Society Publishing House, Iasi, Romania (2016) (in Romanian)
10. MP 025-04: Methodologies of Risk Evaluation and Strengthening Solutions for Heritage Buildings. Directive of the Ministry of Transportation, Construction and Tourism, No. 743, Bucharest, Romania (2004). (in Romanian)
11. De Felice, G., De Santis, S., Lourenco, P., Mendes, N.: Methods and challenges for the seismic assessment of historic masonry structures. Int. J. Archit. Herit. **11**(1), 143–160 (2017). https://doi.org/10.1080/15583058.2016.1238976
12. P100-1/2013: Code of Seismic Design. Part I – Rules for Seismic Design of Buildings. Directive of the Ministry of Regional Development and Constructions, No. 489. Bucharest, Romania (2013). (in Romanian)
13. Lupăşteanu, V., Soveja, L., Lupăşteanu, R., Chingălată, C.: Installation of a base isolation system made of friction pendulum sliding isolators in a historic masonry orthodox church. Eng. Struct. **188**, 369–381 (2019)
14. Barbu (Mogosanu), A.M., Calbureanu, M.X.: Contributions above the rehabilitation of heritage buildings in advanced degradation - precollapse due to the action of environmental factors and strong earthquakes occurred during their lifetime. In: International Multidisciplinary Scientific GeoConference: SGEM, Sofia, vol. 18, no. 1.1 (2018). https://doi.org/10.5593/sgem2018/1.1/s05.093
15. LNCS Homepage. http://www.cultura.ro/sites/default/files/inline-files/LMI-IS.pdf. Accessed 22 Oct 2019
16. Cusutura, S.: Structural Rehabilitation of Braunstain Palace. Dissertation Thesis, Iasi, Romania (2015). (in Romanian)
17. LNCS Homepage. https://www.facebook.com/plugins/post.php?href=https%3A%2F%2Fwww.facebook.com%2F229031573813414%2Fphotos%2Fa.248593048523933%2F1786371734746049%2F%3Ftype%3D3&width=500. Accessed 26 Oct 2019
18. LNCS Homepage. https://en.wikipedia.org/wiki/File%3ARO_IS_Iasi_%2C_Braunstein_Palace.jpg, Argenna [CC BY 3.0] https://commons.wikimedia.org/w/index.php?curid=8755848. Accessed 29 Oct 2019

19. P100-3/2019: Code of Seismic Design. Part III – Provisions for Seismic Assessment of Existing Buildings. Directive of the Ministry of Regional Development and Public Administration No. 2834. Bucharest, Romania (2019). (in Romanian)
20. Lungu, D., Arion, C.: Seismic risk and/or real estate risk for the bucharest heritage buildings. In: 15th World Conference of Earthquake Engineering Proceedings, Lisbon, Portugalia, 10 p. (2012)
21. Bhosale, A., Davis, R., Sarkar, P.: Vertical irregularity of buildings: regularity index versus seismic risk. ASCE-ASME J. Risk Uncertain. Eng. Syst. Part A Civ. Eng. **3**, 04017001 (2017)
22. Przewłócki, J., Zielinska, M.: Analysis of the behavior of foundations of historical buildings. Procedia Eng. **161**, 362–367 (2016)
23. Vicente, R., Lagomarsino, S., Ferreira, T.M., Cattari, S., Mendes da Silva, J.A.R.: Cultural heritage monuments and historical buildings: conservation works and structural retrofitting. In: Costa, A., Arêde, A., Varum, H. (eds.) Strengthening and Retrofitting of Existing Structures. BPR, vol. 9, pp. 25–57. Springer, Singapore (2018). https://doi.org/10.1007/978-981-10-5858-5_2
24. Lupăşteanu, R., Soveja, L., Lupăşteanu, V.: Structural rehabilitation of old masonry churches. a cost-oriented case study. Bull. Polytech. Inst. Jassy. Constr. Architect. Sect. **LXI/LXV**(2), 27–42 (2015)
25. Covatariu, D., Lungu, I., Ţăranu, N., Budescu, M., Toma, I.O.: The influence of joints rehabilitation on structural response of masonry with low-strength mortars. Roman. J. Mater. **43**(3), 251–262 (2013)

# Geometrical Analysis of the Frame Pillars of the Cathedral of Orleans Resulting from 3D Survey

Laurent Josserand and Xavier Brunetaud(✉)

University of Orleans, Polytech Orleans, LaMé, 8 Léonard de Vinci St, 45072 Orleans, France
xavier.brunetaud@univ-orleans.fr

**Abstract.** A collaborative study concerning the cathedral of Orleans in France is in progress. This study involves historical and archeological aspects as well as a global 3D survey. Yet, about 400 laser scanning stations were performed to progressively build the digital replica of the cathedral. After scanning registration, architectural plans (vertical and horizontal) of the gothic cathedral and its roman remains, including basement, were extracted.

First, from these architectural documents, a frame analysis was conducted to improve our knowledge of the construction of the cathedral. This frame analysis highlighted some changes in the architecture project during construction. It also gave us data about various technical aspects of construction like stonework implementation and distance units.

Secondly, the comparison between different horizontal cuts of the central nave pillars allowed us to measure the horizontal displacement of these structural elements. These measurements can help identifying singular behavior and document degraded elements, and can also serve as references for the monitoring of the displacements of the structural elements of the cathedral.

**Keywords:** 3D survey · Laser scanning · Cathedral

## 1 Introduction

At the end of the thirteenth century, the Cathedral of the Holy Cross, in a roman style, is evoked in a state of decay with mentions of cracks and structural collapses. At the same time, in the surroundings, many religious buildings are taken over, completed or built in a new style, more airy, more luminous which has been called the flamboyant Gothic style. The idea of the need to rebuild the cathedral is emerging, more in keeping with the image that one had of the prestigious character of the True Cross relic and of the building that was to house it.

In 1287 the first stone was laid on the new bedside of the Orleans cathedral. As is often the case, work begins at the bedside. Indeed, the erection of a Gothic bedside outside the Romans cathedral has the great advantage of not disturbing the services inside it. Figure 2 shows this new cult building.


A. Rotaru (Ed.): CRIT-RE-BUILT 2019, SSGG, pp. 568–575, 2021.
https://doi.org/10.1007/978-3-030-61118-7_46

The remains of the Romans building are visible in the present cathedral basements. Figure 1 below, although not perfectly realistic, gives an idea of the volumes of this cathedral. In fact, to speak only of the cathedral district, one must imagine Orleans's cathedral surrounded by various buildings, such as canonical houses, at least one barn of which there remain the «cellars of the chapter», and the Episcopal Palace. The bishop gave the location of this palace for the benefit of the future choir of the new building.

**Fig. 1.** Synthetic image [1] of the Orleans cathedral of Sainte Croix in roman style (before the end of the thirteenth century)

The aim of the collaborative research is to improve the understanding of this architectural transformation in various aspects. Among other studies, we want to provide data on the initial architectural project and its changes over time, and an assessment their consequence on the overall stability of the building.

## 2 Method

3D survey is becoming popular to contribute to the documentation of archeological and historical studies. To perform 3D survey, different techniques can be applied, such as laser scanning and photogrammetry [2–10]. For the digitalization of the geometry of complex architectural sites such as cathedral, laser scanning is known to be more robust and precise than photogrammetry, especially for inner parts where laser scanning is rather easy to perform.

To detect singularities in the geometry of architectural elements, different approaches are possible. Direct comparison between different temporal states can be quantified using

**Fig. 2.** Recent photographic view of the Orleans cathedral (personal photograph)

distance maps [11], but this approach cannot be applied here because no previous 3D survey is available for comparison. Comparison between the actual geometry and a theoretically perfect one is also a possible strategy. However, it requires either the simplification of the actual geometry, which generates a conceptual bias, or makes it difficult to compare a complex geometry with a simplified primitive shape due to singularities induced by ornaments or stylistic changes. Manual extraction of geometrical features fitted on point clouds or meshes of the monument, such as symmetrical axis of nave [4], can provide a basis for a global geometrical analysis. This study is based on the latest approach.

First, about 400 different laser scanning stations were positioned both on the exterior and inside the building. Figure 3 shows the position of most laser stations. A Faro Focus 3D was used for this 3D survey. Scanning registration was performed with Recap, with significant proportion of manual positioning before automatic refinement. 3D point clouds were then exported into Autocad for the extraction of slices and length measurements. In this software, a manual vectorization of geometry was driven over the point cloud segmented in 15 cm thick slices (vertical or horizontal depending on the required representation). For the elevation of structural elements such as pillars, this operation was repeated every 2 m to provide sections. On each section, the symmetrical axes were manually identified to obtain the theoretical center of the pillar. Lines connecting the pillars centers were drawn to obtain the general frame of the building.

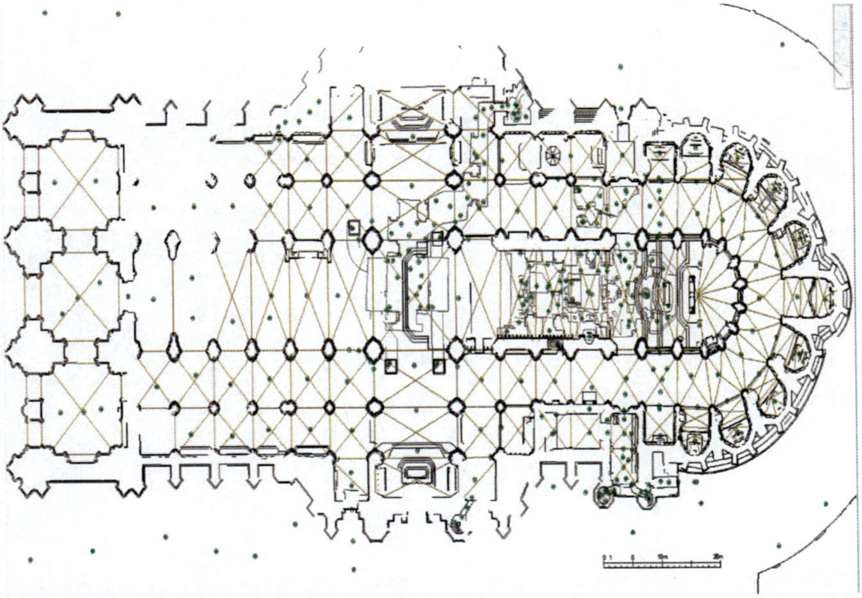

**Fig. 3.** Vectorized drawing of the cathedral showing the actual cathedral, the romans remains and on the other hand the position of most laser stations

## 3 Results and Discussion

### 3.1 Frame Analysis

First of all, we have classified the frame lines by orientation category. This Analysis of structural elements shows two building patterns which differ in their orientation (see Fig. 4). This difference reflects a change in the construction project during construction. More research should be conducted to determine the reason for this difference. Nevertheless, it indicates a chronological stage of the construction site. This step must be close to a stylistic change easily visible in the interiors of the apse. We can note that orientation B also corresponds to the main orientation of the romans remains.

### 3.2 Horizontal Displacements of Pillars

A second line of research was to investigate the origin of cracks which are found essentially at the transept (see Fig. 5). In order to check if structural displacements could be the origin of cracking, the verticality of the structure was analyzed. To achieve this goal, the position of the pillars centers at different heights were compared to the position of the bases. Results of displacement analysis is shown on Fig. 6.

The displacement is measured in the direction of the building center. The thrust of the main arch of the frame, although diminished by the holding of the outer buttresses, undoubtedly explains this lack of verticality. Figure 7 shows the resultant horizontal displacements measured at the pillar head. Most pillars stray under the thrust of the vault,

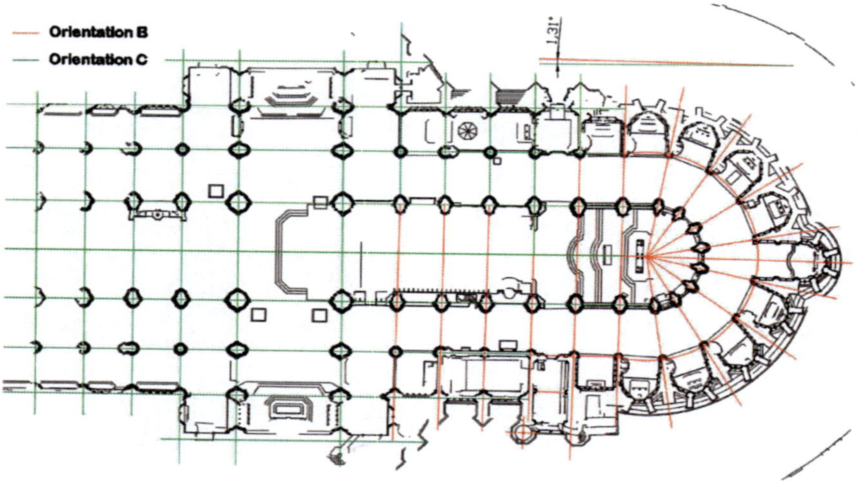


**Fig. 4.** Major building pattern of the cathedral

**Fig. 5.** Observation of joint openings in North and South Transepts, cracks are shown by red ellipses (personal photographs)

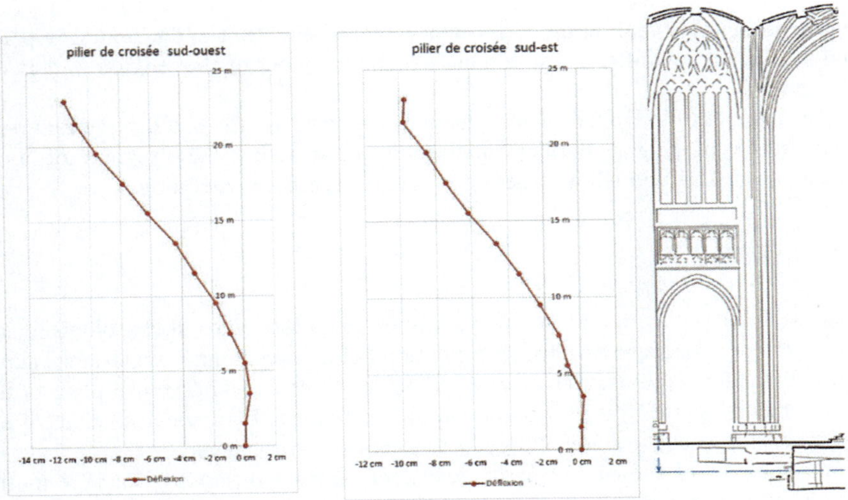


**Fig. 6.** Measurement of the horizontal displacement of the souths pillars with height

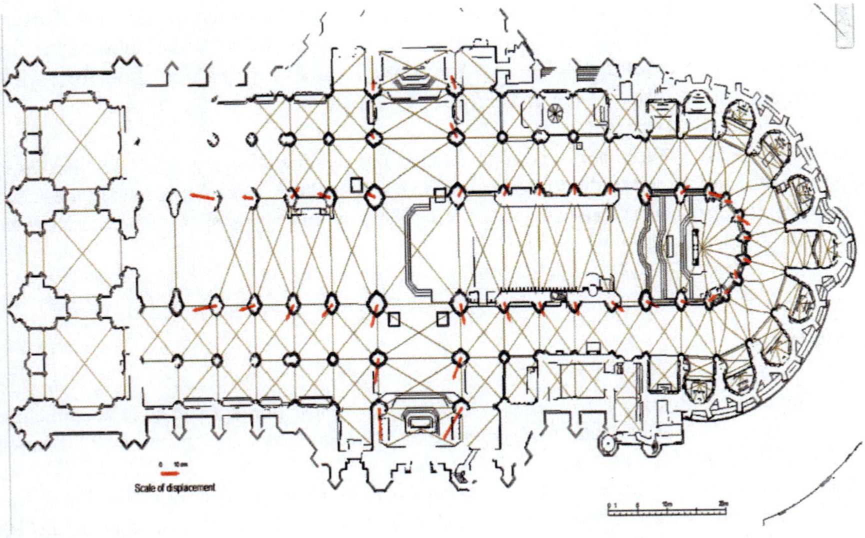


**Fig. 7.** General view of the movements observed at the head of the pillar in relation to its base

but not all. It is possible that successive repairs have impacted the structural behavior of the singular elements. A second hypothesis could lie in the removing sequence of the supporting scaffold. Unfortunately, there is no historical documentation about this sequence.

The temporal evolution of these deformations is also unknown to us since we only have this single measure. However, most of this defect are supposed to occur during the opening of the main vault in the 17th century. Indeed, the vault of the central ship and its pillars were rebuilt after a destruction of the building during the religion wars in the 16th century.

The maximum horizontal displacement is observed on the southwest pillar with 12 cm at 23 m height. Further monitoring could assess the progress of these displacements. This will allows reacting in case of significant evolution.

## 4 Conclusion

Based on a 3D survey, a collaborative study between historians, archaeologists, and civil engineers is in progress. Processing of point clouds resulting from laser scanning allowed extracting architectural documents such as vertical and horizontal plans of the gothic cathedral and its roman remains, as well as a series of horizontal sections of the main pillars to conduct a study of the global stability of the cathedral.

The analysis of the architectural document allowed observing changes in the construction phases that requires further study from the historian point of view.

The analysis of the horizontal displacement of the pillars showed significant displacement at their head, while presenting a vertical progression in relation with the surrounding structural elements. Most pillars stray under the thrust of the vault, but not all. It is possible that successive repairs may have changed structural behavior. This first measurement of horizontal displacement now constitutes a reference state for further monitoring of the stability of the cathedral of Orleans.

**Acknowledgements.** The authors wish to thank the students Julia Wuillequez and Antoine Dupré who participated in this part of the study. We also thank Nathalie Arbaret for managing the access to the cathedral. More generally, we thank the scientific community that shares its research on the cathedral.

## References

1. Josserand, L., Maury, F.: Recherche de la trame de construction du chevet de la cathédrale Sainte Croix d'Orléans. Bulletin de la Société archéologique et historique de l'Orléanais **XXII**(176), 5–14 (2017)
2. Fassi, F., Achille, C., Fregonese, L: Surveying and modelling the main spire of Milan Cathedral using multiple data sources. In: Extended Version of a Paper Presented at the ISPRS Commission V Symposium on "Close Range Image Measurement Techniques" Hosted by Newcastle University, 22nd to 24th June 2010 (2010). https://doi.org/10.1111/j.1477-9730.2011.00658.x
3. Grenzdörffer, G.J., Naumann, M., Niemeyer, F., Frank, A.: Symbiosis of UAS photogrammetry and TLS for surveying and 3D modeling of cultural heritage monuments – a case of study about the cathedral of St Nicholas in the city of Greifswald. In: The International Archives of the Photogrammetry, Remote Sensing and Spatial Information Sciences, 2015 International Conference on Unmanned Aerial Vehicles in Geomatics, 30 August–02 September 2015, Toronto, Canada, vol. XL-1/W4 (2015). https://doi.org/10.5194/isprsarchives-xl-1-w4-91-2015

4. Aita, D., Barsotti, R., Bennati, S., Caroti, G., Piemonte, A.: 3-dimensional geometric survey and structural modeling of the dome of Pisa Cathedral. In: The International Archives of the Photogrammetry, Remote Sensing and Spatial Information Sciences, 3D Virtual Reconstruction and Visualization of Complex Architectures, 1–3 March 2017, Nafplio, Greece, vol. XLII-2/W3 (2017). https://doi.org/10.5194/isprs-archives-xlii-2-w3-39-2017
5. Landes, T., Kuhnle, G., Bruna, R.: 3D modeling of the Strasbourg's cathedral basements for interdisciplinary research and virtual visits. In: The International Archives of the Photogrammetry, Remote Sensing and Spatial Information Sciences, 25th International CIPA Symposium 2015, 31 August 04 September 2015, Taipei, Taiwan, vol. XL-5/W7 (2015) https://doi.org/10.5194/isprsarchives-xl-5-w7-263-2015
6. Ragone, A., Ippolito, A., Liberatore, D., Sorrentino, L.: Emerging technologies for the seismic assessment of historical churches: the case of the bell tower of the Cathedral of Matera, Southern Italy (2017). https://doi.org/10.4018/978-1-5225-0675-1.ch006
7. Brunetaud, X., Stefani, C., Janvier Badosa, S., Beck, K., Al-Mukhtar, M.: Comparison between photomodeling and laser scanning applied to realizing 3D model for digital health record. Eur. J. Environ. Civil Eng. **16**(Suppl. 1), 48–63 (2012). https://doi.org/10.1080/19648189.2012.681957
8. Brunetaud, X., De Luca, L., Janvier-Badosa, S., Beck, K., Al-Mukhtar, M.: Application of digital techniques in monument preservation. Eur. J. Environ. Civil Eng. **16**(5), 543–556 (2012). https://doi.org/10.1080/19648189.2012.676365
9. Al-Omari, A., Brunetaud, X., Beck, K., Al-Mukhtar, M.: Preliminary digital health record of limestone in Al-Ziggurat, Al-Nimrud city, Iraq. J. Cult. Herit. **16**(5), 737–740 (2015). https://doi.org/10.1016/j.culher.2014.11.005
10. Stefani, C., Brunetaud, X., Janvier-Badosa, S., Beck, K., De Luca, L., Al-Mukhtar, M.: Developing a toolkit for mapping and display stone alteration on a web-based documentation platform. J. Cult. Herit. **15**(1), 1–9 (2014). https://doi.org/10.1016/j.culher.2013.01.011
11. Janvier, R., Brunetaud, X., Janvier-Badosa, S., Beck, K., Al-Mukhtar, M.: The potential of laser scanning to describe stone degradation forms. In: 13th International Congress on the Deterioration and Conservation of Stone (STONE 2016), Glasgow, Scotland, 6–10 September 2016 (2016)

# Structural Assessment of Mehmed Pasha Hammam in Prizren, Kosovo

Enea Mustafaraj(✉) and Jetmir Morina

Epoka University, Tirana, Albania
emustafaraj@epoka.edu.al

**Abstract.** Historical structures, built through centuries, designated mainly for religious and recreational purposes have had a significant role in the socio-cultural heritage of Kosovo. Many of these buildings such as mosques, hammams, houses, etc. still present nowadays, were built during Ottoman era (1445–1912) using stone masonry. During their lifespan, these structures have been subjected to several seismic activities, deterioration of construction materials, weather conditions and lack of maintenance.

In this paper, it is presented an approach on structural assessment of Mehmed Pasha Hammam, a heritage structure built during 16$^{th}$ century in Prizren, Kosovo. The methodology is based on visual inspection of the structural condition, identification and documentation of the visible defects, as well as finite element modelling, using SAP2000, to study structural behavior under earthquake loadings.

As a result, structural and non-structural damages were identified, critical areas were highlighted and the global behavior of the structure under seismic loading conditions was simulated. Understanding the structural response of this structure would make it easier to suggest retrofitting solutions, extend the service life and preserve this structure for future generations.

**Keywords:** Stone masonry · Finite element modeling · Structural assessment

## 1 Introduction

Historical structures involve representative values and knowledge for a certain era, place, local or wider community and their traditions, political and social background, as well as methods of construction and technologies used [1].

Preservation of cultural heritage has become an emerging topic in the last decades, in particular the historical monuments. Many of these structures were constructed in the past based on the masters' experience and knowledge when there was neither modern scientific research nor designs standards. During their lifespan they have been exposed to various loading and occupancy conditions, materials deterioration, earthquakes, man-made interventions and other natural disasters. Thus, there is a need to assess their structural stability and determine most critical areas, in order to preserve these structures for the future generations.


A. Rotaru (Ed.): CRIT-RE-BUILT 2019, SSGG, pp. 576–586, 2021.
https://doi.org/10.1007/978-3-030-61118-7_47

However, assessment of seismic vulnerability for historic masonry constructions is a very challenging task and a proper evaluation of their deficiencies requires a particular set of skills due to several uncertainties regarding a material's mechanical properties and geometric characteristics of the structure.

## 2 Materials and Methods

Retrofitting, repair and strengthening of historical structures attempt mitigate the associated hazards from the deterioration of building due to age, as well as those from natural disasters. The primary objective of the repair and strengthening of structures is to ensure improved load-resisting capacities, and extension of their service life.

Having survived natural disasters and other hazards to this day, many unreinforced masonry (URM) buildings have shown that the combination of their respective material and structural properties have resisted the effects of ground-shaking as well as deterioration in time.

Careful planning, as well as the task of implementation of those plans are paramount for every intervention. For the strengthening and repair interventions to be successful, other than meeting the aims in terms of structural performance, they must not alter the original shape, features and architectural appearance of the building, and even in cases where the addition of new structural elements is necessary, they must have similar appearance and features as the original members [1].

Each historical URM structure is unique, thus the process of assessing the seismic vulnerability is a very challenging task as it involves many uncertainties in terms of the characteristics of the materials used, geometrical properties, etc. To perform a correct structural analysis a lot of data is required. Background on the construction history and evolution of the structure, geometrical data, properties of materials, structural details, construction methods as well as crack patterns are all required in order to obtain results as accurately as possible. The combination of results from a series of laboratory and in-situ experiments can be a good way of developing a good representative structural model. However, obtaining all the needed data to properly define the numerical model is very difficult, or even impossible in some cases. For this reason, a simplified and iterative procedures of assessment are needed.

Structural assessment of historical structures has been the subject of many researchers [2–19]. In their studies, combining visual inspections, non-destructive techniques (NDT) and laboratory tests, with various finite element modelling (FEM) approaches, they have successfully conducted structural interventions which have increased structural resistance of historical structures, and eventually have extended their life cycle.

In Kosovo, there is a lack of research studies about structural assessment, strengthening and repair of heritage structures. For this reason, the aim of this paper is to assess the structural performance of a representative structure under static and dynamic loading and to identify the most vulnerable areas.

## 2.1 Case Study – Mehmed Pasha Hammam

Gazi Mehmed Pasha Hammam is located in the urban center of the ancient town Prizren, Kosovo. Together with that of Daut Pasha in Skopje, and Nusret Bey in Sarajevo, were built during the most successful era (1437–1703) of Ottoman architecture in the Balkans [3].

The Hammam dates back to 1573–74 and it is comprised of two main parts; eastern entrance, dedicated for men, and western entrance, dedicated for women (see Fig. 1 and 2). It has two areas with domes, using Turkish tiles, with an octagonal shape.

**Fig. 1.** Location of Mehmed Pasha Hammam.

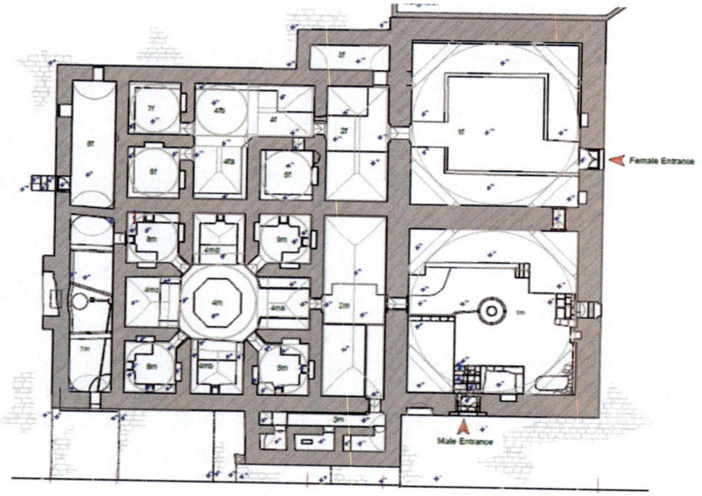


**Fig. 2.** Architectural floor plan of Mehmed Pasha Hammam.

Based on the plaque located on the main entrance of the building, major rehabilitations were done during 1833. The Hammam served until 1926. Around it, different stores were built on the southern and eastern part, whereas on the western part living houses were constructed.

This is one of the many URM historical structures in Kosovo which bear a very important role in the cultural and social heritage of the country, thus the need for further assessments and preservation of these monuments is of great significance.

## 2.2 Material Properties

Evaluation of the mechanical parameters of the masonry materials is a very crucial part for defining a numerical model. Other than the in-situ visual inspection of the structure for cracks, damages and other irregularities, it is of high importance to also collect data regarding mechanical properties of the materials used for construction (masonry) by means of laboratory examinations. Once the correct data is collected, appropriate models and analysis can be applied for evaluation of historical URM structure, and final diagnosis can be extracted, from which conclusions can be drawn with regards to best-suited and feasible repair and strengthening methods.

Due to the special status of the building as a historical monument, destructive testing was impossible to conduct. Thus, the mechanical parameters of bricks and stones were obtained from a literature survey and are presented in Table 1.

**Table 1.** Assumed material properties for the model [3].

| Properties | Brick | Stone |
|---|---|---|
| Unit weight, $\gamma$ (kN/m$^3$) | 17 | 24 |
| Modulus of elasticity, E (MPa) | 2100 | 3000 |
| Compressive strength, [$f_{c,\ allowable}$] (MPa) | 1.54 | 30.5 |
| Tensile strength, [$f_{t,\ allowable}$] (MPa) | 6.42 | 3.55 |
| Void r, [$\upsilon$] | 0.2 | 0.2 |

## 2.3 FEM Modeling of Mehmed Pasha Hammam

FEM modeling of the structure was carried out using SAP2000 v16 software [20], following the macro-modeling approach. This technique is the most used strategy for studying large structures and the effect of global factors. In this strategy, the whole masonry is considered as one continuum element with homogenous material properties, showing elastic behavior. This method is suitable for numerical and for analytical modeling as well, since it does not require detailed modeling into separate discontinued elements.

In order to proceed with the modeling, firstly geometrical data is required. For historical URM structures that are very old, usually laser measurements are performed to obtain the geometry of the structure, due to the lack of data. However, in this study we obtained the architectural drawings from the archives of the municipality of Prizren.

The hammam is a one-story structure with the big domes reaching a maximum elevation point of 11 m. The total area of the whole structure is approximately 450 $m^2$. The roof consists of 2 big, 1 medium-sized and 8 small domes, as well as 2 big vaults and 10 smaller ones.

Due to its complex shape, the modeling of Mehmed Pasha Hammam was quite a challenging task.

The outer and inner walls are made of stone, whereas the whole roofing structure, which includes domes, vaults, pendentives and arches are made of bricks. All structural members are modeled as shell elements with a three- or four-node formulation that combines separate membrane and plate-bending behavior.

In total the model consists of 14462 joints and 15953 shell elements (Fig. 3).

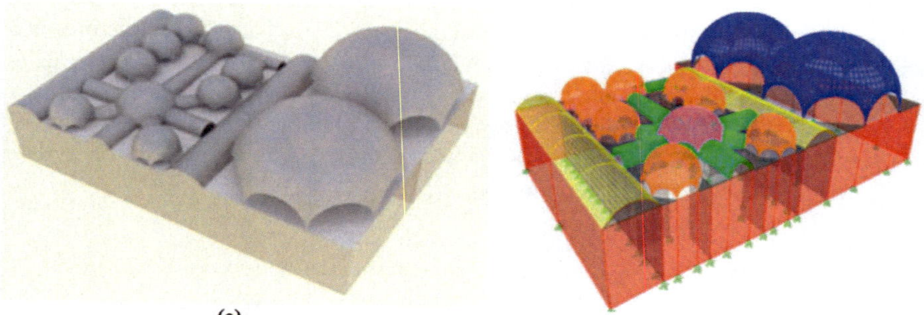

(a) (b)

**Fig. 3.** 3-D model of Mehmed Pasha Hammam: (a) AutoCAD and (b) SAP2000 v.16.

## 3 Results and Discussion

In order to fully assess the URM structure of Mehmed Pasha Hammam both, visual inspections, as well as FEM analysis were carried out.

The former analysis enabled us to identify visual deficiencies of the structure such as cracks, deformations and irregularities, as well as leaning of the structural elements. Whereas, FEM analysis was performed in order to define and examine the behavior of the undamaged structural model under both, static and dynamic loads, as well as identifying the regions of the structure with the highest stress concentrations and distribution of secondary stresses.

### 3.1 Visual Inspection of the Mehmed Pasha Hammam

From the visual assessment of the Mehmed Pasha Hammam it is observed that the most damaged part of the structure is the roof. There are many cracks on the domes and vaults ranging from superficial cracks (plaster cracks, with no direct structural risks), to some severe cracks, mainly found on one of the big domes, and on the big vault. There are visible cracks on the pendentives too, which extend to the vertical load-bearing walls.

The main reasons for these structural cracks are believed to be multiple earthquakes, the amortization of the structure and deterioration due to lack of maintenance.

There are also other deficiencies such as spalling of plaster, presence of molds and vegetation, high level of humidity and water leakage due to improper isolation of the roof. Moreover, due to foundation settlement, there are observed out-of-plane deformations on the vertical walls (Fig. 4).

**Fig. 4.** Cracks on the big vaults Cracks along one of the small vaults. Cracks and molding on one of the small domes, its pendentive and along the vertical wall.

### 3.2 FEM Analysis Results

The FE model of the hammam was analyzed under real-life scenarios under specific loading cases where various parameters such as stress distribution, maximum deflections, period of vibration, etc., were obtained.

From the linear static analysis, the maximum compressive stress values generated are 2.351 MPa (S22 & SMin, respectively) and they are both located on the big dome near the main entrance. The maximum tensile stress value reaches 2.1 MPa (SMaxV) on the big dome near the entrance for men.

In Table 2 it is presented the modal eigenvector results of the free-vibration mode shapes of the structure and the corresponding period.

The response spectrum analysis results considering the earthquake loads, showed the deformed shapes and the vulnerability of the structure when subject to a combination of static and dynamic loads. The two load combinations with respect to response spectrum analysis are:

$$\text{COMB1} = 0.9 * \text{Dead Load} + \text{EQx} + 0.3 * \text{EQy}, \quad (1)$$

**Table 2.** First five modes and their respective modal shapes.

| Mode | Period, T (s) | Modal shapes |
|---|---|---|
| Mode 1 | 0.342 | E-W LATERAL |
| Mode 2 | 0.341 | SQUEEZING |
| Mode 3 | 0.314 | BREATHING |
| Mode 4 | 0.272 | N-S LATERAL |
| Mode 5 | 0.222 | TORSIONAL |

$$\text{COMB2} = 0.9 * \text{Dead Load} + 0.3 * \text{EQx} + \text{EQy} \tag{2}$$

From results summarized in Table 3, it is observed that the worst-case scenario stresses for both tensile and compressive loads, as well as in both, X and Y directions are generated by radial stresses (S22).

**Table 3.** Stress values under response spectrum analysis.

| Seismic loads | S11 (MPa) | S22 (MPa) | S12 (MPa) | S13 (MPa) | S23 (MPa) |
|---|---|---|---|---|---|
| EQx | −2.792 | −6.025 | −1.064 | −2.114 | −2.985 |
| | 2.829 | 3.849 | 1.056 | 1.907 | 3.407 |
| EQy | −2.374 | −5.704 | −1.079 | −1.682 | −1.583 |
| | 2.143 | 2.781 | 1.024 | 1.657 | 1.804 |

With regards to seismic loads is exhibited in the X-direction, the maximum tensile stress value is (3.849 MPa) in S22, reached on the big dome near the inner wall separating men and women areas. While the maximum compressive stress for EQx is 6.025 MPa, in S22, reached on the big dome covering the women's entrance (see Fig. 5 and 6).

In terms of seismic loads in Y-direction, it is observed that the maximum tensile stress value of 2.781 MPa, as well as the maximum compressive stress value of 5.704 MPa (both in S22) are reached on the big dome near the men's entrance.

The maximum displacement caused by seismic load in X-direction reaches a value of 6.064 mm and it is located at the intersection of the big dome (males' entrance side) and its pendentive.

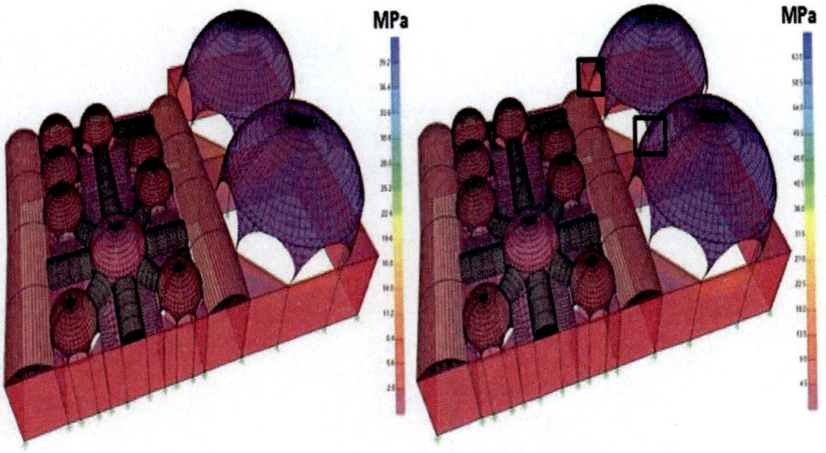


**Fig. 5.** S11 (left) and S22 (right) under seismic loads in X-direction.

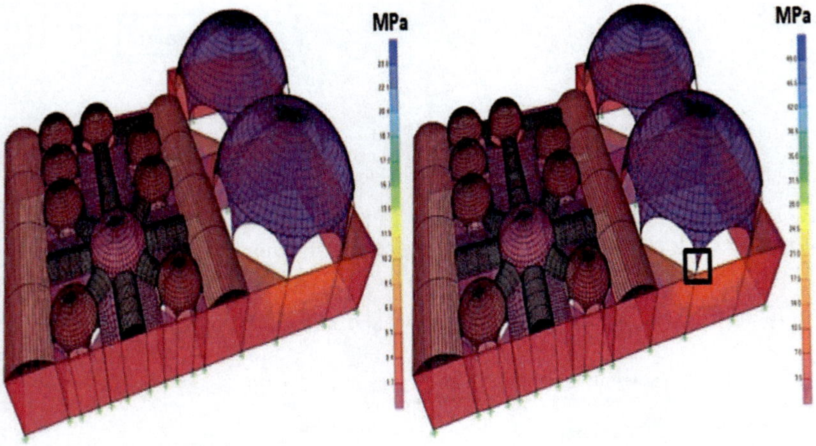


**Fig. 6.** S11 (left) and S22 (right) under seismic loads in Y-direction.

The maximum displacement caused by seismic load in Y-direction and reaches a value of 5.940 mm, which is located at the intersection of the big dome (females' entrance side) and its pendentive, whereas in X-direction the maximum displacement caused by EQy reaches 3.804 mm (see Fig. 7).

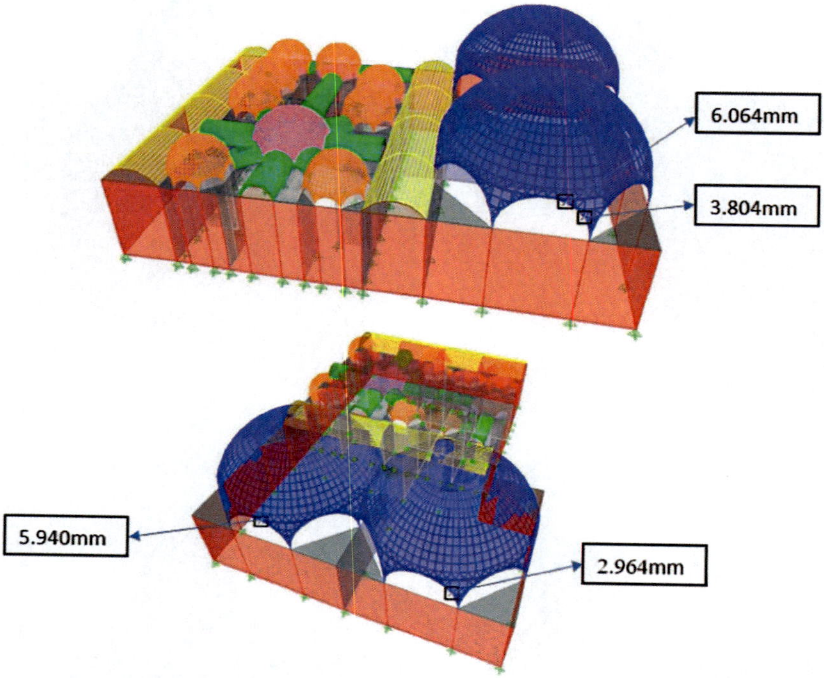


**Fig. 7.** Location of maximum displacements generated by EQx & EQy.

The displacement values for X and Y-directions are summarized in Table 4.

**Table 4.** Maximum displacements generated from response spectrum analysis.

| Seismic loads | x-direction (mm) | y-direction (mm) |
|---|---|---|
| EQx | 6.064 | 2.964 |
| EQy | 3.804 | 5.940 |

## 4 Conclusions

In this paper it is presented a structural assessment of a representative heritage structure of Ottoman Era (1445–1912), the Mehmed Pasha Hammam in Prizren, Kosovo. By means of visual inspection and finite element modelling using SAP2000 v16 software, the structure was analyzed, and structural deficiencies and structural behavior was observed.

From the visual inspection, the structural cracks were identified and addressed accordingly. They indicate that there is a need for intervention and repair. Additionally, the spall of plaster, water leakage, presence of vegetation, broken tiles etc., need to be addressed in order to extend the service life of the structure.

Using FEM analysis, it was defined and examined the behavior of the undamaged model under both, static and dynamic loads, as well as identifying the regions of the structure with the highest stress concentrations and distribution of secondary stresses.

The structure needs retrofitting in order to have a better performance under earthquake loadings.

## References

1. Bucur-Horváth, I., Papa, I., Bulbuk, M., Virag, J.: Historical Constructions - Authenticity and Adaptation to Modern Demands. Structural Analysis of Historic Construction, Taylor & Francis Group, London (2008)
2. Betti, M., Galano, L.: Seismic analysis of historic masonry buildings: The Vicarius Palace in Pescia (Italy). Buildings **2**(2), 63–82 (2012)
3. Morina, J.: Structural assessment of Mehmed Pasha Hammam in Prizren, Kosovo. Epoka University, MSc Thesis (2019)
4. Mustafaraj, E., Yardim, Y.: Determination of structural behavior of an unreinforced Masonry Clock Tower using FEM analysis. Am. J. Eng. Res. **5**(1), 100–104 (2016)
5. Yardim, Y., Mustafaraj, E.: Effects of soil settlement and deformed geometry on a historical structure. Nat. Hazards Earth Syst. Sci. **15**, 1051–1059 (2015)
6. Yardim, Y., Mustafaraj, E.: Structural assessment and restoration of the Leaden Mosque in Berat, Albania. KSU-J. Eng. Sci. **15**(2) (2012)
7. Barushi, K., Mustafaraj, E.: A case study on structural assessment and restoration of King Zog's villa in Durres, Albania. In: 2-BCCCE 2013, Epoka University, Albania, 23–25 May 2013 (2013)
8. Mustafaraj, E., Yardim, Y.: Repair and strengthening of historical structures: Naziresha's Mosque in Elbasan. In: 3rd Annual International Conference on Civil Engineering, Athens, Greece, 10–13 June 2013 (2013)
9. Mustafaraj, E.: Structural assessment of historical buildings: a case study of five Ottoman Mosques in Albania. In: 2-BCCCE 2013, Albania, 23–25 May 2013 (2013)
10. Yardim, Y., Mustafaraj, E.: Selected assessment and retrofitting application techniques for historical unreinforced masonry buildings. In: Asteris, P.G., Plevris, V. (eds.) Seismic Assessment and Rehabilitation of Historic Structures, vol. 1, pp. 257–292. IGI Global (2014)
11. Gulkan, P., Wasti, S.T.: Seismic assessment and rehabilitation of historic structures. Struct. Longevity J. **1**(2), 111–134 (2009)
12. Portioli, F., Mammana, O., Landolfo, R., Mazzolani, F.M.: Seismic assessment of historical Mosques under exceptional earthquakes: a case study in Skopje. In: Urban Habitat Constructions Under Catastrophic Events: Proceedings of the COST C26 Action Final Conference, Naples, pp. 415–420 (2010)
13. Ustundag, C., Sesigur, H., Cili, F.: Seismic evaluation and retrofit of the 16th century Mihrimah Sultan Mosque. In: 14th European Conference on Earthquake Engineering. Ohrid: 14WCEE, p. 308–314 (2010)
14. Gedik, Y.H., Celep, Z.: Earthquake analysis and strengthening of the historical Mehmet Aga Mosque in Istanbul. In: 14th World Conference on Earthquake Engineering, China (2008)
15. Giordano, E., Ferrante, A., Clementi, F., Milani, G., Formisano,A.: Cultural Heritage and earthquake: Tthe case study of San Francesco's church in Amandola (Central Italy), In: IMEKO TC-4 International Conference on Metrology for Archaeology and Cultural Heritage, Florence, Italy (2019)
16. Giaccone, D., Santamaria, U., Corradi, M.: An experimental study on the effect of water on historic Brickwork Masonry. Heritage **3**, 29–46 (2020)

17. Corradi, M., Castori, G., Sisti, R., Borri, A., Pesce, G.L.: Repair of block masonry panels with CFRP sheets. Materials **12**(15), 2363 (2019)
18. Castori, G., Sisti, R., Borri, A., Corradi, M., De Maria, A.: Seismic assessment of the palace of Priors in Perugia. In: Aguilar R., Torrealva D., Moreira S., Pando M.A., Ramos L.F. (eds.) Structural Analysis of Historical Constructions. RILEM Bookseries, vol. 18, pp. 1743–1750. Springer (2019)
19. Castori, G., Corradi, M., De Maria, A., Sisti, R., Borri, A.: A numerical study on seismic damage of masonry fortresses. Bull. Earthq. Eng. **16**(10), 4561–4580 (2018)
20. CSI: SAP2000 Linear and Nonlinear Static and Dynamic Analysis and Design of Three-Dimensional Structures. Computers and Structures (2012)

# Structural Rehabilitation of Historical Masonry Buildings

Mihaela Movilă(✉), Gogonea Mihaela, Ioana Olteanu-Donțov, and Oana-Mihaela Banu

Technical University Gheorghe Asachi of Iasi, Iasi, Romania
mihaela.movila@academic.tuiasi.ro

**Abstract.** The rehabilitation of heritage buildings represents a challenge from two perspectives: from architectural point of view, which involves the conservation, repair and/or replacement of damaged elements and from structural point of view by adopting solutions to increase its bearing capacity with minimum intervention on structure.

Masonry is one of the oldest, traditional construction materials, used for religious and civil architecture in the world. Beside numerous advantages, masonry has some disadvantages on long term, which may lead to severe damage and even unfavorable failure mechanisms. Given the high number of historical masonry buildings, the article makes a review of the specific problems for these types of structures and presents some specific methods of intervention.

The article also presents a case study, in which a heritage building is analyzed. The monument is located in the Moldavia region from Romania. Its main damages and the proposed rehabilitation methods are shown, in order to ensure the structural safety increase, and to obtain a target structural safety level for the structure.

**Keywords:** Masonry · Failure mechanism · Rehabilitation · Heritage · Structural analyze

## 1 Introduction

### 1.1 Heritage

Heritage is an achievement of the mind or the hand of a man who comes out of the ordinary, distinguishing itself by its value or magnitude. In architecture, the heritage term applies to an important building, such as the Egyptian pyramids, the Greek temples and byzantine, roman or gothic churches, renaissance palaces and so on. The entire heritage of the past culture has become the subject of concerns to protect and restore it. As a result of this enlargement, the monument term, was limited only to objects with a unique character.

However, the notion of historical heritage was extended also to urban areas of historical interest, to historical centers and even entire cities, although the great mass of


A. Rotaru (Ed.): CRIT-RE-BUILT 2019, SSGG, pp. 587–599, 2021.
https://doi.org/10.1007/978-3-030-61118-7_48

buildings, within the perimeter of these areas, give the historical and architectural value, compared to monuments with unique character.

Thus, concepts as historical area, historical center and historical city have developed along the years. In Fig. 1 an example of an historical center from Sibiu, Romania is presented [1]. Sibiu is one of the most important cultural centers of Romania and was awarded "the European Capital of Culture" in 2007, along with the city of Luxembourg.

**Fig. 1.** Historical center of Sibiu city [2]

In 2008, Forbes magazine ranked Sibiu as "Europe's 8th-most idyllic place to live". Sibiu city has one of the best-preserved historical sites in the country, many of its medieval fortifications having been kept in excellent state. Its old center has begun the process for becoming a UNESCO World Heritage Site in 2004.

### 1.2 The Masonry

The masonry has a long tradition and, for several hundreds of years, has provided empirically the most rational solution for the construction field. Throughout history, from modest houses up to palaces, churches, fortresses, bridges or aqueducts have been constructed by masonry.

Along the years, specific problems for masonry buildings had been noticed. Researchers worked hard to find appropriate solutions by permanently improving the technological progress. Some of the specific problems for masonry structures are:

- choosing the proper materials to build the masonry (natural stones or artificial);
- to associate the basic elements, so that the continuity of the masonry is ensured;
- achieving horizontal structures of coverage and separation between levels;
- making door and window opening;
- in seismic area, taking horizontal forces in addition to gravitational loads;
- the need to perform functions, other than the bearing, such as protection against climatic factors, but also aesthetic requirements.

In solving most of the above problems, the main limitation was represented by the weak tensile strength (practically negligible) of the masonry, resulting in a series of restrictions on the heights of the buildings, the distances between the supporting walls and the dimensions of the openings. On the other hands, civil engineer's creativity had offered various solutions to these problems, based on the available technical means from various historical eras [3].

Figure 2 presents an old photo of Noyon cathedral, France, with the gothic vault from the main nave. The vault is quadruped, that is, a vault module in 4 parts and covers the area between one 1 bay and one span from the church. The photo on the right gives some information regarding the execution technology used to build this challenging Gothic vault.

**(a)** **(b)**

**Fig. 2.** a.) Old photo of Noyon cathedral, France; b.) Construction stages for the quadruped vault [4]

## 1.3 Damages of Masonry

Masonry structures are vulnerable to water absorption and significant deterioration may occur if the drainage system is not fully functional. Another problem of heritage building is the susceptibility of failure to horizontal force caused by earthquake or wind. Being built in a period when seismic design codes did not exist, historic buildings are vulnerable to any kind of horizontal action and the absence of interventions on the structural system can lead to serious damages. Other structural damages can be produced by problems at the foundation soil which might overcome its bearing capacity [5].

The need to rehabilitate a building listed as heritage is obvious but, the applicable method and technology differs, producing numerous disagreements between the specialists involved. In case of masonry structures, the self-weight and the live loads are taken over by the walls. In other words, taller buildings require thicker walls, proportional to the number of levels. The lateral forces are taken over by the mortar between the brick or stone blocks. In case of masonry without mortar the lateral forces are balanced thanks to frictions between blocks. The erosion of the mortar, freeze-thaw cycles, rainwater not properly drained and the aging of the material cause loss of integrity and bearing capacity in case of masonry structure.

In cold climate areas, some types of masonry structures are more predisposed to degradation than others. Tall structures, with both sides of the masonry exposed to outside temperature, such as towers, chimneys, turrets, show strong damages caused by freeze-thaw cycles. Another factor which might speed up the erosion on the material is represented by the roof absence [6].

The structural damages of masonry structures can be grouped into direct and indirect. The direct degradations occur independently of damages to adjacent structures and causes deformation of the building leading to loss of bearing capacity. The direct causes of masonry degradation can be chemical, biological, but also thermal, such as the freeze-thaw phenomenon [7]. Deformations produced by mechanical actions generally occur due to overloading of the primary system of the structure. Dynamic forces from earthquake, wind or traffic actions produce major damage to the structures which are not designed to withstand such loads.

Indirect degradations are caused by the displacement of adjacent buildings. These damages occur at the foundation area due to soil deformations. Irregular settlement of the building, erosion of the foundation, lowering of groundwater level, produce cracks in the masonry. Major cracks and degradations of the structural system are also caused by extraordinary actions as fires, explosions, floods.

## 2 Heritage Building Assessment

A correct intervention to rehabilitate the structural system of a masonry heritage should have as a starting point a precise diagnosis of the building in order to minimize the interference of the structural intervention with history nature of the architecture.

The structural performance of an old masonry building can be understood and quantified by identifying the construction of the monument, with an analysis of its history and especially with an analysis of the current degradation state. In this direction, the aim is to obtain as many information as possible regarding the structure geometry (which is complex in case of historical monuments with elements such as arches, vaults, pendants, walls with conical opening etc.), characterizing the masonry texture and the cross sections of the elements (in one or more layers, with or without bound between layers), defining physical, chemical and mechanical properties of the material components (brick, stone, mortar) and masonry evaluation like a composite material.

With the same degree of interest, consideration should also be given to studying the area in which the historical monument is located, taking into account the destructive level of the external actions (seismicity of the area, wind and snow loads, humidity, etc.), the interference of the neighboring buildings and the characteristics of the foundation soil [5, 6].

## 2.1 Setting Performance Goals

The M.P. 025-04 methodology defines the term "historical monument", as a construction that has an important "cultural value" that makes it necessary to guarantee its preservation without affecting its own architectural, typological and material characteristics [8].

In this direction, general requirements that ensure the fulfillment of the performance criteria necessary for the structural rehabilitation process are defined:

- designing and executing intervention works in compliance with the Venice International Book;
- all component parts of the structure are considered at assessing the safety level in accordance with „strength and stability requirements" (infrastructure, superstructure, nonstructural elements);
- „safety" includes also the limiting the degradation of cultural assets;
- anti-seismic protection principles and general aspects can be applied, as far as they do not contradict to provisions of the methodology;
- ultimate limit state principles are used for the dimensions and checking of the final cross-sections, so actual codes are considered;
- the seismic performance level pursued through intervention works is established, usually, for the maximum possible earthquake on site (with an average return period of 500 years) and is defined by two criteria:

  1. small damages and sometimes extended ones are acceptable, if repairable for structural and nonstructural elements;
  2. no irreparable damages are accepted at finishes and decorations levels, with cultural value;

- the technical report is mandatory and a minimal content, composed of: architectural and structural surveys, degradation surveys, photographic surveys, technical memory, computing notes, laboratory or non-destructive tests in situ, geotechnical study, proposed plans with the solution of recommended intervention, is necessary.

## 2.2 Evaluation Steps

The first stage of work consists of collecting information through the basic methods:

- available relevant documents;
- visual investigation of the objective, the location and summary of neighboring building;
- on-situ measurements;

- laboratory tests on materials and foundation soil;
- numerical simulation.

The specific failure mechanisms and principal damage causes are identified in this stage.

Based on the interpretation of the information from the analysis stage, the main objectives of diagnosing are set: establishing the causes of the damage, assessing the safety at the current state and deciding on the opportunity, urgency and intervention mode. The absence of these stages in the diagnosis or the superficiality of the process makes the presented solutions arbitrary and most often oversized, thus covering the lack of knowledge when making decisions [9].

The phase of diagnosis of the historical heritage consists of two distinct phases, with an equal role in the elaboration of the final conclusions, thus: qualitative evaluation and evaluation by calculation.

Represents in fact a final conclusion of the diagnostic phase, all the intervention works must satisfy the strength and stability requirements and comply with the specific provisions applicable to historical monuments.

### 2.3 Consolidation Solutions Classification

A classification of such consolidation techniques can be formulated [10] from constructive point of view, into:

- substitution consolidation refers to the replacement of structurally damaged areas, using higher performance materials and elements;
- passive consolidation has as a main objective the improvement of the material properties in the damaged structural areas;
- active consolidation involves the introduction of new structural elements, which can withstand the causes of structure failure;
- reconstruction involves the demolition of the destroyed elements and their restoration to the initial conditions.

Depending on the purpose of the intervention works on the structures, the consolidation techniques can be classified as:

- confinement, which is mainly used to increase the strength of the structural elements by preventing deformations;
- reinforcement, involves introduction of a new materials in the existing masonry, with superior mechanical properties, resulting an increase in structural strength and rigidity;
- the extension, refers to the modification of a section by supplementing it with new materials;
- material replacement, involves the repair and restoration of the damaged elements;
- binding/connecting the different structural elements or subassemblies in order to ensure a compatibility of the deformations;
- anchoring, consists of fixing an element or part of a structure to another supporting structure;

- structural development, implies the general improvement of the characteristics for the elements or of the entire structure;
- pretension, refers to changing the tensions state of an element by means of external loads or by prestressing;
- the isolation, consists of dissipation of the seismic energy and the vibrations through the external systems usually located at the foundation level.

The classification into two categories: reversible and irreversible techniques is the main criterion in choosing intervention techniques. If the interventions are reversible, the materials used impose relatively few restrictions, however, if the interventions are irreversible, compatibility with the original materials and long-term durability must be ensured. These restrictions force a thorough knowledge of the properties of the original materials, so that their characteristics constitute clear criteria for choosing the intervention materials [11].

# 3 Case Study

## 3.1 Building Description

The studied building is located in Paşcani city, in the north-east of Romania. It was built in the XVII$^{th}$ century and is was included on the List of historical monuments in Iasi County in 2004 (Fig. 3). The original plans of the building have not been preserved, although it is assumed that there were no such plans and no other documents regarding its construction were discovered until now.

The construction has an irregular plan view, with the maximum dimensions of approximately 30 × 33 m, if the facades are also measured, and extends on basement (partially), ground floor, floor and attic (Fig. 4 and 5).

In one or more subsequent stages of consolidation, on the vaults from the ground floor, concrete was poured and a floor made of steel profiles and concrete slab was added. The structural system is made of brick walls and stone brick vaulted floors at the ground floor and concrete floors with I steel profiles at the first floor. The thickness of the walls varies from 60 cm to 130 cm. The attic rooms have a wooden floor at the top, which is currently partially collapsed. The stairs are made of reinforced concrete ramps and were probably made with the overcrowding of the vaults on the ground floor and the floor above. The foundation is made of stone masonry connected with clay mortar, having the same depth as the walls, going down to −2.50 m.

## 3.2 Causes of Damage

After a close investigation of damages for the analyzed heritage building, the following causes were emphasized:

### 3.2.1 Repeated Seismic Action

The repeated seismic action at which the structure was subjected in previous years lead to the necessity of consolidating the building. Such works were conducted prior to current

**Fig. 3.** The manor court and Cantacuzino manor throughout the 50's [12]

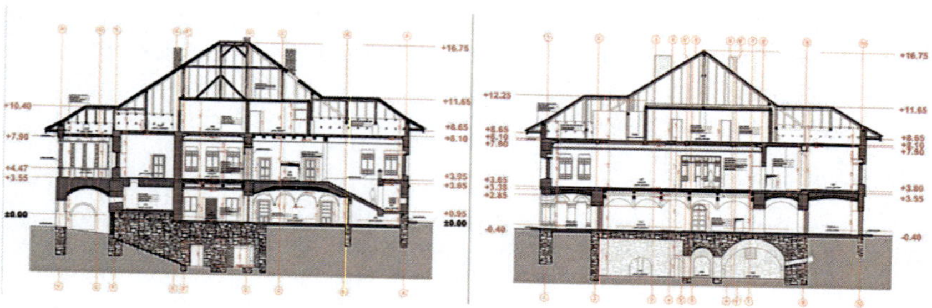

**Fig. 4.** The longitudinal and transversal cross-section

**Fig. 5.** Main façade of Cantacuzino manor

analysis. At present, the construction presents specific degradations to seismic action, as diagonal cracks in the full walls; indicating the tendency of the porches to detach from the main structure;

### 3.2.2 Water Action

The water infiltration due to lack of maintenance work at the roof level lead to the degradation of the steel structure and also to wall exfoliation of the paint and plaster. The missing drainage system produced water infiltration at the ground level producing damage at the ground floor and the basement level. Due to water infiltration at the attic level, the ceiling is partially collapsed and in case of continuing the water infiltration, a chain reaction with imminent danger regarding the stability and the strength of the roof system may occur leading even to global failure mechanism. The different settlements of the foundation soil marked the structural system of the building with visible vertical cracks;

### 3.2.3 Structural Interventions

The introduction of a mortar layer with copped masonry on the walls, the over-concreting of the ground floor vaults, adding a floor of steel profiles and concrete slab above the floor, building an area next to the old porch, filling of several windows and doors openings led to the changes of the building's structural characteristics and the loading capacity. The lack of maintenance work lead to vandalized action by destructive actions which increased building degradation.

## 3.3 Computation Summary

The structure was analyzed according to P100/3-2019 and CR6-2006 [13, 14] codes. The structural safety level for each masonry peer, at shear force, axial force and at bending moment was established. The computation was performed on the main directions, longitudinal and transversal one.

The numerical simulation was performed by ETABS software, which is a finite element program. The structural walls were discretized into surface elements (Fig. 6).

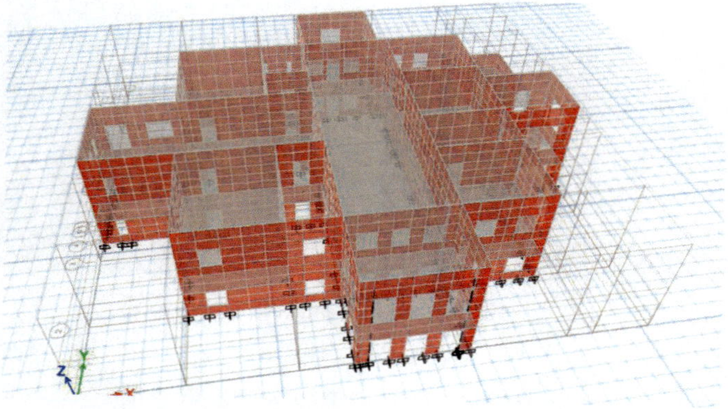

**Fig. 6.** The adopted design model

The program computes the stress state from which, through integration, the stress states result. The ratio between the minimum capable effort able from a section and the external one, represents the structural safety level, of the peer in this particular case.

A 3D model was considered in the numerical simulation. The masonry walls were modeled with shell elements. The efforts in the peers result from integrating stress values. For each wall, the minimum capable shear force and the probable breaking mode (fragile or ductile) in the base section are determined according to the P100-3/2019 code and is compared with the actual efforts resulting from the linear static analysis, obtaining the structural safety level, (Fig. 7).

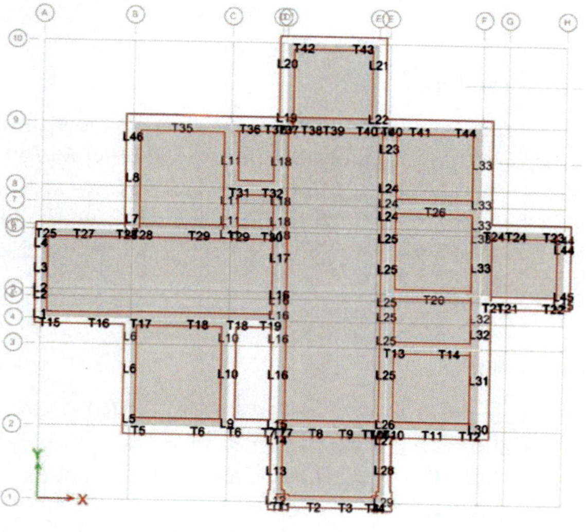


**Fig. 7.** Peers disposal

The structural safety level ($R_{3i}$) for the entire structure, on each direction is computed with:

$$R_{3i} = \frac{\sum_{jd} V_{fd} + \sum_{kf} V_{ff}}{F_b} \tag{1}$$

$\sum_{jd} V_{fd}$ - the sum of the wall's strength capacity with the ductile breaking;
$\sum_{kf} V_{ff}$ - the sum of the wall's strength capacity with the fragile breaking
$F_b$ - base shear force [15] (Table 1).

**Table 1.** The structural safety level of the damaged structure

| Direction | $\sum_{jd} V_{fd}$(kN) | $\sum_{kf} V_{ff}$(kN) | $R_{3i}$ | $R_3 = \frac{R_{3L}+R_{3T}}{2}$ |
|---|---|---|---|---|
| Longitudinal | 15638 | 2829 | 0.181 | 0.173 |
| Transversal | 14990 | 2475 | 0.165 | |

According to P100/3-2019 the structural safety level should be greater than 0.65 for Vrancea seismic region. Taking into consideration this restriction, several consolidation works are proposed in order to increase R3. The load transfer improvement to the walls is realized by fixing the over concreting in a belt included in the walls.

The masonry consolidation by means of vertical straps inserted into drilled channels injected with mortar to ensure the achievement of an advantageous efforts state in case of horizontal action. Also, wall intersections confinement of the masonry is proposed.

Increasing the load bearing capacity of the masonry by reinforcing and pouring a lime mortar coating reinforced with supplementary polymeric grids after the thick layer of plaster with broken brick was removed.

At the foundation level an exterior waterproofing isolation and a clay plug, with a perimeter drainage system, connected to a functional system to efficiently remove water from the building. At the basement level, the water infiltration in the walls will be prevented by waterproofing the masonry with injections of hydro structural resins.

Repairing the cracks by local injectors and local weavings, where necessary.

Existing roof reconstruction with similar wood elements, properly connected in a belt constructed at the upper part of the masonry from the attic.

The numerical simulation was recomputed and in the modal analysis care, the fundamental period of vibration decreased from 0.18 s to 0.16 s, with approximately 10 %. Meanwhile the structural safety level increase from 0.173 to 0.964, with more than 80 %, exceeding the minimum recommended value from the norm.

## 4 Conclusion

The value of a construction considered a historical monument, does not refer only to its exterior or interior appearance, but encompasses all its components, as a result of the

technology specific to the place and time in which it was made. The importance of the conservation of the built heritage puts into discussion the interventions that take place on such constructions, interventions that must not be realized improvably or by applying some norms that are addressed to the new constructions.

Even though the protection of the user's lives is very important, this does not mean neglecting the heritage and stresses that every generation has the architectural heritage only on a temporary basis and is responsible for transmitting it to future generations.

In our country it is necessary to elaborate specific norms for interventions on heritage constructions; in the absence of such regulations at this moment the application of the norms that are addressed to new constructions is practically obligatory also for this type of constructions. It should be noted that detailed rules are very difficult to elaborate, but from the experience of other countries (for example Italy) regulations can be made adaptable to the multitude of situations that can be encountered, to the technological or location characteristics, suggesting a design attitude that it takes into account the specificity of the object. Such regulations may take the form of a "restoration book" which would include the principles of intervention recommended in the existing international books.

The decision regarding the intervention on a building with historical value must take into account the desire to conserve, to emphasize its aesthetic or historical values, without neglecting the appropriate choice of materials or techniques of conservation and consolidation.

Through the materials and the constructive procedures used, it is necessary to keep in mind the preservation of the authenticity of the object that is to be protected. Thus, by this type of interventions the aim is not to obtain a structural safety level similar to the one related to the new buildings, the main intention being that of preservation and only exceptionally of adaptation to the current norms, in situations where the global restauration of the construction is provided.

## References

1. Curinschi Ghe., Restaurarea monumentelor, ed. Tehnică, Bucureşti (1968)
2. LNCS Homepage. http://wikimapia.org/83605/ro/Centrul-Istoric-Hermannstadt. Accessed 03 June 2020
3. Crişan, R.: Construcţii din zidărie şi beton armat, ed. Universitară „Ion Mincu"(2012)
4. LNCS Homepage. https://cursarhitectura.wordpress.com/curs-2-arhitectura-evului-mediu-si-arhitectura-byzantina/. Accessed 03 June 2020
5. Budescu, M. et. al.: Reabilitarea construcţiilor, Ed. Vesper (2001)
6. Soveja Lucian, Teză de Doctorat - Evaluarea şi reabilitarea structurilor istorice din zidărie, Iaşi (2015)
7. Clim, D.A., Groll, L., Groll L.I.: The Main Cause of the Degradation of Historic Buildings, Buletinul Institutului Politehnic din Iaşi, Universitatea Tehnică „Gheorghe Asachi" din Iaşi Volumul 63 (67), Numărul 3, Secţia Construcţii. Arhitectură (2017)
8. MP-025-04 – Metodologie pentru evaluarea riscului şi propunerile de intervenţie necesare la structurile construcţiilor monumente istorice în cadrul lucrărilor de restaurare ale acestora (2004)
9. Petrescu Vlad, Teza de Doctorat - Protecţia clădirilor cu statut de patrimoniu prin metode moderne de intervenţie, Bucureşti (2014)

10. Pleşu, R.: Teză de Doctorat - Contribuţii privind metodele de reabilitare a structurilor de zidărie, Iaşi (2013)
11. Curinschi Gheorghe, Arhitectură, Urbanism, Restaurare, Editura Tehnică (1996)
12. LNCS Homepage. http://asirys.blogspot.ro/2014/01/palatul-cantacuzino-pascani.html. Accessed 03 June 2019
13. P100-3/2019 Cod de proiectare seismică- partea a III-a – Prevederi pentru evaluarea seismică a clădirilor existente (2019)
14. CR6-2013 – Cod de proiectare pentru structure din zidărie (2013)
15. P100-3/2013 Cod de proiectare seismică- partea a I-a – Prevederi de proiectare pentru clădiri (2013)

# Author Index





A. Rotaru (Ed.): CRIT-RE-BUILT 2019, SSGG, pp. 601–602, 2021.
https://doi.org/10.1007/978-3-030-61118-7